W9-BDS-610

Progress in Mathematics
Volume 132

Functional Analysis on the Eve of the 21st Century
Volume II

In Honor of the Eightieth Birthday of I. M. Gelfand

Simon Gindikin
James Lepowsky
Robert L. Wilson
Editors

Birkhäuser
Boston • Basel • Berlin

Simon Gindikin
Department of Mathematics
Rutgers University
New Brunswick, NJ 08903

James Lepowsky
Department of Mathematics
Rutgers University
New Brunswick, NJ 08903

Robert L. Wilson
Department of Mathematics
Rutgers University
New Brunswick, NJ 08903

Library of Congress Cataloging-in-Publication Data

Functional analysis on the eve of the 21st century in honor of the 80th birthday of I. M. Gelfand / [edited] by S. Gindikin, J. Lepowsky, R. Wilson.
p. cm. -- (Progress in mathematics ; vol. 132)
Includes bibliographical references.
ISBN 0-8176-3860-1 (set : acid-free paper). -- ISBN 0-8176-3755-9 (v. 1 : acid-free paper). -- ISBN 0-8176-3855-5 (v. 2 : acid-free paper)
1. Functional analysis. I. Gel'fand, I. M. (Izrail' Moiseevich) II. Gindikin, S. G. (Semen Grigor'evich) III. Lepowsky, J. (James) IV. Wilson, R. (Robert), 1946- . V. Series: Progress in mathematics (Boston, Mass.) ; vol. 132.
QA321.F856 1995 95-20760
515'.7--dc20 CIP

Printed on acid-free paper

Birkhäuser ®

ISBN 0-8176-3755-9 (Volume 1)
ISBN 0-8176-3855-5 (Volume 2)
ISBN 0-8176-3860-1 (set)

ISBN 3-7643-3755-9 (Volume 1)
ISBN 3-7643-3855-5 (Volume 2)
ISBN 3-7643-3860-1 (set)

Typeset and reformatted from disk by TEXniques, Inc., Boston, MA
Printed and bound by Quinn-Woodbine, Woodbine, NJ.
Printed in the U.S.A.

9 8 7 6 5 4 3 2 1

I. M. Gelfand

Contents

Volume II

Volume I

Preface

A four-day conference, "Functional Analysis on the Eve of the Twenty-First Century," was held at Rutgers University, New Brunswick, New Jersey, from October 24 to 27, 1993, in honor of the eightieth birthday of Professor Israel Moiseyevich Gelfand. He was born in Krasnye Okna, near Odessa, on September 2, 1913.

Israel Gelfand has played a crucial role in the development of functional analysis during the last half-century. His work and his philosophy have in fact helped to shape our understanding of the term "functional analysis" itself, as has the celebrated journal *Functional Analysis and Its Applications,* which he edited for many years.

Functional analysis appeared at the beginning of the century in the classic papers of Hilbert on integral operators. Its crucial aspect was the geometric interpretation of families of functions as infinite-dimensional spaces, and of operators (particularly differential and integral operators) as infinite-dimensional analogues of matrices, directly leading to the geometrization of spectral theory. This view of functional analysis as infinite-dimensional geometry organically included many facets of nineteenth-century classical analysis, such as power series, Fourier series and integrals, and other integral transforms.

Quantum mechanics provided a further strong stimulus and source of new ideas for the development of functional analysis. Several brilliant new directions in functional analysis appeared: Banach algebras (Gelfand), operator algebras (von Neumann), infinite-dimensional representations of semisimple Lie groups (Gelfand-Naimark, Bargmann, Harish-Chandra), and the theory of distributions or generalized functions (Sobolev, L. Schwartz). The continued development of quantum physics stimulated the creation of one of the most remarkable ideas in functional analysis, the idea of integration over spaces of functions: Feynman integrals.

Today we can observe new horizons of functional analysis. Dramatic recent developments in theoretical physics — string theory, conformal field theory and topological field theory — are again supplying new problems.

Israel Gelfand's own sense of the relative importance of research directions has played a major role in the development of these new areas. We mention for example his instructive and influential lectures prepared for the International Congresses in Amsterdam, Edinburgh, Stockholm and Nice; in these lectures, he formulated many important problems concerning functional analysis in the broad sense. His personal scientific activity has been distinguished by the coexistence of a very broad spectrum of mathematical interests and also by faithfulness to the ideology of functional analysis. Two remarkable examples are the application of the ideas of infinite-dimensional representations to the study of representations of finite groups, and the development of combinatorics with "infinite-dimensional" background.

It was very difficult to make a selection of topics for this conference, and we decided to follow Gelfand's taste and choose topics in which he has been working actively in recent years or in which he has a very strong interest today. For instance, we have chosen not to include subjects in which Gelfand made fundamental contributions but in which he does not actively work now. For these reasons, we chose the following list of (interrelated) topics for this conference:

(1) Mathematical physics, especially geometric quantum field theory;

(2) Representation theory, particularly, certain problems concerning representations of groups over local fields;

(3) Combinatorics and hypergeometric functions, with emphasis on combinatorial structures underlying various "continuous" constructions;

(4) Noncommutative geometry, quantum groups and geometry.

Support for the conference was generously provided by the National Science Foundation, the A. P. Sloan Foundation and Rutgers University. Eighteen invited mathematical talks were presented at the conference. In addition, President Francis L. Lawrence of Rutgers University awarded Professor Gelfand the honorary degree of Doctor of Science, and on this occasion, Professor Israel M. Singer delivered a tribute to Professor Gelfand. Professor Singer's tribute and the program of the conference are included here. There was great interest in the talks throughout the entire conference, and the lecture hall was constantly filled with people from many countries, including many graduate students and young researchers who were able to attend due to support from the National Science Foundation and the Sloan Foundation.

These two volumes contain papers contibuted by most of the invited speakers. The second of the two volumes contains the somewhat more "geometric" papers, although such a designation is to a certain extent arbitrary, because of the breadth of the papers.

The organizing committee for the conference consisted of: Sir Michael Atiyah, Felix Browder, Alain Connes, Simon Gindikin, Phillip Griffiths, Friedrich Hirzebruch, David Kazhdan, Bertram Kostant, James Lepowsky, George Daniel Mostow, Ilya Piatetski-Shapiro, Mikio Sato, Isadore Singer, Robert Wilson and Edward Witten.

Special thanks are due to Mary Anne Jablonski, who expertly coordinated the conference arrangements at Rutgers University, and to Ann Kostant and the entire staff of Birkhäuser, who have displayed untiring efforts in bringing these volumes to completion.

We believe that this conference gave the mathematical community the opportunity to honor one of the most remarkable mathematicians of our time. We are very happy to see that, as he enters his ninth decade, Israel Gelfand continues his brilliant mathematical life as a young mathematician.

Simon Gindikin
James Lepowsky
Robert Lee Wilson

Functional Analysis on the Eve of the Twenty-First Century

A Conference in Honor of the Eightieth Birthday of Israel M. Gelfand

Rutgers University, New Brunswick, New Jersey

October 24-27, 1993

Sunday, October 24

D. Kazhdan, *Quantization and series of representations of reductive groups*

G. Lusztig, *From modular representations to combinatorics*

I. Frenkel, *A representation-theoretic approach to four-dimensional topology*

C. Moeglin, *Wave front set and unipotent representations for p-adic groups*

S.-T. Yau, *Variational problems of differential geometry*

Monday, October 25

I. Singer, *A tribute to Israel Gelfand*

Presentation of honorary degree of Doctor of Science to Israel Gelfand

B. Kostant, *Minimal unitary representations and the generalized Capelli identity* (joint work with R. Brylinski)

A. Polyakov, *Gravitational dressing*

M. Kontsevich, *Linear algebra of elliptic operators* (joint work with S. Vishik)

I. Singer, *On the quantization of two-dimensional gauge theories*

Tuesday, October 26

M. Kapranov, *Analogies between the Langlands correspondence and topological quantum field theory*

L. Jeffrey, *Equivariant cohomology and pairings in the cohomology of symplectic quotients* (joint work with F. Kirwan)

K. Aomoto, *Connection problem in the q-analog of de Rham cohomology*

A. Zamolodchikov, *Boundary S-matrix and boundary state in two-dimensional integrable quantum field theory* (joint work with S. Ghoshal)

R. MacPherson, *Combinatorial differential manifolds*

Wednesday, October 27

M. Jimbo, *Algebraic analysis of solvable lattice models*

O. Mathieu, *On the cohomology of the Lie algebra of hamiltonian vector fields* (joint work with I. M. Gelfand)

M. Gromov, *Almost flat bundles and applications*

E. Witten, *Physical methods applied to Donaldson theory*

Israel M. Gelfand

Born: September 2, 1913, in Ukraine
Currently Distinguished Professor at Rutgers University
Ph.D. in Mathematics, Moscow State University, 1935
Doctor of Science in Mathematics, Moscow State University, 1940

Awards

State Prize of the USSR, 1953
Wolf Foundation Prize, 1978
Wigner Medal, 1980
Kyoto Prize, 1989
MacArthur Foundation Fellowship, 1994

Memberships

Academy of Sciences of the USSR, Moscow, Corresponding Member, 1953
American Academy of Arts and Sciences, Boston, 1964
Royal Irish Academy, Dublin, 1970
National Academy of Sciences of the USA, 1970
Royal Swedish Academy of Sciences, Stockholm, 1974
Académie des Sciences de l'Institut de France, 1976
Royal Society, London, 1977
Academy of Sciences of the USSR, Moscow, 1984
Accademia dei Lincei, Italy, 1988
Academy of Sciences of Japan, Tokyo, 1989

Honorary degrees

Oxford University, 1973
Université Pierre et Marie Curie (Paris VI)
and Université Paris VII, 1974
Harvard University, 1976
University of Uppsala, 1977
Université de Lyon, 1984
Scuola Normale Superiore, Pisa, 1985
City University of New York, 1988
Kyoto University, 1989
University of Pennsylvania, 1990
New York University, 1992
Rutgers University, 1993

Mathematical Publications of I. M. Gelfand 1987-1995

The list of I.M. Gelfand's papers prior to 1987
appears in Gelfand's Collected Papers Vols I–III, published by Springer-Verlag

Books

1. *Collected papers, Vol. I*, Springer-Verlag, Heidelberg, 1987
2. *Collected papers, Vol. II*, Springer-Verlag, Heidelberg, 1988
3. *Collected papers, vol. III* Springer-Verlag, Heidelberg, 1989
4. *Lectures on Linear Algebra*, Dover Publ., Inc., NY, 1989
5. I. M. Gelfand, S. G. Gindikin (eds.), *Mathematical Problems of Tomography*, Amer. Math. Soc., Providence, 1990
6. I. M. Gelfand, M. I. Graev, I. I. Piatetskii-Shapiro, *Representation Theory and Automorphic Functions*, Academic Press, Boston, 1990
7. I. M. Gelfand, E. G. Glagoleva, A. A. Kirillov, *The Method of Coordinates*, Birkhäuser, Boston, 1990
8. I. M. Gelfand, E. G. Glagoleva, E. Shnol, *Functions and Graphs*, Birkhäuser, Boston, 1990
9. I. M. Gelfand, A. Shen, *Algebra*, Birkhäuser, Boston, 1993; revised edition 1995
10. I. Gelfand, L. Corwin, J. Lepowsky (eds.), *The Gelfand Mathematical Seminars 1990–1992*, Birkhäuser, Boston, 1993
11. I. M. Gelfand, M. M. Kapranov, A. V. Zelevinsky, *Discriminants, Resultants and Multidimensional Determinants* , Birkhäuser, Boston, 1994
12. I. M. Gelfand, J. Lepowsky, M. Smirnov (eds.), *The Gelfand Mathematical Seminars 1993–95*, Birkhäuser, Boston, 1995
13. I. M. Gelfand, M. Smirnov (eds.), *The Arnold-Gelfand Mathematical Seminars*, Birkhäuser, Boston, to appear 1996
14. I. M. Gelfand, T. Fokas (eds.), Memorial Volume for Irene Dorfman, to appear 1996
15. I. M. Gelfand, M. Saul, A. Shen, *Algebra*, Teacher's Edition, in preparation
16. I. M. Gelfand, T. Alexeeyevskaya, *Geometry*, in preparation
17. I. M. Gelfand, A. Borovik, N. White, *Coxeter Matroids*, in preparation
18. I. M. Gelfand, M. Saul, A. Shen, *Calculus*, in preparation
19. I. M. Gelfand, V. S. Retakh, *Quasideterminants, Noncommutative Symmetric Functions and their Applications*, in preparation

Papers

1. (with V. A. Vassiliev and A. V. Zelevinsky) General hypergeometric functions on complex Grassmanian, *Funct. analiz i ego priloz.* (Functional Analysis & Applications), **21**:1 (1987), 23–28
2. (with M. I. Graev) Hypergeometric functions associated with the Grassmanian $G_{3,6}$, *Doklady AN SSSR* **293** (1987), 288–293
3. (with V.V. Serganova) Combinatorial geometries and torus strata on homogeneous compact manifolds, *Uspekhi Mat. Nauk*, **42**:2 (1987), 107–134
4. (with V. V. Serganova) Strata of maximal torus in a compact homogeneous space, *Doklady AN SSSR*, **292**:3 (1987), 524–528
5. (with V. V. Serganova) On the definition of a matroid and greedoid, *Doklady AN SSSR*, **292**:1 (1987), 15–20
6. (with M. Goresky, R.D. MacPherson and V.V. Serganova) Combinatorial geometries, convex polyhedra and Schubert cells, *Advances in Math.*, **63**:3 (1987), 301–316
7. (with M. I. Graev, A. V. Zelevinsky) Holonomic systems of equations and series of hypergeometric type, *Doklady AN SSSR*, **295** (1987), 14–19
8. (with T.V. Alexeevskaya, A. V. Zelevinsky) Distributions of real hyperplanes and the partition function connected with it, *Doklady AN SSSR*, **297**:6 (1987), 1289–1293
9. (with A. V. Varchenko) Heaviside functions of a configurations of hyperplanes, *Funct. analiz i ego priloz.*, **21**:4 (1987), 1–18
10. (with V.A. Ponomarev) Preprojective reduction of the free modular lattice D_r, *Doklady AN SSSR*, **293**:3 (1987), 521–524
11. (with V.S. Retakh and V.V. Serganova) Generalized Airy functions, Schubert cells and Jordan groups, *Doklady AN SSSR*, **298**:1 (1988), 17–21
12. (with A. V. Zelevinsky and M.M. Kapranov) Equations of hypergeometric type and Newton polyhedra, *Doklady AN SSSR*, **300**:3 (1988), 529–534
13. (with M.M. Kapranov and A. V. Zelevinsky) A-discriminants and Cayley-Koszul complexes, *Doklady AN SSSR*, **6** (1989), 1307–1311
14. (with I.S. Zakharevich) Spectral theory of a pencil of third-order skew-symmetric differential operators on S^1, *Funct. analiz i ego priloz.*, **23**:2 (1989), 1–11
15. (with Yu.L. Daletsky, B.L. Tsygan) On a variant of noncommutative differential geometry, *Doklady AN SSSR*, **308**:6 (1989), 1293–1297
16. (with M.M. Kapranov, A. V. Zelevinsky) Projective-dual varieties and hyperdeterminants, *Doklady AN SSSR*, **309**:2 (1989), 385–389
17. (with G.L. Rybnikov) Algebraic and topologic invariants of oriented matroids, *Doklady AN SSSR*, **307**:4 (1989), 791–795
18. (with A. V. Zelevinsky, M. M. Kapranov) Newton polyhedra of principal A-discriminant, *Doklady AN SSSR*, **308**:1 (1989), 20–23

19. (with A. V. Zelevinsky, M.M. Kapranov) Hypergeometric functions and toric varieties, *Funct. anal. i ego priloz*, **23**:2 (1989), 12–26

20. (with M. I. Graev) Hypergeometric functions associated with the Grassmanian $G_{3,6}$, *Matem. Sborn.*, **180**:1 (1989), 3–38

21. (with M. I. Graev) The commutative model of the principal representation of the current group $SL(2, R)$ with respect to a unipotent subgroup, in: *Group Theoretical Methods in Physics*, **1** (1989), Gordon & Breach, 3–22.

22. (with M. I. Graev, A.M. Vershik) Principal representations of the group U_∞, in: *Representations of Lie Groups and Related Topics*, Gordon & Breach, 1990, 119–153

23. (with A. V. Zelevinsky, M.M. Kapranov) Discriminants of polynomials in several variables and triangulations of Newton polyhedra, *Algebra i Analiz*, **2**:3 (1990), 1–62

24. (with M. M. Kapranov, A. V. Zelevinsky) Newton polytopes of the classical resultant and discriminant, *Advances in Math.*, **84**:2 (1990), 237–254

25. (with A. V. Zelevinsky, M. M. Kapranov) Discriminant of polynomials in several variables, *Funct. analiz i ego priloz*, **24**:1 (1990), 1–4

26. (with S. G. Gindikin) Integral geometry and tomography, *Voprosy Kibernetiki*, **157** (1990), 3–7

27. (with M. M. Kapranov, A. V. Zelevinsky) Generalized Euler integrals and A-hypergeometric systems, *Advances in Math.*, **84** (1990), 255–271

28. (with M. M. Kapranov, A. V. Zelevinsky) Hypergeometric functions, toric varieties and Newton polyhedra, in: *Special functions*, Proc. Hayashibara Forum, (1990), 101–121

29. (with M. I. Graev, V.S. Retakh) Γ-series and general hypergeometric function on the manifold of kxn-matrices, Preprint Inst. Prikl. Mat. Akad. Nauk SSSR, **64** (1990)

30. (with M. I. Graev, V.S. Retakh) Hypergeometric functions on strata on small codimensions in $G_{k,n}$, Preprint Inst. Prikl. Mat. Akad. Nauk SSSR, **126** (1990)

31. (with D.B. Fairlie) The algebra of Weyl symmetrised polynomials and its quantum extension, *Comm. Math. Phys.*, **136**:3 (1991), 487–499

32. (with M. I. Graev) The Crofton function and inversion formulas in real integral geometry, *Funct. analiz i ego priloz.*, **25**:1 (1991), 1–6

33. (with I. Zakharevich) Webs, Veronese curves and bi-Hamiltonian systems, *Funct. Analysis*, **99**:1 (1991), 15–178

34. Two Archetypes in the Psychology of Man, *Nonlinear Sci. Today*, **1**:4 (1991), 11–16

35. (with M. I. Graev, V.S. Retakh) Reduction formulae for hypergeometric functions on Grassmanian $G_{k,n}$ and a description of hypergeometric functions on strata of small codimensions, *Doklady AN SSSR*, **318** (1991), 793–797

36. (with M. I. Graev, V.S. Retakh) Hypergeometric functions on the k-th exterior degree of the space $\mathbf{C}^n$ and the Grassmanian $G_{k,n}$ and the connection between them, *Doklady AN SSSR*, **320** (1991), 20–24

37. (with V.S. Retakh) Determinants of matrices over noncommutative rings, *Funct. analiz i ego priloz*, **25**:2 (1991), 13–25

38. (with M. I. Graev, V.S. Retakh) Recent developments in the theory of general hypergeometric functions, in: *Special Differential Equations*, Proc. Taniguchi workshop, 1991, 86–91

39. (with M. I. Graev, V.S. Retakh) Generalized hypergeometric functions associated with an arbitary finite or locally compact continuous field, *Doklady AN SSSR*, **323** (1992), 394–397

40. (with M. I. Graev, V.S. Retakh) Difference and q-analogues of general hypergeometric systems of differential equations, *Doklady AN SSSR*, **325** (1992), 215–220

41. (with B.L. Tsygan) On the localization of topological invariants, *Comm. Math. Phys.*, **146**:1 (1992), 73-90

42. (with O. Mathieu) On the cohomology of the Lie algebra of Hamiltonian vector fields, *J. Funct. Anal.*, **108**:2 (1992), 347–360

43. (with R.D. MacPherson) A combinatorial formula for Pontrjagin classes, *Bull. Amer. Math. Soc.*, **26**:2 (1992), 304–309

44. (with M.M. Kapranov, A. V. Zelevinsky) Hyperdeterminants, *Advances in Math.*, **96**:2 (1992), 226–263

45. (with V.S. Retakh) A theory of noncommutative determinants and characteristic functions of graphs. I, *Funct. analiz i ego priloz.*, **26**:4 (1992), 1–20

46. (with M. I. Graev, V.S. Retakh) General hypergeometric systems of equations and series of hypergeometric type, *Uspekhi Mat. Nauk*, **47**:4 (1992), 3–80

47. (with L.J. Billera, B. Sturmfels) Duality and minors of secondary polyhedra, *J. Comb. Theory B*, **57** (1993), 258–268

48. (with M.M. Kapranov) On the dimension and degree of the projective dual variety: a q-analog of Katz-Kleiman formula, in: *The Gelfand Mathematical Seminars 1990–1992*, L. Corwin, I. M. Gelfand, J. Lepowsky (eds.), Birkhäuser, Boston, 1993

49. (with L. Corwin) Hopf algebra structures for the Heisenberg Algebra. 1, in: *The Gelfand Mathematical Seminars, 1990–1992*, L. Corwin, I. M. Gelfand, J. Lepowsky (eds.), Birkhäuser, Boston, 1993

50. (with M. Smirnov) Nonlocal differentials, in: *The Gelfand-Mathematical Seminars 1990–1992*, L. Corwin, I. M. Gelfand, J. Lepowsky (eds.), Birkhäuser, Boston, 1993

51. (with I. Zakharevich) On the local geometry of a bihamiltonian structure, in: *The Gelfand Mathematical Seminars 1990–1992*, L. Corwin, I. M. Gelfand, J. Lepowsky (eds.), Birkhäuser, Boston, 1993

52. (with A. Fokas) Bi-Hamiltonian structures and integrability, in: *Important developments in solitons theory*, Springer-Verlag, 1993, 259–282

53. (with M. I. Graev, V.S. Retakh) Formulae of reduction for hypergeometric functions connected with the Grassmanian and hypergeometric functions on strata of small codimension in $G_{k,n}$, *Russian J. on Math. Phys.*, **1**:1 (1993)

54. (with M. I. Graev, V.S. Retakh) Hypergeometric functions on $\Lambda^k C^n$ and the Grassmanian $G_{k,n}$, their connections and integral representations, Russian, *J. Math. Phys.*, **1**:3 (1993)

55. (with M. I. Graev) GG-functions, *Doklady RAN*, **328**:6 (1993), 645–648

56. (with M. I. Graev, V. S. Retakh) Q-hypergeometric Gauss equation and its solutions as series and integrals, *Doklady RAN*, **331**:2 (1993), 140–143

57. (with M. I. Graev, V.S. Retakh) (r,s)-hypergeometric functions, *Doklady RAN*, **333**:5 (1993), 567–570

58. (with A. Borovik) Matroids on chamber systems, Publ. LACIM, UQAM, Montreal, **14** (1993), 27–62

59. (with M. I. Graev) Projective representations of the current group $\mathrm{SU}(1,1)^X$, *Funct. analiz i ego priloz.*, **27**:4 (1993), 65–68

60. (with M. I. Graev) Special representations of the group $\mathrm{SU}(n,1)$ and projective unitary representations of the current group $\mathrm{SU}(n,1)^X$, *Doklady RAN*, **332**:3 (1993), 280–282

61. (with A.S. Fokas) Quadratic Poisson algebras and their infinite-dimensional extensions, *J. of Math. Phys.*, **35**:6 (1994), 3117–3131

62. (with A. Borovik) WP-matroids and thin Shubert cells on Tits systems, *Advances in Math.*, **103**:2 (1994), 162–179

63. (with M. Smirnov) Lagrangians satisfying Crofton formula, Radon transforms and nonlocal differentials, *Advances in Math.*, **109**:2 (1994), 188–227

64. (with M. I. Graev, V.S. Retakh and S.A. Spirin) (r,s)-exponents, *Doklady RAN*, **336**:6 (1994), 730–732

65. (with L. J. Corwin, R. Goodman) Quadratic algebras and skew-fields, *Contemp. Math.*, **177** (1994), 217–225

66. (with M. I. Graev) Hypergeometric functions on flag spaces, *Doklady RAN*, **338**:2 (1994), 154–157

67. (with M. I. Graev) Projective non-unitary representations of current groups, Doklady RAN, **338**:3 (1994) 298–301

68. (with I. Zakharevich) The spectral theory for a pencil of skew-symmetrical differential operators of the third order, *Commun. Pure and Appl. Math.*, **47** (1994), 1031–1041

69. (with A.S. Fokas) Integrability of linear and nonlinear evolution equations and the associated nonlinear Fourier transform, *Letters in Math. Physics*, **32**:3 (1994), 189–210

70. (with D. Krob, A. Lascoux, B. Leclerc, V.S. Retakh and J. Tibon) Noncommutative symmetric functions, *Advances in Math.*, **112**:2 (1995), 218-348

71. (with G.L. Rybnikov, D.A. Stone) Projective orientation of matroids, *Advances in Math.*, **113**:1 (1995), 118–150

72. (with V.S. Retakh) A noncommutative Vieta theorem and symmetric functions, in: *The Gelfand Mathematical Seminars 1993–1995*, I. M. Gelfand, J. Lepowsky, M. Smirnov (eds.), Birkhäuser, Boston, 1995

73. (with M. Smirnov) Cocycles on the Gauge Group and the Algebra of Chern-Simons Classes, *The Gelfand Mathematical Seminars 1993–1995*, I. M. Gelfand, J. Lepowsky, M. Smirnov (eds.), Birkhäuser, Boston, 1995

Tribute to I. M. Gelfand for his 80th Birthday Celebration

I. M. Singer

We are here to honor Israel Gelfand and to celebrate the continued vitality of one of the most influential mathematicians of the twentieth century—I dare say, the most outstanding of the last fifty years.

Unfortunately, our society neither understands nor appreciates mathematics. Despite its many applications, despite its intellectual power which has changed the way we do science, mathematicians are undervalued and ignored.

Naturally, its practitioners, its leaders, go unrecognized. They have neither power nor influence. Watching the negative effects popularity causes in other fields, and wincing at the few superficial articles about mathematics, I think it is just as well.

Faced constantly with problems we can't solve, most mathematicians tend to be modest about themselves and their accomplishments. Perhaps that is why we have failed to recognized a giant in our midst. I won't compare Gelfand with other outstanding mathematicians or scientists of the twentieth century; if I did, you would stop listening and start checking for yourselves whether you agree with me. But focus on my point— *we have a giant in our midst.* I turn to other fields to find comparable achievements: Balanchine in dance, or Thomas Mann in literature, or Stravinsky, better still, Mozart in music; but for me, a better comparison is with artists like Cézanne and Matisse. I commend to you the great poet Paul Rilke's letter on Cézanne. He said, "Paul Cézanne has been my supreme example, because he has remained in the innermost center of his work for forty years... which explains something beyond the freshness and purity of his paintings" (of course, for Gelfand, 60 years).

Evoking Matisse is perhaps more apt. A Matisse is breathtaking. No matter what his personal circumstance, he turns to new frontiers with joy and energy. Particularly outstanding is his later work: Jazz, and the remarkable "papier-découpés"—efforts done in his early eighties.

Gelfand too continues to dazzle us with new and profound ideas. His latest book with Kapranov and Zelevinsky is a major work that maps out new directions for decades to come.

In preparing this tribute, I asked many people for topics I should emphasize today. You will be interested in what happened. First, there was little intersection in the subjects my correspondents chose. Second, everyone gave me a five to twenty minute enthusiastic lecture on the essence of Gelfand's contribution—simple, and profound.

Reviewing Gelfand's contributions to mathematics has been an education.

Let me remind you of some of his main work.

1. Normed Rings
2. C^*-Algebras (with Raikov)—the GNS Construction
3. Representations of complex and real semi-simple groups (with Neumark and Graev)
4. Integral Geometry— Generalizations of the Radon Transform
5. Inverse scattering of Sturm Liouville systems (with Levitan)
6. Gelfand-Dickey on Lax operators and KdV
7. The treatises on generalized functions
8. On elliptic equations
9. The cohomology of infinite dimensional Lie algebras (with Fuks)
10. Combinatorial characteristic classes (beginning with MacPherson)
11. Dilogarithms, discriminants, hypergeometric functions
12. The Gelfand Seminar

It is impossible to review his enormous contributions in a few minutes. If I were Gelfand himself, I would orchestrate this occasion, like his seminar, by calling on many of you unexpectedly and demanding a one-sentence synopsis of a particular paper. But rather than intimidate you, I will comment on a few results that affected me.

As a graduate student, one of the first strong influences on me was Gelfand's Normed Ring paper. Marshall Stone had already taught us that points could be recaptured in Boolean algebras as maximal ideals. But Gelfand combined analysis with algebra in a simple and beautiful way. Using maximal ideals in a complex commutative Banach algebra, he represented such algebras as algebras of functions. Thus began the theory of commutative Banach algebras. The spectral theorem and the Wiener Tauberian Theorem were elementary consequences. I was greatly influenced by the revolutionary view begun there.

A natural next step for Gelfand was the study of non-commutative C^*-algebras. He represented such algebras as operator algebras using the famous GNS construction. It seemed inevitable to find unitary representations of locally compact groups using their convolution algebras. The representation theory of complex and real semi-simple Lie groups followed quickly after. What struck me most was the geometric approach Gelfand and his coworkers took. Only recently, it appears this subject has become geometric again.

In 1963, twenty American experts in PDEs were on their way to Novosibirsk for the first visit of foreign scientists to the academic city there. It was in the midst of a Khrushchev thaw. When I learned about it, I asked whether I could be added to the list of visitors, citing the index theorem Atiyah and I had just proved. After reading his early papers, I wanted to meet Gelfand. Each day of my two week stay in Novosibirsk I asked Gelfand's students when he was coming. The response was always "tomorrow." Gelfand never came. I sadly returned to Moscow. When I got to my room at the infamous Hotel Ukraine, the telephone rang and someone said Gelfand wanted to meet me; could I come downstairs. There was Gelfand. He invited Peter Lax and me for a walk. During the walk, Peter tried to tell Gelfand about his work on $SL(2, R)$ with Ralph Phillips. Gelfand tried to explain his own view of $SL(2, R)$ to Peter, but his English was inadequate. (He was rusty; within two days his English was fluent.) I interrupted and explained Gelfand's program to Peter. At the corner Gelfand stopped, turned to me, and said: *"But you are my student."* I replied, "Indeed, I *am* your student." (By the way, Gelfand told me he didn't come to Novosibirsk because he hates long conferences. That's why this celebration lasts only four days.)

Although it is an honor to be a Gelfand student, it is also a burden. We try to imitate the depth and unity that Gelfand brings to mathematics. He makes us think harder than we believed possible. Gelfand and I became close friends in a matter of minutes, and have remained so ever since. I was ill in Moscow, and Gelfand took care of me.

I didn't see him again for ten years. He was scheduled to receive an honorary degree at Oxford, where I was visiting. It was unclear that he would be allowed to leave the Soviet Union to visit the West. I decided not to wait and returned home. A week later, I received a telegram from Atiyah; Gelfand was coming—the Queen had asked the Russian ambassador to intercede. I flew back to England and accompanied Gelfand during his visit, a glorious time. Many things stood out.

But I'll mention only one, our visit to a Parker Fountain Pen store. Those of you who have ever shopped with Gelfand are smiling; it is always an unforgettable experience. Within fifteen minutes, he had every salesperson scrambling for different pens. Within an hour, I knew more about the construction of fountain pens than I ever cared to know, and had ever believed possible! Gelfand's infinite curiosity and the focused energy on details are unbelievable; that, coupled with his profound intuition of essential features is rare among human beings. He is beyond category.

Talking about Oxford, let me emphasize Gelfand's paper on elliptic equations. In 1962, Atiyah and I had found the Dirac operator on spin manifolds and already had the index formula for geometric operators coupled to any vector bundle, although it took another nine months to prove our theorem. Gelfand's paper was brought to our attention by Smale. It enlarged our view considerably, as Gelfand always does, and we quickly realized, using essentially the Bott periodicity theorem, that we could prove the index theorem for any elliptic operator.

I haven't talked about the applications of Gelfand's work to Physics—Gelfand-Fuks, for example, on vector fields of the circle, the so-called Virasoro Algebra, which Virasoro did not in fact define. Although I mentioned Gelfand-Dickey, I haven't stressed its influence very recently on matrix model theory. Nor have I described how encouraging he is and how far ahead of his time he is in understanding the implications of a paper which seems obscure at the time.

Claude Itzykson told me that his now famous paper with Brezin, Parisi and Zuber that led to present-day methods of triangulating moduli space went unnoticed by scientists. The authors received one request for a reprint—from Gelfand.

Ray and I were very excited about our definition of determinants for Laplacian-like operators and its use in obtaining manifold invariants-analytic torsion. The early response in the U.S. was silence; Gelfand sent us a congratulatory telegram.

It has been a great honor to have been chosen to pay tribute to Gelfand on this very special occasion. As you can tell, he means a great deal to me personally.

Among his many special qualities, I will mention only one in closing. He is a magician. It is not very difficult, not very difficult at all, for any of us mere mortals to keep the difference in our ages a constant function of time. But with Gelfand... when I met him 30 years ago, and 20 years ago, I thought Gelfand was older than I. About ten years ago, I felt we were the same age. Now it is quite clear that he is younger; in fact, *much* younger than most in the audience. It is important for us all that Gelfand continue to prosper and to do such great mathematics.

We wish him good health and happiness.

I. M. Singer

Functional Analysis on the Eve of the 21st Century

Volume II

In Honor of the Eightieth Birthday of I. M. Gelfand

Positive Curvature, Macroscopic Dimension, Spectral Gaps and Higher Signatures

M. Gromov

to I.M.G.

Our journey starts with a macroscopic view of Riemannian manifolds with *positive scalar curvature* and terminates with a glimpse of the proof of the homotopy invariance of some *Novikov higher signatures* of non-simply connected manifolds. Our approach focuses on the spectra of geometric differential operators on compact and non-compact manifolds V where the link with the macroscopic geometry and topology is established with suitable index theorems for our operators twisted with *almost flat bundles* over V. Our perspective mainly comes from the asymptotic geometry of infinite groups and foliations.

Contents

1. Scalar curvature Sc (V)

Let $V = (V, g)$ be a C^2-smooth Riemannian manifold where g denotes the Riemannian metric tensor. Then the scalar curvature of V is a function Sc_v on V built in a certain way out of the first and second derivatives of g. In fact there is a unique, up-to-scale, second order differential operator, say $\mathcal{S}$ acting from metrics g to functions $V \to \mathbb{R}$, such that

(a) $\mathcal{S}$ is Diff-equivariant for the natural action of diffeomorphisms of V on metrics and functions, and

(b) $\mathcal{S}$ is linear in the second derivatives of g.

(The existence and uniqueness of $\mathcal{S}$ follows from the fact that the natural representation of the orthogonal group $\mathcal{O}(n)$, for $n = \dim V$, on the space of the curvature tensors $R_{ijk\ell}$ on $\mathbb{R}^n$ has a unique one-dimensional factor). Then one defines

$$\mathrm{Sc}_v(V, g) = \mathcal{S}(g)(v)$$

with the customary normalization condition

$$\mathrm{Sc}(S^2 \times \mathbb{R}^{n-2}) = 2$$

for all $n = \dim V \geq 2$, where S^2 is the unit 2-sphere.

The infinitesimal (and microscopic) meaning of Sc_v is revealed by the following easy formula relating the volumes of the Riemannian ε-ball at $v \in V$ and the unit Euclidean ball B,

$$\mathrm{Vol}\, B_v(\varepsilon) = \varepsilon^n \big(1 - \varepsilon^2 \alpha_n \,\mathrm{Sc}_v + o(\varepsilon^2)\big)\, \mathrm{Vol}\, B$$

where $\alpha_n = (6n)^{-1}$ and $o(\varepsilon^2)$ refers to $\varepsilon \to 0$. For example if $\mathrm{Sc}_v > 0$, then

$$\mathrm{Vol}\, B_v(V, \varepsilon) < \varepsilon^n \,\mathrm{Vol}\, B = \mathrm{Vol}\, B(\mathbb{R}^n, \varepsilon)$$

for all sufficiently small positive $\varepsilon \leq \varepsilon_0(V, v) > 0$. Conversely, if

$$\mathrm{Vol}\, B_v(\varepsilon) \leq \varepsilon^n \,\mathrm{Vol}\, B^n$$

for small ε then $\mathrm{Sc}_v \geq 0$. In other words, positivity of the scalar curvature amounts to V being *volume-wise sub-Euclidean on the microscopic level.*

$1\frac{1}{4}$. Exponential map $\exp_v : T_v(V) \to V$ and curvatures of products

This map is defined by sending each straight ray $\overline{r}$ in $T_v(V) = \mathbb{R}^n$ issuing from zero to the geodesic ray r in V issuing from v in the direction of $\overline{r}$, such that the $\overline{r}$-parametrization of r is (locally) isometric. Thus the ε-balls $B(\varepsilon) \in T_v(V)$ around zero go onto ε-balls in V around v, and so the above expansion formula for $\operatorname{Vol} B_v(V, \varepsilon)$ can be equivalently expressed in terms of the Jacobian of the map $\exp_v$ near zero as follows:

$$\varepsilon^{-n-2} \int\limits_{B(\varepsilon)} (\operatorname{Jac} \exp_v(x) - 1)\, dx \underset{\varepsilon \longrightarrow 0}{\longrightarrow} -\beta_n \operatorname{Sc}_v \qquad (+)$$

$$\text{for } \beta_n = (6n)^{-1} \operatorname{Vol} B \quad \text{and} \quad B = B(\mathbb{R}^n, 1).$$

Now take $V = V_1 \times V_2$ with the metric $g = g_1 \oplus g_2$ and observe that geodesics in V are given by obviously pairing those in V_1 and V_2. In other words the exponential map $\exp_v$ for V from $T_v(V) = T_{v_1}(V_1) \times T_{v_2}(V_2)$, where $v = (v_1, v_2)$, to V is the Cartesian product of the exponents $\exp_{v_1} : T_{v_1}(V_1) \to V_1$ and $\exp_{v_2} : T_{v_2}(V_2) \to V_2$. Therefore,

$$\operatorname{Jac} \exp_v\big(x = (x_1, x_2)\big) = \big(\operatorname{Jac} \exp_{v_1}(x_1)\big)\big(\operatorname{Jac} \exp_{v_2}(x_2)\big)$$

which leads (by an easy computation) to the additivity of the scalar curvature under the Riemannian products.

The scalar curvature of $V = (V_1 \times V_2\ g_1 \oplus g_2)$ *is*

$$\operatorname{Sc} V = \operatorname{Sc} V_1 \oplus \operatorname{Sc} V_2,$$

that is,

$$\operatorname{Sc}_v = \operatorname{Sc}_{v_1} + \operatorname{Sc}_{v_2} \ \textit{for all points } v = (v_1, v_2) \textit{ in } V.$$

Homogeneous examples. It is not hard to compute with (+) that

(a) The unit sphere S^n has constant scalar curvature $n(n-1)$ and the sphere of radius R has $\operatorname{Sc} = R^{-2} n(n-1)$.

(b) In general, the scaled manifold $RV \underset{\text{def}}{=} (V, R^2 g)$ has $\operatorname{Sc}(RV) = R^{-2} \operatorname{Sc} V$.

(c) The hyperbolic space H^n with the sectional curvature -1 has $\operatorname{Sc} = -n(n-1)$.

(d) The Cartesian product of the round ε-sphere S^2 by H^n has

$$\operatorname{Sc}(\varepsilon S^2 \times H^n) = 2\varepsilon^{-2} - n(n-1)$$

which is > 0 for $\varepsilon < \sqrt{2/n(n-1)}$.

(e) Let G be a compact Lie group with a biinvariant metric. Then the scalar curvature is constant ≥ 0 and it is > 0 unless G is a torus. Furthermore, the corresponding metric on each homogeneous space $V = G/H$ also has $\mathrm{Sc} \geq 0$ which is moreover > 0, unless V is a torus. (All this easily follows from the fact that the Riemannian exp equals the Lie-theoretic one for the biinvariant metrics on G.)

Conclusion. *Every compact homogeneous space different from a torus admits an invariant metric with* $\mathrm{Sc} > 0$. (This is also true for those *non-compact* homogeneous Riemannian spaces where the implied isometry group admits a nontrivial compact semisimple factor.)

(f) Every symmetric space V of *non-compact* type has $\mathrm{Sc} \leq 0$ and $\mathrm{Sc} = 0$ implies that V is Riemannian flat (i.e. locally isometric to $\mathbb{R}^n$).

(g) Every connected non-Abelian solvable Lie group G with a left invariant metric has $\mathrm{Sc} < 0$. (Abelian groups are Riemannian flat and have $\mathrm{Sc} = 0$.)

$1\frac{4}{5}$. Collapse with $\mathrm{Sc} > 0$

We shall eventually face the following:

Basic Question. Does the sign of the scalar curvature have any visible macroscopic effect on the geometry of V?

The ultimate "No" for $\mathrm{Sc} < 0$ is asserted by the following *dense h-principle,*

(Lohkamp) *Every Riemannian metric on V can be C^0-approximated by metrics with* $\mathrm{Sc} \leq -1$, *provided* $\dim V \geq 3$.

But what can be expected for $\mathrm{Sc} \geq 0$? Recall that the stronger condition, Ricci > 0, propagates from micro to macro scale. Namely $\mathrm{Ricci}_v > 0$ amounts, microscopically speaking, to the inequality

$$|\operatorname{Jac} \exp_v x| < 1,$$

for all x in a sufficiently small ε-ball in $T_v(V) = \mathbb{R}^n$ around the origin. Remarkably, this inequality, properly reformulated, integrates to the large-scale and implies the following:

Bishop inequality. *Every R-ball in a complete Riemannian manifold with* Ricci > 0 *has volume < volume (Euclidean R-ball).*

Now we want something similar for $\mathrm{Sc} > 0$, but we must be careful in view of our earlier example (d) of the product metric on $\varepsilon S^2 \times H^{n-2}$, say with $\varepsilon = 1/n(n-1)$, which has $\mathrm{Sc} \geq 1$, and yet the volume of the R-ball in this manifold is exponentially growing in R. So the size of V with $\mathrm{Sc}\, V > 0$ cannot be limited merely in terms of the volume. However this product example agrees with the following principle, which will be made precise later on.

The condition $\mathrm{Sc}(V) \geq \varepsilon^{-2}$ *makes* V *look* $(n-2)$*-dimensional on the macroscopic scale* $\gg \varepsilon$*, and as* $\varepsilon \to 0$*, the manifold* V *collapses to something of dimension* $n-2$. Here are some variations of the product example which illustrate this principle.

(i) Take a compact k-dimensional submanifold W_0 in a Riemannian manifold W of dimension $n+1$ and let $W_\varepsilon \subset W$ be the ε-neighbourhood of W_0. To grasp the geometry of W_0 and of its boundary $V_\varepsilon = \partial W_\varepsilon$ near a point $w_0 \in W_0$, we scale W_ε by ε^{-1}, i.e. look at $\varepsilon^{-1} W_\varepsilon$ at w_0. As $\varepsilon \to 0$, this blow-up by ε^{-1} straightens the pair (W, W_0), i.e. this converges to the Euclidean pair $\left(\mathbb{R}^{n+1} \overset{=}{} T_{w_0}(W),\ \mathbb{R}^k = T_{w_0}(W_0)\right)$, and so $\varepsilon^{-1} W_\varepsilon$ metrically converges to the product of the unit Euclidean ball B^{n-k+1} by $\mathbb{R}^k$. Hence the scalar curvature of $\varepsilon^{-1} V_\varepsilon = \partial(\varepsilon^{-1} W_\varepsilon)$ is about $\mathrm{Sc}\,(S^{n-k}) = (n-k)(n-k-1)$ and $\mathrm{Sc}\ V_\varepsilon \approx \varepsilon^{-2}$ for $n-k \geq 2$, which agrees with our principle as ∂V_ε collapses to W_0 for $\varepsilon \to 0$.

(ii) Let us generalize the above by taking a piecewise smooth polyhedron for $W_0 \subset W$. Now the boundary $V_\varepsilon = \partial W_\varepsilon$ has corners, but these can be easily smoothed away without losing much positivity of $\mathrm{Sc}(V_\varepsilon)$. So the smoothed manifolds V_ε collapse to W_0 with $\mathrm{Sc}\ V_\varepsilon$ blowing up as ε^{-2}, provided $\mathrm{codim}\, W_0 \geq 3$. The first interesting case here is that of a connected 1-polyhedron (graph) W_0 in $\mathbb{R}^4$ where the resulting manifolds V_ε are homeomorphic to connected sums of several copies of $S^2 \times S^1$ (see Fig. 1).

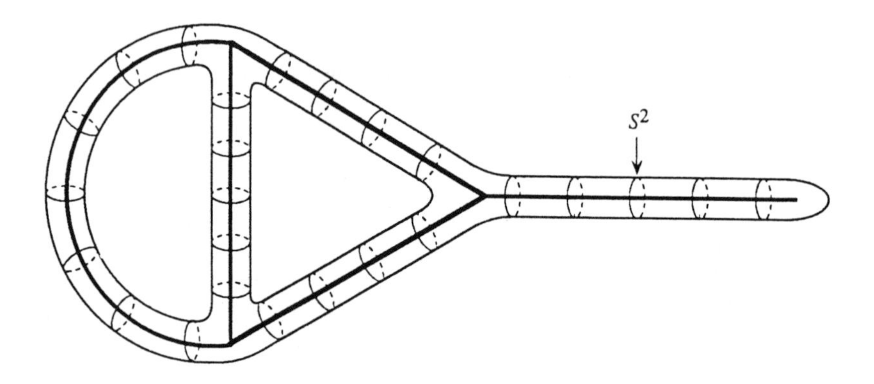

Figure 1

$1\frac{5}{6}$. Surgery for $\mathrm{Sc} > 0$

If we take a framed m-dimensional sphere in an n-dimensional manifold V with positive scalar curvature and do surgery, then the resulting manifold V' admits a rather natural metric with $\mathrm{Sc} > 0$, provided $n - m \geq 3$. In fact, our handle is $S^{n-m-1} \times B^{m+1}$ which can be made very thin, i.e. with $S^{n-m-1} = S_\varepsilon^{n-m-1}$ of small radius ε which has scalar curvature about ε^{-2} and the required metric on V' is obtained out of this by smoothing at the corners; see Fig. 2 below.

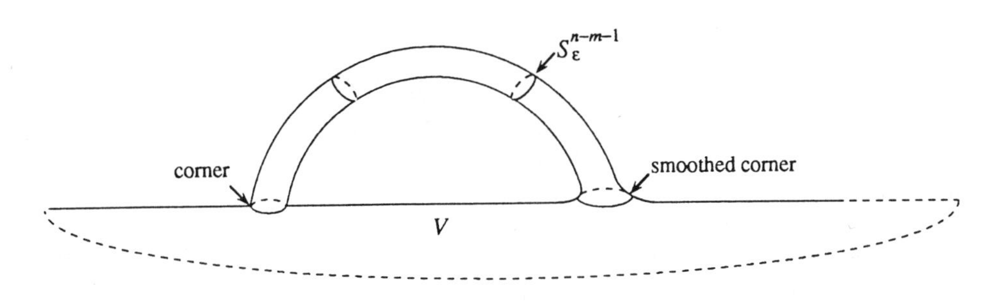

Figure 2

It follows that the existence of a metric of positive scalar curvature is a *spin cobordism invariant* for closed simply connected spin manifolds of dimension ≥ 5 (but definitely not for $\dim V = 4$) and in the non-simply connected case it is *spin bordism invariant* of the classifying map $V \to B\Pi$ for $\Pi = \pi_1(V)$.

Notice that after several surgeries the resulting manifold may turn out diffeomorphic to the original V but with a quite differently looking metric with $\mathrm{Sc} > 0$. In fact, the new metric may be sometimes non-homotopic to the original one in the space of metrics with $\mathrm{Sc} > 0$.

$1\frac{6}{7}$. List of closed manifolds with $\mathrm{Sc} > 0$

This starts with compact symmetric and locally symmetric spaces, e.g. projective spaces over $\mathbb{R}, \mathbb{C}$ and $\mathbb{H}$ and lens spaces; also there are many non-symmetric homogeneous and locally homogeneous spaces with $\mathrm{Sc} > 0$. Furthermore, one may take fibered manifolds $V \to B$ with the above (locally) homogeneous fibers. If we scale the metric in the fibers by ε, then the scalar curvature in the fibers, and *also* in V, blows up roughly as ε^{-2} and, in particular, becomes positive on V. One can slightly generalize by using *foliations* into compact locally homogeneous leaves with $\mathrm{Sc} > 0$. Moreover, one may allow degeneration of leaves as it happens to orbits of compact groups acting on V; whenever the non-degenerate fibers (or orbits) of such a degenerate foliation have metrics with $\mathrm{Sc} > 0$, so does V.

Next one can perform codim ≥ 3 surgeries, thus freely moving V within its spin (co)bordism class for $\dim V \geq 5$.

Finally, one can take smooth minimal hypersurfaces V' in manifolds with $\mathrm{Sc} > 0$ and these V' (or rather $V' \times S^1$) carry metrics with $\mathrm{Sc} > 0$ as well.

Question. Are there metrics with $\mathrm{Sc} > 0$ which are non-homotopic (even better, non-cobordant) to the above in the category $\mathrm{Sc} > 0$?

$1\frac{7}{8}$. Foliations with Sc > 0

Here we deal with foliations endowed with smooth leafwise Riemannian metrics, and $\mathrm{Sc} > 0$ refers to such a metric. It appears that many operations on foliations are compatible with $\mathrm{Sc} > 0$. For example, Reeb's twist of a codimension one foliation around a transversal curve can be made in the category $\mathrm{Sc} > 0$ for dim(leaves) ≥ 3 and the same seems to be true for generalized (Thurston) Reeb twists for higher codimension. So one might think that the existence of a foliation (of dimension ≥ 5) with $\mathrm{Sc} > 0$ is not significantly more restrictive than the existence of a metric with $\mathrm{Sc} > 0$ on all of the manifold. For example, a simply connected parallelizable manifold of dimension n should (?) carry foliations of all dimensions (between 5 and n) with $\mathrm{Sc} > 0$. Conversely, one may think that the existence of such a foliation leads in most cases to a metric with $\mathrm{Sc} > 0$ on the underlying manifold itself. (We shall indicate a proof of this for foliations of codimension one later on.)

Metrics with Sc > 0 *derived from foliations.* Let $\mathcal{F}$ be a smooth foliation on $\mathcal{V}$ with metrics g on $T(\mathcal{F})$ and g' on $T(\mathcal{V})/T(\mathcal{F})$. Choose a complementary (normal) bundle $T' \subset T(\mathcal{V})$ to $T(\mathcal{F}) \subset T(\mathcal{V})$, lift g' to T' where it is still called g' and let $\hat{g}_\lambda = \lambda^2 g' \oplus g$ on $T(\mathcal{V})$. Let us evaluate the curvature of g_λ for large $\lambda \to \infty$. We localize our attention at a single leaf $V \subset \mathcal{V}$ and observe for $\lambda \to \infty$ that the metric $\hat{g}_\lambda$ converges to the canonical metric on the bundle $T'_V \to V$ with the flat connection coming from the $\mathcal{F}$-monodromy. Namely, the space T'_V locally equals $V \times L$ for L being a linear space of dimension $k = \operatorname{codim} \mathcal{F}$ with metrics (Euclidean structures) g'_v on $L_v = v \times L$. These, together with $g|V$, give us a metric on $V \times L$, say $\overline{g}$, which is a *generalized warped product metric.* For example if $k = 1$ (and $L = \mathbb{R}$), one has $\overline{g} = g + \varphi^2 dt^2$ (as $g'_v = \varphi^2(v) g'_{v_0}$) which is the ordinary warped product. In particular, if g' is invariant under the monodromy, i.e. if $(\mathcal{F}, g')$ is transversally Riemannian, then $\overline{g}$ locally is the product metric $g \oplus$ Euclidean.

Conclusion. If $\mathcal{V}$ is a compact transversally Riemannian foliation, then every leafwise metric g with positive scalar curvature gives rise to a metric on $\mathcal{V}$, namely $\hat{g}_\lambda$ which has $\mathrm{Sc}(g_\lambda) > 0$ for large λ.

In the general case, the scalar curvature of the metric $\overline{g}$ on T'_V (and hence of $\hat{g}_\lambda \to \overline{g}$) is of the form $\mathrm{Sc}\,(\overline{g}) = \mathrm{Sc}\,\overline{g} + D_v(g'_v)$ where D_v is a combination of the first and second derivatives of g'_v with respect to the flat connection in T'_v. For example, in the codimension 1 case where $\overline{g} = g + \varphi^2 dt^2$, one has $\mathrm{Sc}(\overline{g}) = \mathrm{Sc}(g) + 2\Delta_g \varphi/\varphi$ for Δ_g being the positive (i.e. $-\sum_i \varphi_{ii}$) Laplace operator on (V, g). In particular, if a codim 1 foliation $\mathcal{F}$ admits a *smooth harmonic transversal measure*, then every leafwise metric with Sc > 0 gives rise to such a metric on $\mathcal{V}$ as "harmonic" amounts to $\Delta_g \varphi = 0$ in the above formula.

Connes' bundle $\mathcal{V}^* \to \mathcal{V}$. This is the bundle associated to the vector bundle $T(\mathcal{V})/T(\mathcal{F}) \to \mathcal{V}$ where the fiber at $v \in \mathcal{V}$ equals the space of Euclidean metrics in the fiber $T_v(\mathcal{V})/T_v(\mathcal{F})$. (Thus metrics g' in $T(\mathcal{V})/T(\mathcal{F})$ are sections $\mathcal{V} \to \mathcal{V}^*$).

Example. Let $\mathcal{V} = \mathbb{R}$ be foliated into points. Then $\mathcal{V}^*$ is the principal $\mathbb{R}^\times_+$-bundle associated to the tangent bundle $T(\mathbb{R})$ (or, more precisely, to the symmetric square of the cotangent bundle of $\mathbb{R}$). Thus $\mathcal{V}^*$ has a natural structure of a principal homogeneous space of the group $\mathrm{Aff}(\mathbb{R})$ which admits an invariant Riemannian metric of constant negative curvature. As this $\mathcal{V}^*$ fibers over $\mathbb{R}$, the fibers are geodesics while natural (horizontal) sections $\mathbb{R} \to \mathcal{V}^*$ (corresponding to translation invariant metrics on $\mathbb{R}$) are horocycles. In the general case of any k, the fiber of $\mathcal{V}^* \to \mathcal{V}$ is the homogeneous space $M = GL_k\mathbb{R}/O(k)$ which admits an invariant metric.

Denote by $\mathcal{F}^*$ the pull-back of the foliation $\mathcal{F}$ to $\mathcal{V}^*$. This has the same codimension k as $\mathcal{F}$, and the bundle $T(\mathcal{V}^*)/T(\mathcal{F}^*)$ is induced from $T(\mathcal{V})/T(\mathcal{F})$. Next, using the flat (monodromy) connection in the bundle $T(\mathcal{V})/T(\mathcal{F})$ along the leaves $V \subset \mathcal{V}$ of $\mathcal{F}$, we lift these leaves to $\mathcal{V}^*$, thus getting a foliation $\widetilde{\mathcal{F}}$ of $\mathcal{V}^*$ refining $\mathcal{F}^*$, where the leaves of $\widetilde{\mathcal{F}}$ project diffeomorphically to those of $\mathcal{F}$. Now *the bundle* $T(\mathcal{V}^*)/T(\mathcal{F}^*)$ *has a canonical metric called* g^*, since every point $v^* \in \mathcal{V}$ is, by definition, a metric in the underlying fiber of $T(\mathcal{V})/(\mathcal{F})$ which is canonically isomorphic to the fiber $T_{v^*}(\mathcal{V}^*)/T(\mathcal{F}^*)$. This metric is *not* transversally invariant for $\mathcal{F}^*$ (e.g. the natural metric on parallel horocycles is not invariant under the normal geodesic shift), but it is invariant under the $\widetilde{\mathcal{F}}$-monodromy.

Now we want to construct some metric on $\mathcal{V}^*$ starting from our g on the leaves of $\mathcal{F}$. This g lifts to the bundle $\widetilde{T} = T(\widetilde{\mathcal{F}}) \subset T(\mathcal{V}^v)$ where it is called $\tilde{g}$. We also have a metric h on the vertical bundle T^M of $\mathcal{V}^* \to \mathcal{V}$ corresponding to some invariant metric on M serving as the fiber of $\mathcal{V}^* \to \mathcal{V}$. What remains to do is to take some $T^* \subset T(\mathcal{V}^*)$ complementary to $T(\mathcal{F}^*) = T^M \oplus \widetilde{T}$ with the metric g^* borrowed from $T(\mathcal{V}^*)/T(\mathcal{F}^*)$ (isomorphic to T^*). This is done with $T' \subset T(\mathcal{V})$ complementary to $T(\mathcal{F})$ and some (Bott) $GL_k(\mathbb{R})$-connection in the Connes M-bundle $\mathcal{V}^* \to \mathcal{V}$ compatible with the flat connection along $\mathcal{F}$. Thus we have $T(\mathcal{V}^*)$ split into $T^* \oplus T^M \oplus \widetilde{T}$ with the metrics g^*, h and $\tilde{g}$ in these three bundles.

To better see what happens, let us temporarily forget about $\widetilde{T}$, i.e. assume the foliation $\mathcal{F}$ is zero dimensional. Then we take $\hat{g}^*_\lambda = \lambda^2 g^* \oplus h$ on $\mathcal{V}^*$ fibered over $\mathcal{V}$ with the fibers M_v, $v \in \mathcal{V}$, and observe that for $\lambda \to \infty$ the space $(\mathcal{V}^*, \hat{g}^*_\lambda)$ converges over each $v \in \mathcal{V}$ to the corresponding M-bundle over $\mathbb{R}^k = T_v(\mathcal{V})$, that is the homogeneous space $M_+ = \mathrm{Aff}(\mathbb{R}^k)/\,\mathrm{maxcomp} = \mathbb{R}^k \times M$ where this (natural) splitting is invariant under $\mathrm{Aff}\,\mathbb{R}^k$, the affine automorphism group of $\mathbb{R}^k$. Notice that this convergence may be (for $k \geq 2$) non-uniform in $m \in M$ albeit M is homogeneous. In fact, if $k \geq 2$ the metric $\hat{g}^*_\lambda$ for *each* λ may easily have unbounded curvature on a fiber M_v. To see this, observe that each fiber M_v is totally geodesic for $\hat{g}^*_\lambda$ since the holonomy of our (Bott) connection is isometric in the fibers. But (for $k \geq 2$) the embedding $M_v \to \mathcal{V}^*$ may be very far from isometric in terms of the distance function because the isometries (typically) have *unbounded* displacements on M as the group $GL_k\mathbb{R}$ is non-Abelian for $k \geq 2$. Therefore, an (isometric) monodromy of M_v around a (short) loop in $\mathcal{V}$ may move points $m \in M_v$ arbitrarily far in M_v. On the other hand, the lift of the loop to a horizontal path from m, say to $m' \in M_v$, may be short and so the distance in $\mathcal{V}^*$ small. It would be quite obvious if the metric g^* on the horizontal subbundle T^* were constant along the fibers, i.e. coming from $\mathcal{V}$.

In our case, g^* may be quite large in certain directions. Yet, as $M = M^0 \times \mathbb{R}$ for $M^0 = SL_k\mathbb{R}/SO(k)$, we always can move in the direction $-\infty$ in $\mathbb{R}$ which makes g^* small but does not change the displacement (or the length of the Killing fields) in the M^0-direction. If $k \geq 2$, this can be achieved in M^0 without the help of the $\mathbb{R}$-factor, but the case of pure M^0 for $k = 2$ is unclear to me. Here we have a surface $\mathcal{V}$ with a given *area* element and take $\mathcal{V}^*_0$ consisting of the metrics of *unit* area, so that $\mathcal{V}^*_0$ fibers over $\mathcal{V}$ with the fiber hyperbolic plane $SL_2\mathbb{R}/SO(2)$. Every $SL_2(\mathbb{R})$ (e.g. a Levi Civita) connection on $\mathcal{V}$ gives a metric to $\mathcal{V}^*$, the geometry of which needs clarification (at least in the mind of the author).

Now we return to $\mathcal{F}$ and look at the metric $\hat{g}^*_{\lambda,\varepsilon} = \lambda^2 g^* \oplus h_\varepsilon \oplus \tilde{g}$ where $h_\varepsilon = \varepsilon^{-2} h$ for a small $\varepsilon > 0$. As $\lambda \to \infty$ the metric $\hat{g}^*_{\lambda,\varepsilon}$ approaches, over each point $v \in V$, the product metric $h^+_\varepsilon \oplus g_V$ on $M_+ \times V$, where h^+_ε is the limit of the above $\hat{g}^*_\lambda = \lambda^2 g^* \oplus h_\varepsilon$ for $\lambda \to \infty$ and V denotes the leaf of $\mathcal{F}$ through $v \in \mathcal{V}$. When ε is small, so are the (absolute values of) the curvatures of h_ε and h^+_ε; so the sign of the scalar curvature of $h^+_\varepsilon \oplus g_V$ is determined by that of g_V.

Summing up, let the metric g on $\mathcal{F}$ have $\mathrm{Sc} \geq \delta^2 > 0$. If $\varepsilon > 0$ is sufficiently small, then for each compact subset $\Upsilon^* \subset \mathcal{V}^*$ there exists $\lambda(\Upsilon^*)$, such that the metric $\hat{g}^*_{\lambda,\varepsilon}$ has $\mathrm{Sc}(\hat{g}^*_{\lambda,\varepsilon}) \geq \delta^2/2$ on Υ^* for $\lambda \geq \lambda(\Upsilon^*)$. Moreover, the geometry of $\hat{g}^*_{\lambda,\varepsilon}$ at each point $v^* \in \Upsilon^*$ is close to the sums $h^+_\varepsilon \oplus g_V$, where V is the leaf of $\mathcal{F}$ through the point $v \in \mathcal{V}$ under v^*. In particular, the local geometry of $\hat{g}^*_{\lambda,\varepsilon}$ is bounded on Υ^* for $\lambda > \lambda(\Upsilon^*)$ (with the implied bound independent of Υ^*).

Remarks (a) The group $\mathbb{R}_+^\times$ naturally acts on $\mathcal{V}^*$ as metrics are multiplied by $\mu \in \mathbb{R}_+^\times$. It is clear that the metrics $\hat{g}^*_{\lambda,\varepsilon}$ have bounded geometry along μ-orbits for $\mu \to \infty$ (but not for $\mu \to 0$) and the above "convergence" of $\hat{g}_{\lambda,\varepsilon}$ to the product metric is uniform on the $[1,\infty)$-orbit of Υ^*.

(b) Here as everywhere throughout this discussion "limit" means "limit of Riemannian manifolds". For example, given a metric g on V, the manifolds $(\mathcal{V}, \lambda^2 g, v)$ converge, for $\lambda \to \infty$, to $\mathbb{R}^k = T_v(\mathcal{V})$, while the metrics $\lambda^2 g$ on a *fixed* $\mathcal{V}$ would diverge; the convergence is achieved by adjusting the coordinate gauge in $\mathcal{V}$.

Example. Let $\operatorname{codim} \mathcal{F} = 1$ and let us look at $\mathcal{V}^*$ over a small coordinate neighbourhood in $\mathcal{V}^*$ locally split as $V^*[0,1]$ for some leaf V in this neighbourhood. Then all metrics $g^*_{\lambda,\varepsilon}$ over such a neighbourhood are bi-Lipschitz to $V \times H'$ where H' is the region in the hyperbolic plane between two asymptotic geodesics. Furthermore, the space $\mathcal{V}^*$ *globally* is obtained, up to bi-Lipschitz equivalence, from the product metric on $\mathcal{V} \times \mathbb{R}$ by modifying it in the T^*-direction by the conformal factor expt (where we assumed $\mathcal{F}$ coorientable).

Notice that the metrics $\hat{g}^*_{\lambda,\varepsilon}$ have in this case locally bounded geometries (provided $\mathcal{V}$ is compact) and one may take $\Upsilon^* = \mathcal{V}^*$. The reason is the commutativity of the group $GL_1\mathbb{R} = \mathbb{R}^\times$.

2. Macroscopic (asymptotic) dimension $\dim_\varepsilon$

A metric space V has the *macroscopic dimension on the scale* $\gg \varepsilon$ *at most* k if, by definition, there exists a k-dimensional polyhedron P and a continuous map $\varphi : V \to P$ such that the fibers $\varphi^{-1}(p) \subset V$ are all ε-small, in the sense that $\operatorname{Diam} \varphi^{-1}(p) \le \varepsilon$ for all $p \in P$. This is expressed in writing by

$$\dim_\varepsilon V \le k$$

and then $\dim_\varepsilon V$ is defined as the supremum of the integers k for which this inequality holds.

The macroscopic dimension can be made ε-free in the following somewhat opposite (mutually dual) cases.

I. The space V is infinite, i.e. has $\operatorname{Diam} V = \infty$ and $\dim_\varepsilon V \le k$ for *some*, possibly large, $\varepsilon < \infty$.

II. Instead of a single space we are given a family V_t, $t \to \infty$, such that $\dim_{\varepsilon(t)} V_t \le k$ for $\varepsilon(t) \underset{t \to \infty}{\to 0}$.

In the first case we say that the *asymptotic* (or macroscopic with unspecified scale) *dimension* $\le k$. The second case can be thought of as a *collapse* of V_t to something k-dimensional for $t \to \infty$.

Example. Let $V = V_0 \times \mathbb{R}^k$, where V_0 is bounded, i.e. $\operatorname{Diam} V_0 = \delta < \infty$. Then, clearly, the asymptotic dimension of V is $\leq k$. In fact, $\dim_\varepsilon V \leq k$ for all $\varepsilon \geq \delta$. Furthermore, the classical dimension theory (compare (B') below) implies that the asymptotic dimension of $\mathbb{R}^k$ is $\geq k$, i.e. every continuous map $\varphi : \mathbb{R}^k \to P$ with $\dim P < k$ has

$$\sup_{p \in P} \operatorname{Diam} \varphi^{-1}(p) = \infty.$$

Thus the asymptotic (or macroscopic) dimension of $V_0 \times \mathbb{R}^k$ is exactly k.

$2\frac{1}{2}$. Uniform contractibility

A metric space V is called *uniformly contractible* if every ball $B_v(R)$ in V is contractible inside some concentric ball $B_v(\rho)$, $\rho \geq R$, where $\rho = \rho(R)$ does not depend on x (but may depend on V). For instance, the Euclidean space is uniformly contractible with $\rho(R) = R$. Furthermore, every contractible space V which admits a proper (e.g. discrete) action of an isometry group with compact quotient is (obviously) uniformly contractible. On the other hand the (contractible) surface in Fig. 3 below is not uniformly contractible.

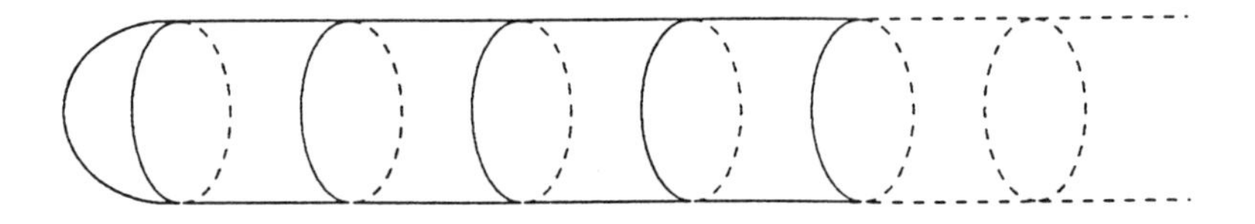

Figure 3

Proposition-Example. *Let V be a complete uniformly contractible Riemannian manifold of dimension k. Then the asymptotic dimension of V equals k. (Observe that the asymptotic dimension of the surface in Fig. 3 equals one.)*

Proof. Let $\varphi : V \to P$ be a map with $\operatorname{Diam} \varphi^{-1}(p) \leq R < \infty$ for all $p \in P$ and let $C_\varphi \supset V$ be the cylinder of this map (i.e. the space obtained by attaching the cylinder $V \times [0,1]$ to P via $v \times 1 \mapsto \varphi(v) \in P$ for all $v \in V$). We assume at this point without loss of generality that the image of φ equals all of P and then retract C_φ on V by appealing to the uniform contractibility of V and an elementary obstruction theory. This retraction, say $q : C_\varphi \to V$, will move each point by a bounded amount, something like $\rho_k = \underbrace{\rho(\rho(\ldots\rho(R)\ldots)}_{k+1}$ and the composition $q \circ \varphi : V \to V$ (recall that φ maps V to P and P sits inside C_φ) is a *proper* map within bounded distance from the identity. It follows, again from the uniform contractibility of V, that $q \circ \varphi$ is *properly homotopic* to $\mathrm{Id} : V \to V$ and therefore, having degree 1, cannot be factored through a map to a polyhedron of dimension $< \dim V$. Thus $\dim_\varepsilon V = \dim V$ for all positive $\varepsilon < \infty$.

Product example. Let V_1 be a compact Riemannian manifold and $V_t = t^{-1}V_1 \times V_2$, where, recall, $t^{-1}V_1 \underset{\mathrm{def}}{=} \left(V_1, t^{-2}g_1\right)$. Then for each $\varepsilon > 0$

$$\dim_\varepsilon V_t = \dim V_2 \quad \text{for } t \geq t(\varepsilon),$$

as V_t collapses to V_2 for $t \to \infty$.

Now we can state a specific conjecture relating macroscopic dimension to the scalar curvature.

Conjecture. Let V be a complete Riemannian manifold with $\mathrm{Sc}(V) \geq \varepsilon^{-2} > 0$. Then the asymptotic dimension of V is at most $\dim V - 2$. In fact one expects $\dim_\delta V \leq \dim V - 2$ for all $\delta \geq c_n \varepsilon$ where one may try to guess the value of c_n by looking at $V = S^n$.

This conjecture looks hard at the present moment (it is proven only for $\dim V = 3$) but still it is not strong enough to capture the full idea of $(n-2)$-dimensionality on the ε-scale for n-manifolds with $\mathrm{Sc} \geq \varepsilon^{-2}$. Namely, the pull-backs $\varphi^{-1}(p) \subset V$ of suitable maps $\varphi : V \to P$, for *known* V with $\mathrm{Sc} \geq \varepsilon^{-2}$, have small (about ε^2) areas as well as diameters and can be ε-small in an even stronger sense. On the other hand the *known* $(n-2)$-dimensionality bounds on V with $\mathrm{Sc}(V) \geq \varepsilon^{-2}$ (see § 5) do not imply (at least not directly) any geometric closeness of V to an actual $(n-2)$-dimensional space.

$2\frac{2}{3}$. Degression to foliations, recurrent dimension and ends of groups

Let $\mathcal{V}$ be a foliated space. It is called *non-recurrent* if each leaf V in $\mathcal{V}$ is a closed subset in $\mathcal{V}$ and there is a neighbourhood Υ of V in $\mathcal{V}$ such that the restriction of our foliation to Υ is Hausdorff, i.e. the space of leaves is Hausdorff.

Basic example. The foliation into the connected components of the fibers of a submersion is non-recurrent.

Next define $\mathrm{recdim}\, \mathcal{V}$ as the minimal number k such that $\mathcal{V}$ can be covered by $k+1$ open subsets where the restriction of the foliation to each subset is non-recurrent. Thus, non-recurrent foliations have recdim $= 0$ according to this definition.

Proposition. *If $\mathcal{V}$ is foliated into n-dimensional manifolds then* $\mathrm{recdim}\, \mathcal{V} \leq n$.

Here one should make some mild assumptions on $\mathcal{V}$, such as paracompactness, and the simplest case is where $\mathcal{V}$ is a smooth foliated *manifold.* Then one may take a sufficiently fine smooth generic triangulation Tr of $\mathcal{V}$ and observe that for a small neighbourhood $\mathcal{U}_0$ of the $(m-n)$-skeleton of Tr for $m = \dim \mathcal{V}$, the induced foliation on $\mathcal{U}_0$ is non-recurrent. Next one takes some open subsets $\mathcal{U}_i$, $i = 1, \dots, n$, such that $\bigcup_{i=0}^{n} \mathcal{U}_i = \mathcal{V}$, where each $\mathcal{U}_i$ for $i \geq 1$ is a union of small disjoint subsets (corresponding to $(m-i+1)$-dimensional simplices in Tr) where the foliation is (obviously) non-recurrent. Q.E.D.

Now let $\mathcal{V}$ be compact with a leafwise Riemannian metric. Then, clearly, each leaf V has

$$\dim_\varepsilon V \leq \operatorname{recdim} \mathcal{V},$$

for all sufficiently *large* ε. Consequently, if all leaves are contractivle, then $\operatorname{recdim} \mathcal{V} = n$.

Problem. Suppose all leaves have $\dim_\varepsilon \leq k$ for some $\varepsilon > 0$. Is $\operatorname{recdim} \mathcal{V} \leq k$ here as well?

A similar problem can be formulated for (e.g. universal) coverings $\widetilde{V}$ of compact manifolds V as follows.

Define $\dim(\widetilde{V}/V)$ as the minimal number k, such that V can be covered by $k+1$ open subsets U_i, $i = 0, \dots, k$, where each connected lift of U_i to $\widetilde{V}$ is relatively compact. (If V is non-compact, one requires $\widetilde{V}$ be "almost trivial" over each U_i meaning that the connected components of the pull-backs of U_i to $\widetilde{V}$ are mapped back to V finite-to-one.) Now one observes that this $\dim \widetilde{V}/V$ bounds the macroscopic dimension of $\widetilde{V}$ (at least for Galois coverings where "finite-to-one" has "finite" $\leq$ const) and one asks oneself if the opposite is true. Here one has the famous Stallings' theorem about ends of groups which refines the implications

$$\dim_\varepsilon \widetilde{V} \leq 1 \Rightarrow \dim \widetilde{V}/V \leq 1$$

as follows.

Stallings' decomposition theorem. *Let $p : \widetilde{V} \to V$ be the universal covering of a closed manifold V. Then there is a closed (possibly disconnected) hypersurface $H \subset V$ such that*

(1) *H admits a compact lift to $\widetilde{V}$;*

(2) *the closure of each connected component of the complement $\widetilde{V} - p^{-1}(H)$ has at most one end (where the compact (!) boundary components of $\widetilde{V} - p^{-1}(H)$ corresponding to lifts of H are not counted for ends).*

Notice that if $\dim_\varepsilon \widetilde{V} \leq 1$ for some $\varepsilon > 0$, then "the one end" condition makes the above components of $\widetilde{V} - p^{-1}(H)$ relatively compact and the decomposition $V = U_\varepsilon(H) \operatorname{cup}(V - H)$ makes $\dim \widetilde{V}/V \leq 1$, where $U_\varepsilon(\dots)$ denotes the ε-neighbourhood of ....

The above manifold version of Stallings' theorem was suggested by Matthew Brin, who also indicated the following proof using minimal surfaces (which he has never published and which I had the pleasure of discovering myself). If $\widetilde{V}$ has more than one end, take a *volume minimizing* hypersurface $\widetilde{H}_1$ in $\widetilde{V}$ separate some of the ends. The idea is that each deck (Galois) transform of $\widetilde{H}_1$, say $\widetilde{H}'_1$, either coincides with $\widetilde{H}_1$ or does not meet $\widetilde{H}_1$ at all, because out of a pair of intersecting hypersurfaces one could easily concoct a third one with volume $< \mathrm{Vol}\, \widetilde{H}_1$ and still separate some ends. Then we take all transforms of $\widetilde{H}_1$, and if the complement contains a component with more than one end, we take the second minimal separating hypersurface, say $\widetilde{H}_2$ inside this component. The transforms of $\widetilde{H}_2$ miss $\widetilde{H}_1$ and we continue until the process stops at some $\widetilde{H}_m$, such that all components of the complement of the Galois transforms of the $\widetilde{H}_i$ are one-ended. Then the projections of these $\widetilde{H}_2$ to V make our $H = \bigcup_{i=1}^{m} H_i$.

End decomposition of foliations. *Let $\mathcal{V}$ be a compact space foliated into smooth manifolds. Then there exists a compact subset $\mathcal{H} \subset \mathcal{V}$ such that*

1. *Intersection of $\mathcal{H}$ with each leaf V in $\mathcal{V}$ consists of a disjoint union of compact subset in V.*

2. *Each connected component of $V - \mathcal{H}$ has at most one end for all leaves V (where the boundary components of $V - \mathcal{H}$ are not counted for the ends).*

Sketch of the proof. Fix a leafwise Riemannian metric in $\mathcal{V}$ and take a hypersurface H_1 in a leaf which is volume minimizing among all end separating hypersurfaces in the leaves or in the monodromy coverings of these. Then take the second such hypersurface (in the complement $\mathcal{V} - H_2$), say H_2, and continue by transfinite induction thus arriving at a closed set $\mathcal{H}_0 = \mathcal{C}\ell \bigcup_{i\in I} H_i \subset \mathcal{V}$ with the following properties.

1. If $V - \mathcal{H}_0$ for some leaf V has a component with more than one end, then the closure of this component contains some $H_i \subset \mathcal{H}_0$. Or, equivalently, a slightly moved H_i separates ends in this components.

2. If H_i and H_j have mutually ε-close points then H_i is Hausdorff δ-close to H_j for some $\delta = \delta(\varepsilon) \to 0$, for $\varepsilon \to 0$.

It follows that for the monodromy covering of each leaf $\mu_V : \widetilde{V} \to V$, all connected components of the pull-back $\mu_V^{-1}(\mathcal{H}_0)$ are compact and the connected components of the complement $\widetilde{V} - \mu_V^{-1}(\mathcal{H}_\varepsilon) \subset \widetilde{V}$ are (at most) one-ended where $\mathcal{H}_\varepsilon$ denotes an ε-neighbourhood of $\mathcal{H}_0$ for $\varepsilon > 0$. It follows that $\mathcal{H} = \mathcal{H}_\varepsilon$ for a small $\varepsilon > 0$ satisfies our requirements. Q.E.D.

Corollary. *If the monodromy covering $\widetilde{V}$ of each leaf V has* $\dim_\varepsilon \widetilde{V} \leq 1$ *for some $\varepsilon > 0$, independent of V, then* $\mathrm{recdim}\, \mathcal{V} \leq 1$.

Example. One knows that every complete simply connected 3-manifold with $\mathrm{Sc} \geq \delta^2 > 0$ has $\dim_\varepsilon \leq 1$ for $\varepsilon \geq 12\pi\delta$. Hence, a compact foliation into such manifolds has $\mathrm{recdim} \leq 1$.

Remark and questions. The geometric (e.g.) smooth nature of the leaves is not relevant in the decomposition theorem and, as in the group theoretic case, one can probably state and prove everything in terms of the corresponding groupoid of the foliation.

It is unclear what should be a higher dimensional version of the decomposition theorem but the above corollary probably generalizes to a similar implication

$$\dim_\varepsilon \widetilde{V} \leq k \text{ for all leaves } V \Rightarrow \mathrm{recdim}\, \mathcal{V} \leq k$$

under suitable restrictions on $\mathcal{V}$. For example, if the universal covering $\widetilde{V}$ of a compact manifold V has $\dim_\varepsilon \widetilde{V} \leq k$, then one may expect the classifying map $V \to B\Pi$ for $\Pi = \pi_1(V)$ to be contractible to the k-skeleton of $B\Pi$ provided Π has no torsion. In particular, if $\dim_\varepsilon \widetilde{V} < n = \dim V$, then the image of the fundamental class $[V]$ in $H_n(B\Pi;\mathbb{Q})$ must be (?) trivial torsion or no torsion. It also appears in many examples that $\dim_\varepsilon \widetilde{V} < n = \dim V \Rightarrow \dim_\varepsilon \widetilde{V} \leq n-2$, i.e. the macroscopic (asymptotic) dimension of the universal covering $\widetilde{V}$ avoids being equal to $\dim V - 1$. (On the homotopy theoretic level this would say that, whenever $[V]$ goes to zero in $H_n(B\Pi)$, V contracts to the $(n-2)$-skeleton in $B\Pi$ rather than to the $(n-1)$-skeleton as the first level obstruction theory predicts. In fact, this may be true, at least for $\dim V \geq 4$, if the torsion is properly taken into account.)

Uncdim *and* brdim. One can modify the definition of recdim by declaring a foliation $\mathcal{F}$ *simple* on $\Upsilon \subset \mathcal{V}$ if it is non-recurrent and the universal covering of each leaf essentially trivializes on U, i.e. the inclusions of the leaves of $\mathcal{F}|\Upsilon$ to those of $\mathcal{F}$ have finite (and uniformly bounded) π_1-images. Then uncdim is defined with decompositions $\mathcal{V} = \bigcup\limits_i \Upsilon_i$ where $\mathcal{F}|\Upsilon_i$ is simple for all i.

Next, for foliations on *non-compact* spaces $\mathcal{V}$, one may use $\Upsilon \subset \mathcal{V}$ on which the leaves are *non-recurrent and bounded* with respect to a given leafwise metric and define brdim accordingly (where "br" stands for bounded recurrency).

Subadditivity of dimension. All these dimensions (and also the asymptotic dimension) are (obviously) subadditive, if $\mathcal{V}$ is covered by $\mathcal{V}_i$, $i = 1, \dots, m$, then

$$\text{"dim"}\mathcal{V} \leq \left(\sum_{i=1}^{m} \text{"dim"}\mathcal{V}_i\right) + m - 1.$$

Monotonicity of dimension. *Let $\mathcal{V}$ be given two foliations, $\mathcal{F}$ and $\mathcal{F}'$ refining $\mathcal{F}$, i.e. the leaves of $\mathcal{F}'$ are contained in those of $\mathcal{F}$. Then*

$$\dim \mathcal{F} - \operatorname{recdim} \mathcal{F} \geq \dim \mathcal{F}' - \operatorname{recdim} \mathcal{F}', \tag{$*$}$$

where "dim" *refers to the dimension of the leaves.*

Proof. Let $\mathcal{V} = \bigcup_{i=0}^{m} \Upsilon_i'$ be the covering with $m = \operatorname{recdim} \mathcal{F}'$ where $\mathcal{F}'$ is non-recurrent on each Υ_i', and consider the continuous map $\varphi_0' : \mathcal{V} \to \Delta^m \subset \mathbb{R}^{m+1}$ corresponding to (a partition of unity associated to) this covering. We approximate φ_0' by a generic smooth map $\varphi' : \mathcal{V} \to \Delta^m$ and consider the partition $\mathcal{P}'$ of $\mathcal{V}$ refining $\mathcal{F}'$ into the connected components of φ' restricted to the leaves of $\mathcal{F}'$. Then we take the quotient space $\mathcal{K} = \mathcal{V}/\mathcal{P}$ with $q : \mathcal{F} \to \mathcal{K}$ denoting the quotient map and look at the image $q(\mathcal{F})$ of $\mathcal{F}$ under q. This is a (rather singular) foliation of the (compact) space $\mathcal{K}$ (with $\dim \mathcal{K} = m + \dim \mathcal{F} - \dim \mathcal{F}'$) of leaf dimension $\dim \mathcal{F} - \dim \mathcal{F}' + m$ and $(*)$ follows from the (easy) inequalities $\operatorname{recdim} q(\mathcal{F}) \leq \dim q(\mathcal{F})$ and $\operatorname{recdim} \mathcal{F} \leq \operatorname{recdim} q(\mathcal{F})$.

Notice that a similar monotonicity is satisfied by uncdim and brdim as well as by asympdim. For example, *no closed aspherical manifold admits a foliation with* uncdim < dim. In particular, *it admits no m-dimensional foliation with $m \geq 2$ where the universal coverings of the leaves have* $\dim_\varepsilon \leq 1$ *for some* $\varepsilon > 0$. (Probably this is true for all $\dim_\varepsilon < m$.)

Corollary. *A closed aspherical manifold supports no 3-dimensional foliation with* Sc > 0.

(This is unknown for foliations of dimension ≥ 4, not even of codimension zero.)

3. Remarks and References on positivity of curvature

(a) *Hierarchy of curvatures.* The curvature tensor can be viewed as a *quadratic form* Q on $\Lambda^2 T(V)$ and the positive definiteness of this Q is one of the strongest curvature positivity conditions studied by geometers. For example, all compact symmetric spaces have $Q \geq 0$ while $Q > 0$ distinguishes the spheres (and real projective spaces). The restriction of Q to *bivectors* in $\Lambda^2 T(V)$ gives us the *sectional curvature* $K(V)$ (and, conversely, Q can be defined as the unique quadratic extension of K from the Grassmannian $\mathrm{Gr}_2 T(V)$, or the set of bivectors in $\Lambda^2 T(V)$, to all of $\Lambda^2 T(V)$). This $K(V)$ is the only known curvature whose positivity has an adequate macroscopic description, which allows, in particular, a comprehensive theory of *singular* spaces with $K \geq 0$ (see [B-G-P] and [Per]). The sectional curvature K, viewed as a function on the Grassmann bundle $\mathrm{Gr}_2 T(V)$, extends to a function, denoted $K_{\mathbb{C}}$, on the complex Grassmann bundle $\mathrm{Gr}_2 \mathbb{C}T(V)$ as follows. First Q extends by complex multilinearity to $\mathbb{C}T(V)$ and then $K_{\mathbb{C}}(\tau)$ for $\tau \in \mathrm{Gr}_2\,\mathbb{C}T(V)$ is defined by

$$K_{\mathbb{C}}(\tau) = K_{\mathbb{C}}(\alpha \wedge \beta) = Q(\alpha \wedge \beta, \overline{\alpha \wedge \beta}),$$

where α et β are two vectors in τ which are orthonormal for the Hermitian extension of the Riemannian metric g of V to $\mathbb{C}T(V)$. Clearly, positivity of $K_{\mathbb{C}}$ mediates between $Q > 0$ and $K > 0$. Next, following Micallef and Moore, one restricts $K_{\mathbb{C}}$ to the subspace $\mathrm{Gr}_2^{\mathrm{isotr}}\,\mathbb{C}T(V) \subset \mathrm{Gr}_2\,\mathbb{C}T(V)$ which consists of those τ on which the $\mathbb{C}$-linear extension of g to (a $\mathbb{C}$-quadratic form on) $\mathbb{C}T(V)$ vanishes. This restricted curvature is denoted by $K_{\mathbb{C}}^{\mathrm{isotr}}$ and the condition $K_{\mathbb{C}}^{\mathrm{isotr}} > 0$ is significantly weaker than $K_{\mathbb{C}} > 0$. Here is the diagram summarizing our curvature positivity conditions.

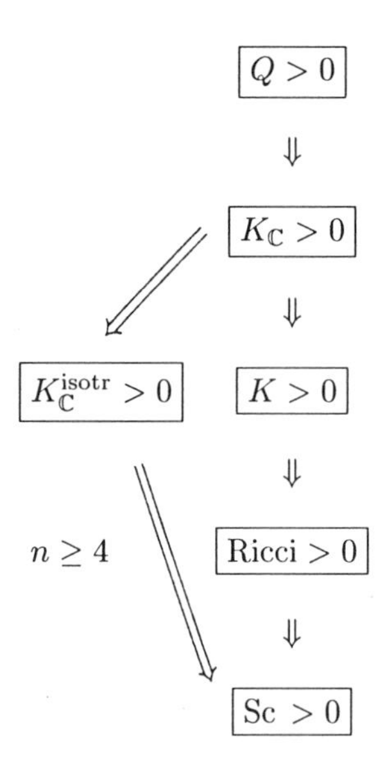

A geometric exposition of these curvatures is given in $[\mathrm{Gro}]_{\mathrm{Sig}}$, where the reader finds further references.

(b) ***On*** $\mathbf{K}_{\mathbb{C}}^{\mathrm{isotr}} > \mathbf{0}$. This condition is vacuous for $n = \dim V \leq 3$ where $\mathrm{Gr}_2^{\mathrm{isotr}}$ is empty but for $n \geq 4$ it implies, according to Micallef and Moore, that V has zero homotopy groups $\pi_2(V), \ldots, \pi_m(V)$ for $m = n/2$ if n is even and $m = (n-1)/2$ if m is odd. On the constructive side, manifolds with $K_{\mathbb{C}}^{\mathrm{isotr}} > 0$ admit 1-dimensional surgery (which was pointed out to me by Mario Micallef) and so the boundaries of ε-neighbourhoods of graphs (i.e. 1-complexes) in W can be smoothed with $K_{\mathbb{C}}^{\mathrm{isotr}} > 0$. More generally, connected sums of spherical spaceforms (with $K = 1$) and copies of $S^{n-1} \times S^1$ can be given metrics with $K_{\mathbb{C}}^{\mathrm{isotr}} > 0$ and, topologically speaking, no other manifold with $K_{\mathbb{C}}^{\mathrm{isotr}} > 0$ is anywhere in sight. Here is the corresponding geometric conjecture:

If $K^{\mathrm{isotr}}_{\mathbb{C}}(V) \geq \varepsilon^{-2}$ and $\dim V \geq 4$ then V is macroscopically 1-dimensional on the scale $\gg \varepsilon$. In particular, the fundamental group $\pi_1(V)$ contains a free subgroup of finite index.

This conjecture would follow (compare (c) below) if one could prove, by extending the method of Micallef-Moore, that every stable minimal disk D in V satisfies $\mathrm{dist}(v, \partial D) \leq \mathrm{const}_n\, \varepsilon$ for all $v \in D$.

(c) If Sc $V \geq \varepsilon^{-2}$ and $\dim V = 3$, then every closed curve γ in V homologous to zero has $\mathrm{Fill\,Rad}\,\gamma \leq 2\pi\varepsilon$, that is γ bounds within its $2\pi\varepsilon$-neighbourhood. This is proven in [G-L]$_{\mathrm{PSC}}$ by looking at the minimal surface in V filling-in γ. (The role of minimal varieties for Sc > 0 was revealed by the earlier work of Schoen and Yau.) Finally, with the bound on Fill Rad γ, one can conclude that

$$\dim_\delta V \leq 1 \quad \text{for } \delta = 12\pi\varepsilon.$$

(See § 10 in [G-L]$_{\mathrm{PSC}}$, Appendix 1 in [Gro]$_{\mathrm{Fil}}$ and [Katz].)

(d) *A dream of* $\dim_\varepsilon \leq$ k *and curvature.* We want to have, for given n and $k < n$, some curvature expression, say $K^{(k)}$, with the usual scaling property, such that

(i) the Cartesian products

$$V = \mathbb{R}^\ell \times S^{n-\ell}, \text{ for } \ell < k, \text{ have } K^{(k)}(V) > 0,$$

(ii) the inequality $K^{(k)}(V) \geq -\varepsilon^2$ implies $\dim_\varepsilon V \leq k$ for all complete n-dimensional Riemannian manifolds (where one should be ready to modify the definition of $\dim_\varepsilon$, e.g. in the spirit of the K-area, if the geometry calls for it).

(iii) The open cone in the space $\mathcal{R}_n$ of the curvature tensors on $\mathbb{R}^n$ defined by the inequality $K^{(k)} > 0$ should be convex or at least connected. (We tacitly assume this cone is $O(n)$-invariant to have our curvature condition meaningful.)

Question. For which n and k does such $K^{(k)}$ (or, equivalently, the corresponding open $O(n)$-invariant cone in $\mathcal{R}_n$) exist? (One is also interested in metrics with $|K^{(k)}| \leq \varepsilon^2$ for $K^{(k)}$ coming from suitable models similar to $\mathbb{R}^k \times S^{n-k}$; these were recently studied by Christophe Margerin using the heat flow in the space of metrics.)

Our optimism is warmed up by the geometry of Euclidean hypersurfaces where $\dim_\varepsilon \leq k$ is linked to an appropriate k-convexity (which we shall explain somewhere else).

(e) *Curvature* h*-principles.* Curvature inequalities can be looked on as particular partial differential relations, and these cannot be integrated to macroscopically visible geometric properties unless a corresponding h-principle fails (see $[\text{Gro}]_{\text{PDR}}$). Recently, Lohkamp proved (using surgery) several powerful h-principles for Sc < 0 and Ricci < 0 thus distroying all (?) hope for a macroscopic geometry for negative scalar and Ricci curvatures. We refer to his papers $[\text{Loh}]_{\text{CLP}}$, $[\text{Loh}]_{\text{GLC}}$, $[\text{Loh}]_{\text{GNR}}$ where he presents and develops his ideas, giving, in particular, the h-principle view on the micro $\Rightarrow$ macro correspondence in Riemannian geometry.

(f) ***On asymptotic dimensions.*** There are several non-equivalent notions of the asymptotic dimension (see, e.g. § 4 in $[\text{Gro}]_{\text{RTG}}$ and $[\text{Gro}]_{\text{AI}}$) but here we emphasized the one directly linked to the *Uryson's width* (see $[\text{Gro}]_{\text{Wid}}$). A quite different notion comes from the idea of the *asymptotic cone* of V, which is an *ultralimit* of εV for $\varepsilon \to 0$. (The idea of ultralimits was injected into geometric context by Van Den Dries and Wilkie, see [VDD-Wi], which I neglected to indicate in $[\text{Gro}]_{\text{AI}}$ where this idea is systematically exploited.) For example, the hyperbolic space H^n has, according to our present definition, asymptotic dimension n. Yet εH^n, $\varepsilon \to 0$, converges to an $\mathbb{R}$-*tree* which is a 1-dimensional space.

(g) Surgery for Sc > 0 provides a non-trivial link between Riemannian geometry and the bordism theory, as exposed in the ICM-talk by Stephen Stolz (see [Sto] and [Ro-St]). Also notice that some surgery is possible for stronger positivity conditions indicated in (d).

4. K-area of a manifold

We want to introduce a certain Riemannian invariant, called K-*area* of V reminiscent of the ordinary area of surfaces with "K" referring to K-theory as well as to the curvature. This K-area is defined by looking at the curvatures of Hermitian vector bundles $X \to V$ endowed with Hermitian connections. Recall that the curvature of X, denoted $\mathcal{R}(X)$, is an End$X$-valued 2-form on V. We equip End$X$ with the *operator norm*, i.e.

$$\|\text{end}\| = \sup_{\|x\|=1} \|x - \text{end}x\|_X,$$

and accordingly, define $\|\mathcal{R}\|$ as $\sup\|\mathcal{R}(\alpha \wedge \beta)\|$ over all orthonormal bivectors $\alpha \wedge \beta$ in V. A relevant feature of this choice of norm is the following strong subadditivity relation, for Whitney sums,

$$\|\mathcal{R}(X \oplus Y)\| = \max\big(\|\mathcal{R}(X)\|,\ \|\mathcal{R}(Y)\|\big),$$

which will become crucially important later on.

Now we define the K-*area* for closed oriented $2m$-dimensional Riemannian manifolds V by maximizing $\|\mathcal{R}(X)\|^{-1}$ over the unitary bundles $X \to V$ for which (at least) one characteristic (Chern) number of X does not vanish. This means that the classifying map of V into the classifying space, say $C\ell_X : V \to BU$ is not homologous to zero, i.e. cannot be contracted to the $(2m-1)$-skeleton of BU. Thus

$$K\text{-area}V = \left(\inf_X \|\mathcal{R}(X)\|\right)^{-1}$$

where the infimum is taken over by the above "homologically significant" bundles X with unitary connections and so this K-area is large, say $\geq \varepsilon^{-1}$, if and only if V admits a "homologically significant" bundle X with small curvature $\|\mathcal{R}(X)\| \leq \varepsilon$.

The definition of the K-area generalizes to open manifolds by sticking to bundles $X \to V$ trivialized at infinity and using the characteristic numbers coming from the cohomology with compact supports. Next one takes care of odd dimensional manifolds by stabilizing

$$K\text{-area}_{\text{st}} V = \sup_k K\text{-Area}\,(V \times \mathbb{R}^k),$$

where one takes those $k \geq 0$ for which $\dim V + k$ is even. Finally, observe that the definition of the K-area extends the homology classes $h \in H_*(V;\mathbb{R})$ by minimizing $\|\mathcal{R}(X)\|$ over those X for which the (classifying) homomorphism $(C\ell_X)_* : H_*(V) \to H_*(BU)$ does not vanish at h (where we may use homology with infinite supports for non-compact manifolds V).

Let us point out at this stage that the K-area is *strictly* positive as every V of dimension $2m$ admits a bundle X with non-zero (top) Chern class $c_m(X) \in H^{2m}(V)$. In fact, one can induce such a bundle over V from a standard bundle over S^{2m} by a map $V \to S^{2m}$ of degree one. Also notice that the K-area scales as the ordinary area,

$$K\text{-area}(\lambda V) = \lambda^2(K\text{-area}\,V).$$

Furthermore, if $V_1 \succ \lambda V_2$, i.e. V *Lipschitz-*λ^{-1} *dominates* V_2 in the sense that there exists a proper λ^{-1}-Lipschitz map $f : V_1 \to V_2$ (where λ^{-1}-Lipschitz amounts to $\|Df\| \leq \lambda^{-1}$) of non-zero degree, then

$$K\text{-area}\,V_1 \geq \lambda^2(K\text{-area}\,V_2). \tag{$*$}$$

Conversely, if V_1 admits a (locally) λ-expanding equidimensional embedding (not a mere immersion!) into V_2 (i.e. an embedding f with $\|D^{-1}f\| \geq \lambda$) then

$$K\text{-area}\ V_1 \leq \lambda^{-2}(K\text{-area}\,V_2). \tag{$**$}$$

In particular every open subset $U \subset V$ has (for the induced metric)

$$K\text{-area}\,U \leq K\text{-area}\,V.$$

It follows from $(*)$ that every *hyper-Euclidean* manifold V, i.e. satisfying $V \succ \mathbb{R}^n$, has K-area $= \infty$.

$4\frac{1}{4}$. K-area for $\pi_1 = 0$

Every compact simply connected manifold V without boundary has K-area$_{\text{st}} < \infty$.

Proof. We recall the following bound on the monodromy M of a unitary connection along the boundary of a disk D,

$$\|M - 1\| \leq 2\sin\left(\frac{1}{2}\|\mathcal{R}\| \cdot \operatorname{area} D\right) \tag{$\square$}$$

where $\|\mathcal{R}\|$ is the sup-norm of the curvature of our fibration $X = (X, \nabla)$ over D (and where the disk D comes along with some Riemannian metric). In fact, the validity of ($\square$) for infinitesimal squares in D (where $2\sin\varepsilon/2 \sim \varepsilon$) follows from the very definition of the curvature and the global inequality ($\square$) is obtained by the (multiplicative) integration of the infinitesimal one. (The role of "sin" is to compare two metrics on the unitary group $U(N)$, where the first one is induced from the metric $\|A - B\|$ on matrices and the second is the corresponding intrinsic *length metric* on $U(N)$. For example, if $N = 1$ and $U(1) = S^1 \subset \mathbb{R}^2$, the first metric is the Euclidean one of $\mathbb{R}^2$ restricted to S^1 and the second one is given by the arc-length on S^1.)

Now we return to $X \to V$, fix a point $v_0 \in V$, join v_0 with every other point $v \in V$ by a minimal geodesic segment γ and transport the fiber X_{v_0} to X_v along γ. If there are two such segments, say γ and γ' between v_0 and v, we obtain two unitary (holonomy or transport) operators, say M_v, $M'_v : X_{v_0} \to X_v$, where the norm $\|M_v - M'_v\|$ can be estimated according to ($\square$) by the area of the minimal disk D filling in the loop $\gamma \circ (\gamma^1)^{-1}$ as follows:

$$\|M_N - M'_v\| \leq 2\sin\left(\frac{1}{2}\|\mathcal{R}\| \cdot \operatorname{area} D\right), \tag{$\square'$}$$

for $\mathcal{R} = \mathcal{R}(X)$ (see Fig. 4 below).

Next let α denote the supremum of the areas of the above minimal disks over all $v \in V$ and all pairs of minimal segments between v_0 and v and suppose that

$$\|\mathcal{R}\| < \pi\alpha^{-1}/3 \quad \text{and so} \quad \|M_v - M'_v\| \leq \delta < 1$$

for all v, γ and γ'. Then, for every v, the convex combinations of the operators $M_v, M'_v, M''_v, \ldots$ corresponding to different segments are all *non-singular*, which allows us to smooth the multivalued correspondence $v \mapsto \{M_v, M'_v, \ldots\}$ to a

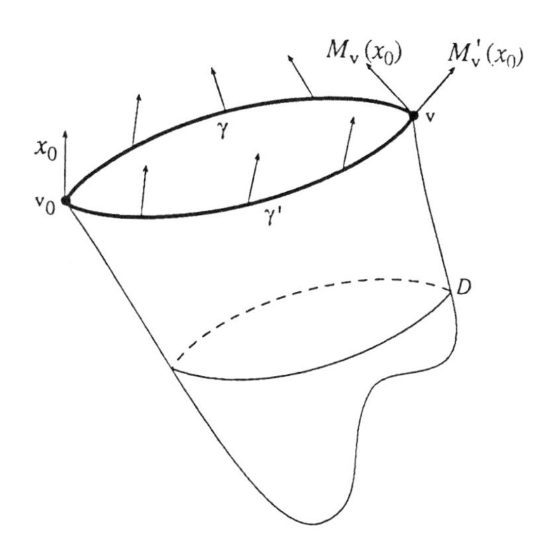

Figure 4

continuous field of *non-singular* operators, say $\overline{M}_v : X_{v_0} \to X_v$ for v running over V. Thus every ε-flat bundle with $\varepsilon < \pi\alpha^{-1}/3$ is trivial and so

$$K\text{-area}\, V \leq 3\alpha/\pi.$$

Furthermore, every ε-flat bundle over $V \times W$ with arbitrary W can be induced by the above argument from a bundle over W which implies the desired bound

$$K\text{-area}_{\,\mathrm{st}} V \leq 3\alpha/\pi < \infty. \qquad (\Delta)$$

Remark. The above argument also works if V has non-empty *connected* boundary. Yet the unit segment $[0,1]$ has infinite stable K-area. (We suggest to the reader to figure out what happens for $\dim V > 1$.)

$4\frac{1}{3}$. K-area under homotopies

The relations K-area $V = \infty$ and $K\text{-area}_{\mathrm{st}} V = \infty$ are homotopy invariants of V for all (possibly non-simply connected) compact manifolds V. (Compact manifolds with infinite K-area do exist as indicated in (iii) below.) This is also true (and obvious) for (possibly infinite) coverings $\widetilde{V}$ of V. For example if V_1 and V_2 are homotopy equivalent *compact* manifolds then their universal coverings satisfy

$$K\text{-area}\, \widetilde{V}_1 = \infty \Longleftrightarrow K\text{-area}\, \widetilde{V}_2 = \infty.$$

$4\frac{1}{2}$. K-area in examples

(i) *Every connected surface of genus zero (i.e. embeddable into S^2) has*

$$K\text{-area}_{\text{st}} = \text{const} \cdot \text{area},$$

where const $\leq \frac{3}{2\pi}$ by (Δ) since $\alpha = \text{area}/2$ for the 2-spheres.

Exercise (suggested to me by Richard Montgomery). Show that, in fact, const $= \frac{1}{2\pi}$. In particular, no bundle X over S^2 with $c_1(X) \neq 0$ has curvature smaller than the Hopf bundle. (It would be nice to find a *sharp* bound on $K\text{-area}_{\text{st}}\, V$ for all (simply connected) V in terms of areas of suitable surfaces in V.)

(ii) *The rectangular solid $V = [0, \ell_1] \times [0, \ell_2] \times \ldots \times [0, \ell_n]$, where $\ell_1 \leq \ell_2 \leq \ldots \leq \ell_n$ has*

$$\text{const} \cdot \ell_1 \ell_2 \leq K\text{-area}\, V \leq \text{const}'\, \ell_1 \ell_2.$$

Thus the inequality $K\text{-area}\, V \leq \varepsilon^2$ makes V "area-wise" ε-close to the $(n-2)$-dimensional space $[0, \ell_3] \times \ldots \times [0, \ell_n]$.

Our next example is more surprising.

(iii) *Every connected surface V of positive genus has K-area $= \infty$.* To see this we first observe the following:

$4\frac{3}{5}$. Push-forward inequality and K-area$^+$

Let $f : \widetilde{V} \to V$ be a finitely sheeted covering which is trivial at infinity (and at the boundary) of V, i.e. the sheets are disconnected at infinity. Then

$$K\text{-area}\, V \geq K\text{-area}\, \widetilde{V}. \qquad (\widetilde{\star})$$

Proof. A bundle $\widetilde{X} \to \widetilde{V}$ goes down to the bundle X over V with the fiber

$$X_v = \bigoplus_{\tilde{v}} \widetilde{X}_{\tilde{v}}$$

where the sum is taken over all $\tilde{v} \in f^{-1}(v)$. If $\widetilde{X}$ has a non-zero Chern number then so does X and, clearly, $\|\mathcal{R}(X)\| = \|\mathcal{R}(\widetilde{X})\|$ while the implied trivialization of $\widetilde{X}$ at infinity gives us such a trivialization on V. Q.E.D.

Remarks. (a) One can also lift bundles from V to $\widetilde{V}$ which gives us the opposite inequality, and hence, the equality

$$K\text{-area}\, V = K\text{-area}\, \widetilde{V}. \qquad (\star\star)$$

(b) It is quite useful to have ($\star$) and a suitable version of ($\star\star$) for infinite coverings as we shall see later on.

In fact our present definition of the K-area is provisional. Ultimately one should work with *virtual bundles* (K^0-classes) $\kappa = [X_1]-[X_2]$ where the bundles X_1 and X_2 are joined by a homomorphism $F : X_1 \to X_2$ which is a unitary connection preserving isomorphism outside a compact subset in V and where, furthermore, one should allow *infinite dimensional* X_1 and X_2 with *Fredholm* F (see $9\frac{1}{6}$). Alternatively, one can refine the notion of K-area by minimizing $\|\mathcal{R}(X)\|$ over smaller classes of bundles X, e.g. by requiring non-vanishing of a specific Chern number, (say, $c_m(X)[V]$) and/or limiting the rank of X (but the latter makes the K-area finite for all V of finite volume as indicated below).

Now we can prove **(iii)** by considering finite (connected) coverings $\widetilde{V}$ of V with arbitrarily large area (and, hence, K-area) satisfying triviality assumption at infinity. (If V has genus zero then every covering $\widetilde{V} \to V$ trivial at infinity has disconnected sheets all over V.)

***Remark on rank* X.** It is crucial here that we do not limit rank X in the definition of K-area as the K-area with a priori bounded rank X is finite for all *compact* (possibly non-simply connected) V. In fact, if

$$(\operatorname{Vol} V)\|\mathcal{R}(X)\|^{-m} \leq C^{-1}(m,r)$$

for $2m = \dim V$, $r = \operatorname{rank} X$ and some (universal) constant $C > 0$ (polynomial in m and r), then (by Chern-Gauss-Bonnet) all Chern numbers of X vanish.

K-area$^+$. If we follow the above suggestion and define the K-area with virtual bundles $\kappa = [X_1] - [X_2]$, where X_1 and X_2 are (actual) unitary bundles isomorphic at infinity, where κ has a non-zero Chern number and where we minimize $\max(\|\mathcal{R}(X_1)\|, \|\mathcal{R}(X_2)\|)$, the result, denoted K-area$^+$, may be significantly greater than the K-area for *open* manifolds V. For example, *every* surface V with *infinite* fundamental group has K-area$^+(V) = \infty$. In fact, the *K-area$^+$ satisfies the push-forward inequality*

$$K\text{-area}^+ V \geq K\text{-area}\, \widetilde{V} \qquad (\widetilde{\star}^+)$$

for all finite coverings $\widetilde{V} \to V$.

But there is a price to pay: the K-area$^+$ is not monotone for equidimensional embeddings. For example, K-area$^+(S^2) = K$-area$(S^2) < \infty$ while non-simply connected open subsets $U \subset S^2$ have K-area$^+ U = \infty$ according to the above.

Examples (continued). **(iv)** Let V_1 be S^2 minus three disks and V_2 be the torus minus a disk. Then $V_1 \times \mathbb{R}$ is (almost obviously) diffeomorphic to $V_2 \times \mathbb{R}$ but it is infinitely smaller than $V_2 \times \mathbb{R}$ "K-area-wise". This is very much similar to the behaviour of the *stable symplectic area* (sometimes called "width" or "capacity") and, in fact, our K-area can be brought into the symplectic ambience as we shall see later on.

(v) *The n-torus T^n has*

$$K\text{-area}\, T^n = \infty.$$

In fact, the self-mapping (endomorphism) $t \mapsto 2t$ provides a (2^n-sheeted) covering of T^n by $2T^n$ (i.e. T^n with the doubled metric) which implies by $(\widetilde{\star})$ that

$$K\text{-area}\, T^n \geq K\text{-area}\, 2T^n = 4(K\text{-area}\, T^n).$$

Hence, K-area $T^n = \infty$ as it is > 0.

(v′). *Let V be a closed manifold with non-positive sectional curvature. If the fundamental group $\pi_1(V)$ is residually finite (which is known to be the case if V is locally symmetric, for example) then K-area $V = \infty$.*

Proof. The residual finiteness of V implies that for each $R > 0$ there exists a finite covering $\widetilde{V} \to V$ where every loop of length $\leq 2R$ at some point $\tilde{v} \in \widetilde{V}$ is contractible. Then the exponential map at $\tilde{v}$ gives us an expanding embedding of the Euclidean R-ball $B(R) = RB(1)$ into $\widetilde{V}$ and so K-area $V \geq$ K-area $\widetilde{V} \geq R\big(K\text{-area}\, B(1)\big)$ which makes K-area $V = \infty$ for $R \to \infty$.

***About* K-*area*$^+$.** The above also applies to complete non-compact manifolds V and shows that K-area$^+ V = \infty$ (while the K-area may be finite, e.g. for surfaces of genus zero).

(v″) ***Remark.*** It is likely that "most" (even among compact) manifolds V with negative curvature admit no non-trivial *finite* covering $\widetilde{V}$. But the above can be generalized to infinite (e.g. universal) coverings with a suitable class of *infinite dimensional* (virtual) bundles mentioned earlier.

(vi) ***Questions.*** It is unclear if our currently used K-area appealing to finite dimensional bundles X is infinite for all V with $K(V) \leq 0$. Moreover, there is no known example of a *closed* aspherical manifold of finite K-area. In fact one would like to have K-area $\widetilde{V} = \infty$ for the universal coverings $\widetilde{V}$ of closed aspherical manifolds V and also for more general π_1-*essential* V for which the classifying map $V \to B\Pi$, $\Pi = \pi_1(V)$, sends the fundamental class $[V]$ to a non-zero element in $H_n(B\Pi; \mathbb{Q})$, $n = \dim V$. One can even aspire to prove that a (suitable) K-area $V = \infty$ whenever some (e.g. universal) covering $\widetilde{V}$ of V has $\dim_\varepsilon V \geq \dim V - 1$ for all *sufficiently large* ε. No counterexample has been found so far.

(vii). ***Distinguishing strict* (K $<$ 0) *and non-strict* (K $\leq$ 0) *negativity of the sectional curvature by the* K-*area.*** Take the R-ball

$B = B(R)$ in a complete simply connected manifold V with $K(V) \leq 0$. Such a ball admits a proper R^{-1}-Lipschitz map onto S^n of degree one (where "proper" means constant on ∂B and where S^n is normalized to have the interior diameter 1) and so

$$K\text{-area}\, B \geq \text{const}_n\, R^2, \tag{+}$$

provided $n = \dim V$ is even (since S^n supports a bundle X with $c_m(X) \neq 0$, $m = n/2$). Now we claim that (+) can be substantially improved for $K(V) \leq -1$. Namely, for $R \geq 2$ we have

$$K\text{-area}\, B \geq \text{const}_n \exp R. \tag{$\times$}$$

In fact the $(R-1)$-sphere in such a V is exponentially large and admits a Lipschitz-$\exp -R$ map to the unit S^{n-1} of degree one. The suspension of such a map gives us a proper *exponentially area contracting* map $B = B(R) \to S^n$. Q.E.D.

An exponential inequality similar to (X) remains valid whenever $K(V) \leq -\kappa < 0$ as is seen with the scaling $V \mapsto \kappa V$, but if $K(V)$ vanishes somewhere the situation radically changes. For example,

Let V be a symmetric space of $\mathbb{R}$-rank ≥ 2 (which amounts for symmetric spaces not to have $K < 0$). Then

$$K\text{-area}\, B \leq \text{const}\, R^2. \tag{$-$}$$

Proof. We may scale the ball B to $R = 1$ and then, for $\mathbb{R}$-rank $\geq 2, B$ can be "swept over" by unit flat 2-disks which give us a universal bound on the K-area. Q.E.D.

Remark. If $\mathbb{R}$-rank $V = r$, there is a proper 1-Lipschitz map $B = B(R) \to S^n$ of degree one which exponentially (in R) contracts the volumes of all $(r+1)$-dimensional submanifolds in B. But the volume contraction on r-submanifolds cannot be stronger than R^{-r} as $B(R)$ can be "swept over" by flat R-balls.

$4\frac{2}{3}$. $\mathcal{R}$-norm on $\mathbf{K}^\circ$

Our definition of the K-area should have been, logically speaking, preceded by the notion of the $\mathcal{R}$-*norm* ($\mathcal{R}$ for curvature) on the even K-theory which assigns to each $\kappa \in K^\circ(V)$ the minimal (infimal) number R, such that κ can be represented as a (formal) difference of two bundles with unitary connections, say $\kappa = [X] - [Y]$, where $\max(\|\mathcal{R}(X)\|, \|\mathcal{R}(Y)\|) \leq \mathcal{R}$. This assignment, denoted $\kappa \mapsto \|\mathcal{R}(\kappa)\|$, defines a positive function $K^\circ(V) \to \mathbb{R}_+$ encoding significantly more geometric information about V than the K-area. Part of this information is homotopy invariant, namely the subgroup $K_0^\circ \subset K^\circ$ consisting of κ with $\|\mathcal{R}(\kappa)\| = 0$, clearly is a homotopy invariant for compact manifolds V. Moreover, the equivalence class of the set of subgroups $K_\varepsilon^\circ = \{\kappa \mid \|\mathcal{R}(\kappa)\| \leq \varepsilon\}$ for $\varepsilon \to 0$ is also a homotopy invariant, where two subsets (of subgroups in K°) are declared equivalent if they differ only by finitely many members. (These K_0° and $\{K_\varepsilon^\circ\}$ depend, as we know, only on the image of K°(classifying space of $\Pi = \pi_1(V)$) in $K^\circ(V)$ but little is known about K_0° and K_ε° for general groups Π. I do not even see an immediate example where the set $\{K_\varepsilon^\circ\}_{\varepsilon>0}$ is infinite and, hence, is not equivalent to $\{K_0^\circ\}$. The most optimistic individual would equate $K_0^\circ(V)$ (at least in $K^\circ \otimes \mathbb{Q}$) with the above image of $K^\circ(B\Pi)$ in $K^\circ(V)$. I would not take this seriously for general *non-residually finite* groups but the residually finite case (especially, where $B\Pi$ is a finite complex) leaves room for hope.)

Next one is tempted to use a suitable L_p-norm of $\mathcal{R}(X)$ and thus distinguishing *L_p-bundles* over non-compact Riemannian manifolds V where, by definition, this norm is finite. If we have such an L_p-bundle over $V \times [0,1]$ then the restrictions to $V = V \times 0$ and $V = V \times 1$ should be declared L_p-equivalent which leads us to a definition of a (nonreduced) group $L_pK_0(V)$. This group carries a norm coming from our L_p-norm on $\mathcal{R}(X)$ and one can reduce it by dividing by the subgroup consisting of κ with $\|\kappa\|_{L_p} = 0$. Furthermore, one may tensor with $\mathbb{R}$ thus getting a Banach space which is Hilbert for $p = 2$.

The first question arising here is the comparison with the L_q-cohomology. For example, when a given L_q-cohomology class of degree $2d$ can be represented as Chern of an L_p-bundle with $p = qd$? (This is easy for degree 2 as every exact 2-form serves as the curvature of some unitary line bundle.) The most interesting manifolds where we want to know $L_pK_0(V)$ are those with cocompact isometry groups and especially contractible ones (e.g. of non-positive curvature) but the easiest (and yet interesting) examples are provided by disjoint unions of compact manifolds.

Examples. (a) Let V be an even dimensional hyperbolic space, say H^{2m} and $\varphi : H^{2m} \to S^{2m}$ be an injective conformal map. Then the pull-back X of every bundle Y over S^{2m} has $\|X\|_{L_p} = \|Y\|_{L_p} < \infty$ for $p = m$, by the obvious conformal invariance of the L_q-norm on d-forms for $q = d/2m$. Now we may take a bundle Y over S^{2m} for which the m-th Chern form has non-zero integral over $\varphi(H^{2m}) \subset S^m$ which clearly makes our X non-trivial in the reduced $L_mK^0(H^{2m})$. (But I am afraid these elements in L_mK° can be generated by line bundles which would make them less interesting.)

(b) Let $V = \mathbb{R}^{2m}$ and $p = m$. Then, by the conformal invariance, every bundle over S^{2m} with non-zero top Chern class gives us a non-zero element κ in the reduced $L_pK^\circ(\mathbb{R}^{2m})$. This k definitely does not come from line bundles as the reduced L_m-cohomology of $\mathbb{R}^{2m}$ vanishes (by an elementary and well known argument, see $[\text{Gro}]_{\text{AI}}$ for instance). But κ can be made with a compact support on $\mathbb{R}^{2m}$ which is somewhat disappointing as, in fact, every orientable $2m$-dimensional V obviously admits a κ with compact support (and hence in L_p for all p) which is non-zero in the reduced L_pK° for all $p \leq m$ having non-trivial top Chern class in H^* with compact supports.

$4\frac{3}{4}$. K-area of symplectic manifolds

Let ω be a non-singular 2-form on V. A Riemannian metric g is called *adapted* to ω if at each point $v \in V$ there is a g-orthonormal coframe, say $x_i, y_i \in T_v^*(V)$, $i = 1, \ldots, m$, for $2m = \dim V$, such that $\omega_v = \sum_{i=1}^{m} x_i \wedge y_i$. Equivalently, g is adapted if the operator A_ω defined by $\langle A_\omega x, y\rangle_g = \omega(x,y)$ has all eigenvalues of the absolute values one.

Now we set

$$K\text{-area}\,(V,\omega) = \sup_g K\text{-area}\,(V,g)$$

over all metrics g adapted to ω. If $dw = 0$ (and thus ω is symplectic) and the cohomology class $[\omega]$ is integral, then ω serves as the curvature of a complex line bundle ℓ (we disregard here the usual $2\pi i$-coefficient) and if V is a closed manifold we have non-zero Chern number $c_1^m[V] = \int_V \omega^m$. Thus we see that K-area $(V,\omega) \geq 1$ in this case. What is less obvious is the opposite inequality,

$$K\text{-area}\,(V,\omega) \leq \text{const} < \infty,$$

for certain symplectic manifolds (V,ω) which is non-trivial even for the "smallest" manifold of all, the unit ball in $(\mathbb{R}^{2m}, \omega = \sum_{i=1}^{m} dx_i \wedge dy_i)$. In fact, the K-area is finite for those V which can be "swept over" by *rational pseudo-holomorphic curves* and, consequently, for open subsets in such V. Here is the simplest example.

Let $V = \mathbb{C}P^m$ with the standard symplectic form ω. Then

$$K\text{-area}\,(V,\omega) \leq \text{const}_0 < \infty.$$

Proof. If g is an adapted metric, then one defines a rational pseudo-holomorphic curve in V as a g-harmonic map $f : S^2 \to V$ whose g-area equals ω-area

$$\int_{S^2} f^*(\omega) = g\text{-area } f(S^2). \tag{$*$}$$

One knows (see $[\text{Gro}]_{\text{PHC}}$) that (at least for generic g) there exists a smooth $(m-2)$-dimensional family of these curves, i.e. a smooth map $F : P \times S^2 \to$

$\mathbb{C}P^m$, where P is a closed manifold, such that $\deg F \neq 0$ and all (pseudo-holomorphic) spheres $F_p : S^2 \to \mathbb{C}P^m$ are homotopic to the projective line and thus have by $(*)$ *unit* g-areas. Now every bundle $X \to \mathbb{C}P^m$ with small curvature lifts to a bundle $\widetilde{X}$ over $P \times S^2$ with small curvatures on the S^2-fibres which makes $\widetilde{X}$ inducible from some bundle over P since these fibres are simply connected and have unit area. As $\dim P < 2m$, the classifying map $P \times S^2 \to BU$ contracts to the $(2m-2)$-skeleton which implies that the bundles $\widetilde{X}$ and X are "homologically insignificant". Q.E.D.

Conjecture. The K-area is finite for all split symplectic manifolds $V = V_0 \times S^2$.

This follows from $[\mathrm{Gro}]_{\mathrm{PHC}}$ for many V_0 and the general case is feasible in view of [La-McD].

5. Scalar curvature and K-area

Let us explain how to bound the K-area in terms of the scalar curvature, under the following assumption on our Riemannian manifold V.

V is oriented and spin. Recall that the special orthogonal group $\mathcal{SO}(n)$ for $n \geq 2$ has a unique non-trivial (i.e. connected) double cover called $\mathrm{Spin}(n) \to \mathcal{SO}(n)$. Consider the oriented frame bundle of V, say Fr where each fiber equals $\mathcal{SO}(n)$, and recall that a *spin structure* on V is a double cover $\widetilde{\mathrm{Fr}} \to \mathrm{Fr}$ which restricts to a *non-trivial* (i.e. connected) cover over each fiber of $\mathrm{Fr} \to V$. Then V is called *spin* if it admits a spin structure. A necessary and sufficient condition for this is the *vanishing* of the *second Stiefel-Whitney class* $w_2 \in H^2(V; \mathbb{Z}/2\mathbb{Z})$. One knows that $w_2 = 0$ if and only if the restriction of the tangent bundle $T(V)$ to an arbitrary (immersed) surface S in V is trivial (where we assume $\dim V \geq 3$ as every oriented surface is spin anyway). In particular, every *parallelizable* manifold V (i.e. with $T(V)$ trivial) is spin. For example, the n-torus is spin. More generally, *stably parallelizable* manifolds (which means parallelizability of $V \times \mathbb{R}$) are spin. Thus every (immersed) oriented hypersurface in $\mathbb{R}^{n+1}$ (e.g. S^n) is spin. On the other hand the complex projective space $\mathbb{C}P^m$ is spin if and only if m is odd. Finally observe that every V contains a submanifold $\Sigma \subset V$ of codimension two (representing the Poincaré dual of w_2), such that $V - \Sigma$ is spin.

Now comes one of the central statements of this paper slightly reformulating our old result with Blaine Lawson.

$5\frac{1}{4}$. K-area inequality

Every complete Riemannian spin manifold of dimension n with $\mathrm{Sc}\, V \geq \varepsilon^{-2}$ *satisfies*

$$K\text{-}\operatorname{area}_{\mathrm{st}} V \leq \mathrm{const}_n\, \varepsilon^2. \tag{K}$$

Proof. Everything hinges upon (the existence of) a remarkable differential operator on V called the *(Atiyah-Singer)-Dirac operator.* If $n = 2m$, this operator acts between (smooth sections of) two vector bundles over V, called *spin bundles* S_+ and S_- and the (Dirac) operator is denoted $D_+ : C^\infty(S_+) \to C^\infty(S_-)$. These S_+ and S_- are unitary bundles of $\mathbb{C}$-ranks 2^{m-1} which are built in a canonical way out of the tangent bundle $T(V)$ and D_+ is a first order (elliptic) operator (algebraically) constructed out of the covariant derivative on S_+ corresponding to the Levi-Civita connection ∇ on $T(V)$. (More precisely, there exist two irreducible complex 2^{m-1}-dimensional representations of the group(!) Spin $(2m)$ and S_+ and S_- are the corresponding vector bundles associated to the principal bundle $\widetilde{\mathrm{Fr}} \to V$. In other words, the bundle S_+ is always defined on small neighbourhoods U_i of V, but the gluing isomorphisms over $U_i \cap U_j$ between $S_+|U_i$ and $S_+|U_j$ are defined only up to $\pm$ sign and making coherent choices requires the spin structure.

Similarly, the bundle S_-, in general, is defined up to $\pm$ sign and it becomes an honest bundle in the presence of spin. Notice that $\pm$ sign ambiguity is the same for S_+ and S_- and so the tensor product $S_+ \otimes S_-$ is globally defined on V even if V is non-spin and the same applies to all even tensor products of these bundles, such as $S \otimes S_-$, $S_+ \otimes S_+ \otimes S_+ \otimes S_-$, etc. This formally follows from irreducibility of the underlying representations of Spin$(2m)$). Two crucial properties of D_+ are as follows.

I. Atiyah-Singer theorem. *The index of D_+ on a closed manifold V equals certain non-zero rational combinations of the Pontryagin numbers of V, the so-called Todd genus $\widehat{A}[V]$.*

Recall that

$$\mathrm{Ind}\, D_+ \underset{\mathrm{def}}{=} \dim \mathrm{Ker}\, D_+ - \dim \mathrm{Ker}\, D_+^*,$$

where $D_+^* : C^\infty(S_-) \to C^\infty(S_+)$ is the adjoint operator which is a "twin" of D_+ and is denoted D_-. It is convenient to bring the two together and form the sum

$$D = D_+ \oplus D_- : C^\infty(S_+ \oplus S_-) \to C^\infty(S_+ \oplus S_-).$$

Clearly, D is self-adjoint and $D^2 > 0 \Rightarrow \mathrm{ind}\, D_+ = 0$.

This index theorem for D_+ is not very interesting for dim V non-divisible by 4 where there are no Pontryagin numbers and consequently ind $D_+ = 0$. On the other hand, one can construct (this is easy but not quite trivial) a *spin* manifold V of a given dimension $4k$ with $\widehat{A}[V] \neq 0$. Also recall that $\widehat{A} = \widehat{A}(p_1, p_2, \ldots)$ is a certain universal formal power series in variables p_i (where each p_i is given degree i) which starts as follows

$$\widehat{A} = 1 - \tfrac{1}{24}\,p_1 + \tfrac{1}{2^7\cdot 3^2\cdot 5}\,(-4p_2 + 7p_1^2) + \tfrac{1}{2^{10}\cdot 3^3\cdot 5\cdot 7}\,(16p_3 - 44p_2p_1 + 31p_1^3) + \dots .$$

If V is a manifold, then p_i is substituted by the *Pontryagin classes* $p_i = p_i(V) \in H^{4i}(V)$ (see $7\frac{1}{2}$) and $\widehat{A} = \widehat{A}_V$ becomes a (non-homogeneous) cohomology class in V (where all terms of degrees $> \dim V/4$ vanish). Thus $\widehat{A}[V]$ denotes the value of $\widehat{A}$ on the fundamental class of V, i.e. the Pontryagin number corresponding to the k-th grade term of $\widehat{A}$ for $k = \dim V/4$. (Notice that the zero grade term of $\widehat{A}$ is non-zero; it is $1 \in H^0(V)$ coming from the map of V to the one-point space.)

II. ***Bochner-Lichnerowicz formula.***

$$D^2 = \Delta_S + \tfrac{1}{4}\,\mathrm{Sc}\,. \tag{BL}$$

Here Δ_S denotes the Bochner (positive coarse) Laplacian acting on the spin bundle $S = S_+ \oplus S_-$ and $\frac{1}{4}\,\mathrm{Sc}$ denotes the multiplication operator $s \mapsto (\frac{\mathrm{Sc}}{4})s$ on *spinors* s, i.e. sections of S. Recall that Δ_S is defined with the Levi-Civita covariant derivative ∇ on S by $\Delta_S = \nabla^*\nabla$ which is equivalent to the integral identity $\langle \Delta_S s_1, s_2\rangle = \langle \nabla s_1, \nabla s_2\rangle$, for smooth spinors with compact support, where the scalar product is defined as usual by

$$\langle \alpha, \beta\rangle = \int_V \langle \alpha(v), \beta(v)\rangle dv.$$

Thus $\Delta_S \geq 0$ and the kernel of Δ_S consists of *parallel spinors*, i.e. those satisfying $\nabla s = 0$.

Remark. It is not surprising at all that $D^2 - \Delta_S$ is a *zero order* operator expressible in terms of the curvature tensor of V. This follows from simple symmetry considerations which apply to all "natural" second order operators over Riemannian manifolds. (For example, the Hodge-Laplacian on the i-form differs from the corresponding Bochner Laplacian by a certain operator on $\wedge^i(V)$ concocted out of $R_{ijk\ell}$, which reduces for $i = 1$ to Ricci acting on 1-form). However, it takes the exceptional symmetry of spinors to make this (zero order) operator a scalar and then the scalar curvature inevitably (?) enters the game. Yet the geometry behind this simple linear algebra remains obscure. (Of course, the reader may complain that it could not be otherwise as we had given *no definition* of D. But, in fact, D is essentially uniquely defined as the square root of $\Delta_S + \frac{1}{4}\,\mathrm{Sc}$ and in any case, we need for the time being only the sheer existence of D with the above properties. A decisive plunge into the algebra of spinors around D is unavoidable, however, for extending Lichnerowicz' approach to more general Dirac type operators as in [Wit] and [Min].)

$5\frac{1}{3}$. Lichnerowicz theorem

Every closed spin manifold V with $\mathrm{Sc}\, V > 0$ *has* $\widetilde{A}[V] = 0$.

Indeed, for all spinors s,

$$\langle D^2 s,\, s\rangle = \langle \Delta_S s + \left(\tfrac{\mathrm{Sc}}{4}\right) s,\, s\rangle = \langle \nabla s,\, \nabla s\rangle + \langle \left(\tfrac{\mathrm{Sc}}{4}\right) s,\, s\rangle = \|\nabla s\|^2_{L_2} + \int_V \left(\tfrac{\mathrm{Sc}}{4}\right) \|s\|^2,$$

where the latter sum is strictly positive for all $s \neq 0$. In other words, the relations $\Delta_S \geq 0$ and $\mathrm{Sc} > 0$ imply that $D^2 = \Delta_S + \frac{1}{4}\,\mathrm{Sc} > 0$ and so $\widehat{A}[V] = \mathrm{Ind}\, D = 0$.

Corollary. *There exist closed manifolds of all dimensions $n = 4k$, $k = 1, 2, \ldots$, admitting no metrics with* $\mathrm{Sc} > 0$.

In fact, closed spin manifolds of dimensions $4k$ with $\widehat{A} \neq 0$ do exist as was indicated earlier. (But one does not find any of them among compact homogeneous spaces as these have $\mathrm{Sc} > 0$ unless they are flat.)

The spinor power of Lichnerowicz' theorem cannot be matched by traditional devices of Riemannian geometry even if one strengthens the condition $\mathrm{Sc} > 0$ to $K > 0$. The simplest example where the theorem applies is the famous *K_3-surface*, which is, topologically speaking, a 4-manifold V presented by a non-singular complex surface in $\mathbb{C}P^3$ of degree 4, say given by the following equation in the homogeneous coordinates in $\mathbb{C}P^3$,

$$x^4 + y^4 + z^4 + w^4 = 0.$$

This 4-manifold V is simply connected (by the Lefschetz theorem) and admits a (Kähler) metric with Ricci $= 0$ by Yau's solution to the Calabi conjecture. Furthermore, it is spin while $\widehat{A}[V] \neq 0$ and so no metric on V has $\mathrm{Sc} > 0$; yet no known elementary geometric argument rules out Ricci > 0 or even $K > 0$.

However, for all its beauty the Lichnerowicz theorem tells us nothing whatsoever about the geometry of V with $\mathrm{Sc}(V) > 0$ nor about the simple-minded topology, such as $\pi_1(V)$ for example, but only about the esoteric Todd genus. (Notice that $\widehat{A} = -\frac{1}{8}$ (signature) for 4-manifolds V, which is not so esoteric.) To overcome this drawback we should use the full power of $\mathrm{ind}\, D_+$, which is not just a number but an element of the *K_0-homology* of V, that is a (linear) *function* on the (set of) vector bundles $X \to V$. (Eventually, $\mathrm{ind}\, D_+$ will be extended to the K-theory of a suitable C^*-algebra incorporating the fundamental group $\pi_1(V)$.) Namely, if X comes along with a linear connection, there is a natural extension of D_+ to a first order operator from $C^\infty(S_+ \otimes X)$ to $C^\infty(S_- \otimes X)$ which is also denoted D_+ and is uniquely (and correctly) defined by the following property. If x is a smooth section of X horizontal at some point $v \in V$, i.e. having $(\nabla_X x)(v) = 0$, then

$$D_+(s \otimes x)(v) = (D_+(s) \otimes x)(v)$$

for all spinors s. For example, if ∇_X is flat and so, locally,

$$S_+ \otimes X = \underbrace{S_+ \oplus S_+ \oplus \ldots \oplus S_+}_{r}, \$$$

for $r = \operatorname{rank} X$, then D_+ on $S_+ \otimes X$ locally equals the direct sum of r copies of D_+. The resulting operator D_+ on $S_+ \otimes X$ is elliptic for all X and the index gives us the desired function on bundles

$$X \mapsto (\text{index of } D_+ \text{ on } S_+ \otimes X)$$

which is (obviously) additive for the Whitney sums of bundles.

$\mathbf{I_X}$. ***Atiyah-Singer theorem for*** $\mathbf{D_+}$ ***on*** $\mathbf{S_+ \otimes X}$. *The index of this (twisted)* D_+ *satisfies*

$$\operatorname{Ind} D_+ = \left(\widehat{A}_V \smile ch_X\right)[V], \tag{Twind}$$

where ch_X *is the Chern character of* X *which is a polynomial in the Chern classes* $c_i = c_i(X)$ (defined below).

$\mathbf{II_X}$. ***Twisted Bochner-Lichnerowicz.*** *Assume that* (X, ∇_X) *is unitary, take the adjoint* D_- *of the (twisted)* D_+ *and set* $D = D_+ + D_-$. *Then*

$$D^2 = \Delta + \frac{1}{4} \operatorname{Sc} + \mathcal{R}_0, \tag{TwiBL}$$

where Δ *is the Bochner Laplacian on* $S \otimes X$ *and* $\mathcal{R}_0$ *is a symmetric bundle endomorphism of* $S \otimes X$ *(i.e. a self-adjoint operator of zero order) and where the pointwise norm of* $\mathcal{R}_0$ *everywhere bounded by the curvature of* ∇_X,

$$\|\mathcal{R}_0\| \leq \operatorname{const}_n \|\mathcal{R}(X)\|, \; n = \dim V. \tag{$\star$}$$

Remark. If X (i.e. ∇_X) is flat, which means $\mathcal{R}(X) = 0$, then the twisted BL trivially follows from the untwisted one. In fact, the full (TwiBL) (including ($\star$)) can be probably derived from scaling considerations.

Now we are ready to prove the implication

$$\operatorname{Sc} V \geq \varepsilon^{-2} \Rightarrow K\text{-}\operatorname{area} V \leq \operatorname{const}_n \varepsilon^2. \tag{$*$}$$

To do this, we must show that every bundle X, where the curvature $\mathcal{R}(X)$ is small compared to $\inf_{v \in V} \operatorname{Sc}_v(V)$, necessarily has all Chern numbers zero. Observe that if the curvature of X is small, i.e. $\leq \delta \operatorname{Sc} V$, then the curvature of every associated bundle X' is also small and then, according to (TwiBL), the operator D^2 on $S \otimes X'$ is *strictly positive* as

$$D^2 = \Delta + \tfrac{1}{4}\operatorname{Sc} + \mathcal{R}'_0$$

where the needed inequality $\mathcal{R}'_0 < \frac{1}{4}\operatorname{Sc}$ is ensured (as explained below) by our bound $\|\mathcal{R}(X)\| \leq \delta \operatorname{Sc} V$ for a suitably small $\delta > 0$. The positivity of D^2 makes the index of D zero, then according to (Twind)

$$\left(\widehat{A}_V \smile ch_{X'}\right)[V] = 0, \tag{0'}$$

and this relation for *all* X' makes *all* Chern numbers of X zero. Actually, we do not need all associated bundles X' but only a finite number of them, depending on $\dim V$, where the needed X' are certain tensor products of copies of X itself and its exterior powers, $\wedge^1 X = X$, $\wedge^2 X, \ldots$. Then indeed, there exists a positive $\delta = \delta_n$, such that the bound $\|\mathcal{R}(X)\| \leq \delta_n \operatorname{Sc}$ makes $\mathcal{R}'_0 \leq \frac{1}{4}\operatorname{Sc}$ for X' from this (finite!) set of bundles.

$5\frac{3}{8}$. Algebraic conclusion of the proof of the K-area inequality

What remains to show is

$$(0') \Rightarrow \text{ vanishing of the Chern numbers of } X.$$

The proof of this we start in the simplest (and essential) case where $\widehat{A}_V = 1$ (e.g. V is stably parallelizable) and $(0')$ reduces to the identity $chX' = 0$. We denote the i-th grade component of $ch \in H^*(V)$ by $ch_i \in H^{2i}$, so that $ch = ch_0 + ch_1 + ch_2 + \ldots$ and recall that ch is defined for (complex) line bundles ℓ by

$$ch_\ell = \exp c_1 = 1 + c_1 + \tfrac{1}{2}c_1^2 + \tfrac{1}{6}\,c_1^3 + \ldots$$

where c_1 is the first Chern class of ℓ. Next, ch extends to sums of line bundles with the following

Additivity.

$$ch_{X \oplus Y} = ch_X + ch_Y,$$

for the Whitney sums of arbitrary bundles. This uniquely defines ch for all bundles X (since they can be *formally* split into sums of line bundles) as a universal polynomial in the Chern classes $c_i = c_i(X) \in H^{2i}(V)$,

$$ch = r + c_1 + (\alpha_2 c_2 + \beta_2 c_1^2) + (\alpha_3 c_3 + \beta_3 c_2 c_1 + \gamma_3 c_1^3) + \ldots$$

where $r = \operatorname{rank} X$. An important feature for us of $ch_i = \alpha_i c_i + \ldots$, is the non-vanishing of the coefficients α_i for all i. Thus every Chern number, i.e. a homogeneous polynomial in c_i (evaluated on $[V]$) is expressible as a polynomial in ch_i. Then we recall that ch is multiplicative for tensor products of bundles

$$ch_{X \otimes Y} = ch_X \smile ch_Y$$

and finally we bring into the picture the *Adams operations* $\psi_k(X)$, $k = 1, 2, 3, \ldots$, which are certain universal combinations of the exterior powers $\wedge^i(X)$ of X. The advantage of $\psi_k(X)$ over $\wedge^i(X)$ is the following simple formula for ch of ψ_k which is best expressed with the notation $ch(t) = \sum_{i=0}^{\infty} ch_i t^i$,

$$ch_{\psi_k(X)}(t) = ch_X(kt)(= \sum_{i=1}^{\infty} ch_i k^i t^i)$$

for all $k = 1, 2, \ldots$.

Trivial Algebraic Lemma. *Consider a formal power series $a(t) = \sum_{i=0}^{\infty} a_i t^i$ and take the products $b(t) = \prod_{\mu=1}^{\nu} a(k_\mu t)$ for all strings of positive integers $k_1, \ldots, k_\nu$. Then the coefficients b_i of these products at each t^i (which are homogeneous polynomials in a_j for $j \leq i$ such as $\alpha_i a_i + \beta_i a_{i-1} a_1 + \gamma_i a_{i-2} a_1^2 + \ldots$) span, (for all integer strings $(k_1, \ldots, k_\nu)$) the space of all homogeneous polynomials in a_j of degree i (where $\deg a_1^{d_1} a_2^{d_2} a_3^{d_3} \ldots \underset{\text{def}}{=} d_1 + 2d_2 + 3d_3 + \ldots$).*

Thus we arrive at the following chain of implications. (The tensor products of the exterior powers of X have $ch_i = 0$ for a given i) $\Rightarrow$ (the tensor products of all $\psi_k(X)$ have $ch_i = 0$) $\Rightarrow$ (all homogeneous polynomials in ch_j of degree i equal zero) $\Rightarrow$ (all homogeneous polynomials in c_j of degree i equal zero).

Here the first implication is based on the fact that the Adams operations $\psi_k(X)$ are polynomials in the exterior powers of X, the second one follows from the above lemma and the third one appeals to nonvanishing of the coefficients α_j in the polynomials $ch_j = \alpha_j c_j + \ldots$ mentioned earlier. Now we see that indeed $\left(\widehat{A} = 0 \text{ and } (0')\right) \Rightarrow$ vanishing of the Chern numbers of X by applying the above to $i = \frac{1}{2} \dim V$ and then the general case (where $\widehat{A} \neq 0$) follows by observing that

$$\left(\widehat{A} \smile (ch)^k\right) [V] = 0 \text{ for all } k) \Rightarrow ch[V] = 0,$$

(since $\widehat{A}$ starts from a non-zero term in the degree zero as was emphasized earlier) and that the tensor power $Y = \underset{k}{\otimes} X$ has $ch_Y = ch_X^k$. This concludes the proof of the inequality K-area $V \leq \text{const}_n \, (\inf \operatorname{Sc} V)^{-1}$ for closed manifolds

V with $\mathrm{Sc}\, V > 0$ and the non-compact case follows from the relative version of the index theorem which is well adjusted to the notion of K-area$^+$ and implies that

$$K\text{-area}^+(V) \leq \mathrm{const}_n\, \varepsilon^2 \qquad (K^+)$$

for $\mathrm{Sc}\, V \geq \varepsilon^2$ (see $6\frac{4}{5}$). Finally one may stabilize by passing to $V \times \mathbb{R}^k$ and apply the index theorem for families. This gives the required inequality

$$K\text{-area}_{st} V \leq \mathrm{const}_n (\inf \mathrm{Sc}\, V)^{-1}$$

for $\mathrm{Sc}\, V \geq 0$ and a similar (stable) inequality for K-area$^+_{st}$. Q.E.D.

$5\frac{4}{9}$. Spin problem, aspherical manifolds and extremal metrics

Here is an application of (K).

Let V be a closed manifold which admits a metric g_0 with sectional curvatures $K_{g_0} \leq 0$. Then V admits no metric g with $\mathrm{Sc} > 0$, *provided it is spin and the fundamental group $\pi_1(V)$ is residually finite.*

Proof. According to (v') in $4\frac{3}{5}$ such a V with $K(V, g_0) \leq 0$ has K-area $= \infty$ which is incompatible with the above bound on K-area by $(\mathrm{Sc}\, V)^{-1}$ for $\mathrm{Sc}\, V > 0$.

Remark. The residual finiteness if $\pi_1(V)$ is non-essential as is seen with the generalized K-area using infinite dimensional bundles (see $4\frac{3}{5}$ and $9\frac{1}{6}$). Alternatively, one may apply the K-area inequality to the universal covering $(\widetilde{V}, \tilde{g})$ since $\mathrm{Sc}\, \tilde{g} = \mathrm{Sc}\, g$. Thus we have the implications

$$\mathrm{Sc}\, g \geq \varepsilon^{-2} \Rightarrow \mathrm{Sc}\, \tilde{g} \geq \varepsilon^{-2} \overset{(K)}{\Rightarrow} K\text{-area}_{st}(\widetilde{V}, \tilde{g})$$
$$\leq \mathrm{const}_r\, \varepsilon^2 < \infty \Rightarrow K\text{-area}(\widetilde{V}, \tilde{g}_0) < \infty,$$

where the third implication is explained in $4\frac{1}{3}$. But the K-area of $(\widetilde{V}, \tilde{g}_0)$ is obviously infinite as the exponential map gives us an expanding embedding $\mathbb{R}^n \to \widetilde{V}$. Also notice that $\widetilde{V}$ (being contractible) is always spin which shows the redundancy of the spin condition as well.

Questions. (A) Are there closed aspherical (and, more generally, π_1-essential as in (vi) of $4\frac{3}{5}$) manifolds V admitting a metric with $\mathrm{Sc} > 0$?

Schoen and Yau anounced "No" for $\dim V = 4$ and they expressed a belief that a technical refinement of their argument will work for all n; see [Sch] and [Yau].

(B) Does the K-area inequality (K) extend to *all* complete (*non-spin*!) manifolds V with $\mathrm{Sc}\, V \geq 0$?

We shall indicate a modified version of (K) with a certain $K_{\surd}$-area in place of the K-area but this $K_{\surd}$-area inequality will give us no topological restriction on such a V. On the other hand, the minimal surface techniques of Schoen-Yau (see $5\frac{2}{3}$) lead to topologically significant geometric inequalities similar to (K) for all (not even necessarily complete) manifolds V with $\mathrm{Sc}\, V \geq 0$ but these inequalities do not seem to imply (or follow from) our (K).

(C) When does the K-area inequality become sharp and what are the corresponding *extremal* Riemannian manifolds V ?

Possible definitions of extremality. We assume here V has $\mathrm{Sc}\, V \geq 0$ and consider all possible Riemannian manifolds V' and proper maps $f : V' \to V$ such that

$$\mathrm{Sc}_{v'}\, V' \geq \mathrm{Sc}_{f(v')}\, V \text{ for all } v' \in V.$$

We insist on our maps f having non-zero degree in a suitable sense. It may be just non-zero degree in the sense of f_* on H^n_{comp} or something more general, such as

(i) the map f is spin (e.g. both V and V' are spin) and has non-zero $\widehat{A}$-degree, i.e. the pull-back of a regular value, $f^{-1}(v) \subset V'$, has $\widehat{A} \neq 0$.

(ii) There exists an almost flat $\kappa \in K^0(V')$ (i.e. representable by bundles with arbitrarily small $\|\mathcal{R}\|$) such that $ch\kappa$ does not vanish on $f^{-1}(v)$. (An instance of that is the projection $V' = V_1 \times V \to V$, where K-area $(V_1) = \infty$.)

Now V is called *length extremal* if it admits no map $f : V' \to V$ as above which is *strictly* contracting, i.e. having $\mathrm{Lip}\, f < 1$. Moreover, one may require the implication

$$\mathrm{Lip}\, f \leq 1 \Rightarrow f \text{ is a Riemannian submersion.}$$

Another possibility, more in the spirit of the K-area, is to call V *area extremal* if there is no $f : V' \to V$ strictly area contracting, i.e. strictly contracting the area of the surfaces in V'. (Here one should be careful with the equality case, especially for $n = \dim V = 2$, as surfaces admit plenty of area preserving non-isometric maps).

Llarull theorem (see $[\mathrm{Lla}]_{\mathrm{ShEs}}$, and $[\mathrm{Lla}]_{\mathrm{Scn+4k}}$). *The spheres S^n for $n \geq 2$ are area extremal in the spin category.*

Llarull's proof uses a sharp TwiBL formula for some (twisting) bundle $X \to S^n$ (which accidentally is the spinor bundle). He states in his paper only the $\widehat{A}$-degree theorem, but his argument also applies to such manifolds as $S^n \times T^k \to S^n$. This will be used in $5\frac{5}{6}$ to prove some semicontinuity of Sc under C^0-limits of metrics.

Notice that by transitivity of degree, Llarull's theorem implies that the product manifolds $S^n \times$ (complete flat) are also area extremal in the spin category, i.e. where the comparison manifolds $V' \to S^n \times V_1$ are spin and where the extremality, in the case of non-compact (flat) V_1, is understood in the (slightly weaker) sense of non-existence of area contracting maps $f : V' \to S^n \times V_1$ with non-zero degree where $\mathrm{Sc}\, V' \geq \sigma^2 > n(n-1) = \mathrm{Sc}\, S^n$. (Beware of manifolds V' of *positive* curvature admitting proper contracting maps $V' \to \mathbb{R}^k$ of degree one!)

Next, since the curvature term in (TwiBL) is additive for the products $(V_1 \times V_2,\ X_1 \otimes X_2)$, Llarull's computation also yields (this seems obvious but I did not honestly check it) that Cartesian *products of spheres (possibly of non-equal radii) are also area extremal.* Furthermore, Min-Oo recently proved the extremality of the compact symmetric spaces which are spin and have non-zero Euler characteristic. In fact, one may expect all compact symmetric spaces to be area extremal. Also some non-symmetric homogeneous spaces may be extremal (but the 3-sphere squeezed along Hopf's circles is not extremal).

Can one produce extremal manifolds by the following maximization process? Start with (V_0, g_0) where $\mathrm{Sc}\, g_0 > 0$ and start enlarging g_0 without making the scalar curvature smaller. One may hope that there is some limit manifold (V, g), possibly non-homeomorphic to V_0 but admitting a suitable contracting map $V \to V_0$ and having $\mathrm{Sc}\, V \geq \mathrm{Sc}\, V_0$ and being extremal. An important point here is to show that the scalar curvature is semicontinuous, i.e. it cannot jump down in the limit but this is not known in the sharp form (compare $5\frac{5}{6}$). However, this maximization obviously works in the category of homogeneous spaces and then one asks if the resulting homogeneously extremal manifolds are external.

(D) The K-area inequality is unlikely to be sharp unless the scalar curvature (function) $\mathrm{Sc} = Sc_v$ is constant on V. But for non-constant Sc one may improve (K) by conformally scaling $g \mapsto g' = \mathrm{Sc} \cdot g$ and observing that the proof of (K) yields

$$K\text{-area}_{st}(V, g') \leq \mathrm{const}_n, \qquad (K')$$

which is significantly sharper than K for (strongly) variable Sc.

$5\frac{1}{2}$. K-area and the spectrum for Inf Sc $= -\sigma < 0$

Let V be a closed Riemannian spin manifold and let us bound $|\widehat{A}[V]|$ in terms of the spectrum of the Bochner Laplacian Δ_S on spinors. We use Hermann Weyl's variational principle for Δ_S and observe with the BL-formula $D^2 = \Delta_S + \frac{1}{4}\mathrm{Sc}$ that the harmonic spinors yield eigenfunctions of Δ_S in the spectral interval

$[0, \frac{1}{4}\sigma]$ for $-\sigma = \inf \mathrm{Sc}\, V$. Then, by the index theorem,

$$|\widehat{A}[V]| = |\operatorname{ind} D| \leq \operatorname{rank} \ker D \leq \#\operatorname{spec} \Delta_S[0, \frac{1}{4}\sigma],$$

where $\#\operatorname{spec} \Delta_S[0, \lambda]$ denotes the number of eigenvalues of Δ_S in the segment $[0, \lambda]$. To make this bound interesting we should relate $\operatorname{spec} \Delta_S$ to more significant geometric invariants of V and we invoke at this stage the following *Kac-Feynman-Kato inequality* connecting the eigenvalues λ'_i of Δ_S with the eigenvalues λ_i of the ordinary (positive) Laplace operator Δ on V,

$$\sum_{i=0}^{\infty} \exp -\lambda'_i t \leq (\operatorname{rank}_{\mathbb{R}} S) \sum_{i=0}^{\infty} \exp -\lambda_i t \qquad \text{(KFK)}$$

which holds for all $t > 0$ and where $\operatorname{rank}_{\mathbb{R}} S = 2^n$ for $n = \dim V$. (Such an inequality is valid for the Bochner Laplacian in an arbitrary bundle.) Thus we have

$$|\widehat{A}[V]| \exp -\frac{1}{4}\sigma t \leq 2^n \sum_{i=0} \infty \exp -\lambda_i t$$

for $-\sigma = \inf \mathrm{Sc}\, V$ and all $t > 0$. Here we notice that $\lambda_0 = 0$ and so the inequality (KFK) provides a non-trivial information on λ_i only for $\widehat{A}[V] \geq 2^{\dim V}$ and relatively small σ. In fact, since the splitting $S = S_+ \oplus S_-$ is parallel, the KFK-inequality applies to S_+ and S_- separately which yields

$$|\widehat{A}[V]| \exp -\tfrac{1}{4}\sigma t \leq 2^{n/2} \sum_{i=0}^{\infty} \exp -\lambda_i t = 2^{\frac{n}{2}} \left(1 + \sum_{i=1}^{\infty} \exp -\lambda_i t\right), \qquad (\widehat{A}\text{-exp})$$

(where we assume V is connected and so $\lambda_0 = 0$ has multiplicity one). The advantage of $\lambda_i = \lambda_i(\Delta)$ over $\lambda'_i = \lambda_i(\Delta_S)$ from our (possibly naive) geometric viewpoint is the fact that λ_i are continuous in the space of metric g with the C^0-topology as follows from the variational principle for the quadratic form

$$f \mapsto \int \langle \Delta f(v),\ f(v) \rangle_g dv = \int \langle df(v), df(v) \rangle_g dv$$

involving no derivatives of g, while the corresponding form for Δ_S (as well as for D^2) uses the first derivatives of g entering via the Levi-Civita connection. (This remark and (BL) imply, as was pointed out by Lohkamp, that the integral $\int \psi(v)\, \mathrm{Sc}_v(g) dv$, for an arbitrary function ψ on V, only depends on the first derivatives of g, although $\mathrm{Sc}\, g$ involves the second derivatives. The same applies to other Bochner curvatures, e.g. Ricci on 1-forms, and implies the C^1-closeness of the upper bounds on Sc and Ricci; see $[\mathrm{Loh}]_{\mathrm{GLC}}$ and compare p. 24 in $[\mathrm{Gro}]_{\mathrm{PDR}}$.) We see, consequently, that *for an arbitrary closed spin manifold*

(V, g_0) *with* $|\widehat{A}(V)| > 2^{\frac{\dim V}{2}}$, *there exists a constant* $\sigma_0 = \sigma_0(g_0) > 0$ *such that* g_0 *admits no* C^0*-approximation by metrics* g *with* $\operatorname{Sc} g \geq -\sigma_0$.

As another corollary of ($\widehat{A}$-exp) one obtains the following bound on $\widehat{A}(V)$ in terms of $\delta = \operatorname{Diam} V$ and $-\rho^2 = \inf \operatorname{Ricci} V$,

$$\widehat{A}(V) \leq 2^{\frac{n}{2}} + (\delta\rho)^n (\operatorname{const}_n)^{1+\delta\rho},$$

where one uses besides ($\widehat{A}$-exp) the inequality

$$\lambda_i \geq \delta^{-2} C_n^{1+\delta}\, i^{2/n}, \quad i = 1, 2, \ldots, \qquad (\lambda_i\text{-Ricci})$$

valid for all closed n-manifolds V with Ricci $V \geq -1$ (and where the general case of Ricci $\geq -\delta^2$ follows by scaling).

Now we apply the KFK-inequality to the Dirac operator with coefficients in X with curvature bounded by some constant, say $\|\mathcal{R}(X)\| \leq R_X$, and conclude to the following twisted $\widehat{A}$-exp-inequality.

$$|\widehat{A}_V \smile ch_X[V]| \exp(-\tfrac{1}{4}\sigma - C_n R_X)t \leq 2^{n/2} \operatorname{rank}_{\mathbb{R}} X \left(1 + \sum_{i=1}^{\infty} \exp -\lambda_i t\right) \qquad (\widehat{A}\text{-}ch\text{-exp})$$

(where, recall, $1 + \sum\limits_{i=1}^{\infty} \exp -\lambda_i t = \operatorname{Trace} \exp - \Delta t$).

In order to make this inequality useful, one needs unitary bundles X with $R_X \lesssim \sigma$ and large (comparable to Vol V) $ch_X[V]$ while rank X should be small. Then one obtains a lower bound on $\sum\limits_{i=1}^{\infty} \exp \lambda_i t$ for certain (large) values of t, which leads to the following.

Non-approximation example. *Let* (V, g_0) *be a closed Riemannian manifold which admits a map of non-zero degree into a closed manifold* W *with negative sectional curvature. Then, if the covering* $\widetilde{V}$ *of* V *induced by the universal covering of* W *is spin* (e.g. $V \to W$ *is a homotopy equivalence), there exists a constant* $\sigma_0 = \sigma_0(g_0) > 0$, *such that* g_0 *admits no* C^0*-approximations by* C^2*-metrics* g *on* V *with* $\operatorname{Sc} g \geq -\sigma_0$.

Sketch of the proof. Assume to start with that $\pi_1(W)$ is residually finite; let $\widetilde{W}_j$, $j = 1, 2, \ldots$, be finite coverings of W approximating the universal covering and let $\widetilde{V}_j$ be the corresponding coverings of V. One knows that the heat flow exponentially decays on the universal coverings $\widetilde{V}$ and $\widetilde{W}$ since the fundamental group $\pi_1(W)$ is non-amenable and so the heat flow on $\widetilde{V}_j$ satisfies

$$\operatorname{Trace} - \widetilde{\Delta}_j t \leq \operatorname{const}(\exp -at) \operatorname{Vol} \widetilde{V}_j$$

for a fixed $a = a(V) > 0$ and $1 \leq t \leq t(j)$ where $t(j) \to \infty$ for $j \to \infty$. (The equivalences non-amenability $\Leftrightarrow$ exponential heat decay $\Leftrightarrow \lambda_0 > 0$, can be traced to Kesten's work on random walk in groups and, probably, to the original work by von Neumann; this was brought to the attention of geometers by Robert Brooks.) Observe that for every $\varepsilon_0 > 0$ there exists j_0 such that W_{j_0} admits an ε_0-contracting map to S^n, $n = \dim W$, and so the same is true for $\widetilde{V}_j$ with $\varepsilon_0' = \text{const}\,\varepsilon_0$.

To simplify the notations we assume $\varepsilon_0' = \varepsilon_0$ and $\widetilde{V}_{j_0} = V$, so all $\widetilde{V}_j$ cover $V_{j_0} = V$. Now, if $n = 2m$, we get an ε_0-flat bundle X over V with $c_m(X) \neq 0$ by pulling to V a standard non-trivial bundle over S^n of rank n. This bundle goes up to all $\widetilde{V}_j \to V$ where it has $c_m \geq \delta \operatorname{Vol} \widetilde{V}_j$ with $\delta = \delta(V) \approx (\operatorname{Vol} V)^{-1}$. Next we assume V is spin and to make it even easier, let $\widehat{A}_V = 1$. Then the above inequality ($\widehat{A}$-*ch*-exp) applied to $\widetilde{V}_j$ reads

$$\delta' \operatorname{Vol} \widetilde{V}_j \exp(-\tfrac{1}{4}\sigma - C_n\varepsilon_0)t \leq n2^n \operatorname{Trace} -\widetilde{\Delta}_j\, t \leq \text{const}'(\exp -at) \operatorname{Vol} \widetilde{V}_j,$$

where $\delta' = \delta/n!$ and $t \leq t(j)$. This implies

$$\exp(-\tfrac{1}{4}\sigma - C_n\varepsilon_0)t \leq C_n'\delta^{-1} \exp -at,$$

for all $t > 1$ as j can be taken arbitrarily large and so

$$\tfrac{1}{4}\sigma + C_n\varepsilon_0 \geq a. \tag{$*$}$$

Therefore, if $\frac{1}{4}\sigma_0 \underset{\text{def}}{=} a - C_n\varepsilon_0 > 0$, we obtain the (non-trivial) inequality $\sigma \geq \sigma_0 > 0$ for the lower bound $-\sigma$ of $\operatorname{Sc} V$. Finally we observe that we could have chosen $\varepsilon_0 > 0$ arbitrarily small and that the constant $a = a(V)$ is C^0-continuous in the metric on V. Hence ($*$) applies to small perturbations g of the original metric g_0 in V with $\operatorname{Sc} g \geq -\sigma$. This concludes the proof in the presence of the finite coverings $\widetilde{V}_j$ and the general case needs a similar argument in the universal covering $\widetilde{V}$ in the spirit of $9\frac{1}{9}$ and $9\frac{1}{6}$.

Remarks. All we needed of the curvature condition $K(V) < 0$ is (a) $\pi_1(W)$ is non-amenable; (b) the universal covering $\widetilde{W}$ has infinite K-area.

It seems likely that the above remains true for complete non-compact manifolds V (replacing the above $\widetilde{W}$). Namely let

(a) the heat flow on V exponentially decays with the rate given by the lowest eigenvalue $\lambda_0 > 0$ of Δ;

(b) every R-ball in V has K-area $\geq cR^2$, for some $c > 0$;

(c) V has local geometry bounded by ρ, or at least Ricci $V \geq -\rho > -\infty$ to avoid a major pathology;

Then, probably, $\inf \operatorname{Sc} V \leq -\sigma$ for some $\sigma = \sigma(\lambda_0, c, \rho, \dim V) > 0$.

Approximation problems. *Let a smooth metric g be a limit of g_i. Is then*

$$\operatorname*{Inf}_V \operatorname{Sc} g \geq \operatorname*{Inf}_i \operatorname*{Inf}_V \operatorname{Sc} g_i \,?$$

Even better, if

$$\operatorname{Sc}_v g \geq \liminf_{i\to\infty} \operatorname{Sc}_v g_i$$

for all $v \in V$? Or, maybe

$$\int_V \operatorname{Sc}_v g \, dv_g \geq \liminf_{i\to\infty} \int_V \operatorname{Sc}_v g_i \, dv_{g_i}$$

under some extra conditions on g_i? (See [Loh]$_{\mathrm{GLC}}$ for a comprehensive discussion of this problem and $5\frac{5}{6}$ for partial results.)

$5\frac{2}{3}$. Remark and references on scalar curvature, minimal subvarieties and asymptotically standard manifolds

There are two competing methods in the study of $\operatorname{Sc} > 0$.

I. *Minimal hypersurfaces, splitting and symmetrization.* If $V_1 \subset V$ is a smooth stable minimal hypersurface in $V = (V, g)$ with $\operatorname{Sc} g > 0$, then $\widehat{V}_1 = V_1 \times \mathbb{R}$ admits an $\mathbb{R}$-invariant metric $\widehat{g}_1$ with $\operatorname{Sc} \widehat{g}_1 > 0$ which in the quotient space $V_1 = \widehat{V}_1/\mathbb{R}$ equals the restriction of g to V_1. (Recall that "minimal" means critical for the functional $V_1 \mapsto \operatorname{Vol}_{n-1} V_1$ and "stable" is implied by V_1 being a local minimum for Vol_{n-1}.) Then with a suitably minimal $V_2 \subset V_1$ one obtains an $\mathbb{R}^2$-invariant metric $\widehat{\widehat{g}}_2$ on $\widehat{\widehat{V}}_2 = V_2 \times \mathbb{R}^2$ with $\operatorname{Sc} \widehat{\widehat{g}}_2 > 0$ etc, which eventually leads to strong topological and geometrical restrictions on (V, g) similar to (but yet seemingly different from) the K-area inequality (compare $5\frac{5}{6}$).

The positive curvature splitting with minimal surfaces (in a somewhat different form) was introduced by Schoen and Yau about 15 years after the appearance of the Lichnerowicz spinor paper (see [Sch-Ya]$_{\mathrm{EIMS}}$, [Sch-Ya]$_{\mathrm{SMPS}}$). They also applied their method to non-compact asymptotically flat manifolds and resolved the positive mass and action conjectures of the general relativity (see [Sch-Ya]$_{\mathrm{PM}}$, [Sch-Ya]$_{\mathrm{PA}}$). Further modifications and application of the minimal surface techniques appear in [FC-Sch], [G-L]$_{\mathrm{PSC}}$, [Gro]$_{\mathrm{FPP}}$, [Sch] and [Ya], where the reader finds further references).

II. *Twisted* BL-*formula.* The original (untwisted) spinor method of Lichnerowicz-Atiyah-Singer was further developed by N. Hitchin in 1972, in his theses where he showed, in particular, that *some exotic* 9-*spheres admit no metrics with* $\operatorname{Sc} > 0$. The twist idea was introduced by G. Lusztig, also in 1972, who was concerned with the Hirzebruch formulae rather than with $\operatorname{Sc} > 0$. Namely, he looked at the cohomology of a manifold V with coefficient in a flat bundle X over V and observed that in the presence of a parallel quadratic (possibly indefinite) form Q on X, one could pair the middle dimensional cohomology, $H^m(V;X) \otimes H^m(V;X) \to \mathbb{R}$ for $2m = \dim V$, and for m even speak of the signature $\sigma(V;X,Q)$. He then identified this signature with the index of the signature operator on V suitably twisted with (X,Q) and expressed the index in terms of the characteristic classes of the $O(p,q)$-bundle (X,Q) (where (p,q) is the type of Q) according to the Atiyah-Singer theorem thus generalizing the classical formula of Rochlin-Thom-Hirzebruch for the ordinary signature $\sigma(V) = \sigma(V; \mathrm{Triv}^1, Q = x^2)$ (see $7\frac{1}{4}$, $7\frac{1}{2}$, $8\frac{1}{2}$).

Lusztig also proved a similar signature formula for *families* of flat S^1-bundles which, as was observed in $[\text{G-L}]_{\text{SSC}}$, admitted the $\widehat{A}$-version yielding non-existence of metrics with $\mathrm{Sc} > 0$ on tori. Then the twisting was applied in $[\text{G-L}]_{\text{SSC}}$ and $[\text{G-L}]_{\text{PSC}}$ to *almost* flat bundles over *sufficiently large* manifolds V (all having K-area $= \infty$ in our present terminology) and the relevant (macroscopic) concepts of largeness were further investigated in $[\text{Gro}]_{\text{LRM}}$. It is worth noticing here that the existence of a flat G-bundle over V with non-trivial characteristic class in $H^n(BG;\mathbb{R})$ for a connected Lie group G and $n = \dim V$ (e.g. $O(p,q)$-bundle X with a non-trivial Pontryagin number of $X_+ - X_-$) makes a suitable covering of V rather large. In particular, the stable K-area of such a V is infinite as we shall see later on.

The KFK-inequality has been apparently known to physicists from time immemorial (at least it was known to Jürg Fröhlich who explained it to me around 1980; also see [H-S-U]) and it nicely fits with the (λ_i-Ricci)-inequality (proven in $[\text{Gro}]_{\text{PL}}$) as was observed in $[\text{Gro}]_{\text{VBC}}$ (see p. 86 there) and in [Gal]. It is tempting to sharpen the KFK inequality by replacing the bound on $\Sigma\exp\text{-}\lambda_i t$ by a similar bound on individual eigenvalues λ_i. Such inequalities are implicitly present in the formulae (9), (9′) and (10) in $[\text{Gr}]_{\text{LRM}}$ but now I believe I erred at that point and one should rewrite (9)-(10) with $\Sigma \exp \lambda_i\, t$ (as in our inequality ($\widehat{A}$-*ch*-exp)) properly replacing $N(\lambda)$ in $[\text{Gro}]_{\text{LRM}}$. (I have not tried to find a counterexample to (9)-(10) of $[\text{Gro}]_{\text{LRM}}$). Fortunately, this does not essentially damage the geometric message contained in these formulae.

There is still one case where the individual eigenvalue bound is possible, namely that for λ_0, which was exploited for $\mathrm{Sc} < 0$ by Ono and later by Mathai who proved something similar to the above non-approximation example with a special regard to non-amenability (see [Ono], [Math], [Hij] and also [Bera] for general information on Bochner formulae and λ_i).

The KFK-inequality and the BL-formula $D^2 = \Delta_S + \frac{1}{4}\mathrm{Sc}$ suggest that the positivity of the operator $\Delta + \mu\,\mathrm{Sc}$, where Δ is the (positive) Laplacian on functions, must have, for large μ, comparable effect on D to that of the positivity of Sc. In fact the lowest eigenvalue of this operator, denoted $\lambda_0(\mu)$, should play the same role as $\inf \mathrm{Sc}$. This can be justified for $\mu = \frac{1}{2}$ (and hence for all $\mu \geq \frac{1}{2}$) by observing that the ($\mathbb{R}$-invariant) metric $\hat{g} = g + \varphi^2 dt^2$ on $\widehat{V} = V \times \mathbb{R}$ has $\mathrm{Sc}\,\hat{g} = \mathrm{Sc}\,g + \frac{2}{\varphi}\Delta\varphi$. If we take the lowest (and hence, non-vanishing) eigenfunction of $\Delta + \frac{1}{2}\,\mathrm{Sc}$ (on $V = (V, g)$) for φ, we get $\mathrm{Sc}\,\hat{g} = 2\lambda_0(\frac{1}{2})$. Then we observe that $(\widehat{V}, \hat{g})$ is (at least) as large as (V, g) since $V = \widehat{V}/\mathbb{R}$. For example,

$$K\text{-area}_{\mathrm{st}}\widehat{V} \geq K\text{-area}_{\mathrm{st}}V,$$

at least for *compact* V (which actually was tacitly assumed here anyway). To see that we go further, to the Riemannian product $\widehat{\widehat{V}} = \widehat{V} \times \mathbb{R}$ and observe that the standard (virtual) bundle Z on $\mathbb{R}^2$ with compact support representing the generator in the K-theory of $\mathbb{R}^2/\infty$ can be made arbitrarily ε-flat and then it lifts to a bundle $\widehat{\widehat{Z}}$ on $\widehat{\widehat{V}}$, also as flat as we wish (since V is compact and so the projection $\widehat{\widehat{V}} \to \mathbb{R}^2$ is Lipschitz). Then every bundle X on V, after lifting to $\widehat{\widehat{V}}$ and tensoring with $\widehat{\widehat{Z}}$, gives us a comparably flat bundle on $\widehat{\widehat{V}}$. Hence *our K-inequality for V remains valid with $2\lambda_0(\frac{1}{2})$ in place of* $\inf \mathrm{Sc}$.

The geometric role of $\lambda_0(\mu)$ for $\mu < \frac{1}{2}$ is not so clear but the topology of V feels positivity of $\Delta + \mu\,\mathrm{Sc}$ up to $\mu = \frac{1}{4}$. Namely we have the following

Observation. *If $\Delta + \mu\,\mathrm{Sc} > 0$ for some $\mu \geq \frac{1}{4}$ then the Cartesian product with the torus, $V \times T^k$ for some k, admits a metric with $\mathrm{Sc} > 0$. In particular, this implies according to* [G-L]$_{\mathrm{SSC}}$ *that $\widehat{A}(V) = 0$, provided V is spin.*

Proof. If $\Delta + \frac{1}{4}S$ is strictly positive then so is $\Delta + \mu_N\,\mathrm{Sc}$ for $\mu_N = \frac{N-1}{4(N-2)}$ and large $N = \dim V + k$ (we assume here V is a closed manifold) and let φ be the first (positive!) eigenfunction of $\Delta + \mu_N\,\mathrm{Sc}$. Then we conformally change the product metric on $V \times T^k$ by $\tilde{g} = (g \oplus \text{flat}) \mapsto \tilde{g}_1 = \tilde{\varphi}_{\frac{4}{N-2}}\tilde{g}$, where $\tilde{\varphi}$ is the obvious lift of φ from V to $V \times T^k$, and recall (see [Ka-Wa], [BerBe]) that $\mathrm{Sc}\,\tilde{g}_1 > 0$. (Probably, a combination of the Schoen-Yau successive splitting technique with surgery could deliver a metric g_1 with $\mathrm{Sc}\,g_1 > 0$ on V itself for $\dim V \geq 5$.)

Question. What is the geometric (and topological) significance (if any) of the strict positivity of $\Delta + \mu\,\mathrm{Sc}$ for $\mu < \frac{1}{4}$?

III. *Comparison between* I *and* II. There are two basic advantages of the minimal surface techniques over the spinors. First of all, one does not need the underlying manifold V to be spin. For example, no direct Dirac operator argument rules out $\mathrm{Sc} > 0$ on the connected sum $T^4 \# \mathbb{C}P^2$. Second of all, minimal surfaces work in (sufficiently large) non-complete manifolds where one has a problem with the Dirac operator. On the other hand, whenever the Dirac method applies it delivers finer geometric (and topological) information although in no serious case the results obtained by one method can be completely recaptured by the other. (This indicates a dark invisible mass of deep hidden structure showing two little tips, minimal and spinor.)

One may try to extend the Dirac operator techniques to non-spin manifolds V by removing a suitable codimension 2 submanifold W for which the complement $V_0 = V - W$ is spin and proving an appropriate relative index theorem for V_0. (Compare [Cho]$_{1,2}$. It would be even better to give V_0 a complete metric of positive scalar curvature whenever V and W possess such metrics.) Also, one may look for a relative index theorem for "sufficiently large" manifolds with "far away" boundary.

IV. ***Connes' theorem on foliations with*** **Sc** > 0. Let $\mathcal{V}$ be a closed foliated manifold with leafwise Riemannian metric having $\mathrm{Sc} > 0$. Alain Connes proved in [Con]$_{\mathrm{CCTF}}$ the following generalization of Lichnerowicz' theorem,

If the leafwise tangent bundle is spin then $\widehat{A}(V) = 0$.

Connes' proof relies on his rather sophisticated version of the index theorem integrating the "along the leaves" analysis transversally to the leaves. We shall indicate in $9\frac{2}{3}$ a more elementary approach and also explain how a twisted version of Connes' theorem suggests a conjectural bound on the recurrency dimension of foliation (see $2\frac{2}{3}$), namely

$$\mathrm{recdim} \leq \dim(\mathrm{leaves}) - 2$$

for $\mathrm{Sc} > 0$.

V. ***More on*** **Sc** $> -\sigma$. If a closed manifold V admits no metric with $\mathrm{Sc} \geq 0$, then $\mathrm{Inf}\,\mathrm{Sc}$ is expected to be small for small metrics on V. For example, if $V = (V, g_0)$ is locally symmetric with Ricci< 0, then every metric g on V which is smaller than g_0 (or, maybe, just having $\mathrm{Vol}(V,g) \leq \mathrm{Vol}(V, g_0)$) is likely to have $\inf \mathrm{Sc}\, g \leq \mathrm{Sc}\, g_0$. Some result of this kind is proven for $\dim V = 3$ in [Gro]$_{\mathrm{FPP}}$ using minimal surfaces and we indicate in $5\frac{5}{6}$ (following a hint by Rick Schoen) a similar approach with stable soap bubbles. (See [Gro]$_{\mathrm{VBC}}$ and [B-C-G]$_{\mathrm{ER}}$ for such results with Ricci instead of Sc and [B-C-G]$_{\mathrm{VE}}$ for bounds on $\mathrm{Sc}\, g$ where g is (conformally) close to g_0.)

VI. ***Spinors without the index theorem.*** E. Witten observed among other things in [Wit] that the Bochner-Lichnerowicz formula alone rules out certain metrics with $\mathrm{Sc} > 0$. For example, let g be a metric with $\mathrm{Sc}\, g \geq 0$ on $\mathbb{R}^n$ which is Euclidean outside a compact subset. It is not hard to show that D_g^2 is positive and D_g is L_2-invertible and so for each spinor s_0 on $\mathbb{R}^n$ parallel at infinity there exists an L_2-spinor φ, such that $D_g\varphi = D_g s_0$. Then the spinor $s = s_0 - \varphi$ is g-harmonic and asymptotically parallel, and the BL-formula shows that s is g-parallel over all of $\mathbb{R}^n$ which eventually implies g is flat (earlier proved for small n by Schoen and Yau with minimal surfaces, see [Kazd] for details and further references).

Min-Oo extended Witten's method to the hyperbolic space H^n instead of $\mathbb{R}^n$ and proved, for example, that *every complete non-compact connected spin manifold V with $\mathrm{Sc}\, V \geq -n(n-1)$, $n = \dim V$, which is isometric to H^n at infinity, is isometric to H^n* (see $[\mathrm{Min}]_{\mathrm{SCR}}$).

In other words, one cannot modify H^n on a compact subset without pushing scalar curvature down somewhere. Similarly, the hemisphere with the constant sectional curvature is "scalar curvature rigid" according to $[\mathrm{Min}]_{\mathrm{SCRH}}$

Let us explain this for the *Ricci* curvature by looking at a family of parallel horospheres in the perturbed space H^n; see Fig. 5 below.

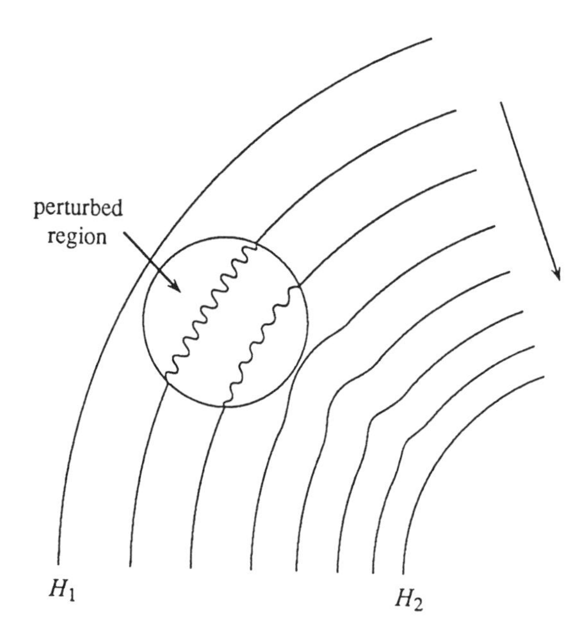

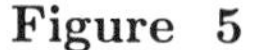

Figure 5

If the Ricci curvature of the perturbed manifold is $\geq -n+1 = \mathrm{Ricci}\,(H^n)$, then the perturbed horospheres are more mean-convex than the original ones (of mean curvature $n-1$) and they also have smaller $(n-1)$-volumes. Therefore if we normally project a perturbed horosphere after it has passed the perturbed region onto a non-perturbed one (H_2 on Fig. 5) we obtain a map with smaller Jacobian than in the non-perturbed case. This gives us a map of (a part of) a starting horosphere (H_1 on Fig. 5) to H_2 which is standard at infinity and which contracts more than its regular share inside. Clearly, this implies that there was no extra contraction at all and that the horospheres did not change while passing through the perturbed region and so there was no any perturbation to start with.

A modification of this argument (see iv in $5\frac{5}{7}$) applies to rather general manifolds and gives a non-trivial upper bound on Ricci in compact regions in terms of the ambient geometry (where such a bound should be sharp, I guess, for symmetric spaces). On the other hand the moving horospheres can be deformed to μ-bubbles of $5\frac{5}{6}$, which leads to the following version of the Min-Oo theorem.

Let V be a complete connected (possibly non-spin) manifold with $\mathrm{Sc} \geq -n(n-1)$, $n = \dim V$, *having an end isometric to that of H^n (and possibly having other infinite ends). Then it is isometric to H^n provided $n \leq 7$.* (The dimension restriction is due to possible singularities of μ-bubbles, but according to an unpublished result by Schoen and Yau these singularities are irrelevant in so far as we deal with the scalar curvature and so the conclusion of the theorem holds for all n.)

This theorem (as well as the original version by Min-Oo) tells us that the filling of the round sphere in H^n by the ball is the best possible in terms of the lower bound on Sc which brings in the following general discussion.

$5\frac{5}{7}$. Topological and Riemannian filling problems

We say that a closed oriented n-dimensional Riemannian manifold is *filled in* by an oriented manifold W if W is a compact manifold with boundary $\partial W = V$. The original Riemannian metric g on V can be always extended to some h on W and then we say that (W, h) *fills in* (V, g) *in the Riemannian category.* We assign some measure of geometric and/or topological complexity to W, try to find a filling W with minimal possible complexity, and use this minimal (infimal) complexity among all fillings W as a Riemannian invariant of V. We start with a couple of topological versions of this problem.

I. Find W filling in V with a *minimal possible Morse number*, i.e. with a Morse function on W vanishing on V and having the minimal possible (Morse) number of critical points.

This problem makes sense in each cobordism (and also bordism) theory and it can be successfully attacked in many cases by traditional surgery techniques (albeit this may be rather subtle already for estimating the Morse numbers of manifolds realizing given homology classes in *non-simply connected* spaces; see $8\frac{1}{2}$). Namely, the Browder-Novikov theory seems to imply the following filling estimate.

Every V of dimension $n \geq 5$ with trivial characteristic numbers admits a filling W such that the Morse number $M(W)$ is bounded by

$$M(W) \leq \mathrm{const}_n\big(\mathrm{rank}\,\pi_1(V) + \sum_{i=0}^{n} b_i(V)\big)$$

where $\operatorname{rank}\pi_1(V)$ *denotes the minimal number of generators of* π_1*'s of the connected components of* V *and* b_i *are the Betti numbers with suitable coefficients.*

Idea of the proof. As V is oriented one can kill π_1 by $k = \operatorname{rank}\pi_1$ surgeries, thus making V simply connected (even with smaller k equal the minimal number of elements *normally* generating π_1). Then, on the level of the Poincaré complexes, one can construct the required "small" filling and next it can be given a smooth structure with the Browder-Novikov theorem. The easiest case is that of a framed odd dimensional manifold which can be brought by at most $M(V)$ surgeries to a homotopy sphere V_0 fillable by W_0 with $M(W_0) \leq \text{const}_n$. In general, however, I can immediately see only a $\mathbb{Q}$-version of the filling, i.e. filling W_i with the required bound on $M(W_i)$, not of V itself but of iV for some positive integer $i \leq i_0(n)$. All this equally applies to a realization of a given relative bordism class of a simply connected space but non-simply-connectedness is quite a different matter.

II. Instead of the Morse numbers $M(V)$ and $M(W)$ one may use a stronger invariant carrying more topological information. Here we use $N(V)$, the minimal possible number of simplices for a p.l. (or smooth) triangulation of V. Then we define the *filling number* $FN(V)$ as the minimum of $N(W)$ over all fillings W of V. (If V does not bound, we still may define $NF(V)$ by subtracting from V a combination of some standard generators of the cobordisms in question.) There are at most finitely many manifolds V with $N(V) \leq N_0$ (by the standard smoothing theory) and so every smooth invariant of V, including $FN(V)$, admits a bound in terms of $N(V)$. The problem is to find, qualitatively speaking, the best bounds (where in the definition of $FN(V)$ we may or may not insist that the triangulation of W extends a given one of V). For example, every Pontryagin number P of V is bounded by some function $F_P\left(N(V)\right)$ and we shall see presently that

$$F_P(N) \ll \underbrace{\exp\exp\ldots\exp N}_{\dim V}.$$

But it seems quite realistic to expect just the linear bound $F_P(N) \leq \text{const}_P\, N$, or at worst, a polynomial one, $F_P(N) \leq \text{const}_P\, N^{C_P}$. A similar problem comes up with the function $FN(V)$ where also there is a huge gap between known multi-exponential bounds and the expected linear ones.

II$'$. *Locally bounded fillings.* Let us measure the local complexity of a triangulated space by $N\text{Loc}(\text{Tr}\,V)$, the maximal number of neighbours a simplex may have. Then fix two numbers N_1 and N_2 much larger than N_1 and try to fill in a p.l. triangulated manifold $\text{Tr}\,V$ with $N\text{Loc}(\text{Tr}\,V) \leq N_1$ by a p.l. triangulated W with $N\text{Loc}\,\text{Tr}\,W \leq N_2$. (If one does not like p.l. category one may think of smooth triangulations.) An elementary argument shows, that if V bounds at all, then it bounds some $\text{Tr}\,W$ with $N\text{Loc}\,\text{Tr}\,W \leq N_2$ provided N_2 is sufficiently large compared to N_1. What is more amusing here is a possibility to extend this to (oriented) *pseudo-manifolds* (i.e. spaces built of n-simplices where every $(n-1)$-face has exactly two adjacent n-simplices). Namely, every *oriented pseudomanifold V of dimension $n > 0$ can be filled in by a pseudomanifold W (with boundary $\partial W = V$), such that $N \mathrm{Loc}\, W \leq F(N \mathrm{Loc}\, V)$ for some universal function $F = F_n(N)$.*

To grasp the idea, first let V be a manifold. Then it bounds a pseudomanifold W with some standard singularities, namely cones over the generators in the corresponding (here oriented) cobordism group. The number of these is finite by Thom's theorem and so their complexities are bounded.

Next, suppose V has only isolated singularities. These are cones over certain manifolds, say $U_1, \dots, U_j$, which are, in totality, bound V minus the cones and have $N(U_i) \leq N_1$, $i = 1, \dots, j$. If two of them, say U_1 and U_2 together, bound some manifold $V_{1,2}$, this can be chosen with $N(V_{1,2})$ bounded by a constant (depending on N_1) and we can eliminate such a pair of singularities by cobordism $W_{1,2}$ between $V = V_1$ and V_2 where $W_{1,2}$ satisfies some local bound and V_2 has by two singularities (corresponding to U_1 and U_2) less than $V = V_1$. If one could divide all U_i into such pairs (or just groups with an a priori bounded number of members) one would construct step by step our $W_k = W_{1,2} \,\mathrm{cup}\, W_{2,3} \,\mathrm{cup} \dots$. Of course, such a grouping is not possible in general, but it becomes so after introducing extra singularities, also added one by one in pairs, isomorphic to the cones over $\pm U_i$ which are added in small groups the way we wanted to subtract them. This (after a little thought) gives us a cobordism W_j from V to a non-singular V_j to which the previous argument applies. Notice that these steps from V to V_j do not need any cobordism theory (but this will enter again if we look for a bound on the function $F_n(N_1) \geq N_2$).

Now, suppose the singularity is supported on the k-skeleton and then make some modification over the k-simplices to push the singularity to the $(k-1)$-skeleton. Over each open k-simplex Δ_k the singularity is ($\Delta_k \times$ cone over U_i) where U_i is a $(n-k-1)$-dimensional manifold and where the number of isomorphism classes of these U_i is bounded in terms of N_1. If some finite combination (with a priori bounded number of members) of U_i bounds a manifold, we can by surgery eliminate these U_i, and again, in the general case, such a grouping is preceded by adding to V some (possibly very large) number of "standard" V_ν's, where each has $N(V_\nu) \leq \mathrm{const}_{N_1}$, such that their $U_{i\nu}$ match our U_i. Again, this is a matter of elementary algebra (with no any topology being used). Thus, by elementary induction, every V with bounded singularities can be made non-singular by a cobordism with bounded singularities and the resulting non-singular space, say V', can be filled in by a W' with bounded singularities according to Thom's theorem.

Now comes the true problem: *estimate the number of simplices of these W with bounded singularities in terms of the number of simplices in V.* This is related to a similar problem stated earlier where we did not require the local bounds (but insisted on V and W being manifolds rather than pseudomanifolds) since one can achieve such bounds by induction on skeletons as follows. Suppose our V has a bound on the links of the simplices of codimensions $1, 2, \ldots, n-k-1$, and we want to achieve it over the k-simplices Δ. Here again the singularity is ($\Delta\times$ cones over U_i) where each U_i is locally but not globally bounded. If each U_i can be filled in by a locally bounded $\operatorname{Fill} U_i$ with $N(\operatorname{Fill} U_i) \leq F(N_i)$, we could make the singularities of V smaller by adding only a $F(N(V))$ number of simplices. Thus the locally unbounded filling problem (of estimating $N(W)$) reduces to the locally bounded one.

Low dimensional examples. (1) The circle triangulated into N segments can be filled in by a triangulated disk with at most 7 triangles at every vertex and with at most $10^4 N$ triangles. (Of course we need fewer, but I take 10^4 to be safe without much thinking; in any case, the proof is left to the reader.)

(2) Every oriented surface V triangulated into N simplices bounds a 3-manifold (in fact a handle body) divided into (at most) $10^{100} N$ simplices.

To prove this one may assume, by the above, the triangulation of V has at most 14 triangles at every vertex. Then the corresponding locally bounded problem can be solved in the following geometric setting (while a purely combinatorial proof is left to the reader).

II$''$. Let V be a Riemannian manifold and $\operatorname{Loc} V$ denote $\sup(|K(V)| + (\operatorname{inj}\operatorname{Rad} V)^{-1})$. First we want to fill in V by W with $\operatorname{Loc} W \leq \operatorname{const} \operatorname{Loc} V$ for some fixed (possibly huge) const $= \operatorname{const}_n$, such that near the boundary $\partial W = V$ the manifold W has product geometry, i.e. that of $V \times [0, \varepsilon]$ with $\varepsilon = (\operatorname{Loc} V)^{-1}$, where the injectivity radius of W is measured only ε-far from ∂W. If V bounds some W, then it is easy to give W a metric with such properties but what is more interesting is to have such a W with a bound $\operatorname{Vol} W \leq F(\operatorname{Vol} V)$ where the best function F would be the linear one giving the bound $\operatorname{Vol} W \leq \operatorname{const}_n \operatorname{Vol} V$. Notice that every V with $\operatorname{Loc} V \ll 1$ can be triangulated into $N \approx \operatorname{Vol} V$ simplices with $N \operatorname{Loc} \operatorname{Tr} V \lesssim 1$. Conversely, such a triangulation can be smoothed (if it is smoothable) to a metric with $\operatorname{Loc} \lesssim N \operatorname{Loc}$ and $\operatorname{Vol} \lesssim N(\operatorname{Tr})$. Thus the combinatorial problem concerning locally bounded pseudo-manifold filling is equivalent to its Riemannian counterpart.

Filling-in surfaces. Let first V be diffeomorphic to S^2. Then it admits, by a theorem of Alexandrov, a convex isometric embedding into the hyperbolic space H^3 with given constant curvature $-\kappa < \inf K(V)$. Furthermore, one knows that the local geometry of this embedding (i.e. its second quadratic form) is bounded by that of V and by κ and the volume of the convex body $W < H^3$ is bounded by $\operatorname{const} \operatorname{Area} V$ for all $\kappa < -1$. This W (slightly modified near the boundary) linearly (!) solves our Riemannian filling problem for V

diffeomorphic to S^2.

Next, an arbitrarily oriented surface with bounded geometry can be cut into pairs of pants by controlled cuts which reduces the problem to the case of S^2. Thus, *every oriented surface V with a bound on* $\operatorname{Loc} V$ *admits a filling W with another bound on* $\operatorname{Loc} W$ *such that* $\operatorname{Vol} W \leq \operatorname{const} \operatorname{Area} V$. *Consequently, the locally bounded pseudomanifolds filling problem is linearly solvable in dimension two* as we have claimed.

Remark. One could replace the Alexandrov embedding theorem by the Riemann mapping theorem for S^2. On the other hand one could construct the filling for general surfaces directly by using Laborie's isometric embedding theorem.

II$'''$. Let us give a homotopy theoretic version of the above filling problem where we concentrate on the simplest case of maps $S^{m+n} \to S^m$. If such a map f has $\operatorname{Lip} f \leq \lambda$, it can be regularized (smoothed) so that the pullback of some regular value, say $V = f_{\mathrm{reg}}^{-1}(s) \subset S^{m+n}$, will have (local and global) geometry controlled by λ as follows, for example, from Yomdin's quantitative transversality theorem. Conversely, if we have a framed n-manifold $V \subset S^{m+n}$ with local control over the geometry, the corresponding map $S^{m+n} \to S^m$ is Lipschitz-controlled. Thus the volume-controlled Riemannian filling problem translates in this case to the Lipschitz extension problem. This may be used to prove the above mentioned multi-exponential bound where the appearance of $\exp\exp\ldots$ is due to the use of iterated loop spaces or Postnikov systems. (For example, contracting a map $S^k \to \mathbb{C}P^m$, $3 \leq k \leq 2m$, involves an exponential distortion in the course of the lift to S^{2m+1}. Yet the above 2-filling argument allows sometimes to eradicate this exp, e.g. for the map $S^{m+2} \to S^m$.)

Having failed to prove the linear bound for the above filling, one may look for obstruction and an obvious one is the μ-invariant of V. But this can be linearly bounded by $\operatorname{Vol} V$ (see [Ch-Gr]$_{\mathrm{CN}}$) with the heat flow serving as a kind of linear analytic filling.

III. *Filling without curvature.* A given Riemannian metric g on $V = \partial W$ can be extended to an h on W with arbitrarily small volume but then necessarily the distance function dist_h becomes smaller on V than dist_g. In fact, if $\operatorname{dist}_h = \operatorname{dist}_g$ on V, then (almost) obviously

$$\operatorname{Vol}(W,h) \geq C(V,g).$$

What is less obvious is the existence of h with $\operatorname{dist}_h |V = \operatorname{dist}_g$ satisfying

$$\operatorname{Vol}(W,h) \leq \operatorname{const}_n (\operatorname{Vol} V, g)^{\frac{n+1}{n}},$$

which generalizes the isoperimetric inequality of Federer-Fleming and which is proven in [Gro]$_{\mathrm{FRM}}$ using ideas borrowed from the classical Plateau problem.

Unfortunately, one has a poor understanding of this const_n. For example, one does not know if the filling of the equator $S^n \subset S^{n+1}$ by a hemisphere is the best possible. (Other natural candidates for extremal examples, where the above inequality may become sharp, use a distance function on V related to a Riemannian metric not on V but on an ambient space. For example, one may take a sphere V in a, say symmetric, space X and ask if there is a filling W of V with $\text{dist}_W |V \geq \text{dist}_X |V$ and having smaller volume than the ball in X bounded by V.)

Filling radius. Besides $\text{Vol}\, W$, an important characteristic of a filling is $\text{inRad}\, W \underset{\text{def}}{=} \sup_{w \in W} \text{dist}(w, \partial W = V)$ (where recall the distance in W defined as the infimum of lengths of curves between the points in question where the curves may touch the boundary at some points.) Next one defines $\text{Fil Rad}\, V$ as the infimum of $\text{inRad}\, W$ over all fillings W with $\text{dist}_W |V = \text{dist}_V$. This is related to $\dim_\varepsilon V$ in the following obvious way, $\dim_\varepsilon V \leq n-1 \Rightarrow \text{Fil Rad}\, V \leq \varepsilon$ and so Fil Rad can be vaguely thought of as a distance from V to something lower dimensional. For example, suppose we are given a λ-Lipschitz map $V \to V_1$ where V_1 has the following local contractibility property: *every ball of radius* $\delta \leq 1$ *in* V_1 *is contractible within the concentric* 2δ*-ball.* Suppose that $\lambda\, \text{Fil Rad}\, V \leq \delta_n$ for a small positive $\delta_n \approx 2^{-n}$. Then f admits a continuous extension to W. This is done by sending each $w \in W$ to the nearest point $v \in V$ and then to $f(v) \in V_1$. But the nearest point v may be not unique. What we do is to choose an ε-fine triangulation of W and making some choices for the vertices $w_i \mapsto v_i \to f(v_i)$. The distance between $f(v_i)$ and $f(v_j)$ for adjacent w_i and w_j cannot exceed $2\,\text{inRad} + \varepsilon$ and so we have a short path between $f(v_i)$ and $f(v_j)$ in V_1. Then the boundary of each triangle in W goes to a closed curve of length $6\,\text{inRad}\, W + 3_\varepsilon$ which can be filled in in V_1 because of the local contractibility of V_1. This gives us an extension of f to the 2-skeleton of W. Then we extend to the 3-skeleton, etc. (compare $[\text{Gro}]_{\text{FRM}}$).

Examples. $\text{Fil Rad}\, V \geq \delta_n' \,\text{Inj Rad}\, V$ for some $\delta_n' > 0$.

If $\lambda \leq \delta_n (\text{Fil Rad}\, V)^{-1}$ for the above $\delta_n > 0$, then every λ-Lipschitz map $V \to S^n$ for $n = \dim V$ has zero degree.

Notice that the condition $\text{dist}_W |V = \text{dist}_V$, albeit crucial, could have been slightly relaxed in the above argument which will become relevant presently.

IV. ***Fillings with lower bounds on curvature.*** Let again (V, g) be a closed Riemannian manifold but now with an additionally given quadratic form $\mathcal{S}$ which we want to serve as the second fundamental form of a filling W of V. Then we pick up some curvature $K^?$, e.g. sectional curvature, Ricci curvature, scalar curvature, and try to *maximize* $K^?_-(W) \underset{\text{def}}{=} \inf_W K^?(W)$ over all fillings. Then the supremum of $K^?_-(W)$ over all fillings W of $V = (V, g, \mathcal{S})$ becomes an invariant of $(V, g, \mathcal{S})$ which we want to evaluate and also we want to understand the geometry of the extremal and nearly extremal fillings W.

(The classical calculus of variation suggests maximizing or minimizing some integral curvature characteristics of W, but even for such strong functionals as $\int_W |K(W)|^\alpha dw$ we know yet too little even to make a conjecture.)

There are two natural choices of $\mathcal{S}$ for this purpose. The first one is where V appears as a closed hypersurface in a standard (e.g. symmetric) space and bounds some domain W_0. Here g and $\mathcal{S}$ are the induced metric and the second fundamental form of V in W_0, and the basic question is whether W_0 is extremal for a particular curvature function $K^?$, i.e. if some filling W of $(V, g, \mathcal{S})$ may have $K^?$ greater than that of W_0. Another useful choice of $\mathcal{S}$ is $\mathcal{S} = \lambda g$ for some constant λ where our $\sup \operatorname{Fill} K^?(V, g, \lambda g) \underset{\text{def}}{=} \sup_{\{W\}} K^?_-(W)$ becomes an invariant of g alone for each choice of λ.

The basic result, motivating our setting, is the Schoen-Yau-Witten theorem claiming that the domains $W_0 \subset \mathbb{R}^n$ are extremal for $K^? = \mathrm{Sc}$, i.e. one cannot find W filling ∂W_0 (with g and $\mathcal{S}$ induced from $\mathbb{R}^n$) without having $\operatorname{Inf} \mathrm{Sc}\, W \leq 0$, and this generalizes, according to Min-Oo, to domains in the hyperbolic space $K = -\operatorname{const}$ as well as in the hemisphere ($K = \operatorname{const}$).

***The case* $\mathbf{K}^? = \mathbf{K}$.** The Gauss theorema egregium expresses $K(W)$ on $V = \partial W$ in terms of $K(g)$ and $\mathcal{S}$ which gives us for $n \geq 2$ an a priori upper bound on $\inf_W K(W)$ (unpleasantly) limiting possibilities of the filling. Yet more precise evaluation of $\sup \operatorname{Fill} K(V, g, \lambda g)$ in terms of the global geometry of (V, g) remains interesting especially for manifolds V of positive curvature.

$\mathbf{K}^? =$ Ricci *and scalar curvature.* Here some $(V, g, \mathcal{S})$ may have fillings W with arbitrarily large Ricci.

Example. Take (V, g) to be a flat torus and $\mathcal{S} = \lambda g$ for some $\lambda < 0$. Then this admits fillings W with $\operatorname{Ricci} W \geq \rho$ for arbitrarily large ρ. To see this realize $V = T^n$ as the boundary of a neighbourhood of the zero section of a real line bundle W over $V_0 = T^n/\mathbb{Z}_2$, say $U_\varepsilon \subset W \supset V_0$. We may give V_0 flat metric and make it totally geodesic in W simultaneously making all sectional curvatures of W on the bivectors $(\tau, \nu) \in T(W)|V_0$ equal $\kappa > 0$, where ν is the unit normal to V_0 in W and τ are tangent to V_0. If κ is large so is $\operatorname{Ricci} W$ while the second fundamental form of $\partial U_\varepsilon = V$ can be adjusted with ε to be λg. Notice that the implied involution on $V = T^n$ can be made for $n \geq 2$ orientation reversing as well as free which makes W orientable. Yet there is something not quite convincing about this example, and probably it can be ruled out by some mild restriction.

Proposition-Example. *Let W be a compact manifold with convex boundary $V = \partial W$ (i.e. $\mathcal{S} \geq 0$) such that the mean curvature of V satisfies $M(V) \geq \mu > 0$ and $\operatorname{Ricci} W \geq -\rho^2$ for $(\mu^2/n-1) - \rho^2 = \delta > 0$. Then*

(a) $$\operatorname{inRad} W \leq a\mu^{-1} \ for\ a = \mu^2/\delta.$$

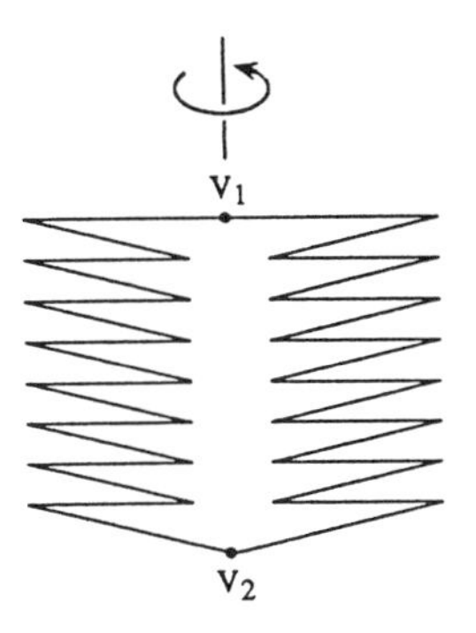

Figure 6

(b) *There exists a constant* $C = C(g, \delta^{-1}) = C(\inf \operatorname{Ricci} V, \delta^{-1})$ *such that every two points in* V *with* $\operatorname{dist}_W(v_1, v_2) \leq 1$ *have* $\operatorname{dist}_V(v_1, v_2) \leq C$.

Sketch of the proof. The first claim follows from the following standard differential inequality for the mean curvatures of the equidistant hypersurfaces $V_t \subset W$,

$$M'_t \geq (n-1)^{-1}(M_t)^2 - \rho^2,$$

which forces M_t to go to ∞ in time $\leq a\mu^{-1}$ (see $[\text{Gro}]_{\text{Sig}}$ for an elementary discussion). Notice that one does not need convexity of $V = \partial W$ at this stage.

Next, to prove (b), we look at the distance function $\operatorname{dist}_W(v_1, v_2)$ and observe that its Laplacian on $V \times V$ tends to be quite negative for large δ. In particular, $\operatorname{dist}_W(v_1, v_2)$ cannot have a local minimum (v_1, v_2) for $v_1 \neq v_2$ if $\delta > 0$. Moreover, the standard minimization argument (relying on the Omori-Yau maximum principle if we want to use $\inf \operatorname{Ricci} V$) shows that there are positive ε_1 and ε_2 depending on (V, g) and δ such that every two points v_1 and v_2 in V can be brought ε_2-closer in W by an ε_1-move in V, and so they cannot be too far apart in V if they are close in W.

Corollary. *If* $M(V) \to \infty$ *then necessarily* $\inf \operatorname{Ricci} W \to -\infty$ *for all fillings of* $(V, g, \mathcal{S} \geq 0)$ *with* g *kept fixed (and* $M =$ *Trace* $\mathcal{S}$*).*

Proof. If M becomes much larger than $-\operatorname{Inf} \operatorname{Ricci}$, then V behaves as if it had a very small filling radius. In particular, every λ-Lipschitz map $V \to S^n$ extends to W if $M(V)$ is large compared to λ and $\operatorname{Inf} \operatorname{Ricci} W$ is not too small, which is, certainly, impossible for large enough λ.

Questions. (a) One can probably significantly relax (if not totally remove) the convexity condition albeit the following example makes one feel unconfortable.

The surface of revolution of the curve in Fig. 6 can be smoothed with positive mean curvature $M \geq \varepsilon^2 > 0$ while the filling is flat.

At the same time the V-distance between v_1 and v_2 can be forced to go to $+\infty$ with $\mathrm{dist}_W(v_1, v_2) \leq$ const. However, this does not constitute a counter-example as the intrinsic geometry of V becomes rather unruly with Ricci $V \to -\infty$.

(b) Can one replace in the above corollary Inf Ricci W by Inf Sc W? (Maybe with $S \to \infty$ strengthening $M \to \infty$.)

$5\frac{3}{4}$. $K_{\surd}$-area for non-spin manifolds

The major geometric impact of our bound on the K-area by $(\inf \mathrm{Sc})^{-1}$ for $\mathrm{Sc} > 0$ is the following.

Rough area bound. *Let g_0 be an arbitrary Riemannian metric on V. Then there exists a positive constant $C = C(V, g_0) > 0$ such that every complete metric g which is areawise greater than g_0 has*

$$\inf_{v \in V} \mathrm{Sc}_v\, g \leq C$$

(where "areawise greater" means that every smooth surface S in V has $\mathrm{Area}_g\, S \geq \mathrm{Area}_{g_0}\, S$).

This bound has been established so far only for *spin* manifolds V and now we want to prove it for all V.

First approach. Every V admits an S^m-bundle $W \to V$ for all large m where the global space W is a spin manifold. (For example, if V is orientable then the total space of the unit tangent bundle is spin.) If V comes with an areawise large metric g, then our sphere bundle can be given a connection with small curvature and the fiberwise (unit spherical) metric adds up with g to a metric $\overline{g}$ on W with $\mathrm{Sc}\,\overline{g} \approx \mathrm{Sc}\, S^m + \mathrm{Sc}\, g$, where, recall, $\mathrm{Sc}\, S^m = m(m-1)$. Furthermore we can enlarge the fibers by scaling them by a suitably large R (yet with R^{-1} not too small compared to the curvature of our connection) which makes the corresponding metric on W, say $\overline{g}_R$, areawise large for g being areawise large (as $\overline{g}_R$ is area-wise monotone in g) and having $\mathrm{Sc}\,\overline{g}_R \approx \frac{m(m-1)}{R^2} + \mathrm{Sc}\, g$ (where we need R^{-1} comparatively large). This largeness of $\overline{g}_R$ signifies, in particular, a lower bound on K-area W and then our spin result applied to W implies the desired conclusion for our (non-spin) manifold V.

Remark. One may use here more general (non-spherical) fibrations $W \to V$ with compact homogeneous fibers such as $\mathbb{C}P^m$ or $\mathrm{Gr}_p\mathbb{R}^m$ handily coming along with vector bundles which can be used in the definition of the K-area.

Second approach. If V is non-spin, the spinors are defined up to $\pm$ sign and form, what we call $\frac{1}{2}$-*spin bundles* S_+ and S_-. Now, instead of the ordinary bundles X we use $\frac{1}{2}$-spin bundles X which have the same $\pm$ ambiguity as S_+ and S_-. Then the tensor products $S_+ \otimes X$ and $S_- \otimes X$ are ordinary vector bundles and we have the (twisted) Dirac operator $D : C^\infty(S_+ \otimes X) \to C^\infty(S_- \otimes X)$. We define the *$K_{\surd}$-area* using these $\frac{1}{2}$-spin bundles X requiring as earlier that some Chern number of X does not vanish (noticing that the notions of a Chern number and of the curvature make perfect sense for these X) and observe that the *proof of the K-area inequality now applies to non-spin manifolds V and shows that if* $\mathrm{Sc}\, V \geq 0$, *then*

$$K_{\surd}\text{-area}_{\mathrm{st}} V \leq \mathrm{const}_n (\inf \mathrm{Sc}\, V)^{-1}. \qquad (K_{\surd})$$

Remark. The $K_{\surd}$-area has the functorial properties similar to those of the K-area but only for *spin* maps $f : V_1 \to V_2$ which respect the second Stiefel-Whitney class, i.e. having $f^*\big(w_2(V_2)\big) = w_2(V_1)$. For example,

$$K_{\surd}\text{-area} V_1 \geq \lambda^{-2} K_{\surd}\text{-area} V_2 \qquad (*)_{\surd}$$

whenever there exists a spin λ-Lipschitz map $V_1 \to V_2$ of non-zero degree (compare $(*)$ in §4), and other properties of the K-area (see §4) similarly extend to the $K_{\surd}$-area. Unfortunately, we are unable to compare the $K_{\surd}$-area of a non-spin manifold with that of S^n or $\mathbb{R}^n$ and the inequality $(K_{\surd})$ leads to no topological restriction on V with $\mathrm{Sc}\, V > 0$ if the universal covering of V is non-spin. (Most topological restrictions for $\mathrm{Sc} > 0$ without the spin assumption follow by the techniques of minimal varieties of Schoen and Yau, but it is less clear how to recapture the geometric aspects of the K-area by these techniques.)

$5\frac{4}{5}$. Symplectic manifolds and positive scalar curvature

Let (V, ω) be a symplectic manifold of dimension $n = 2m$ and g a Riemannian metric on V. Then ω can be diagonalized with respect to g at each point $v \in V$, e.g. $\omega_v = \sum_{i=1}^{m} a_i x_i \wedge y_i$ for a g-orthonormal coframe x_i, y_i at v. Thus g is symplectically characterized by m numbers $|a_i|$, and so every system of intervals $\mathcal{I} = \{I_i \subset [0, \infty)\}$, $i = 1, \dots,$ gives a class of metrics $G_{\mathcal{I}}$ on (V, ω) characterized by $a_i = a_i(g) \in I_i$ for $g \in G_{\mathcal{I}}$. In particular, we have the class $G_1 = G_1(\omega)$ of *adapted* metrics g where each I_i reduces to the single point 1. Two other important classes are $G_>$, where $|a_i| \leq 1$ and $G_<$ where $|a_i| \geq 1$. Notice that each class $G_{\mathcal{I}}$ is invariant under the symplectic automorphism group of the tangent bundle $\big(T(V), \omega\big)$ which is a huge extension of the group $\mathrm{Sympl}\,(V, \omega)$ of symplectic automorphisms of V.

Now every metric invariant $g \mapsto \mathrm{inv}(V, g)$ gives us a function on $G_{\mathcal{I}}$ (invariant under $\mathrm{Sympl}\,(V, \omega)$) from which we may hope to extract symplectic invariants. For example, inf and sup of $\mathrm{inv}(V, g)$ over $g \in G_{\mathcal{I}}$ are invariants of (V, ω). Furthermore, one may use all of the Morse landscape of the function $G_{\mathcal{I}} \underset{\text{inv}}{\longrightarrow} \mathbb{R}$ as a (symplectic) invariant of (V, ω) (and a physicist would try $\int_{G_{\mathcal{I}}} \exp(-\lambda \operatorname{inv}(g)) dg)$.

A single example we have met so far was $\operatorname{inv}(g) = K\text{-area}(g)$ in $4\frac{3}{4}$ and here we look at the scalar curvature $\operatorname{Sc} g$.

If (V, ω, g) is a closed *Kähler* manifold, then one knows (Chern?) that

$$\int_V \operatorname{Sc}(g) dv = \mathbf{I}_0(\omega) \underset{\text{def}}{=} \alpha_m \left(c_1(V) \smile [\omega^{m-1}]\right) [V] \tag{$*$}$$

where $\alpha_m = 4\pi/(m-1)!$, and this was recently extended by David Blair to quasi-Kähler manifolds, i.e. for g adapted to ω as follows:

$$\int_V (\operatorname{Sc} g + \tfrac{1}{4} \|\nabla J\|^2) dv = \mathbf{I}_0(\omega) \tag{$\substack{*\\*}$}$$

where J is the almost complex structure naturally associated to ω and g (defined by $g(x, Jy) = \omega(x, y)$) and ∇ is the covariant derivative of g (see [Bla]). Thus $\int_V (\operatorname{Sc} g) dv \le \mathbf{I}_0(\omega)$ with the equality exactly for Kähler metrics g. This suggests the following three (symplectic) invariants of (V, ω):

(1) $\mathbf{I}_1(\omega) = \sup\limits_g \int_V \operatorname{Sc} g \, dv - \mathbf{I}_0(\omega)$,

(2) $\mathbf{I}_2(\omega) = \left(\sup\limits_g \inf\limits_{v \in V} \operatorname{Sc}_v g\right) \operatorname{Vol} V - \mathbf{I}_0(\omega)$,

(3) $\mathbf{I}_3(\omega) = \left(\sup\limits_g \operatorname{Sc} g\right) \operatorname{Vol} V - \mathbf{I}_0(\omega)$,

where in (1) and (2) g runs over all adapted metrics and in (3) over the adapted metrics with *constant* scalar curvature.

Notice that $\mathbf{I}_3 \le \mathbf{I}_2 \le \mathbf{I}_1 \le 0$ and the basic question is whether (or when) the vanishing of $\mathbf{I}_i$, for a given $i = 1, 2, 3$, implies the existence of an adapted Kähler metric g on (V, ω) (and it would be useful to understand the Euler-Lagrange equation for the function $g \mapsto \int\limits_V \|J_g\|^2 d_g v$ on the space of adapted metrics g, where the solutions generalize Kähler metrics by saying that J and ω are in a certain sense g-harmonic). Also observe that $\mathbf{I}_3$ may be, a priori, equal to $-\infty$, if (V, ω) admits no adapted metric g with $\operatorname{Sc} g$ constant, but I guess the existence of an adapted g with $\operatorname{Sc} g = -$ (large const) must follow for $m \ge 2$ from a suitable h-principle. On the other hand it may be hard to decide when (V, ω) admits an adapted metric with $\operatorname{Sc} > 0$, as besides the topological restrictions disregarding ω, one must take into account the inequality $\mathbf{I}_0(\omega) > 0$ (which also appears in the context of Floer homology and seems to be quite restrictive).

Now let us modify our $\mathbf{I}_i$ by allowing metrics g from a larger class, namely $G_> \supset G$. Notice that the condition $g \in G_>(\omega)$ is equivalent to $\|\omega\|_g \leq 1$ which fits well into the K-area discussion. Define

$$\mathbf{I}_1^>(\omega) = \sup_g \mathrm{Vol}(V, g_0)\big(\mathrm{Vol}(V, g)\big)^{-1} \int_V \mathrm{Sc}\, g\, dv - \mathbf{I}_0$$

where g runs over G and g_0 is some metric from G_1. Clearly $\mathbf{I}_1^>(\omega) \geq \mathbf{I}_1(\omega)$ and, for all we know, it may be $+\infty$ (as it happens if we disregard ω and observe that every smooth manifold V of dimension ≥ 3 admits a metric g with arbitrarily large average $\mathrm{Vol}^{-1} \int \mathrm{Sc}\, g$, obtained by adding spherical bubbles to a given (V, g_0)). Trying to prevent this, one may modify $\mathbf{I}_1^>$ to $\mathbf{I}_1^>(\omega, \sigma_0, \lambda)$ where g runs over the metrics in $G_>(\omega)$ with $\mathrm{Sc}\, g \geq -\sigma_0$ and $\mathrm{Vol}(V, g) \leq \lambda \, \mathrm{Vol}(V, g_0)$ for $g_0 \in G_1$. (Every $g \in G_>$ has $\mathrm{Vol}(V, g) \geq \mathrm{Vol}(V, g_0)$ with the equality iff $g \in G_1$). Similarly, we modify (2) and (3) by enlarging G_1 to $G_>$ and replacing $\mathrm{Vol}\, V$ factor by $\mathrm{Vol}(V, g_0)$ for some $g_0 \in G_1$.

It follows from the K-area inequality (or $K_{\surd}$-area for non-spin manifolds) that $\mathbf{I}_2^>(\omega) < \infty$ for all (V, ω), i.e. one cannot make $\mathrm{Sc}\, g$ everywhere large keeping "$g \geq \omega$", i.e. $\|\omega\|_g \leq 1$, and, moreover, one expects here sharp inequalities of this kind. For example, let (V, ω, g_0) be a compact symmetric Kähler manifold. Then one may think that every metric $g \in G_>$ has $\inf \mathrm{Sc}\, g \leq \mathrm{Sc}\, g_0$. This may be approached by a detailed analysis of the Bochner formula for the Dirac operator twisted with the line bundle corresponding to ω (compare [Lla] and yet unpublished work by Min-Oo).

$5\frac{5}{6}$. Soap bubbles for Sc $> -\sigma$

Rick Schoen once said to me, about 5 years ago, that soap bubbles could be applied to the geometry of $\mathrm{Sc} \geq -\sigma$ as (and even more) efficiently as minimal surfaces. We were talking at the moment about the foliated Plateau problem in hyperbolic 3-manifolds (see p. 73 in $[\mathrm{Gro}]_{\mathrm{FPP}}$) and I was not ready to appreciate Rick's remark. But now we shall follow Rick's suggestion and look at such bubbles and see that they indeed provide a flexible tool for the study of $\mathrm{Sc} > -\sigma$. For example, we shall prove that *the hyperbolic metric of constant sectional curvature and* $\mathrm{Sc} = -n(n-1)$ *cannot be, even locally, approximated by metrics with* $\mathrm{Sc} \geq -\sigma > -n(n-1)$.

Usually, *soap bubbles* refer to surfaces of constant mean curvature. Here we use a more general (well known) notion of a *μ-bubble* where μ is a real function on a Riemannian manifold V. We look at a hypersurface W in V bounding some domain $W^+ \subset V$ and set

$$V\ell_\mu W^+ = \int_{W^+} \mu(v) dv.$$

Actually, one should think of μdv as an n-form on V for $n = \dim V$ and $V\ell_\mu W^+$ should be regarded as a 1-form on the space of the hypersurfaces W in question. Here we allow W to have a non-empty boundary and then $V\ell_\mu W^+$ is defined up to an additive constant (i.e. as a 1-form) on the space of W's with a boundary $\partial W \subset V$ *kept fixed.* Then we consider the function(al) $W \mapsto \mathbb{R}$ given by

$$W \mapsto A(W) - V\ell_\mu W^+$$

for $A(W) = \mathrm{Vol}_{n-1} W$ (thought of as "area") and define μ-bubbles as critical points (i.e. hypersurfaces $W \subset V$) of this function.

Examples. (a) If $\mu = 0$, these bubbles are the ordinary minimal subvarieties, which have (at their non-singular points) zero mean curvature.

(b) Let $V = \mathbb{R}^n$ and $\mu(v) = (n-1)\|v\|^{-1}$. Then the μ-bubbles are exactly the concentric spheres $W_t \subset \mathbb{R}^n$ of radii $t \in \mathbb{R}_+$ around the origin. Not-accidentally these have constant mean curvatures, this is because the levels of μ have constant mean curvatures, $M\big(\mu^{-1}(t)\big) = (n-1)t^{-1}$. Here our function $W_t \mapsto A(W_t) - V\ell_\mu W_t^+$ is (clearly) constant ($=0$) in t (where W_t^+ is the ball bounded by the sphere W_t and where the mean curvature of W_t is $+(n-1)$ with our sign convention). In fact each sphere W_t provides the (non-strict) global minimum for the function $W \mapsto A(W) - V\ell_\mu W^+$.

We want to show that in general μ-bubbles W have mean curvature $M(W) = \mu | W$ and then to compute the second variation (derivative) of $A - V\ell_\mu$. Such a variation at W is defined with a normal field $\varphi(W)\nu$ for a unit normal field ν looking outward (of W^+) as in Fig. 7 below.

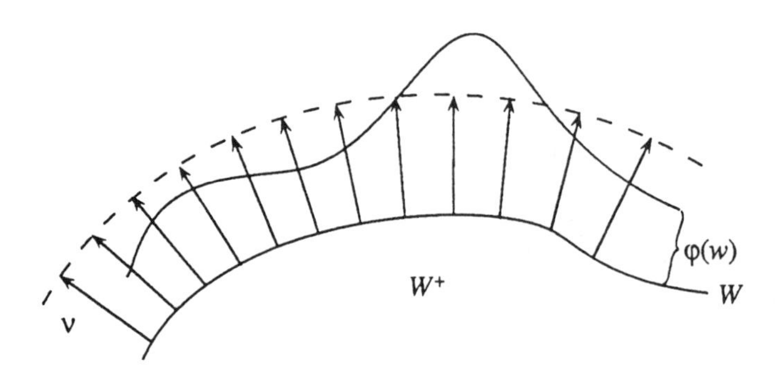

Figure 7

One knows that the first variation (derivative) of $A = A(W)$ at W is

$$A' = \int_W M(w)\varphi(w)dw$$

where M denotes the mean curvature of W, and

$$V\ell'_\mu = \int_W \varphi(w)\mu(w)dw.$$

Thus

$$d(A - V\ell_\mu) = 0 \Rightarrow \int \big(M(w) - \mu(w)\big)\varphi(w) = 0$$

for all functions φ on W which implies $W(w) = \mu(w)$ as we mentioned earlier. Next, one knows that

$$A'' = \int_W \left(\|d\varphi\|^2 - (\operatorname{Trace} \mathcal{S}^2 - M^2 + \operatorname{Ricci}_V(\nu,\nu))\varphi^2(w)\right) dw,$$

where $\mathcal{S}$ is the *shape operator* corresponding to the *second quadratic form* of $W \subset V$, and so Trace $\mathcal{S}^2 = \sum_{i=1}^{n-1} \lambda_i^2$ for the principal curvatures λ_i of W, while $M = \sum_{i=1}^{n-1} \lambda_i$. Furthermore

$$V\ell''_\mu = \int_W \big(\mu(w)M(w) + \mu'_\nu(w)\big)\varphi^2(w)dw,$$

where μ'_ν is the ν-normal derivative of μ. So

$$(A - V\ell_\mu)'' = \int_W (\|d\varphi\|^2 + R\varphi^2)dw$$

for

$$R = -\operatorname{Trace} \mathcal{S}^2 + M^2 - \operatorname{Ricci}(\nu,\nu) - \mu M - \mu'_\nu.$$

This can be related to the scalar curvature by (following Schoen and Yau) substituting

$$\operatorname{Ricci}(\nu,\nu) = \tfrac{1}{2}\big(\operatorname{Sc} V - \operatorname{Sc}(V|W)\big)$$

where $\operatorname{Sc}(V|W)$ is obtained, at each point $w \in W \subset V$, by summing up the sectional curvatures of V over an orthonormal frame in $T_w(W) \subset T_w(V)$, and

$$-\tfrac{1}{2}\left(\operatorname{Trace}\mathcal{S}^2 - M^2\right) = -\sum_{i,j=1}^{n-1} \lambda_i\lambda_j = \tfrac{1}{2}\big(\operatorname{Sc}(V|W) - \operatorname{Sc} W\big)$$

which makes

$$R = -\tfrac{1}{2}\operatorname{Trace}\mathcal{S}^2 + \frac{1}{2}M^2 + \tfrac{1}{2}(\operatorname{Sc}W - \operatorname{Sc}V) - \mu M - \mu'_\nu.$$

In particular, if W is μ-critical, then the second variation becomes

$$\int_W (\|d\varphi\|^2 - \tfrac{1}{2}(\operatorname{Trace}\mathcal{S}^2 - \mu^2 + 2\mu'_\nu - \operatorname{Sc}W + \operatorname{Sc}V)\varphi^2)dw,$$

where W has constant mean curvature $M = \mu$. Finally we observe that Trace $\mathcal{S}^2 \geq (n-1)^{-1}M^2$ and so

$$(A - \mu V\ell)'' \leq \int_W \|d\varphi\|^2 - \tfrac{1}{2}\left(\frac{n\mu^2}{n-1} + 2\mu'_\nu - \operatorname{Sc}W + \operatorname{Sc}V\right)\varphi^2 dw.$$

It follows that if W locally *minimizes* $A(W) - V\ell_\mu(W)$ then

$$\int_W \|d\varphi\|^2 - \tfrac{1}{2}\left(\frac{n\mu^2}{n-1} + 2\mu'_\nu - \operatorname{Sc}W + \operatorname{Sc}V\right)\varphi^2 dw \geq 0 \qquad (*)$$

for all functions φ on W.

Example. Let V be a *warped product*, $V = W \times \mathbb{R}$ with the metric $g = a^2(t)h + dt^2$ for some metric h on W and a positive function $a(t)$ satisfying $a(0) = 1$. Then

$$A(W \times t) = a^{n-1}(t)A_0,$$

and $\mathcal{S}$ on $W \times t$, viewed as the second quadratic form, equals $(a'/a)h$,

$M = \operatorname{Trace}\mathcal{S} = (n-1)a'/a$,
$-\operatorname{Ricci}(\nu,\nu) = \operatorname{Trace}\mathcal{S}^2 + M' = (n-1)\left[(a'/a)^2 + a''/a - (a')^2/a^2\right] = (n-1)a''/a.$

Each $W \times t$ here has *constant* mean curvature $M = M(t)$ and so it is μ-critical for $\mu(t) = M(t)$. In fact it is even (non-strictly) locally minimal and so the second variation of $A - V\ell_\mu$ is non-negative vanishing exactly at the (constant) normal field ν since $V\ell_\mu(W \times t)$ for $\mu = M = (n-1)a'/a$ equals

$$A_0 \int_{-\infty}^t \mu(\tau)a^{n-1}(\tau)d\tau = \int_0^t (n-1)a'(\tau)a^{n-2}(\tau)d\tau = A(W \times t).$$

Notice that the principal curvatures of $W \times t$ are all equal,

$$\lambda_1(t) = \lambda_2(t) = \ldots = \lambda_{n-1}(t) = \lambda = a'/a$$

and so Trace $\mathcal{S}^2 = (n-1)^{-1}M^2$. Consequently $(*)$ becomes equality for $\varphi =$ const in this case (which checks up with the equality

$$R = -\tfrac{1}{2}\left(\tfrac{n\mu^2}{n-1} + 2\mu' - \mathrm{Sc}(W \times t) + \mathrm{Sc}\,V\right) = 0,$$

obtained by a straightforward computation i.e. by subsituting $\mu = (n-1)a'/a$,

$$\mathrm{Sc}(W \times t) = \mathrm{Sc}(V|W \times t) + (n-1)(n-2)\lambda^2,$$

and $\mathrm{Sc}\,(V|W \times t) = \mathrm{Sc}\,V + 2\,\mathrm{Ricci}$ for $\mathrm{Ricci} = (n-1)a''/a$).

***Warping* W *with* S^1.** Now we look at a different kind of a warped metric on $W \times S^1$ defined with a metric h on W and a positive function f on W by

$$\widehat{g}_f = h + f^2 d\,s^2$$

which has

$$\mathrm{Sc}(\hat{g}_f) = \mathrm{Sc}(h) + (2\Delta f)/f$$

where Δ is the positive Laplacian on W (see p. 157/369 in [G-L]$_{\mathrm{PSC}}$).

We apply this warping procedure to a stable (e.g. locally minimal) μ-bubble $W \subset V$, i.e. where $(*)$ holds for all functions φ on W vanishing on the boundary. Then there exists a function f on W vanishing on ∂W and satisfying

$$\Delta f + Rf = \lambda_0 f$$

for some $\lambda_0 \geq 0$, where

$$R = -\tfrac{1}{2}\left(\tfrac{n\mu^2}{n-1} + 2\mu'_\nu - \mathrm{Sc}\,W + \mathrm{Sc}\,V\right),$$

and so

$$\mathrm{Sc}\,\hat{g}_f = \mathrm{Sc}\,W + 2(\lambda_0 - R) = 2\lambda_0 + \tfrac{n\mu^2}{n-1} + 2\mu'_\nu + \mathrm{Sc}\,V \geq \tfrac{n\mu^2}{n-1} - 2|\mu'_\nu| + \mathrm{Sc}\,V.$$

Approximation corollary. *Let the metric $g = a^2(t)h + dt^2$ on $V = W \times \mathbb{R}$ be C^0-approximated on a fixed band $V_\delta = W \times [0, \delta]$ by metrics g_ε with $\mathrm{Sc}\,g_\varepsilon \geq \mathrm{Sc}\,g + \sigma_0$.*

Then there exist functions f_ε on W such that the warped product metric $\hat{g}_\varepsilon = h_\varepsilon + f_\varepsilon^2 ds^2$ on $W \times S^1$ has $\mathrm{Sc}\,\hat{g}_\varepsilon \geq \mathrm{Sc}\,W + \sigma_0 - \varepsilon'$ for $h_\varepsilon \underset{C^0}{\to} h$ and $\varepsilon' \to 0$ with $\varepsilon \to 0$.

(One should regard warped metrics on $W \times S^1$ as kinds of generalized metrics on W and so this corollary reduces dimension in the C^0-approximation problem by metrics with $\mathrm{Sc} \geq \sigma$, compare §12 in [G-L]$_{\mathrm{PSC}}$.)

Proof. We slightly perturb the function $\mu = \mu(t) = M(t)$ in order to make some $W \times t_0 \subset W \times [0, \delta]$, with small t_0 eventually going to zero, strictly μ-minimal. Then (V, g_ε), which is $C^0 - \varepsilon$-close to (V, g), also has, for small ε, a μ-minimal bubble, say W^ε-close to $W \times t_0$, in fact W^ε is non-singular and C^1*-close* to $W \times t_0$ (by an easy argument). Then we warp the induced metric h_ε on W^ε with $f = f_\varepsilon$ as above. Q.E.D.

Non-approximation conclusion. *A metric g of constant sectional curvature near a point $v_0 \in V$ cannot be C^0-approximated by g_ε with* $\mathrm{Sc}\, g_\varepsilon \geq \sigma >$ $\mathrm{Sc}\, g$.

Proof. The metric g near a point is a warped product in polar coordinates, $g = a^2(t)h + dt_2$ where (W, h) is a small round sphere in V around v_0. By the above corollary, an approximation g_ε with $\mathrm{Sc}\, g_\varepsilon \geq \mathrm{Sc}\, g + \sigma_0$ with $\sigma_0 > 0$ would give rise to a warped metric $\hat{g}_\varepsilon$ on $W \times S^1$ with $\mathrm{Sc}\, \hat{g}_\varepsilon \geq \mathrm{Sc}\, h + \sigma_0'$ which is incompatible with the version of Llarull's theorem stated in $5\frac{4}{9}$; as such $W \times S^1$ comes along with the contracting map to the sphere W_t of the radius t slightly less than that of W and yet with $\mathrm{Sc} < \mathrm{Sc}\, \hat{g}_\varepsilon$. Q.E.D.

Remark. One may be relieved to learn that Llarull's theorem (based on Dirac) can be excluded and the proof rendered purely Plateau. To show this, let us think of $\hat{g}_\varepsilon$ as a generalized metric on W approximating the original (spherical) one. Then one may assume, by induction on dimension, that the scalar curvature of $\hat{g}_\varepsilon$ cannot essentially exceed that of h. The details of the argument here are similar to those in §12 of [G-L]$_{\mathrm{PSC}}$ and left to the reader. Notice that all μ-minimal varieties in the present case are non-singular, being C^1-close* to round spheres, and so one does not have to limit $\dim V \leq 7$ as in [G-L]$_{\mathrm{PSC}}$.

Approximation for non-constant sectional curvature. One can apply the above argument to a very small and narrow spherical band around a point $v_0 \in V$ with a suitable $\mu(v) = \mu(\mathrm{dist}(v, v_0)$ and obtain a certain upper bound on $\mathrm{Sc}\, g_\varepsilon$ in terms of the infinitesimal geometry of g at v_0. For example, *if the sectional curvature at v_0 satisfies $K_{v_0}(V) \leq \kappa_0$, then $\mathrm{Sc}\, g_\varepsilon \leq n(n-1)\kappa_0 + \varepsilon'$ with $\varepsilon' \to 0$ for $\varepsilon \to 0$* (which recaptures the above constant curvature result where $\mathrm{Sc}_{v_0}\, g = n(n-1)\kappa_0$). The proof is similar to the above and left to the reader.

On global effects of $\mathrm{Sc} > -\sigma$. Consider a compact Riemannian manifold V with two boundary components, say W_1 and W_2, and take a function μ on V such that $\mu | W_1 \geq -M(W_1)$ and $\mu | W_2 \leq M(W_2)$, where the mean curvatures are signed with the exterior normal field. Then there exists a minimal μ-bubble W between W_1 and W_2 with implied W^+ being the band between W_1

---

* Grisha Perelman explained to me that this "C^1-closeness" is a non-trivial matter and so our "Remark" should be called a "conjecture."

and W, since W has $M(W) = \mu|W$ and cannot touch neither W_1 nor W_2 by the maximum principle. If W is non-singular (which is always the case for $\dim V \leq 7$) then we can make the warped product metric $\hat{g}$ on $W \times S^1$ with

$$\mathrm{Sc}\,\hat{g} \geq \alpha(V,\mu) = \inf_V \left(\mathrm{Sc}\,V + \tfrac{n}{n-1}\,\mu^2 - 2\|d\mu\|\right).$$

Now, suppose we know a priori that $\mathrm{Sc}\,\hat{g} \leq \sigma_0$. (For example, the topology of V may prevent every W separating the ends from having positive scalar curvature on $W \times S^1$, e.g. if the homology class $[W]$ has infinite K-area, then $\mathrm{Sc}\,\hat{g} \leq 0$, or there exists an area contracting map of V to the round sphere S_r^{n-1} and by Llarull's theorem $\mathrm{Sc}\,\hat{g} \leq r^{-2}(n-1)(n-2)$.) Then we conclude that

$$\inf_V \left(\mathrm{Sc}\,V + \tfrac{n}{n-1}\mu^2 - 2\|d\mu\|\right) \leq \sigma_0, \tag{$\star$}$$

for *every* function μ on V satisfying the above boundary relations.

Here is a specific example. Let V be homeomorphic to $W \times [1,2]$, where W has infinite K-area (and so $W \times S^1$ admits no metric with $\mathrm{Sc} > 0$). Let $\sup_{W_1} -M(W_1) = M_1^-$ be not too large, i.e. W_1 is not too concave in the M-sense, while $M(W_2) \geq M_2^+ > 0$ (i.e. W_2 is mean convex). Furthermore, suppose that $\mathrm{dist}(W_1, W_2)$ is large so it is easy to make up μ with small $\|d\mu\|$ and given behavior near W_1 and W_2. Then we have the bound

$$\mathrm{Sc}\,V \leq -\tfrac{n}{n-1}\,\mu^2 + 2\|d\mu\|$$

where the second term can be made small for $\mathrm{dist}(W_1, W_2)$ large while the best for the first term is given by $\mu = M_2^+$. Thus we can have a bound of Sc of the form

$$\inf \mathrm{Sc}\,V \leq -\tfrac{n}{n-1}\,(M_2^+)^2 + \text{small term}.$$

This looks crude but it may be sharp in some cases. For example if we start with the warped product metric $g = e^{2t}h + dt^2$ on V with constant sectional curvature -1, the above shows that every metric g' on V which equals g near the boundary and has $\mathrm{dist}_{g'}(W_1, W_2) \geq \mathrm{dist}_g(W_1, W_2)$ necessarily satisfies $\mathrm{Inf}\,\mathrm{Sc}\,g' \leq \mathrm{Sc}\,g$ with the equality only for g' isometric to g. This still works where W is non-compact, e.g. $W = \mathbb{R}^{n-1}$ in the above example with $(V = \mathbb{R}^{n-1} \times \mathbb{R},\ e^{2t}h + dt^2)$ being the hyperbolic space, provided the metric g' is sufficiently standard (e.g. equals g) at infinity, which implies the version of the Min-Oo theorem stated in VI of $5\frac{2}{3}$.

There are further applications of μ-bubbles to $\mathrm{Sc} \geq -\sigma$ but these deserve a separate paper.

6. Index and the spectrum

Observe that the BL-formula $D^2 = \Delta_S + \frac{1}{4}\mathrm{Sc}$ bounds the spectrum of the Dirac operator D on a complete manifold V from below by

$$\inf \operatorname{spec} D^2 \geq \frac{1}{4} \inf_V \mathrm{Sc}_v(V) \qquad (*)$$

and so every upper bound on $\inf \operatorname{spec} D^2$ in terms of the macroscopic geometry of V implies a similar bound on $\mathrm{Sc}\, V$. Now we focus on the spectrum of D (and of D^2) rather than on the scalar curvature and try to relate this spectrum directly to the geometry of V. Notice that $\operatorname{Spec} D$ (unlike $\operatorname{spec} d + d^*$) is not immediately linked to the coarse macroscopic geometry of $V = (V, g)$ as the construction of D essentially uses the first derivatives of g. Yet we shall see below, following Vafa and Witten, that a suitable macroscopically visible largeness of V leads to an upper bound on $\inf \operatorname{spec} D^2$ (which amounts, for a compact V, to an appearance of an eigenvalue λ_0 of D small in the absolute value) similar to the bound of $\inf_V \mathrm{Sc}\, V$ by the K-area of V. Moreover, we shall obtain such bounds for all geometric operators D, where the most interesting D's are Hodge's $d + d^*$ and Dolbeault's $\overline{\partial} + \overline{\partial}^*$.

$6\frac{1}{4}$. *K*-length and ε-straightness

We want to introduce an invariant of a unitary bundle $X = (X, \nabla)$ over a Riemannian manifold V measuring the deviation of X from being a *straight*, i.e. *trivial flat* bundle. Recall that our K-area concerns the deviation of X from a flat but not necessarily trivial bundle by measuring the curvature $\mathcal{R}(X)$. Now we want to integrate $\mathcal{R}$ to some quantity $\mathcal{P}$ recording the parallel transport of the connection and thus measuring non-straightness of (X, ∇). What we do in practice is comparing ∇ with a trivial connection in a larger trivial bundle $X^0 \supset X$ as follows. First, more generally, let X^0 be an arbitrary unitary bundle containing X and let ∇^0 be a connection on X^0. Then the difference $\nabla^0 - \nabla$ on X^0 is a 1-form on V with values in $\mathrm{End} = \mathrm{End}(X_0) \supset \mathrm{Hom}(X \to X^0)$ (with the inclusion induced by the normal projection $X_0 \to X$). Thus we may speak of *the operator norm* in End_v and in $\mathrm{Hom}(T_v(V) \to \mathrm{End}_v)$ for all $v \in V$, denoted by $\|\nabla^0 - \nabla\|_v$ and $\|\nabla^0 - \nabla\| = \sup_{v \in V} \|\nabla^0 - \nabla\|_v$.

Definitions. The *non-straightness* of (X, ∇), denoted by $\|\mathcal{P}_N(X)\|$, is

$$\|\mathcal{P}_N(X)\| = \inf_{X^0} \|\nabla^0 - \nabla\|$$

where "inf" is taken over all straight (i.e. trivial flat) bundles $X^0 = (X^0, \nabla^0)$ of rank N and all unitary embeddings $X \hookrightarrow X_0$.

The *K-length$_N$* of V is

$$K\text{-length}_N(V) = \inf_X \ \|\mathcal{P}_N(X)\|^{-1}$$

where X runs over all "homologically substantial" unitary bundles $X = (X, \nabla)$, with the same meaning of "homologically substantial" as earlier in §4, namely, non-vanishing of some Chern number.

This definition of the K-length is meaningful for $\dim V$ even. If $\dim V$ is odd, we stabilize by passing to $V \times \mathbb{R}$, where (as for non-compact manifolds in general) we restrict to bundles X trivialized at infinity.

Our major concern will be limiting the K-length from below, i.e. constructing sufficiently straight homologically significant bundles X over V. This can be done (as in bounding from below the K-area) by exhibiting sufficiently contracting maps $f : V \to S^n$, for $n = \dim V$, of non-zero degree and, for n even, pulling back to V a standard bundle over the unit sphere S^n with non-zero top Chern class. (Notice that for the K-area purposes f needs be only sufficiently *area* contracting but here we need contraction in all directions.)

It is convenient at this stage to introduce the *hypersphericity* radius of V, denoted Rad V/S^n as the maximal (suprimal) number R, such that V admits a Lipschitz-R^{-1} map $V \to S^n$ of non-zero degree, where, if V is non-closed, each component of the boundary and/or infinity of V must go to a single point in S^n (and where these points may be different for different components of the boundary/infinity).

Now we use all these notations just to express the indicated above lower bound on the K-length in writing,

$$K\text{-length}_N V \geq \text{const}_n \ \text{Rad} \ V/S^n \qquad \text{for} \quad N \geq n \ . \tag{$\star$}$$

This is proven with a non-trivial complex vector bundle X_0 of $\mathbb{C}$-rank $n/2$ over an even dimensional sphere S^n which has $\|\mathcal{P}_n(X_0)\| \leq \text{const}_n < \infty$ and which pulls back to V under an R^{-1}-contracting map to a bundle X with $\|\mathcal{P}_N(X)\| \leq R^{-1} \ \text{const}_n$ for all $N \geq n$.

Remark about reversing ($\star$). An elementary argument for surfaces V ($n = 2$) shows that

$$\text{Rad} \ V/S^2 \geq \text{const}(K\text{-length}_N V) \tag{$\bar{\star}_2$}$$

for all N and some const ≥ 0.01. In particular K-length$_N V$ is essentially independent of N for $N \geq 4$. I do not know if this is true for $n \geq 3$ but one can show that the stabilized "Rad" can be bounded from below by $\text{const}_N \cdot$ length. For example, if V is even dimensional, then

$$\text{Rad}(V \times \mathbb{R})/S^{n+1} \geq \text{const}_N \cdot (K\text{-length}_N V) \qquad \text{for every} \quad N = 1, 2, \ldots . \tag{$\bar{\star}_n$}$$

This follows from Serre's theorem on inducing rational cohomology classes from odd dimensional spheres. In our case we start with an odd dimensional non-torsion homology class $h \in H_{n+1}((\mathrm{Gr}_k\,\mathbb{C}^N) \times S^1)$ for which, according to Serre's theorem, there exists a Lipschitz map $\varphi : (\mathrm{Gr}_k\,\mathbb{C}^N) \times S^1 \to S^{n+1}$ non-vanishing on h. We take the circle S^1 of a very large length $L = L_N$ so that the best (i.e. infimal) Lipschitz constant of our map becomes a function of k and N, say $s_{k,N}(h)$, and notice that $\sup\limits_{h \in H_{n+1}} s_{k,N}(h) = s_{k,N,n+1} < \infty$ as the group $H_{n+1} = H_{n+1}((\mathrm{Gr}_k\,\mathbb{C}^N) \times S^1))$ is finitely generated. We observe that if V supports a homologically significant bundle X of rank k with $\|\mathcal{P}_N(X)\| \leq \ell^{-1}$, then V admits a ℓ^{-1}-contracting map f into the Grassmannian $\mathrm{Gr}_k\,\mathbb{C}^N$, with the metric induced from the operator norm (metric) for the imbedding $\mathrm{Gr}_k\,\mathbb{C}^n \to \{\text{operators}\}$ sending each k-plane $\tau \subset \mathbb{C}^N$ to the normal projection operator $\mathbb{C}^N \to \tau$, such that $h = f_*[V] \neq 0$ (compare below). Now the relevant map $V \times S^1 \to S^{n+1}$ comes by composing φ with $f \times 1 : V \times S^1 \to (\mathrm{Gr}_k\,N) \times S^1$.

Problem. Evaluate the (Serre) constants $s_{k,N,n+1}$ for $N \to \infty$ (and possibly $n \to \infty$ and $k \to \infty$). This seems interesting already for $\mathbb{C}P^N \times S^1$.

This problem arises any time when the algebraic topology provides homotopically interesting maps between standard manifolds but gives us no *realistic* bound on the Lipschitz constants of these maps. (Serre's type arguments evaluate these constants by something like $\underbrace{\exp\exp\ldots\exp}_{n} N$ if not worse, compare $5\frac{5}{7}$).

K-length and mappings to Grassmannians. If V admits a ℓ^{-1}-contracting map f to $\mathrm{Gr}_k\,\mathbb{C}^N$ then the pullback of the canonical rank k bundle over $\mathrm{Gr}_k\,\mathbb{C}^N$ to V, say $X \to V$, has $\mathcal{P}_N(X) \leq \ell^{-1}$. In fact, maps $V \to \mathrm{Gr}_k\,\mathbb{C}^N$ correspond to embeddings $X \hookrightarrow X^0$ where X^0 is the trivial bundle over V of rank N. The trivial connection ∇^0 on X^0 induces ∇ on X by $\nabla = P\nabla^0$ for the normal projection $P : X^0 \to X$.

Conversely, starting from a connection ∇ on X with small K_N-length, one has, by definition, an embedding $X \to X_0$, where clearly the induced connection, say ∇_1 on X, is close to ∇ and this ∇_1 is induced from $\mathrm{Gr}_k\,\mathbb{C}^N$. In fact, such an embedding $X \to X_0$ can be often achieved whenever X has small curvature $\mathcal{R}$ by constructing N sections of X (or rather of the dual bundle X^*) with small covariant derivatives. For example, let V be covered by open subsets U_i, $i = 1, \ldots, N_0$, such that every loop of length $\leq \delta$ in the ε-neighbourhood of each U_i bounds a disk of area A in the 2ε-neighbourhood of U_i where δ, ε and A are certain positive constants satisfying $\delta \geq 2(\mathrm{Diam}\,U_i + 2\varepsilon)$ (e.g. $\varepsilon \approx \mathrm{Diam}\,U_i \approx \delta/4 \leq 1$ and the ε-neighbourhood of each U_i is bi-Lipschitz to the Euclidean $(\delta + \varepsilon)$-ball with the implied Lipschitz constant $\approx A^{\frac{1}{2}}$). Then, assuming $A\|\mathcal{R}\|$ is small, say $\leq \exp -k$, one can construct over the ε-neighbourhood of each U_i an almost parallel k-frame, and these, bumped down to zero near the boundaries of these ε-neighbourhoods, will give us an embedding $X \to X^0$ for rank $X^0 = N = kN^0$ with controlled derivatives, namely bounded roughly by $\mu(\varepsilon^{-1} + A\|\mathcal{R}\|)$, where μ denotes the multiplicity of the covering by $U_\varepsilon(U_i)$. In particular, K-area + local geometry bound K-length.

$6\frac{1}{2}$. Differential operators twisted with almost straight bundles

We want to compare the twisted Dirac operators in (X^0, ∇^0) and (X, ∇). In fact, we do this for an arbitrary first order operator D acting between two unitary bundles, and, to save notations, we assume this is the same bundle, say S, and $D : C^\infty(S) \to C^\infty(S)$. We recall the *principal symbol* $\sigma = \sigma(D)$ of D, that is an End$S$-valued 1-coform (vector) on V defined as follows. Take a 1-form ℓ on V and a section s of S. To find σ_v for a given $v \in V$ we take a smooth functions f with $df(v) = \ell(v)$ and set $\sigma_v(\ell \otimes s) = (Df)(s)(v)$. In other words, the endomorphism $\sigma_v(\ell)$ maps s to $(Dfs)(v)$. We denote by $\|\sigma\|$ the operator norm, i.e. $\sup\limits_{\|\ell\|\leq 1, \|s\|\leq 1} \|\sigma(\ell \otimes s)\|$ and use the following twist D_X of D with an arbitrary bundle (X, ∇)

$$C^\infty(S \otimes X) \overset{1\otimes\nabla}{\longrightarrow} C_\infty(S \otimes (L \otimes X)) = C^\infty((S \otimes L) \otimes X) \overset{\sigma\otimes 1}{\longrightarrow} C^\infty(S \otimes X)\,,$$

$$\underbrace{\qquad\qquad\qquad\qquad\qquad\qquad\qquad\qquad}_{D_X}$$

where L denotes the cotangent bundle of V. (This agrees with our earlier twist for the Dirac operator where there is no zero order term.) Observe that this twist does not increase the norm of the principal symbol. In fact, $\|\sigma(D_X)\| = \|\sigma(D)\|$ for rank $X > 0$. Thus all twisted Dirac (as well as Hodge and Dolbeaut) operators have their symbols bounded by a fixed constant const_n (which actually does not depend on $n = \dim V$ either with our choice of $\|\sigma\|$).

Now we compare D_{X^0} and D_X by taking the difference $D_{X^0} - D_X$ which is a homomorphism $X \to X^0$ obtained by comparing $\nabla^0 - \nabla$ with the symbol of D and abiding the bound $\|D_{X_0} - D_X\| \leq n\|\sigma(D)\| \cdot \|\nabla^0 - \nabla\|$. To see this we write $\nabla' = \nabla^0 - \nabla$ and $x \overset{\nabla'}{\mapsto} \sum\limits_{i=1}^n \ell_i \otimes \nabla'_i(x)$ for an orthonormal basis ℓ_i in L. Similarly, write, $\sigma(s \otimes \ell_i) = \sigma_i(s)$ and then compose $\nabla' = (\nabla'_1, \ldots, \nabla'_n)$ and $\sigma = (\sigma_1, \ldots, \sigma_n)$ as follows:

$$\sum_{\mu,\nu} a_{\mu\nu}\; s_\mu \otimes x_\nu \;\overset{1\otimes\nabla'}{\mapsto}\; \sum_{\mu,\nu,i} a_{\mu\nu}\; s_\mu \otimes \ell_i \otimes \nabla'_i(x_\nu) \;\overset{\sigma\otimes 1}{\mapsto}$$

$$\sum_{\mu,\nu,i} a_{\mu\nu}\; \sigma_i(s_\mu) \otimes \nabla'_i(x_\nu) = \sum_{i=1}^n \sum_{\mu,\nu} a_{\mu\nu}\; \sigma_i(s_\mu) \otimes \nabla'_i(x_\nu).$$

The norms of the operators σ_i and ∇'_i are bounded by those of σ and ∇' (by the definition of $\|\sigma\|$ and $\|\nabla'\|$) and so we have

$$\left\| \sum_{\mu,\nu} a_{\mu\nu}\, \sigma_i(s_\mu) \otimes \nabla'_i(x_\nu) \right\| \leq \|\sigma\| \cdot \|\nabla'\| \cdot \sum_{\mu,\nu} a_{\mu\nu}\, s_\mu \otimes x_\nu$$

since the norm of the tensor product of operators is submultiplicative ($\|A \otimes B\| \leq \|A\| \cdot \|B\|$, as is seen with an orthonormal basis which remains orthogonal under A and a similar basis for B), which yields the required bound by summing over $i = 1, \ldots, n$.

We specify the above to a Dirac type operator D (i.e. Dirac, Hodge, Dolbeaut, possibly twisted with an unitary bundle) and come to the following conclusion, that *$D'_X = D_{X^0} - D_X$ is a zero order operator, i.e. a homomorphism $X \to X_0 \supset X$ satisfying*

$$\|D'_X\| \leq \delta' = \text{const}_n \|\nabla'\| \tag{+}$$

for $\nabla' = \nabla^0 - \nabla$ where $\text{const}_n = n\times$ (universal constant)).

Spectral Corollaries. (1) *If V is a complete Riemannian manifolds, then the bottom of the spectrum of $|D_{X_0}|$ is bounded from below by that of $|D_X|$ as follows:*

$$\inf \text{spec}|D_{X_0}| \leq \inf \text{spec}|D_X| + \delta'$$

for the above $\delta' = \text{const}_n \|\nabla'\|_n$ *and* $\nabla' = \nabla^0 - \nabla$ (where, recall spec |self-adjoint operator| = |spec(operator)|).

(2) *Let V be a closed Riemannian manifold. Then the number of the eigenvalues of D_{X_0} in every interval $[a,b]$ is bounded from below by the number of the eigenvalues of D_X in the interval $[a',b']$ for $a' = a + \delta'$ and $b' = b - \delta'$ (where $[a',b']$ agreed to be empty for $a' > b'$). This is expressed in writing by*

$$\#\ \text{spec}\ D_{X_0}[a,b] \geq \#\ \text{spec}\ D_X[a',b']. \tag{$*$}$$

(3) *Let V admit a discrete co-compact isometric action of a group Γ which lifts to X and X_0 and commute with D_X and D_{X_0}. Then the above $(*)$ remains valid with the (von Neumann) Γ-dimension (of the space corresponding to* spec $\in [a,b]$*) instead of the ordinary dimension (= #* spec$[a,b]$*). That is*

$$\dim_\Gamma \text{spec}\ D_{X_0}[a,b] \geq \dim_\Gamma \text{spec}\ D_X[a',b'] \tag{$*$}_\Gamma$$

(which is equivalent to $(*)$ for finite groups Γ, where $\dim_\Gamma \text{spec} \ldots = |\Gamma|^{-1} \#\ \text{spec} \ldots$*).*

The proof of (1), (2) follows by the following elementary perturbation argument which automatically extends to the Γ-case of (3) (see $9\frac{1}{9}$). To prove (1) we observe, for an arbitrary self-adjoint operator $\mathcal{D}$ on a Hilbert space $\mathcal{X}$ and $\lambda \geq 0$, that

$$\text{infspec}|\mathcal{D}| \geq \lambda \Leftrightarrow \|\mathcal{D}(\mathrm{x})\| \geq \lambda\|\mathrm{x}\| \quad \text{for all } \mathrm{x} \in \mathcal{X}. \tag{+}$$

First we apply this to $\mathcal{D} = D_X$ and $\lambda = \text{infspec } D_X + \varepsilon$ thus obtaining a vector x for which $\|D_X(x)\| < \lambda\|x\|$. Then we apply D_{X_0} to this x (recall that D_{X_0} is defined on a larger space than D_X) and see that $\|D_{X_0}(x)\| < (\lambda + \delta')\|x\|$ by the triangle inequality as $\|D_{X_0} - D_X\| \leq \delta'$ on the domain of D_X. Then by applying (+) to D_{X_0} we conclude to the inequality $\inf \text{spec}|D_{X_0}| \leq \lambda + \delta'$ which yields (1) for $\varepsilon \to 0$.

Next, in order to study the spectrum in a given segment $[a, b]$, we apply (+) to $\mathcal{D} - c$ for some $c \in [a, b] \cap \text{spec } \mathcal{D}$ and see that a perturbed operator, $\mathcal{D}_0 = \mathcal{D} + \mathcal{D}'$, necessarily has a spectrum point in $[a, b]$ if $\|\mathcal{D}'\| \leq \min(a - c, c - b)$ (where we may have $\mathcal{D}_0$ defined on a larger space than $\mathcal{D}$). In other words, a δ'-perturbation of $\mathcal{D}$ moves each eigenvalue by at most δ'. This implies (2) as the spectra of D_X and D_{X_0} are discrete in the compact case and (3) also follows with necessary Γ-provisions.

$6\frac{3}{4}$. When an operator D over a large manifold V has many eigenvalues near zero

We want to apply the above corollary to bound from above $\inf \text{spec } D$ where D is a twisted (which incudes "untwisted") Dirac, Hodge or Dolbeaut operator. This is done by using an auxiliary bundle X such that D twisted with X has non-zero index and thus $\text{spec} D_X \ni 0$. (Here we assume $n = \dim V$ even and D splits into $D_+ \oplus D_-$ with $\text{ind}\, D$ actually referring to $\text{ind}\, D_+$). We try to choose this X as straight as possible, i.e. with a unitary embedding into a straight (trivial flat) bundle X_0 such that $\|\nabla^0 - \nabla\|$ is small. Then by the above (1) the zero mode of D_X gives a λ-mode of $D_{X_0} = \underbrace{D \oplus \ldots \oplus D}_{N}$ for $\lambda \leq \delta' = \text{const}_n \|\nabla^0 - \nabla\|$, which also serves as a λ-mode of D itself. If V is large, in the sense of having large K-length or, even better (see $\star$), large hyperspherical radius, we may choose such an X with small δ'. Summing up we come to the following:

spectral inequalities:

$$\inf \text{spec } D \leq \text{const}_n (K\text{-length}_N\, V)^{-1} \tag{$\star\star$}$$

$$\inf \text{spec } D \leq \text{const}_n (\text{Rad } V/S^n)^{-1} \tag{$\star\star\star$}$$

where V is a complete Riemannian manifold and $N = 1, 2, \ldots$ is an arbitrary integer.

Explanations. The above argument works, strictly speaking, if V is a *closed* even dimensional manifold with $D = D_+ \oplus D_-$ such that the index of the operator D_+ twisted with X is given by the formula

$$\text{ind} = A_D \smile \text{ch}_X[V], \qquad (*)$$

where A_D is a polynomial in p_i with *non-zero* term of degree zero. This is the case, for example, for the Dirac signature (i.e. Hodge's $D = d + d^*$ with the splitting $D = D_+ + D_-$ according to the eigenvalues of the Hodge $*$-operator) and Dolbeaut. If V is *complete non-compact* we may assume without loss of generality that $\inf \operatorname{spec} D > \sigma_0 \geq 0$ (otherwise there is nothing to prove). In this case the operator D, twisted with a bundle X trivial at infinity, is Fredholm and still satisfies the essential part of $(*)$ by the following noncompact version of the Atiyah-Singer index theorem.

$6\frac{4}{5}$. Relative index theorem

Let V be a complete Riemannian manifold where our (self-adjoint) operator D is positive at infinity in the sense that there exists a compact subset $K \subset D$, such that the vanishing $s \mid K = 0$ implies

$$\|D(s)\|_{L_2} \geq \sigma_0 \|s\|_{L_2}$$

for all L_2-section s in the domain of D and a fixed (depending on D and K) constant $\sigma_0 > 0$. Let V' be another manifold which is identified with V at infinity (say, outside K) and let D' be an operator over V' identical with D at infinity. Then the operators D and D' are Fredholm and the difference of the indices of D_+ and D'_+ is given by the usual formula

$$\text{ind } D_+ - \text{ind } D'_+ = (A_D - A_{D'}) \; [V \operatorname{cup} V'], \qquad (**)$$

where A_D and $A_{D'}$ are the Atiyah-Singer polynomials (in characteristic classes) associated to D and D'.

This theorem applies, in particular, to two twisted operators over the same manifold V, say $D = D_X$ and $D' = D_{X'}$, where the bundles X and X' are identified at infinity, for example, where X is a bundle trivialized at infinity and X' is the trivial bundle with $\operatorname{rank} X' = \operatorname{rank} X$ (identified with X outside some $K \subset V$). Next, the theorem yields the spectral bounds $(\star\star)$ and $(\star\star\star)$ also for *odd* dimensional V by passing to $V \times \mathbb{R}$ or $V \times S^1$ for a sufficiently long circle S^1. (But one gets by far more mileage from the relative odd index theorem; see $6\frac{8}{9}$ and $6\frac{11}{12}$.) Actually, it is worth stabilizing also for even dimensional V

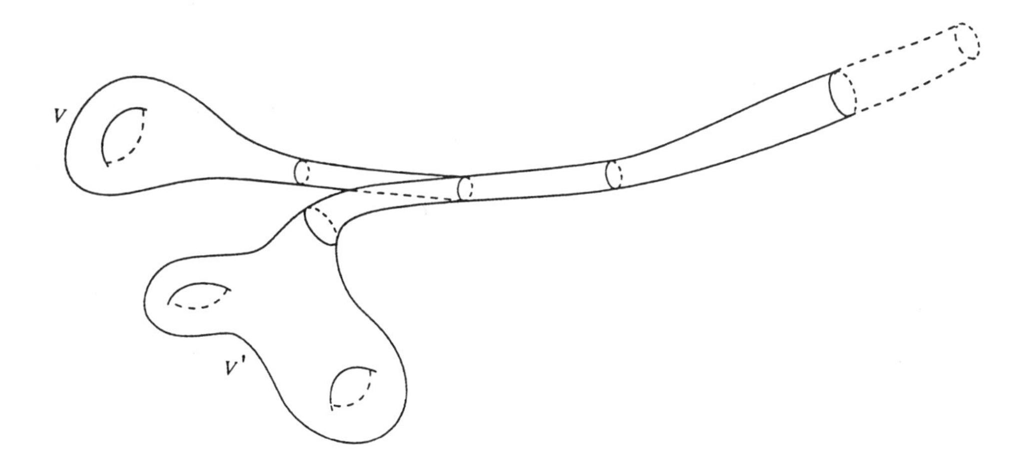

Figure 8

as the K-length of $V \times \mathbb{R}^M$ as well as the hypersphericity radius may go up with increase of M (albeit it is unclear by how much). Then, in order to avoid the dependence of our constants on M, it is better to use the index theorem for families (rather than the ordinary index theorem applied to $V \times \mathbb{R}^M$) where $t \in \mathbb{R}^M$ is our parameter. Thus we set

$$K\text{-length}_{\text{st}} V = \sup_{N,M} \; K\text{-length}_N \,(V \times \mathbb{R}^M)$$

and stabilize $(\star\star)$ to

$$\inf \operatorname{spec} D \leq \operatorname{const}_n \,(K\text{-length}_{\text{st}})^{-1}. \qquad (\star\star)_{\text{st}}$$

Similarly we define $\operatorname{Rad}_{\text{st}} V/S^n$ as $\sup\limits_M \operatorname{Rad}(V \times \mathbb{R}^M)/S^{n+M}$ and stabilize $(\star\star\star)$ by substituting $\operatorname{Rad}_{\text{st}}$ for Rad. (If V is non-compact it is less restrictive to use maps $V \times \mathbb{R}^M \to S^{n+M}$ which are locally constant at infinity on each individual slice $V \times t$, $t \in \mathbb{R}^M$, rather than locally constant at infinity on $V \times \mathbb{R}^M$. Similarly, in the definition of K-area$_{\text{st}}$, we may use bundles X trivialized at infinity of each $V \times t$ where the trivialization may move with $t \in \mathbb{R}^M$. The relative index theorem *for families* works perfectly in this situation while the corresponding individual index theorem does not seem to apply to such X on $V \times \mathbb{R}^M$).

On the proof of the relative index theorem. This can be traced to the original work by Atiyah and Singer where they discuss the *excision* property of the index homomorphism. The above formulation copies that in

[G-L]$_{\mathrm{PSC}}$ (where we limited ourselves with Blaine Lawson to Dirac operators on complete manifolds V with $\mathrm{Sc}\, V > 0$ at infinity), and the proof of [G-L]$_{\mathrm{PSC}}$ can be adapted to the present situation. A more conceptual argument is given in [Ang]. Here we sketch yet another proof (closely following that in [Roe]$_{\mathrm{RIT}}$ where the author works with a finer index valued in K_0 of a suitable algebra) which clarifies the "excision" aspect of the relative setting. Namely, we observe that the right hand side of (**) makes sense *without* assuming D is positive at infinity and we want to give an operator-theoretic expression replacing $\mathrm{ind}\, D_+ - \mathrm{ind}\, D'_+$ for general (non-positive at infinity) operators D. Heuristically, we rewrite

$$\begin{gathered}\mathrm{ind}\ D_+ - \mathrm{ind}\ D'_+ = (\dim\ker D_+ - \dim\ker D_-) - (\dim\ker D'_+ - \dim\ker D'_-) = \\ (\dim\ker D_+ - \dim\ker D'_+) - (\dim\ker D_- - \dim\ker D'_-) = \\ \mathrm{ind}\ \delta_+ - \mathrm{ind}\ \delta_-\end{gathered}$$

for suitable *Fredholm* operators δ_+ and δ_- (where, recall $D_- = (D_+)^*$ and $(D'_-) = (D_+)^*$). We want $\delta_\pm$ to act from ker $D_\pm$ to ker $D'_\pm$ and for this we need an operator connecting the domains of D and D'. We use for this purpose the identification between (V, S, D) and (V', S', D') at infinity and take some operator Φ from sections of S to those of S' which is given at infinity by this identification and which is zero over some compact region. More precisely, we take a smooth (cut-off) function φ on V which equals 1 at infinity and which has $\mathrm{supp}\,\varphi$ inside the region where V is identified with V'. Then $\Phi(s)$ is defined for all section $s : V \to S$ in three steps.

1. Multiply s by φ and restrict the product φs to $u = \mathrm{supp}\,\varphi \subset V$.

2. Take the section $(\varphi s)'$ corresponding to φs over $U' \subset V'$ indentified with U.

3. Extend $(\varphi s)'$ by zero on $V' - U'$.

Now we compose Φ restricted to Ker D_+ with the orthogonal projection P'_+ onto Ker D'_+, call this composition $\delta_+ : \ker\ D_+ \to \ker\ D'_+$ and similarly define $\delta_- : \ker\ D_- \to \ker\ D'_-$.

Example. Suppose we have a single manifold V and δ maps $\ker D$ into itself by $D \circ \varphi$, first by multiplying with a function φ equal 1 at infinity and then by normally projecting to ker D in the L_2-space of sections of X. Such a δ is Fredholm since the product φs is L_2- close to s for every $s \in \ker\ D$ which is (relatively) ε-small in a neighbourhood $U \subset V$ containing $K = \mathrm{supp}(\varphi - 1)$, i.e. satisfying $\int_U \|s\|^2\, dv \leq \varepsilon \int_V \|s\|^2\, dv$, and since, for every $\varepsilon > 0$ and relatively compact $U \subset V$, there exists a subspace $L \subset \ker\ D$ of *finite* codimension such that all $s \in L$ satisfy this inequality (because the restriction operators from U_1 to $U_2 \subset\subset U_1$ are compact on ker D as D is elliptic).

In general, however, where $X \neq X'$, the operator δ_+ is not always Fredholm. In fact, it is Fredholm if zero is an isolated point in the spectra of D and D', but may be not otherwise. To remedy this we must regularize the projection operator $P'_+ : L_2(S') \to \ker D'_+$ by another operator $Q'_+ = \psi(D')$ for a suitable (spectral) function $\psi : \mathbb{R} \to \mathbb{R}$ replacing the Dirac δ-function concentrated at the zero point (of the spectrum) which defines $P'_+ = \delta(D'_- \circ D'_+)$ (where, recall, S' is split, $S' = S'_+ \oplus S'_-$ and $D = D'_+ \oplus D'_-$ with $D'_+ : S'_+ \to S'_-$ and $D'_- : S'_- \to S'_+$ being mutually adjoint operators). Notice that if $\psi(1) = 1$, then, formally,

$$\text{ind } D'_+ = \text{Trace } P'_+ - \text{Trace } P'_- = \text{Trace } \psi(D'_- \circ D'_+) - \text{Trace } \psi(D'_+ \circ D'_-),$$

since the operators $D'_- \circ D'_+$ and $D'_+ \circ D'_-$ have equal spectra apart from zero, and so by "tracing" $\psi(D_- \circ D_+), \ldots, \varphi(D'_+ \circ D'_-)$ one may recapture $\text{ind } D_+ - \text{ind } D_-$.

Now the idea is to choose ψ so that the operators $\psi(\ldots)$

(a) will have ***finite propagation***, i.e. their Schwartzian kernels $K(v_1, v_2)$ vanish for $\text{dist}(v_1, v_2) \geq \text{const}$; and

(b) will be ***locally traceable***, which means for positive operators A (and only such will be needed) that $\varphi A \varphi$ is in the trace class for all continuous functions φ with compact supports (which is equivalent under (a) to tracebility of either φA or $A \varphi$).

Now, as pointed out by Roe, these properties are satisfied for the functions ψ for which the *Fourier transforms* $\widehat{\psi}$ *have compact supports* and so such ψ are readily available.

One can define for such ψ

$$\text{Tr}_+ \, \psi \underset{\text{def}}{=} \text{Trace } \psi(D_- \circ D_+) - \text{Trace } \psi(D'_- \circ D'_+)$$

since the operators $\psi(D_- \circ D_+)$ and $\psi(D'_- \circ D'_+)$ coincide outside a compact set and hence their difference is traceable (where the relevant ψ is positive, but in fact any ψ with compact supp $\widehat{\psi}$ will do). To make the above precise one should bring the operator to a single Hilbert space as earlier. In fact, it is convenient here first to split $\psi(D_- \circ D_+) = \mathcal{A}_1 + \mathcal{A}_2$, where $\mathcal{A}_1 = \varphi_1 \, \psi(D_- \circ D_+)$ and $\mathcal{A}_2 = \varphi_2 \, \psi(D_- \circ D_+)$ and where φ_1 and φ_2 are smooth non-negative functions such that φ_1 has (large) compact support. Then $\mathcal{A}_1$ will be of trace class, and if $\text{supp}\, \varphi_1$ is sufficiently large, then

$$\Phi \, \mathcal{A}_2 \, \Phi^* = \mathcal{A}'_2 \underset{\text{def}}{=} \varphi_2 \, \psi(D'_- \circ D'_+),$$

which allows us to define

$$\text{Tr}_+ \, \psi \underset{\text{def}}{=} \text{Trace } \mathcal{A}_1 - \text{Trace } \mathcal{A}'_1.$$

Similarly, we define

$$\mathrm{Tr}_- \ \psi = \mathrm{Trace}\ \varphi_1\ \psi(D_+ \circ D_-) - \mathrm{Trace}\ \varphi_1\ \psi(D'_+ \circ D'_-)$$

and set

$$\mathrm{ind}\left([D_+] - [D'_+]\right) \underset{\mathrm{def}}{=} \mathrm{Tr}_+\ \psi - \mathrm{Tr}_-\ \psi.$$

Excision Proposition. *If* $\psi(1) = 1$ *and* supp $\widehat{\psi}$ *is compact then the above "index" satisfies* $(**)$, *i.e.*

$$\mathrm{ind}\left([D_+] - [D'_+]\right) = (A_D - A_{D'})\ [V \,\mathrm{cup}\, V']. \tag{exc}$$

Proof. This (excision) formula is local and immediately follows, for example, from the local version of the Atiyah-Singer formula. (Notice that it makes sense and remains true for non-complete manifolds as well.)

Finally, let us derive $(**)$ from (exc) for *Fredholm* operators D and D'. To do this we choose a sequence ψ_i weakly converging to the δ-function and use the continuity of the trace. This yields

$$\lim_{i\to\infty}\ \mathrm{Tr}_+\ \psi_i = \mathrm{Trace}\,\delta(D_-\ D_+) - \mathrm{Trace}\,\delta(D'_-\ D'_+) = \dim\ker D_+ - \dim\ker D'_+,$$

and similarly

$$\lim_{i\to\infty}\ \mathrm{Tr}_-\ \psi_i = \dim\ker D_- - \dim\ker D'_-.$$

Q.E.D.

Examples and applications. (1) *Let K-length*$_{\mathrm{st}}$ $V = \infty$. *Then* $0 \in$ spec D. *In particular, if* $\mathrm{Rad}_{\mathrm{st}}\ V/S^n = \infty$, *e.g. if V is hyper-Euclidean (i.e. $V \succ \mathbb{R}^n$, which means the existence of a proper Lipschitz map $V \to \mathbb{R}^n$ of non-zero degree) then* $0 \in$ spec D.

Notice that the above geometric criteria (K-length $= \infty$, Rad $= \infty$, "hyper-Euclidean") are very robust. In particular, if V appears as a (infinite) covering of a compact manifold $\overline{V}$, these properties are homotopy invariants of $\overline{V}$. Thus, for example, if $(\overline{V}, \overline{g})$ is a closed manifold admitting a metric $\overline{g}_0$ of non-positive sectional curvature and V is the universal covering of $(\overline{V}, \overline{g})$, then V is (obviously) hyper-Euclidean and, consequently, D on V contains zero in the spectrum.

Problem. Let V be a covering of a closed manifold $\overline{V}$. Find a (most general) homotopy condition on $\overline{V}$ (and on $\pi_1(V) \subset \pi_1(\overline{V})$) which would ensure the presence of zero in the spectrum of D on V.

Remark. If D is Hodge's $d + d^*$ then the inclusion $0 \in \operatorname{spec} D$ on X is a homotopy invariant of $\overline{V}$ as this (inclusion) is equivalent to non-vanishing of the non-reduced L_2-cohomology of V (and in all known examples this cohomology does not vanish). On the other hand, if D is Dirac, then the presence of zero in the spectrum may depend on a particular metric. Actual examples are known for compact V (e.g. $V = S^3$, see [Hit]) but no one seemed to work it out for infinite coverings. For example, let $(\overline{V}, \overline{g})$ be a closed Riemannian manifold which admits a metric $\overline{g}_0$ with Sc $\overline{g}_0 > 0$. Can an infinite (say cyclic) covering of $(\overline{V}, \overline{g})$ have zero in the spectrum of the Dirac operator? (One may ask similar questions for Dolbeault's $\partial + \partial^*$.)

(2) Let $\widetilde{V}_i$ be a sequence of finite k_i-sheeted coverings, where $k_i \to \infty$ for $i = 1, 2, \ldots$, of a closed manifold V. The problem is to find an asymptotic bound on inf spec $|\widetilde{D}_i|$ for $i \to \infty$ where $\widetilde{D}_i$ stands for the lifts to $\widetilde{V}_i$ of a given D on V. Here again, the asymptotics of inf spec $|\widetilde{D}_i|$ is a topological (even homotopy) invariant of V (and $\pi_1(\widetilde{V}_i) \subset \pi_1(V)$) for $D = d + d^*$ but not for general D. Yet we seek a bound on inf spec $\widetilde{D}_i$ in topological terms for all our D. The asymptotics of the metric invariants of $\widetilde{V}_i$ we used above, Rad $\widetilde{V}_i/S^n$, K-length $\widetilde{V}_i$, etc. are all homotopy invariants and can be sometimes nicely computed with a suitable metric. For example, if V admits a metric of negative curvature and all loops in $\widetilde{V}_i$ at some point $v_i \in \widetilde{V}_i$ of length $\leq \ell_i$ are constructible, then, obviously, Rad $\widetilde{V}_i/S^n \geq \ell_i$ and consequently

$$\text{inf spec } |\widetilde{D}_i| \leq \text{const}_V \ \ell_i^{-1}. \tag{+}$$

This estimate is qualitatively sharp for the coverings of the tori T^n given by $t \mapsto it$ (where $\ell_i \sim i$) but not for general coverings of T^n.

Questions. (a) Which sequences of coverings $\widetilde{V}_i$ of T^n (determined by the subgroups $\pi_1(\widetilde{V}_i) \subset \mathbb{Z}^n = \pi_1(T^n)$) have $\inf \operatorname{spec} \widetilde{D}_i \underset{i \to \infty}{\to} 0$ for the Dirac operators $\widetilde{D}_i$ lifted from $V = T^n$ with an arbitrary (non-flat) Riemannian metric? (Notice that the Hodge Laplace operators on $\widetilde{V}_i$ have the spectra accumulating at zero on form of each degree for all sequences of coverings with the numbers of sheets k_i going to infinity).

(b) Let $V = H/\Gamma$ where H is the Heisenberg group and Γ is a co-compact lattice. This V admits standard coverings $\widetilde{V}_i \to V$ corresponding to dilations of H. The problem is to bound $\inf \operatorname{spec} \widetilde{D}_i$ by something better than ℓ_i^{-1}. (It is easy to see that ℓ_i^{-1} works here since the balls $B(R)$ in H have Rad $B(R)/S^n \approx R$.)

(c) Let V be homotopy equivalent to an arithmetic variety S/Γ where S is symmetric space of non-compact type and $\widetilde{V}_i$ correspond to a sequence of congruence subgroups $\Gamma_i \subset \Gamma$. Again we want to bound $\inf \operatorname{spec} \widetilde{D}_i$, say for the

Dirac operator D by something better than ℓ_i^{-1}, or to see in examples that ℓ_i^{-1} is the best general bound.

Finally we notice that the ideology behind the Novikov conjecture suggests that inf spec $\widetilde{D}_i \to 0$ whenever $\ell_i \to \infty$ provided the classifying map $V \to B\Gamma$ for $\Gamma = \pi_1(V)$ sends the fundamental homology class of V to a *non-zero* element in $H^n(B\Gamma;\mathbb{R})$. But even in the cases where this is known one yet has to find a good upper bound on inf spec D_i. (The bound inf spec $\widetilde{D}_i \leq$ const ℓ_i^{-1} seems plausible for linear (sub)groups Γ as they act on products of Bruhat-Tits buildings and for similar reason for hyperbolic and related groups, such as the *mapping class group* where the negative curvature argument requires some caution.)

$6\frac{5}{6}$. Lower bounds on the number of eigenvalues

We introduce a new invariant of a closed Riemannian manifold V, denoted maxch(V, N), $N = 1, 2, \ldots$, as the maximum of the absolute values of Chern numbers of all complex bundles X over satisfying $\|\mathcal{P}_N(X)\| \leq 1$ (where $\|\mathcal{P}_N\|$ measures non-straightness of an optimal realization of X in the trivial bundle of rank N; see $6\frac{1}{4}$). This "maxch" will be applied to (the metric of) V scaled by some $\ell > 0$, and so maxch$(\ell V, N) \geq d \Leftrightarrow$ there exists an X over V with $\|\mathcal{P}_N(X)\| \leq \ell$ and having some Chern number at least d. (Our old friend K-length$_N$ corresponds to the minimal ℓ for which maxch$(\ell V, N) \geq 1$.) Similarly we refine Rad V/S^n by defining $\max\deg(\ell V/S^n)$ as the supremum of degrees of ℓ-Lipschitz maps $V \to S^n$. This "maxdeg" is increasing in ℓ and

$$\max\deg(\ell V/S^n) = s_n\, \ell^n \text{ Vol } V + o(\ell^n) \quad \text{for} \quad \ell \to \infty$$

for some universal constant s_n (see [G-L-P]). The two "maxes" are related by the obvious inequality

$$\text{maxch}(\ell V, N) \geq \text{const}_n \ \max\deg \ell V/S^N$$

for all even $n = \dim V$, all $N \geq 2n$ and some universal const$_n > 0$. (This is proven by pulling a standard non-trivial bundle from S^n to V.)

Finally we extend the definition of maxch(V, N) to n odd by setting

$$\text{maxch}(\ell V, N) \underset{\text{def}}{=} \text{maxch}(\ell V \times S^1, N)$$

for the unit circle S^1 and we notice that, typically,

$$\max\deg \ell V/S^n \approx \max\deg(\ell V \times S^1)/S^{n+1}.$$

Length-spectrum Estimate. *Let V be a closed Riemannian manifold and D a geometric differential operator (i.e. twisted or untwisted Dirac, Hodge's*

$d+d^*$, or Dolbeaut $\partial+\partial^*$). Then the number of the eigenvalues of D in each segment $[-a,a]$ satisfies

$$\#\ \mathrm{spec}\ D[-a,a] \geq \delta_n\ N^{-1}\ \mathrm{maxch}(\gamma_n a V, N), \qquad (*)$$

for all N and some universal positive constants δ_n and γ_n. Consequently

$$\#\ \mathrm{spec}\ D[-a,a] \geq \delta'_n \max\deg(\gamma'_n\ aV/S^n).$$

Corollary. *For every closed Riemannian manifold (V,g_0) there exists a constant $\delta=\delta(g_0)>0$ such that for each metric $g_1\geq g_0$ the* corresponding operator $D=D_{g_1}$ *has*

$$\#\ \mathrm{spec}\ D[-a,a] \geq \delta\ a^n - 1 \qquad (\#)$$

for all $a\geq 0$. Moreover, (#) remains valid for every manifold (V_1,g_1) admitting a contracting map $V_1\to V_0$ of non-zero degree.

(No such bound is possible for the ordinary Laplace operator on functions; see $6\frac{12}{13}$).

Proof of $(*)$. By slightly tinkering at X (and transforming it to the tensor product of suitable exterior powers of X, compare $5\frac{3}{8}$) we arrive at the situation where the top dimensional term in the Chern character of the modified X becomes of order d and so the twisted operator D has index about d. Then $(*)$ in $6\frac{1}{2}$ gives us a bound on the spectrum of D twisted with the trivial bundle of rank N (in fact, slightly greater than N as we have modified X) which is the same thing as N times the spectrum of D. We leave filling in the details to the reader.

$6\frac{7}{8}$. Evaluation of "maxdeg" for "simple" manifolds

This does not come up as readily as one might expect, yet several examples are available.

Tori and beyond. Take the flat torus T^n, where the shortest closed geodesic has length L. Then, clearly, $\max\deg \ell T^n/S^n \approx \ell^n \mathrm{Vol}\ T^n$ for $\ell L\geq$ const_n (while $\max\deg\ \ell T^n/S^n = 0$ for $\ell L < 2\pi$). A similar estimate applies to approximately flat tori such as finite coverings $\widetilde{V}_i$ of a fixed torus V. Namely, $\max\deg \ell\ \widetilde{V}_i/S^n \geq \delta_V\ \ell^n\ \mathrm{Vol}\ \widetilde{V}_i$ for $\ell L\geq \mathrm{const}_V$ which implies the following lower bound on the number of the eigenvalues of D lifted from V to $\widetilde{V}_i$

$$\#\ \widetilde{D}_i\ [-a,a] \geq a^n\ \mathrm{Vol}\ \widetilde{V}_i$$

for $a \leq \widetilde{L}_i^{-1}$ where $\widetilde{L}_i$ denotes the length of the shortest non-contractable closed curve in $\widetilde{V}_i$. This is qualitatively sharp as T^n-invariant operator D (e.g. non-twisted Dirac and $d + d^*$ on the flat torus T^n) have $\#$ spec $\widetilde{D}_i\ [-a, a] \approx a^n$ Vol T^n for all $a \leq \ell$.

Next, look at a more general situation where V is an arbitrary manifold (not homeo T^n anymore) and $\widetilde{V}_i$ are finite Galois coverings converging to the universal covering of V, i.e. $\widetilde{L}_i \to \infty$. What we keep of T^n is the existence of a map $f : V \to T^n$ of non-zero degree (which amounts to the presence of n cohomology classes in $H^1(V)$ with non zero cup-product). Then we observe with pleasure that for *every* $\ell > 0$ (where small ℓ's are the ones we are after)

$$\max \deg \ell\ \widetilde{V}_i/S^n \geq \text{const}_V\ \ell^n\ \text{Vol}\ \widetilde{V}_i$$

for all $i \geq i_0(\ell)$ *and, hence*

$$\lim_{i\to\infty} \inf(\#\ \text{spec}\ \widetilde{D}_i\ [-a, a])/\text{Vol}\ \widetilde{V}_i \geq \delta_V\ a^n \qquad (\widetilde{\#})$$

for some $\delta_V > 0$.

In fact, the pertinent maps $\widetilde{V}_i \to S^n$ come from composing $\widetilde{V}_i \to \widetilde{T}_i^n \to S^n$ and by the same token $(\widetilde{\#})$ remains valid for all sequences of finite Galois covering $\widetilde{V}_i$ converging to $\widetilde{V}$ lying over the covering induced (by f) from the universal covering $\mathbb{R}^n \to T^n$, (such as the maximal Abelian covering of V, for example). And $(\widetilde{\#})$ also extends to this (infinite) covering $\widetilde{V}$ by

$$\dim_\Gamma\ \text{spec}\ \widetilde{D}\ [-a, a] \geq \delta_V\ a^n$$

for Γ being the Galois group of $\widetilde{V}$ (see $9\frac{1}{9}$). The key case here is that of the universal covering $\widetilde{V}_{\text{univ}} \to V$, i.e. $\Gamma = \pi_1(V)$, and the simplest non-Abelian example is V, a surface of genus ≥ 2 (which does admit the required map f to T^2 of non-zero degree).

Nilmanifolds. Let $V = G/\Gamma$ where G is a simply connected nilpotent Lie group and Γ is a co-compact discrete subgroup. Take a sequence of finite Galois coverings $G/\Gamma_i = \widetilde{V}_i \to V$ converging to the universal covering $\widetilde{V}_{\text{univ}} = G$ (which amounts to $\bigcap_i \Gamma_i = 1$) and try to construct ℓ-contracting maps $f_i : \widetilde{V}_i \to S^n$, $n = \dim V$, with possibly large degrees $\deg f = \int_{\widetilde{V}_i} \text{Jac}\, f_i$, where "large" here means close to ℓ^n Vol $\widetilde{V}_i$ and where ℓ is small eventually converging to zero. We recall that large metric balls $B(R)$ in G have

$$\text{Vol}\ B(R) \approx R^h$$

for some integer $h \geq n = \dim V$ called the *exponent* of G (which equals the Hausdorff dimension of the limit εG, $\varepsilon \to 0$). It is not hard to show that such a ball admits a proper Lipschitz map onto the Euclidean R-ball where the implied Lipschitz constant is independent of R. (This can be seen by looking at the limit $\lim\limits_{\varepsilon\to 0} \varepsilon G$). One takes a maximal system of disjoint R-balls in $\widetilde{V}_i$ which are the same as in G for $i \geq i_0(R)$ and by ℓ-contracting each of them to S^n with $\ell \approx R^{-1}$ one obtains maps $f_i : \widetilde{V}_i \to S^n$ with $\deg f_i \approx \ell^h$ Vol $\widetilde{V}_i$ for every fixed $\ell > 0$ and $i \geq i_0(\ell)$. If $h > n$ (which happens for all non-Abelian G), this is rather inefficient for small ℓ as the average Jacobian of such f_i is about ℓ^h rather than ℓ^n but no improvement is possible (even if we stabilize to $V \times \mathbb{R}^k$ and/or use maxch) as follows from the (Carnot-Caratheodory) geometry of $\lim\limits_{\varepsilon\to 0} \varepsilon G$ (see 1.4.E$'$ in [Gr]$_{\rm CCS}$). Now, our lower bound on the spectrum of $\widetilde{V}_i$ reads

$$\# \text{ spec } \widetilde{D}_i \; [-a,a] \geq \delta_V \; a^h \text{ Vol } \widetilde{V}_i \quad \text{for each} \quad a \in \;]0,1] \quad \text{and} \quad i \geq i_0(a) \;\; (\widetilde{\#}_h)$$

and this generalizes as in the Abelian case to finite (as well as infinite Galois) covering of manifolds V admitting maps to G/Γ of non-zero degree.

The bound $(\widetilde{\#}_h)$ is hardly sharp. For example, if D is the untwisted Hodge's $d + d^*$ on the 3-dimensional Heisenberg manifold (which has $h = 4$), then

$$\dim_\Gamma \text{ spec } \widetilde{D} \; [-a,a] \approx a^2$$

as is proven by John Lott in [Lot] who also established the lower bound for this $\dim_\Gamma$ by $a^{\frac{n+1}{2}}$ for the n-dimensional (with $h = n+1$) Heisenberg group. But it is still conceivable that some of our $\widetilde{D}$'s (e.g. twisted or perturbed $d + d^*$) have significantly less of the spectrum in $[-a,a]$ than Lott's $a^{\frac{n+1}{2}}$ for small a. In fact, the ordinary Laplace on *functions* has $\#$ spec $\Delta^{\frac{1}{2}}$ $[0,a] \approx a^h$ as follows from the isoperimetric inequality on G (proven by Pansu for the Heisenberg groups and by Varopoulos for general nilpotent groups), but this does not tell us much of what we want as Δ is not a square of any of our D's. Also, the spectrum of $D = d + d^*$ on *all* forms is bounded from *above* by

$$\# \text{ spec } \widetilde{D}_i \; [-a,a] \leq a^\alpha \text{ vol } \widetilde{V}_i$$

for all i, all sufficiently small $a > 0$ and some $\alpha > 0$, say $\alpha = 1/n^2$, which follows from the cohomological interpretation of spec $d + d^*$ near zero (see $6\frac{10}{11}$) and the proof is the easiest for V admitting expanding endomorphisms.)

(3) ***Solvmanifolds.*** Let G be a simply connected non-nilpotent solvable Lie group, $V = G/\Gamma$ for a co-compact discrete subgroup $\Gamma \subset G$, and $\widetilde{V}_i = G/\Gamma_i$, $i = 1, 2, \ldots$, are finite coverings converging to $\widetilde{V}_{\rm univ} = G$, i.e. $\bigcap\limits_i \Gamma_i = \text{id}$ as earlier. It is not hard to construct, for every (small) $a > 0$ and all $i \geq i_0(a)$, a-contracting maps $\widetilde{V}_i \to S^n$ of degrees $\geq$ (Vol $\widetilde{V}_i$)$/ \exp a^{-1}$ which gives us the

following lower bound on the spectrum of our operator D lifted to $\widetilde{V}_i$,

$$\# \operatorname{spec} \widetilde{D}_i \, [-a, a] \geq (\exp ca^{-1})^{-1}$$

for some $c = c(V) > 0$, every $a > 0$ and $i \geq i_0(a)$. But this does not look sharp, not even in a most generous qualitative sense. In fact, one knows much here for the ordinary Laplace operator $\widetilde{\Delta}$ on functions on $\widetilde{V}_{\text{univ}} = G$, where the spectral density near zero (or equivalently, the rate of decay of the random walk on G) has been investigated by Varopoulos and one can descend, if one so wishes, to $\widetilde{V}_i$ (see [Var-Sa-Co]).

On the other hand, it is conceivable that a-contracting maps $\widetilde{V}_i \to S^n$ for small $a > 0$ and large i are necessarily exponentially non-efficient, i.e. have average Jacobians $\approx \exp a^{-1}$ (rather than a^{-n} as for the flat manifolds) and their $|\text{degrees}| \leq \operatorname{Vol} \widetilde{V}_i / \exp a^{-1}$, but I could not prove it already in the first interesting case of 3-dimensional solvmanifolds V. These are fibered by 2-tori (corresponding to $\mathbb{R}^2 = [G, G] \subset G$) which are exponentially distorted when lifted to $G = \widetilde{V}_{\text{univ}}$ or roughly so in $\widetilde{V}_i$ and a-contracting maps $\widetilde{V}_i \to S^3$ can be perturbed to $(\exp -a^{-1})$-contracting ones *along these tori*. This makes the bound $|\text{degree}| \leq (\exp -a^{-1})/(\operatorname{Vol} \widetilde{V}_i)$ quite plausible. (What is wrong with this argument is a possible exponential strech of the perturbed maps in the direction transversal to the tori.)

(4) ***Algebraic manifolds.*** Let V be complex algebraic submanifold in $\mathbb{C}P^N$ of real dimension $n = 2m$ and of algebraic degree δ. Then obviously

$$\operatorname{maxch}(\ell V, N) \geq \delta^m \tag{$*$}$$

for all $\ell \geq 10$. In fact, this is true for every submanifold $V \subset \mathbb{C}P^N$ with the induced metric which is homologous to d times the linear subspace in $\mathbb{C}P^N$. Of course this example is tailor made for $(*)$; what remains unclear, however, is a similar lower bound on $\max \deg \ell V / S^n$ (or, at least on $\max \deg \ell (V \times S^1)/S^{n+1}$) for a fixed ℓ independent of N and δ.

One sees with $(*)$ that D has, for large d, about d/N eigenvalues in the segment $[-a, a]$ for a (large) fixed a independent of D and d.

Question. Can one have a lower bound on $\#$ spec D for (more) general Kähler (or quasi-Kähler) manifolds in terms of their complex (quasi-complex) structure and the cohomology class of the (symplectic) structure form ω? (See $[\text{Gro}]_{\text{MIK}}$ for some information.)

Exercise. Bound from below $\operatorname{maxch}(\ell V; N)$ for all (large) ℓ using self-mappings $\mathbb{C}P^N \to \mathbb{C}P^N$ of growing degrees. Then bound from below $\#$ spec D $[-a, a]$ for large a and generalize this to homologically significant submanifolds V in a fixed W (with constants depending on W and the homology class $[V] \in H_n(W)$ but not on the actual geometric position of V in W).

(5) ***Manifolds of negative curvature.*** If V is a closed manifold with $K(V) \leq 0$, then, as we mentioned earlier, one expects the spectral density of $\widetilde{D}$ on $\widetilde{V}_{\text{univ}}$ be higher near zero than that in $\mathbb{R}^n$ which would imply a similar lower bound for $\#$ spec $\widetilde{D}_i$ $[-a, a]$ for finite covering $\widetilde{V}_i$ approximating $\widetilde{V}_{\text{univ}}$ whenever such coverings exist. The corresponding geometric problem concerning these coverings is the existence of a-contracting map $\widetilde{V}_i \to S^n$ of degrees $\geq a^{-n}$ Vol $\widetilde{V}_i$ for small $a \to \infty$ and $i \geq i_0(a)$. The latter as we know is possible if V admits a map of positive degree to the torus T^n and then we have

$$\dim_\Gamma \text{spec } \widetilde{D} \; [-a, a] \geq a^n \qquad (**)$$

as expected. Notice that if $\widetilde{V}_{\text{univ}}$ is a symmetric space then the von Neumann dimension $\dim_\Gamma$ spec $\widetilde{D}$ $[-a, a]$ is independent of Γ, but to prove $(**)$ we need Γ with a particular property. Furthermore, once we know $(**)$, we have a bound similar to $(**)$ for $\widetilde{V}_i$ corresponding to $\Gamma_i \subset \Gamma$ where this Γ may be different from the one used to prove $(**)$. Of course, for symmetric spaces and $\widetilde{D}$ associated to the "symmetric" metric, one can compute the spectral density via the harmonic analysis of the corresponding Lie group (if one is an adept in the representation theory). But the above still seems to have some independent merit (as, for example, it applies to "non-symmetric" Γ-invariant $\widetilde{D}$ on $\widetilde{V}$).

We shall prove in $9\frac{3}{4}$ that

$$\dim_\Gamma \text{spec } \widetilde{D} \; [-a, a] \geq a^\alpha$$

with $\alpha > 0$ for some V with $K(V) < 0$, where we also shall discuss $\max \deg \ell\widetilde{V}/S^n$ in the foliated framework.

$6\frac{8}{9}$. Vafa-Witten in odd dimension

The idea to use the twisted index theorem for lower spectral bounds is due to Vafa and Witten (see [Va-Wi]) who emphasize in their paper the fact that the *twisted Dirac operator admit a lower spectral bound independent of the twist*, i.e. of the implied bundle with connection. (A year earlier, a similar idea fleetingly appeared on the top of p.200/412 of $[\text{Ros}]_{C^*\text{APS}}$ where the author worked with the Dirac operator twisted with some C^*-algebra module.) Here (as in $[\text{G-L}]_{\text{PSC}}$ and $[\text{Gro}]_{\text{LRM}}$) we are more interested in the effect of the macroscopic geometry of the underlying manifold V on the spectrum of the untwisted Dirac on spinors as well as Hodge's $d + d^*$ on forms.

Notice that the ordinary index theorem is essentially vacuous if $\dim V$ is odd and we had to stabilize V to $V \times \mathbb{R}$. But Vafa and Witten use in their original paper the odd dimensional index with values in $K_1(V)$ (see $6\frac{11}{12}$) which allows them, for $\dim V$ odd, to bound from below the gaps in the spectrum of D everywhere on $\mathbb{R}$, not only at zero. Their main result (brought to our geometric framework) reads,

Odd VW. *Let V be a closed odd dimensional Riemannian manifold and D be either (twisted or untwisted) Dirac operator (for which V must be spin) or Hodge's $d + d^*$. Then the number of the eigenvalues in every interval $[a, b]$ of length $c = b - a$ is bounded from below by certain geometric invariant of V, $\mathrm{Inv}_c V$,*

$$\# \text{ spec } D[a, b] \geq \mathrm{Inv}_c V \ , \qquad (\#_{\text{odd}})$$

where $\mathrm{Inv}_c V$ has the following properties.

(I) *For every V and $c \geq c_0(V)$ this invariant is positive in fact $\mathrm{Inv}_c(V) \geq 1$ for $c \geq c_0(V)$ and moreover $\mathrm{Inv}_c(V) \gtrsim c^n$, $n = \dim V$, for large c. That is*

$$\mathrm{Inv}_c(V) \geq \mathrm{const}_V \ c^n \ , \qquad (*)$$

for $c \geq c_0(V)$ and some $\mathrm{const}_V > 0$. (Notice that $\mathrm{Inv}_c V$ appearing in $(\#_{\text{odd}})$ is independent of the implied twist as emphasized by Vafa and Witten.)

(II) *The invariant $\mathrm{Inv}_c V$ is monotone increasing in (the Riemannian metric of) V for every fixed c (and, of course, it is monotone increasing in c). Moreover, if $V_1 \to V_2$ is distance decreasing map of non-zero degree, then $\mathrm{Inv}_c V_1 \geq \mathrm{Inv}_c V_2$ for each $c \geq 0$. Furthermore $\mathrm{Inv}_c V$ is scale invariant, $\mathrm{Inv}_{\lambda a} \lambda^{-1} V = \mathrm{Inv}_c V$ for all $c, \lambda > 0$ (when $\lambda^{-1}(V, g) = (V, \lambda^{-2} g)$). (Consequently $\mathrm{Inv}_c V$ is C^0-continuous in the Riemannian metric g of V.)*

(III) *Let V admit a Lipschitz-λ^{-1} map onto S^n of degree $\geq d > 0$. Then*

$$\mathrm{Inv}_c V \geq d \quad \text{for} \quad c \geq \gamma_n \lambda$$

where $\gamma_n > 0$ is a universal constant.

(III') *Let $U(N)$ be the unitary group with the operator norm metric and let us fix the standard generators $h_1, \dots, h_k$ in the cohomology group $H^n(U(N))$ which are independent of N for large $N \geq n$. Then, if V admits a Lipschitz-λ^{-1} map $f : V \to U(N)$ and $c \geq \gamma_n \lambda^{-1}$,*

$$\mathrm{Inv}_c V \geq \delta_n \ N^{-1} \max_{i=1,\dots,k} \langle f^*(h_i), [V] \rangle \qquad (**)$$

where $\delta_n > 0$ is a universal constant.

The Vafa-Witten method was succintly exposed by M. Atiyah in $[\mathrm{At}]_{\mathrm{EDO}}$ and further developed and applied to geometric problems in a variety of papers, especially by Steven Hurder for (finite and infinite) coverings and foliations (see $[\mathrm{Hur}]_{\mathrm{CGF,EIOI,EIT,ETF}}$) and by John Roe in his coarse (macroscopic) index theory on complete manifolds (see $[\mathrm{Roe}]_{\mathrm{CCIT,PNM}}$). We shall return to this later on in this paper but yet mention here that the basic bound on the spectral gaps of D which claims that

each segment $[a, a+c]$ *for* $c \geq c_0(V)$ *contains some spectrum of* D,

remains valid for all *complete* odd dimensional manifolds V by the odd-dimensional version of the relative index theorem.

It is worth emphasizing that the idea of the Vafa-Witten method consists in reducing solution of an *inequality*, say $\|Dx\| \leq \lambda\|x\|$ (equivalent to $\infty\mathrm{spec}|D| \leq \lambda$) to an *equation* $D'x = 0$ for some auxiliary operator D'. A similar reduction was earlier used for lower spectral bound on S^2 with $\bar{\partial}$ in place of D' (see [Her]) where the direct link with VW-method is not quite clear (see $[\mathrm{Gro}]_{\mathrm{MIK}}$ for further information and references. Also see [M-M] for a VW-style application of $\bar{\partial}$ to a lower bound on the Morse index of minimal spheres in manifolds with $K_{\mathbb{C}}^{\mathrm{isotr}} > 0$ and see [Dem] for an interplay of the spectrum with an asymptotic Riemann-Roch theorem leading to Demailly-Morse inequalities for holomorphic vector bundles with controlled curvatures). The application of solutions of the Cauchy-Riemann equation to solving geometric inequalities extends to the non-linear domain (e.g. for bounding the symplectic area with pseudo-holomorphic curves, see $4\frac{3}{4}$) and it would be interesting to delinearize VW for more general operators (e.g. in Donaldson theory).

$6\frac{9}{10}$. Spectral gaps for general geometric operators

Let D be a positive selfadjoint differential (or pseudifferential) operator of order r on a closed manifold D. Then the number of eigenvalues in a large interval $[0, a]$ is about $a^{\frac{n}{r}}$ (this is an elementary exercise) with the error term for $a \to \infty$ of order $a^{\frac{n-1}{r}}$ (proven by Hörmander using the wave equation). That is, in writing,

$$\#\ \mathrm{spec}\ D[0,a] = C_D\ a^{\frac{n}{r}} + \mathcal{O}\left(a^{\frac{n-1}{r}}\right)\ . \qquad (\#_D)$$

This trivially implies a bound on the gap in the spectrum of $D^{\frac{1}{2}}$ (i.e. the set $\left\{\lambda_i^{\frac{1}{2}}\right\}$, $\lambda_i \in \mathrm{spec}\ D$) which reads

$$\#\ \mathrm{spec}\ D^{\frac{1}{r}}\ [a,b] \geq 1 \quad \text{for} \quad b-a \geq \mathrm{const}_D \quad \text{and all} \quad a \geq 0\ ,$$

and, moreover,

$$\#\ \mathrm{spec}\ D^{\frac{1}{2}}\ [a,b] \geq C_D(b-a)\ a^{n-1} \quad \text{for} \quad b-a \geq \mathrm{const}_D\ .$$

If D has a topological twist to it, such as being Dirac or a power of such operator, then the VW-theorem gives a bound on the above const_D and hence on gaps in spec $D^{\frac{1}{2}}$) in terms of C^0-geometry of V; now we want to indicate some geometric (and shamefully weak) bounds on const_D for more general operators D.

We start with the simplest case where D is the ordinary Laplace operator Δ acting on functions on V. In order to bound # spec $\Delta^{\frac{1}{2}}[0,a]$ from below by j one should produce, according to the minimax principle, j mutually orthogonal non-zero functions f_i, $i=1,\dots,j$, satisfying $\|df\|_{L_2} \leq a\|f\|_{L_2}$. A naive (yet often efficient) way to do it is to find j disjoint balls B_i in V of radius $\varepsilon \approx a^{-1}$ and take $\text{dist}(v, V-B_i)$ for $f_i(v)$. If the volumes of the concentric halfballs $\frac{1}{2}\, B_i$ satisfy

$$\text{Vol}\ \frac{1}{2}\ B_i \geq \delta\ \text{Vol}\ B_i$$

then, clearly

$$\|f_i\|_{L_2} \geq \frac{1}{2}\ \varepsilon\ \sqrt{\delta\ \text{Vol}\ B_i}$$

while

$$\|df_i\|_{L_2} = \sqrt{\text{Vol}\ B_i}$$

which makes

$$\|df_i\|_{L_2} \geq \frac{1}{2}\ \varepsilon\ \delta^{\frac{1}{2}}\ \|f_i\|_{L_2}\ .$$

For example, let V have the Ricci curvature bounded from below, say Ricci $\geq -\rho^2$. Then every pair of concentric balls satisfies *Bishop's inequality*

$$\text{Vol}\ B(\varepsilon/2) \geq 2^{-n(1+\varepsilon\rho)}\ \text{Vol}\ B(\varepsilon)$$

and each $B(\varepsilon)$ has

$$\text{Vol}\ B(\varepsilon) \leq \text{const}_n\ \varepsilon^n\ 2^{n\varepsilon\rho}\ .$$

In particular, if Ricci ≥ -1, then for each $\varepsilon \leq 1$, V contains about Vol V/ε^n disjoint balls B_i with Vol $\frac{1}{2}\ B_i \geq \delta_n$ Vol B_i and thus

$$\#\ \text{spec}\ \Delta^{\frac{1}{2}}\ [0,a] \geq \text{const}_n\ a^n\ \text{Vol}\ V \tag{Ri}$$

for all $a \geq 1$ (and one has a similar *upper* bound on # spec $\Delta^{\frac{1}{2}}$ for Ricci ≥ -1, see [Gro]$_{\text{PL}}$).

Now, recall that Ind + BL + KFK yields a similar bound with the scalar curvature instead of Ricci and a suitable K-area of V instead of the volume where, unfortunately, the bound on the number of eigenvalues λ_i of Δ in $[0, a]$ is replaced by an average bound (see $5\frac{1}{2}$). For example, if V is a connected spin manifold with $|\widehat{A}\text{-genus}| > 2^{\frac{n}{2}}$, $n = \dim V$, and $Sc \geq -\sigma$, then we do not even need the K-area, as

$$\sum_{i=1}^{\infty} \exp -\lambda_i \, t \geq \left(2^{-\frac{n}{2}} \, |\widehat{A}[V]| \exp -\frac{1}{4} \, \sigma \, t\right) - 1$$

for all $t > 0$ (we count from $i = 1$ as $\lambda_0(\Delta) = 0$; compare $\widehat{A}$-exp in $6\frac{1}{2}$). No simple minded construction with distance functions can ever deliver an estimate of this kind! (Yet the above elementary construction gives us the bonus of test functions f_i which are not just orthogonal but have disjoint supports).

Next, we drop our assumption on $\widehat{A}[V]$ and bring in the K-area in the following simplified form. Suppose V admits a smooth map $\varphi : V \to S^n$ of degree $\geq d = d(A)$ which is area expanding at most by A, i.e.

$$\text{area } \varphi(\Sigma) \leq A \cdot \text{area } \Sigma$$

for all smooth surfaces Σ in V. Then, as we know, the Dirac operator twisted with a suitable $\mathbb{C}$-bundle of $\mathbb{R}$-rank n has ind $\approx d$ and so the corresponding Bochner Laplacian has # spec $[0, a] \geq d$ for $a \approx A + \sigma$. Hence, we obtain with KFK the bound

$$\sum_{i=0}^{\infty} \exp \lambda_i \, t \geq \left(n \, 2^{\frac{n}{2}}\right)^{-1} \, d \exp -\alpha_n(A/\sigma) \, t \, ,$$

for some universal $\alpha_n > 0$, all $t > 0$, all $A \geq 0$ and $d = d(A)$. (As we allow maps $V \to S^n$ with larger and larger A we shall also have $d \to \infty$ and then the above estimate becomes better and better for $t \to 0$ which corresponds to producing higher eigenvalues λ_i of Δ.)

The intermediate steps of the above proofs giving bounds on spectra of the Bochner Laplacians are also quite interesting. In the first case the purely topological condition $|\widehat{A}(V)| = N \geq 2^{\frac{n}{2}} + 1$ implies the existence of a unitary bundle $Z = (Z, \nabla)$ over V of $\mathbb{R}$-rank $2^{\frac{n}{2}}$ where the Bochner Laplacian has at least N eigenvalues below $\frac{1}{4}\sigma$ for $-\sigma = \inf \text{ Sc } V$, which is equivalent to the presence of N mutually orthogonal non-zero sections Z_i $i = 1, \ldots, N$ of Z satisfying $\|\nabla Z_i\|_{L_2} \leq \frac{1}{2} \sqrt{\sigma} \, \|Z_i\|_{L_2}$. Moreover, this Z is (spin) associated to the tangent bundle of V thus having the curvature bounded in terms of that of V. But even without knowing the true identity of Z (which, in fact, S_+ or S_-), we gain non-trivial information about the geometry of V.

Next, in the twisted case, we assume nothing about $\widehat{A}[V]$ and yet obtain a Z of rank $n2^{\frac{n}{2}}$ having about d eigenvalues below $\lambda \approx A + \sigma$ for the above A and d (where the curvature of Z is about $A|K(V)|$). This is again a non-vacuous property of (V, g) as KFK prevents bundles of $\mathbb{R}$ rank k over V from having $> k$ very small eigenvalues of the Bochner Laplacian.

A geometric bound on the gap in* spec $\Delta^{\frac{1}{2}}$ *away from zero. The C^0-continuity of spec $\Delta^{\frac{1}{2}}$ in the metric is non-uniform and a small deformation may create large gaps (an ε-perturbation of a metric roughly corresponds to composing Δ with $1 + A_\varepsilon$ where $\|A_\varepsilon\| \approx \varepsilon$). It seems to be unknown if the gaps are uniformly bounded on C^0-Riemannian manifolds (and I do not know the minimal smoothness of (V, g) needed for the Hörmander method) but we shall now establish such a bound for odd dimensional C^2-manifolds (in fact we only need $C^{1\cdot 1}$) using VW.

***Gap bound for* $\Delta^{\frac{1}{2}}$.** *Let V be a closed odd dimensional Riemannian manifold with the sectional curvature and the injectivity radius bounded by one, i.e.*

$$|K(V)| \leq 1 \quad \textit{and} \quad \text{Inj Rad } V \geq 1$$

(where the bound Inj Rad ≥ 1 *is equivalent in our case to the absence of closed geodesics of length* < 2*). Then the gaps in* spec $\Delta^{\frac{1}{2}}$ *are bounded by a constant* const_n, $n = \dim V$, *and moreover*

$$\# \text{ spec } \Delta^{\frac{1}{2}} \ [a, b] \geq C_n (b - a)^n \text{ Vol } V$$

for a universal constant $C_n > 0$, *all* $a \geq 0$ *and* $b \geq a + \text{const}_n$.

Proof. We are going to reduce the gap bound for $\Delta^{\frac{1}{2}}$ on functions to a similar bound for $d + d^*$, or equivalently $\Delta_H^{\frac{1}{2}}$ on forms where $\Delta_H = (d + d^*)^2$ is the Hodge Laplacian. First we switch to the Bochner Laplacian Δ_B on forms related to Δ_H by the Bochner formula $\Delta_H = \Delta_B + R$ where R is an endomorphism of $\Lambda^*(V)$ made of the curvatures of V. Our bound $|K(V)| \leq 1$ gives a bound on R and so the spectral gaps of Δ_B are bounded by those for Δ_H plus a constant majorizing $\|R\|$.

Next we observe that our bounds on $|K|$ and Inj Rad (trivially) provide a bound on the straightness $\|\mathcal{P}_N(T(V))\|$ of the tangent bundle $T(V)$ with Levi-Civita's ∇ and hence every associated bundle of V (compare $6\frac{1}{4}$). In particular, the bundle $\Lambda^*(V) = \bigoplus_{i=0}^{n} \Lambda^i \, T(V)$ admits a unitary embedding into the trivial bundle of rank N^* such that the Levi-Civita connection ∇_Λ in $\Lambda^*(V)$ differs from the trivial connection by some const_n and where also $N^* \leq N(n) \approx 4^n$.

We recall that ∇ acts from $H_1 = C^\infty(\Lambda^*(V))$ to $H_2 = C^\infty(\Lambda^*(V) \otimes T^*(V))$ and $\Delta_B = \nabla_\Lambda^* \, \nabla_\Lambda$ which is essentially equivalent to $\Delta_B = \nabla_\Lambda^2$ for a suitable unitary correspondence between H_1 and H_2. This shows that the gaps in spec $\Delta_B^{\frac{1}{2}}$ = "spec ∇_Λ" majorize, up to the above const$_n$, those of spec $\Delta^{\frac{1}{2}}$ = "spec d" for the differential d on functions. Q.E.D.

Commentaries. (a) Our bounds on gaps in spec $\Delta^{\frac{1}{2}}$ extends to *complete* non-compact manifolds V.

(b) It is unclear how to make the above argument work for even dimensional manifolds V.

(c) I suspect that our gap bound for Δ (for both, odd and even n) can be recaptured by the wave equation techniques (which must be obvious for true analists) but the VW-method may still provide additional leverage.

(d) Let X be an arbitrary bundle over V with curvature $\mathcal{R}(X)$ bounded in norm by one. Then the Whiterey sum $\underset{N}{\oplus} X$ for some $N \leq N(n)$ admits a unitary section z with $\|\nabla z\| \leq C_n$ (as we assume $K(V) \leq 0$, Inj Rad $V \geq 1$). It follows that the spectral gaps of Bochner's $\Delta_X^{\frac{1}{2}}$ are bounded, up to some const$_n$, by those of $\Delta^{\frac{1}{2}}$ and so our estimates extend to the Bochner Laplacian on X.

(e) It seems likely that the spectral gap bound for $\Delta^{\frac{1}{2}}$ remains valid for complete manifolds V having $|K(V)| \leq 1$ and Inj Rad$_v$ $V \geq 1$ at a single point $v \in V$ (as this is so for Dirac and $d+d^*$ by VW and the relevant eigenfunctions, probably, sufficiently localize near v).

(e') (Pointed out to me by Misha Shubin.) If V contains an actual flat Euclidean unit ball B (not just an approximate one as in (e)) then $\Delta^{\frac{1}{2}}$ indeed admits a universal gap bound, because $\Delta^{\frac{1}{2}}$ has approximate λ-eigenfunctions for all λ, namely $f_\lambda = \varphi \exp \lambda iu$ for a linear function u on B and a smooth bump function φ on B. Such an f_λ has $\|\lambda^{-1}\, \Delta f_\lambda - \lambda f\| \leq$ const $\|f\|_{L_2}$ from which (an independent of λ) bound on gaps in spec $\Delta^{\frac{1}{2}}$ follows by an obvious perturbation argument.

$6\frac{10}{11}$. On Dirac and Hodge

The VW lower spectral estimate equally applies to the Dirac operator D on spinors (if V is spin) and to Hodge's $d+d^*$ on differential forms. (Notice that $(d+d^*)^2$, unlike D^2 splits into the direct sum of $n+1$ operators, $(d+d^*)^2 = \sum_{i=0}^{n} \Delta_i$ acting on $C^\infty(\Lambda^*(V)) = \bigoplus_{i=0}^{n} C^\infty(\Lambda^i(V))$ and so the VW-theorem for $d+d^*$, when applies, predicts small eigenvalues of *some* of Δ_i, $i = 0, \ldots, n$ without saying of which one. Typically, one expects the largest spectrum for Δ_i with $i = \frac{n}{2}$ for n even and $i = \frac{n-1}{2}, \frac{n+1}{2}$ for n odd.) But the flavour of this is somewhat different in the two cases since the spectrum of $d+d^*$ is continuous in the C^0-topology on the space of Riemannian metrics (as explained below) while spec$\,D$ is only C^1-continuous. Thus the VW bound for $d+d^*$ is an internally C^0-theorem relating two geometric invariants, spec $(d+d^*)$ and the size of V (encoded into the K-length) while in the case of Dirac VW shows D to be more geometric than is apparent from its definition. (This suggests some C^0-stabilization of the eigenvalues of D by taking $\lim_{\varepsilon\to 0}\sup \lambda_i(D_\varepsilon)$ for the ε-perturbations of the metric of V in the C^0-topology, where one may wonder how often this limsup equals $\lambda_i(D)$, compare (4) below.)

To clarify the geometric (and topological) significance of spec $d+d^*$ we observe that it is determined only by $d : \Lambda^*(V) \to \Lambda^*(V)$, which is purely topological, and the L_2-norm in $\Lambda^*(V)$ which (C^0-continuously) depends on the metric. Then, apart from the atom at zero corresponding to the cohomology, we have two quadratic forms (norms) on each Im $d_{i-1} \subset \Lambda^i(V)$, the first induced from the original L_2-norm on $\Lambda^i(V)$ and the second is the quotient norm for the surjection $d_{i-1} : \Lambda^{i-1}(V) \to \operatorname{Im} d_{i-1}$. Then our # spec $d+d^*[-a,a]$ equals rank H^* plus the dimension of a maximal linear subspace Φ in $\operatorname{Im} d$, such that

$$\|\varphi\|_{\text{first}} \le a\|\varphi\|_{\text{second}} \quad \text{for all} \quad \varphi \in \Phi\ .$$

Now, clearly if V and V' are λ-bi-Lipschitz equivalent, then the L_2-norm on forms changes by at most λ^n and so

$$\begin{aligned}\#\ \mathrm{spec}_V\, d+d^*[-\lambda^{-n}a,\lambda^{-n}a] &\le \#\ \mathrm{spec}_{V'}\, d+d^*[-a,a]\\ &\le \#\ \mathrm{spec}_V\, d+d^*[-\lambda^{n}a,\lambda^{n}a].\end{aligned}$$

As we pass to coverings $\widetilde{V}_i$ we see that the asymptotic of (# $\mathrm{spec}_{\widetilde{V}_i}\, d + d^*[-a,a])/\mathrm{Vol}\, \widetilde{V}_i$ for $i \to \infty$ and $a \to 0$ is, in a natural sense, a bi-Lipschitz (even homotopy) invariant of V.

In fact it is more helpful to use infinite coverings of V, such as the universal covering $\widetilde{V}$ where the atomic spectrum at zero defines *the reduced L_2-cohomology*, Ker $\widetilde{d} \mid L_2/$Closure Im $\widetilde{d}(L_2)$, the basic homotopy invariant of V, and where the spectrum of $\widetilde{d+d^*}$ *near zero* contains an essential (homotopy) information on the *non-reduced* L_2-cohomology Ker $\widetilde{d} \mid L_2/\operatorname{Im} \widetilde{d}(L_2)$ (see [No-Sh], [Gr-Sh], [Lot] and references therein).

Furthermore, one may speak of spec $d+d^*$ and spec $\widetilde{d+d^*}$ for quite general (singular) spaces V whenever the cohomology is built with a set of simplices (cells) carrying a measure providing an L_2-structure on the cochain level (e.g. where V is triangulated and d is the boundary operator). Probably, the VW-method straightforwardly extends to conical spaces of Cheeger and to Lipschitz manifolds with the index theory developed by Teleman and Sullivan and a suitable lower bound on # spec $d+d^*[-a,a]$ may survive on most unhospitable singular metric spaces.

Dirac on singular spaces is another story where one should keep the (singular) scalar curvature away from $-\infty$ (compare [Cho]), as in *Alexandrov's spaces* with $K \geq -\text{const}$. But $\text{Sc} \geq -\text{const}$ allows by far more intricate spaces which can be fractalized, for example, by taking iterated connected sums with $\text{Sc} > 0$.

Question. Can one bound the spectra of $d + d^*$ and/or Dirac on the universal covering $\widetilde{V}$ of V in terms of the K-area instead of the K-length? Notice that almost flat bundles $\widetilde{X} \to \widetilde{V}$ are almost straight on arbitrarily large *compact* parts of $\widetilde{V}$ but this seems to fall short of what is needed for a proof. More specifically, let V be a compact (homologically) symplectic aspherical manifold. Do then $d + d^*$ and D on $\widetilde{V}$ have zero in their spectra?

(4) ***Inv_c as a norm on bordisms.*** The Inv_c of the above (at the bingin-ning of $6\frac{8}{9}$) odd VW (as well as of the even one) can be defined axiomatically as the largest number good enough to serve ($\#_{\text{odd}}$) (or the corresponding bound on spec $[0, c]$ for n even). Ultimately, for each $c > 0$ and $a \geq 0$ one defines a kind of a norm on the (spin if D is Dirac) bordisms (and thus homology) of a metric space W by taking

$$\inf_V \ \# \ \text{spec} \ D_V[a, a + c]$$

where V runs over all Riemannian manifolds admitting a distance decreasing maps $V \to W$ representing a given bordism class of W (where one should restrict to $a = 0$ for n even). One can do a similar thing with $\text{Sc}\,V$ instead of the spectrum (see below) and for $D = d + d^*$ one may use singular spaces V. This may bring geometricly tasteful fruits, but I could not go so far beyond a few rather obvious foundational observations.

(5) ***Scalar curvature and spec D.*** Since the Dirac operator equals $\Delta_S + \frac{1}{4}\ \text{Sc}$ the role of inf Sc for $\text{Sc} \geq 0$ is somewhat similar to that of inf spec $|D|$. For example one could define a "norm" on bordisms of the above W as the minimal (infimal) δ for which a bordism class is representable by a contracting (i.e. 1-Lipschitz) map $V \to W$ where $\text{Sc}\ V \geq \delta^{-2}$. In fact, one could use here *area* contracting maps instead of merely contracting ones which are by far more numerous and geometricly appealing as they pertain to the dimension two rather than one (see the K-area inequality $6\frac{1}{4}$). On the other hand spec D beats scalar curvature by the sheer abundance of invariants hidden in it, not only inf spec $|D|$ but also $\#$ spec $D[a, b]$ etc. But the last word on the curvature is yet to be said.

$6\frac{11}{12}$. Odd index theorem

If V is an odd dimensional manifold then the ordinary index of every elliptic operator D is zero; yet there is a non-trivial index defined as an element of the *odd K-homology* of V which assigns to each map $f : V \to GL_N\,\mathbb{C}$, representing a K^1-class, an integer, $\text{ind}_f\, D$, defined as follows. Take some (e.g. trivial)

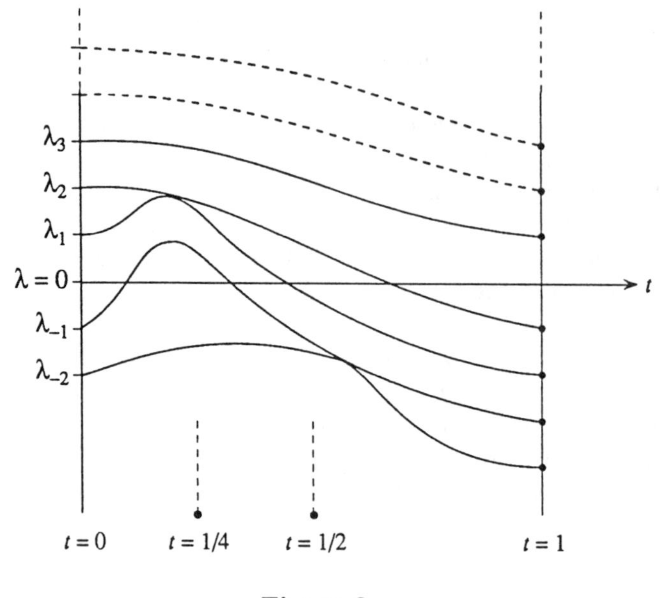

Figure 9

connection ∇_0 in the trivial bundle $X_0 \to V$ of rank N and let $\nabla_1 = f_*(\nabla_0)$, where f is regarded as a fiberwise automorphism of $X_0 \to V$. We consider two twisted operators, D twisted with (X_0, ∇_0) and with (X_0, ∇_1) acting on the same space, namely the sections of $S \otimes X_0$, where S is the original bundle (implied by the definition of D), denoted D_0 and D_1. Moreover, as we can take convex combinations of connections, $\nabla_t = t\nabla_1 + (1-t)\nabla_0$, we can twist D with ∇_t thus obtaining a 1-parameter family of elliptic operators D_t and we want to assign an integer to such a family. This we shall do where D is a *selfadjoint* operator. In this case the operators D_t are also selfadjoint and Fredholm, as we assume at this stage V is compact. Then one can define the *spectral flow* of the family D_t, as follows. Assume the spectrum of D_0 contains no zero and then count how many eigenvalues of D_t cross zero from left to right as t moves from 0 to 1 as in Fig. 9

Here $\lambda_i = \lambda_i(D_0)$ and these moves as $\lambda_i(t) = \lambda_i(D_t)$. (Notice that D_1 has the same spectrum as D_0 since they are conjugated by f.) The spectral flow at the moment $t = \frac{1}{4}$ is $+1$, at $t = \frac{1}{2}$ it is -1, and finally for $t = 1$ it is -2. In general terms, whenever we have a discrete subset in $\mathbb{R}$ moving with t, say $\Lambda_t \subset \mathbb{R}$, $t \in [0,1]$, such that Λ_0 and Λ_t contain no zero, then there is a well defined flow of points from Λ_t through zero. This flow is also defined for non-zero points $\lambda \in \mathbb{R} - \Lambda_0 \,\mathrm{cup}\, \Lambda_1$ and if $\Lambda_0 = \Lambda_1$ the result is independent of λ.

Next, consider the space $\mathcal{D}$ of self-adjoint Fredholm operators on a Hilbert space $\mathcal{H}$. Then non-invertible operators, i.e. having zero in the spectrum form a hypersurface say $\Sigma_0 \subset \mathcal{D}$. The singular locus Σ_0' of this hypersurface has codimension two (not one!) in Σ_0 (which is seen with a finite dimensional reduction where this is more or less obvious. Say, Σ_0 in the space of 2×2 symmetric matrices is given by the equation $a^2 - bc = 0$ where the only singularity is at $a = b = c = 0$. More generally, symmetric matrices with two zero eigenvalues have codimension 3). Thus Σ_0 form a codimension one cycle in $\mathcal{D}$ which has a natural coorientation (the direction of the spectral flow from negative to positive). If the Hilbert space in question is finite dimensional, Σ_0 divides the space $\mathcal{D}$, consisting of *all* symmetric operators, into the components, corresponding to the signature = (number of positive eigenvalues) – (the number of the negative eigenvalues). But if $\mathcal{H}$ is infinite dimensional one may have a closed curve in $\mathcal{D}$ meeting Σ_0 transversally at a single point, i.e. a family D_t with the spectral flow one, for example operators D_t with the spectra $\Lambda_t = \mathbb{Z} + t$, $t \in [0,1]$. Such a curve represents a non-trivial homology class in $H_1(\mathcal{D})$ detected by its intersection with Σ_0, where instead of Σ_0 one could take $\Sigma_\lambda \subset \mathcal{D}$ consisting of the operators $D \in \mathcal{D}$ containing $\lambda \in \operatorname{spec} D$ in-so-far as all D_t in question do not have λ in their essential spectra, i.e. if $D_t - \lambda$ are Fredholm. In particular, one can always use sufficiently small λ as the Fredholm property of D_t implies that for $D_t + \lambda$ if $|\lambda| \leq \varepsilon$. (The difference between the topology of $\mathcal{D}$ in the finite and infinite dimensional spaces is due to the fact that in the infinite dimensional case removing non-Fredholm operators makes the remaining part, i.e. $\mathcal{D}$, non-contractible.)

Now we return to our differential operators D_t acting an (sections of) $S \times X_t$ and observe that here, strictly speaking, the path does not close up as $D_0 \neq D_1$; however, $\operatorname{spec} D_0 = \operatorname{spec} D_1$ (since D_1 is equivalent to D_0 via f) and so the spectral flow is well defined. In fact it is better to think of D_t as acting on a variable space $\mathcal{H}_t$ of section of $S \otimes X_t$ as follows. The automorphism $f : X_0 \to X_0$ defines a vector bundle, say $X \to V \times S^1$ obtained by glueing $X_0 = X \to V \times 0$ with $X_t = f(X_0) \to V \times 1$ according to f and one takes a family of connections ∇_t on $X_t = X \mid V \times t$ with t now running over the circle S^1. The spectral flow makes perfect sense in this situation (which could have been reduced to the case of a fixed $\mathcal{H}$ by Kuiper's theorem claiming the contractibility of the infinite dimensional unitary group) and defines

$$\operatorname{ind}_f D \underset{\text{def}}{=} \text{spectral flow of } D_t \ .$$

Observe that this makes sense whenever the operators D_t are Fredholm (as well as selfadjoint). In particular, if V is a complete (possibly non-compact manifold) and D^2 is positive at infinity (see $6\frac{4}{5}$) then D is Fredholm (i.e. $\lambda = 0$ is not a point of the essential spectrum; here, as everywhere in the index theory, we do not care if D is bounded or not as we are concerned with the spectrum near zero) and, furthermore, if the map $f : V \to GL_N \mathbb{C}$ has compact support, then all D_t are equal at infinity to the Whithey sum of N-copies of D and so also Fredholm. (In fact, D is Fredholm $\Leftrightarrow D^2$ is positive at infinity; see [Ang].) Thus $\mathrm{ind}_f\, D$ is defined for D^2 positive at infinity and it satisfies the following:

(Relative) index formula.

$$\mathrm{inf}_f D = (A_D \smile \mathrm{ch}_f)[V]\ , \qquad (\mathrm{ind}_f)$$

where A_D is the same even cohomology class as in the ordinary index formula, e.g. $A_D = \widehat{A}_V$ for the Dirac operator D and $A_D = L_V$ for the signature operator. What is relevant for our applications is that the zero degree term in A_D is non-zero for the above operators. Next, ch_f is the pull-back under f of some universal polynomial in the standard (odd) generators in $H^*(GL_N\ \mathbb{C})$ which has a non-trivial component of each degree. Since f has compact support, so does ch_f and one can evaluate the cup product $A_D \smile \mathrm{ch}_f$ on the fundamental class [V].

This formula for compact V is due to Atiyah, Patodi and Singer and the non-compact case follows by readjusting the corresponding even argument (compare $6\frac{4}{5}$). In fact the odd case can be reduced to the even one with the (non-selfadjoint!) operator $\widetilde{D} = D_t + \frac{\partial}{\partial t}$ acting on sections of $S \otimes X$ over $V \times S^1$ as explained in $[\mathrm{At}]_{\mathrm{EDO}}$ for compact V. The pertinent points here are the following.

(1) Since f has compact support, the bundle $X \to V \times S^1$ is trivialized at infinity. Furthermore, if $\mathrm{Triv} \overset{N}{\to} V \times S^1$ is the trivial bundle, the corresponding operator D_{Triv} has zero index (essentially, because $\frac{\partial}{\partial t}$ has zero index over S^1) and so $\mathrm{ind}\, \widetilde{D}$ fits into the relative framework of $6\frac{4}{5}$.

(2) $A_{\widetilde{D}}$ equals the pull-back of A_D under the projection $V \times S^1 \to V$ while $\mathrm{ch}\, X$ equals the S^1-suspension of $\mathrm{ch}\, f$.

Finally we observe that this formula is as good as the even one for the Vafa-Witten type estimates. In fact it is better as it applies to $D - \lambda$ for all λ not in the essential spectrum of D (e.g. for *all* λ if V is compact) and yields odd VW as we stated in $6\frac{8}{9}$.

Remark. There is yet another way to define ind_f using *Toeplitz operators* as follows (compare [Ba-Do]). Let λ be *not* in the spectrum of D and let $\mathcal{H}_\lambda^-$ be the spectral space of D twisted with (the trivial of rank N) bundle X_0 corresponding to spec $< \lambda$. The Toeplitz operator T_λ associated to f is defined with the spectral projection P_λ^- on $\mathcal{H}_\lambda^-$ by $h \mapsto P_\lambda^- \circ f(h)$ for all $h \in \mathcal{H}_\lambda^-$. One knows this operator is Fredholm and one can show that $\mathrm{ind}\, T_\lambda = \mathrm{ind}_f(D - \lambda)$ (which is well known in the compact case). This definition nicely fits into the π-invariant and foliated frameworks (see §$9\frac{2}{3}$) where Toeplitz operators were extensively studied by S. Hurder in $[\mathrm{Hur}]_{\mathrm{CGF,EITF}}$.

$6\frac{12}{13}$. Large manifolds with no small eigenvalues of the Laplacian

It seems, intuitively, as if every sufficiently large Riemannian manifold (V, g) must have a small $\lambda_1 = \lambda_1(\Delta)$. For , if g is the metric on the sphere dominating the standard metric g_0 by $g \geq \mu^2 g_0$ one may expect $\lambda_1(\Delta_g) \lesssim \mu^2$. In fact, this so for $\dim V = n = 2$ by a theorem of Hersch, and also, for all n, but with the Dirac or Hodge instead of Δ by the VW-theorem, but we shall exhibit counter examples for Δ and all $n \geq 3$ (compare [CdV]).

There exists metrics $g \geq g_0$ on S^n, $n \geq 3$, with arbitrarily large $\lambda_1(\Delta_g)$.

Sketch of the proof. First we start with large metrics having large λ_1 on manifolds non-diffeomorphic to S^n. Namely, we recall that the congruence coverings $\widetilde{V}_i$ of every compact arithmetic variety V have $\lambda_1(\widetilde{V}_i) \geq \text{const} > 0$ for $i \to \infty$, while $\widetilde{V}_i$ converge to (quite large) universal covering $\widetilde{V}_{\text{univ}}$ of V. (If the fundamental group $\pi_1(V)$ is Kazhdan's T, one may use any sequence of finite coverings converging to $\widetilde{V}_{\text{univ}}$.) Observe that such varieties exist for all dimensions $n \geq 2$, for example those of the form $H^n/\Pi = \mathcal{O}(n)\backslash\mathcal{O}(n,1)/\Pi$ where H^n is the hyperbolic space and Π torsionless arithmetic subgroup in $\mathcal{O}(n,1)$.

Now we want to change the topology of such a $V^0 = \widetilde{V}_i$ with large $\lambda_1(\Delta)$ by a suitable geometric surgery (as in $1\frac{5}{6}$). To make it clear, we suppose $n \geq 4$ and show how to kill the fundamental group of V^0 without introducing small eigenvalues. We assume without loss of generality that V^0 is orientable and so the usual surgery kills π_1. Geometrically, this surgery consists in attaching disks D to some loops in V^0 and then taking boundaries of slightly thickened disks, which are $\partial(D \times B_\varepsilon^{n-1}) = D \times S_\varepsilon^{n-2}$. If we want to keep the spectrum large, we must have D with large λ_1, and these are readily available; just take hyperbolic disks with curvature $\leq -C$ for C large and with boundaries isometric to the circles we kill. If ε is sufficiently small, such surgery does not bring small eigenvalues since the meeting place of D (carrying the geometric essence of the handle for small ε) is of codimensions > 1 in V^0 and so the smallest positive eigenvalue of V^0+ handle is no smaller than that of V^0 or of the handle (with the zero boundary condition) $\pm\varepsilon$.

Similarly, one can make all surgeries along spheres of codimension ≥ 2 except for connected sums (but with possible 1-handles attached to connected manifolds). Therefore if V^0 is orientably bordant to zero, it can be moved to S^n for $n \geq 5$, by surgeries of codimension ≥ 2, since one may choose V^0 stably parallelizable (such V^0 exist, e.g. of constant negative curvature, by a theorem of Deligne and Sullivan).

One can obviously organize the surgeries so that a metric ball B in $V^0 = \widetilde{V}_i$ of large radius R remains intact (as we can choose i as large as we want) and then our sphere V^1 obtained from V^0 by surgeries also contains B. It follows that the metric g_1 on V^1 is larger than the spherical metric g_0 of the intrinsic diameter R as B can be compressed on (S^n, g_0) minus a little ball, and the rest of V^1 compresses to this small ball.

Let us indicate how to make the above work for all topological types of V for $n \geq 3$. This is done by removing from V^0 a small ε-neighbourhood of the $(n-2)$-skeleton of some triangulation Tr of V^0 and glueing in such a neighbourhood in the manifold V^1 with the desired topology. We notice that the toplogy of $V - U_\varepsilon(\mathrm{Tr}^{n-2})$ is essentially independent of V for $n \geq 3$, as this is a handle body with a 1-dimensional spine where the number of the handles can be easily adjusted by changing the triangulation (and where we assume V is orientable to avoid minor troubles). Thus, topologically speaking, we can replace $U_\varepsilon(\mathrm{Tr}^{n-2}\ V^0)$ by $U_\varepsilon(\mathrm{Tr}^{n-2}\ V^1)$ with some diffeomorphism

$$\partial U_\varepsilon(\mathrm{Tr}^{n-2}\ V^1) \leftrightarrow \partial U_\varepsilon(\mathrm{Tr}^{n-2}\ V^0)\ ,$$

where one should be aware of the fact that such a diffeomorphism may have (and usually has) a very large metric distortion going to ∞ with $\varepsilon \to 0$. What remains to do is to indicate a good metric on $U_\varepsilon(\mathrm{Tr}^{n-2}\ V^1)$ extending from the boundary the one induced by the embedding $\partial U_\varepsilon(\mathrm{Tr}^{n-2}\ V^0) \hookrightarrow V^0$. What we do is a fast shrinking of this boundary (as if by filling with a hyperbolic ball) with a simultaneous drift from the metric of V^0 to that of V^1, followed by filling the result by $\delta U_\varepsilon(\mathrm{Tr}^{n-2}\ V^1)$ with small $\delta > 0$ matching the preceding shrinking. (We suggest the reader would fill in the details.)

Thus every closed orientable n-manifold V with $n \geq 3$ admits a metric $g = g_R$, for every given $R > 0$, such that

(1) *V contains an isometric copy of the hyperbolic R-ball for a given R.*

(2) $\lambda_1(\Delta_g) \geq 1$.

Remarks. (a) Probably, it is not hard to remove the orientability assumption.

(b) It seems that one can freely move topology with this kind of surgery (using Tr^k for $k \approx \frac{n}{2}$) without changing non-zero part of the small spectrum of $d + d^*$ apart from $m = n/2$ for n even and $m = \frac{n\pm1}{2}$ for n odd. However, the starting manifolds $\widetilde{V}_i$ cause a problem here. (The only way I see how to control the spectrum on forms of positive degrees is with Bochner-Matsushima type formulae, but these do not seem to cover all m's, but only the range $m \leq \sqrt{n}$.)

(c) In order to replace (1) by $g \geq R^2\, g_0$ one should find triangulations Tr of $\widetilde{V}_i$ (for large i) with metrically large Tr^{n-2}. Here is a related quintessential problem. Can one generate $H^m(\widetilde{V}_i;\mathbb{Q})$ for large i and odd m by the pull-backs $f^*[S^m]^{\mathrm{co}}$ of *distance contracting* maps $f : \widetilde{V}_i \to S^m$ with the implied contraction (i.e. $(\mathrm{Lip} f)^{-1}$) going to ∞ for $i \to \infty$?

(d) It would be nice to make the above construction more elementary by chasing away arithmetic varieties. In fact, it is easy to construct large graphs with large λ_1 (e.g. starting from cubical graphs as in [Gro]$_{\mathrm{FRM}}$) but thickening them to large manifolds does not look obvious (despite 9.2.A in $[\text{Gro}]_{\text{FRM}}$ which I now regard with suspicion.

7. Invariance and non-invariance of the tangent bundle and Pontryagin classes

Can one change the tangent bundle $T(V)$ of a manifold V by modifying its smooth structure while keeping the homotopy type of V intact ? If "yes", in how many ways? "No" is known for the spheres S^n for all n. "Yes" is obvious for many *open* manifolds V. Namely if V and V' are total spaces of two different vector bundles X and X' of the same rank) over some V_0, then the tangent bundles $T(V)$ and $(T(V')$ differ as much as X and X' do, while V and V' are homotopy equivalent being contractible to the same V_0.

In earlier times one could smugly believe in the homotopy rigidity of the smooth structure and, consequently, of the tangent bundle of a *closed* manifold V. After all this had been known for surfaces V, where the essential invariant of $T(V)$, the *Euler class*, i.e. the "algebraic" number of zeros of a generic section (vector field) $V \to T(V)$, *is* a homotopy invariant being equal to the Euler characteristic of V. But as dimension goes up, there appear too many different possibilities for $T(V)$ to be contained by the homotopy type of V. For example, one can show there are infinitely many manifolds $V_1, V_2, \ldots$, all homotopy equivalent to $S^2 \times S^4$ but with quite different tangent bundles, distinguished by their first *Pontryagin classes* $p_1(T(V_i)) \in H^4(S^2 \times S^4) = \mathbb{Z}$, namely with $p_1(T(V_i)) = Mi$ for some (large) fixed integer $M \neq 0$ and $i = 0, 1, 2, \ldots$, where $V_0 = S^2 \times S^4$ and where non-vanishing of p_1 signifies non-triviality of the restriction of the tangent bundle $T(V_i)$ to S^4.

More precisely, the implied homotopy equivalence $S^2 \times S^4 \to V_i$ sends $S^4 = s \times S^4$ into V_i and the "restrictions" means "pull-back" under this map $S^4 \to V_i$. Notice that the non-vanishing of $p_1(V_i)$ precludes any embedding or immersion $S^4 \to V_i$, non-homologous to zero albeit the generator of $H_4(S^2 \times S^4) = \mathbb{Z}$ can be represented by a smooth submanifold. To see this, we compose the homotopy equivalence in the opposite direction, $V_i \to S^2 \times S^4$, with the projection $S^2 \times S^4 \to S^2$ thus obtaining a map $V_i \to S^2$. We make this map smooth by a small perturbation and take the pull-back W of a generic, and hence regular, point $s \in S^2$. This $W \subset V_i$ is a smooth 4-manifold whose fundamental class [W] generates $H_4(V_i)$ and whose *signature*, according to the *Rochlin-Thom-Hirzebruch theorem* (see below), equals $\frac{1}{3}\langle p_1(T(V_i)), [W]\rangle$. In particular, if $p_1 \neq 0$, this signature is also non-zero which prohibits S^4 from serving for W.

$7\frac{1}{4}$. Recollection on signature $\sigma(V)$

Let V be an oriented $4k$-dimensional manifold (possibly non-compact and with boundary) and observe that the intersection index between $2k$-cycles in V is symmetric, $Z_1 \frown Z_2 = Z_2 \frown Z_1$, (it is antisymmetric for $\dim V = 4k+2$), and hence defines a quadratic form on the real vector space $H_{2k}(V;\mathbb{R})$. We assume this space is finite dimensional, say of rank b, and we bring the intersection form to $\sum_{i=1}^{b_+} x_i^2 - \sum_{i=1}^{b_-} y_i^2$. (If V is a *closed* manifold, then this form is non-singular, by the Poincaré duality, and so $b = b_+ + b_-$). The difference $\sigma = b_+ - b_-$ is called the *signature* (of the form and) of V. If V is a closed manifold, the signature $\sigma(V)$ is a homotopy invariant of V (since the intersection on cycles is Poincaré dual to the cup-product on cocycles) and it is not "just an invariant" but *the invariant* which can be matched in the beauty and power only by the Euler characteristic. If V is non-closed, $\sigma(V)$ is an invariant under *proper* homotopy equivalences. Here is what one should know about σ.

(1) $\sigma(V) = -\sigma(V)$ where $-V$ means the reversing the orientation of V. This is obvious.

(2) $\sigma(V_1 \coprod V_2) = \sigma(V_1) + (V_2)$. (So obvious it is hard to not forget to mention it).

(3) ***Cobordism invariance.*** If V, a closed manifold, bounds an oriented $(4k+1)$-manifold W then $\sigma(V) = 0$. The intersection is, obviously, zero on the kernel of the inclusion homomorphism $I_* : H_{2k}(V) \to H_{2k}(W)$ and the orthogonal complement of this kernel for the intersection form on $H_*(V)$, say $\ker^\perp$, is contained in ker itself by the Poincaré duality in W and the intersection vanishes on $\ker^\perp$ as well as on ker. Hence $\sigma(V) = 0$ by obvious linear algebra. It follows that $\sigma(V)$ is a *cobordism invariant* (as well as a homotopy invariant) of V. Namely if V and $-V'$ make a boundary of some W, then $\sigma(V') = \sigma(V)$. For example, if V is an oriented connected sum, $V = V_1 \# V_2$, then $\sigma(V) = \sigma(V_1) + \sigma(V_2)$.

(4) ***Multiplicativity.*** If $\widetilde{V}_1 \to V$ is a finite d-sheeted covering of V, then $\sigma(\widetilde{V}) = d\sigma(V)$, provided V is a *closed* manifold. (Amazingly, there is no direct homological approach to this multiplicativity. The original argument appeals to Thom's cobordism theory with a possible shortcut to the bare essentials, the Serre finiteness theorem for the stable homotopy groups (see $7\frac{8}{9}$). The second proof depends on the Atiyah-Singer index theorem. The latter was originally established using cobordisms but now there are several independent proofs, some K-theoretic and some purely analytic, but none truly elementary).

(5) ***Cartesian multiplicativity.*** $\sigma(V_1 \times V_2) = \sigma(V_1)\sigma(V_2)$. (It follows from the multiplicativity of the signature under tensor product of quadratic forms).

(6) ***Novikov Additivity.*** Let V be cut into two pieces, say V_1 and V_2, by a closed hypersurface S lying in the interior of V. Then

$$\sigma(V_1) + \sigma(V_2) = \sigma(V).$$

The $2k$-homology of V is built of those of V_1 and V_2 and of the intersection of the kernels of the inclusion homomorphisms $i_1 : H_{2k-1}(S) \to V_1$ and $i_2 : H_{2k-1}(S) \to V_2$. Since the intersection form is invariant for the inclusions of V_1 and V_2 into V, the $(2k-1)$-homology of V coming from V_1 and V_2 has $\sigma = \sigma(V_1)+\sigma(V_2)$. On the other hand, the intersection form obviously vanishes on $\operatorname{Im} H_{2k}(S) \hookrightarrow H_{2k}(V)$ and consequently, by the Poincaré duality in S, the intersections $\ker i_1 \cap \ker i_2 \to H_{2k}(V)$ does not contribute to the signature of V.

Examples. (a) $\mathbb{C}P^2$ has signature 1 as $H_2(\mathbb{C}P^2) = \mathbb{Z}$ with positive self-intersection of $\mathbb{C}P^1 \subset \mathbb{C}P^2$ generating $H_2(\mathbb{C}P^2)$. Consequently $\mathbb{C}P^2\#\mathbb{C}P^2$ has $\sigma = 2$ and so it is neither homotopy equivalent nor cobordant to $S^2 \times S^2$, which has the same Betti numbers but zero signature.

(b) Let V be the total space of an oriented vector bundle X of rank $2k$ over a closed connected $2k$-dimensional manifold V_0. Then the signature of V (obviously) equals sign $e(X)$, where "e" stands for the Euler number defined as the self-intersection number of V_0 in V realized as the zero section. Thus $e(X)$ is a proper homotopy invariant of (the total space of) X and, as one knows, this is the only numerical invariant (characteristic number) with this property.

$7\frac{1}{2}$. Pontryagin classes, L-classes, signature theorem, and so on...

Every real vector bundle $X \to V$ can be induced by a continuous map $\alpha : V \to Gr_r\mathbb{R}^\infty$, for $r = \operatorname{rank} X$, from the canonical r-bundle over the Grassmann manifold $Gr_r\mathbb{R}^\infty$ and the isomorphism class of X is determined by the homotopy class of α. As we *stabilize* X by adding trivial bundles, we embed $Gr_r\mathbb{R}^\infty \subset Gr_{r+1}\mathbb{R}^\infty \subset \cdots$ and take the union, called $Gr\mathbb{R}^\infty = BGL$, the *classifying space of the stabilized linear group* $GL = GL(\infty) = \bigcup_{r=1}^\infty GL(r)$. The sole purpose of this stabilization is to remove the Euler class and if $\operatorname{rank} X > \dim V$ the stabilization is unnecessary.

The non-torsion part of the cohomology of BGL is a polynomial ring which can be polynomially generated by certain distinguished classes $p_i \in H^{4i}(BGL;\mathbb{Z})$, $i = 1,2,\ldots$, called the (universal) *Pontryagin classes.* The pull-backs of these to V under the classifying map $\alpha : V \to BGL$ are the *Pontryagin classes* of X, denoted $p_i(X) \in H^*(V;\mathbb{Z})$. If V is a closed oriented $4k$-manifold, one extracts numerical invariants out of (the cohomological invariants) $p_i = p_i(X)$ by taking their various products of total degree $4k$ and evaluating on the fundamental class of V, namely, $p_1^k[V], p_1^{k-2}p_2[V], \ldots, p_k[V]$. These are called the *Pontryagin numbers of* X, and for $X = T(V)$, the *Pontryagin numbers of* V. The totality of the Pontryagin numbers encodes the homology class $\alpha_*[V] \in H_{4k}(BGL;\mathbb{Q})$. In particular, if the Pontryagin numbers vanish, this class is zero which means that a "multiple of V" can be homotoped to the $(4k-1)$-skeleton of (some triangulation of) BGL.

In general, one may pair (products of) p_i's with the homology classes in V, and the resulting numbers encode the $\mathbb{Q}$-*information* on our (stabilized) bundle X. More precisely, we say that two bundles X_1 and X_2 over V are $\mathbb{Q}$-*equivalent* if there is an integer $M > 0$ such that MX_1 is *stably equivalent* to MX_2, where $MX = \underbrace{X \oplus X \oplus \cdots \oplus X}_{M}$ and "stably equivalent" means "equivalent after adding trivial bundles of suitable ranks". One knows that *two bundles are* $\mathbb{Q}$-*equivalent if and only if they have equal rational Pontryagin classes* where "rationalization" means passing to $H_*(V;\mathbb{Q}))$ which is equivalent to having equal *numbers* $p_i(h)$ for all $h \in \bigoplus_j H_{4j}(V)$. (This implies equality of all $\Pi_\mu(h)$ for the products Π_μ of p_i). *And there exists an integer* $M_0 = M_0(V) > 0$ *such that for arbitrary* $p' \in H_{4i}(V)$, $i = 1, 2, \ldots$, *the multiples* $M_0 p'_i$ *can be realized as Pontryagin classes of some* $X \to V$. All this follows from Serre's finiteness theorem (see $7\frac{8}{9}$).

Another consequence of this theorem is the *finiteness of the number of proper homotopy equivalence classes of stable vector bundles over* V. This means, in particular, that *there exists an integer* $M_1 = M_1(V)$ *such that every vector bundle* $X \to V$ *or* $\operatorname{rank} r > \dim V$ *stably equivalent to* $M_1 Y$ *for some* $Y \to V$ *is proper fiberwise homotopy equivalent to the trivial bundle* $V \times \mathbb{R}^r$. Such proper equivalence implies the homotopy equivalence of the corresponding sphere bundle S_X to $V \times S^{r-1}$ which is more attractive being a closed manifold for closed V, while, by the above, the Pontryagin classes of X, and hence of S_X, may be taken almost at will. For example, if all $p_i(V) = 0$, then $p_i(S_X)$ equal the pull-backs of $p_i(X)$ (for the projection $S_X \to V$) and these $p_i(X) \in H_i(V)$ can be chosen multiples of arbitrary classes $p'_i \in H_{4_i}(V)$, $i = 1, \ldots$. Thus the Pontryagin classes of $V \times S^{r-1}$ can be easily varied by varying the smooth structure within the fixed homotopy class of $V \times S^{r-1}$. This agrees with (but does not formally imply) our earlier example of $S^2 \times S^4$ where the pertinent bundle has rank 3 over S^4 which is not the stable range but where Serre's theorem still applies.

L-classes. There is nothing sacred about the generators p_i of $H^*(BGL)$. In fact we prefer another set of polynomial generators of the *rational* cohomology of BGL, denoted $L_i \in H^{4i}(BGL : \mathbb{Q})$, which are uniquely characterized by the following condition.

Let V be an oriented $4k$-dimensional manifold which is the Cartesian product of some complex projective spaces, and $\alpha : V \to BGL$ the classifying map for the tangent bundle $T(V) \to V$. Then

$$L_k(\alpha_*[V]) \underset{\text{def}}{=} \sigma(V), \tag{$*$}$$

i.e. $L_k(\alpha_*[V]) = 1$ if all $\mathbb{C}P^j$-factors of V have j even and $L_k(\alpha_*[V]) = 0$ if some j are odd. This indeed correctly defines L_i, since the classes $\alpha_*[V] \in H_{4k}(BGL)$ form a rational basis in this H_{4k} for all $\mathbb{C}P^j$-product manifolds V as an elementary computation (of Pontryagin numbers of these V's) shows. The first L_i can be easily computed in terms of p_i,

$$L_1 = \tfrac{1}{3}p_1, \ L_2 = \tfrac{1}{45}(7p_2 - p_1^2), \ L_3 = \tfrac{1}{945}(62p_3 - 13p_2p_1 + 2p_1^3), \ldots,$$

but then it becomes a mess; yet, one can show that $L_i = \ell_i p_i + \cdots$ where $\ell_i \neq 0$ for all i and so the p_i's can be rebuilt out of L_i's. (Actually, we could start with L_i defined by (*) without ever mentioning p_i but we paid our respect to the custom).

Signature theorem. *Every closed oriented $4k$-manifold V has*

$$\sigma(V) = L[V],$$

where

$$L[V] \underset{\text{def}}{=} L_k[V] \underset{\text{def}}{=} L_k(\alpha_*[V]) \qquad (**)$$

for the classifying map $\alpha : V \to BGL$.

Proof. According to Thom's cobordism theory (which can be reduced in our case to Serre's finiteness again), two manifolds V_1 and V_2 are $\mathbb{Q}$-*cobordant*, i.e. MV_1 is cobordant to MV_2, where MV denotes the disjoint union of M copies of V if (and, obviously, only if) their classifying maps are $\mathbb{Q}$-homologous, i.e. $\alpha_1[V_1]$ equals $\alpha_2[V_2]$ in $H_{4k}(BGL;\mathbb{Q})$. Since $H_{4k}(BGL)$ is spanned by products of $\mathbb{C}P^j$'s, every V is $\mathbb{Q}$-cobordant to a disjoint union of products of $\mathbb{C}P^j$'s and their inverses (i.e. with reversed orientations) and, hence obviously by linearity, (with the properties 1-3 and 5 of σ) (**) follows from (*). (Everything here but "hence obviously" is due to Thom with the final "hence obviously" furnished by Hirzebruch. Apparently, what Thom missed was "linearity", i.e. the implication

$$Ma = Mb \Rightarrow a = b$$

in the vector space of linear functions on the cobordism group $\mathcal{C}ord_{4k}$ as he aimed at the actual generators of $\mathcal{C}ord_{4k}$ not only those over $\mathbb{Q}$ provided by the products of $\mathbb{C}P^j$'s).

Multiplicativity corollary (see (4) in $7\frac{1}{4}$). *If $\widetilde{V} \to V$ is a finite k-sheeted covering then $\sigma(\widetilde{V}) = k\sigma(V)$.*

In fact, $L[V]$ is multiplicative as, obviously, $\alpha_*[\widetilde{V}] = k\alpha_*[V]$ in $H_*(BGL)$. Q.E.D.

Browder-Novikov theorem. We saw earlier how one could vary Pontryagin classes (or, equivalently L-classes of $V \times S^r$ and this extends to all closed *simply connected* manifolds V of dimension ≥ 6, where, according to BN, the signature formula $L(V) = \sigma(V)$ is the only homotopy restriction on the $\mathbb{Q}$-type of the stable tangent bundle $T(V)$. Namely, *there is an integer* $M = M(V)$ *such that for arbitrary (integer) classes* $L'_i \in H_{4_i}(V)$, $i = 1, \ldots, k-1$, *where* $4k - 3 \geq \dim V \leq 4k$, *one can find* V' *homotopy equivalent to* V *and having* $L_i(V') = L_i(V) + ML'_i$, $i = 1, \ldots, k-1$, where this equality refers to the identification between $H^*(V')$ and $H^*(V)$ for the implied homotopy equivalence $V' \leftrightarrow V$ and where $L_i(V) \underset{\text{def}}{=} \alpha^*(L_i)$ for the classifying map $\alpha : V \to BGL$. But one cannot vary $L_k(V)$ for $\dim V = 4k$ as it must abide (**)). The Browder-Novikov proof consists of the reduction of this by surgery to Serre's finiteness theorem. Practically all $\mathbb{Q}$-finiteness of the number of homotopy restrictions on $T(V)$ apart from the equalities $L_k(T(V)) = \sigma(V)$ and $e(T(V)) = \chi(V)$ are derived from Serre's theorem. And this theorem, in a certain precise sense, is less elementary than the derivation arguments.

$7\frac{3}{4}$. On the invariance of L_i and the Novikov conjecture

Now the stage is set for a discussion on the homotopy invariance of the classes $L_i(V) \in H^{4i}(V;\mathbb{Q})$ (or equivalently of "rationalized" classes p_i) of *non-simply connected* manifolds V. For example, let all homotopy of V come from the fundamental group $\Pi = \pi_1(V)$, i.e. V be a closed *aspherical* (also called $K(\Pi;1)$ and/or $B\Pi$) manifold which means contractibility of the universal covering $\widetilde{V}$ of V. Then one may conjecture, following Novikov, that the tangent bundle is uniquely determined in the $\mathbb{Q}$-sense by the homotopy type of V, i.e. by the fundamental group. That is, every map between two such manifolds, $V \to V'$, which is isomorphic on π_1's sends $L_i(V) \leftarrow L_i(V')$. Take for example the n-torus T^n for V. This manifold is parallelizable and so all characteristic classes vanish.

According to the conjecture, this must be true for every n-manifold V' homotopy equivalent to T^n, all L_i and p_i must be zero. (Since $H^*(V') = H^*(T^n)$ has no torsion, vanishing of p_i in $H^*(V;\mathbb{Q})$ implies vanishing in $H^*(V;\mathbb{Z})$ and the Euler class is zero anyway being equal to $\chi(V') = \chi(T^n)$). To see this from another angle, let $\widetilde{V}'_{\text{univ}} \to V'$ be the universal covering of V' viewed as a principal bundle with the group $\Pi = \pi_1(V') = \mathbb{Z}^n$ for the fiber and let $X \to V'$ be the associated $\mathbb{R}^n$-bundle for the standard action of $\mathbb{Z}^n$ on $\mathbb{R}^n$. (X equals $\widetilde{V}'_{\text{univ}} \times \mathbb{R}^n/(\text{diagonal action of } \mathbb{Z}^k)$ and it naturally projects to V'). Then the conjecture claims that X (turned into a vector bundle by choosing a "zero" section $V' \to X$) is $\mathbb{Q}$-equivalent to $T(V')$ i.e. has the same L-classes.

A similar interpretation is possible for all aspherical V. Namely, we take the fibration $X \to V'$ associated to $\widetilde{V}'_{\text{univ}} \to V'$ with the fiber $\widetilde{V}_{\text{univ}}$ for the Galois action of $\Pi = \pi_1(V) = \pi_1(V')$ on $\widetilde{V}_{\text{univ}}$ and (conjecturally) claim that X is $\mathbb{Q}$-equivalent to $T(V)$ (which means here a fiberwise diffeomorphism between the fibrations $MX \oplus \text{Triv}$ and $MT(V') \oplus \text{Triv}'$). (The universal covering $\widetilde{V}_{\text{univ}}$ does not even have to be diffeomorphic to $\mathbb{R}^n$, but this is recovered by adding the trivial bundle). In fact, we would rather exclude V from this altogether, as we want to reconstruct (the $\mathbb{Q}$-type of) $T(V')$ (as well as $T(V)$) functorially out of Π alone. (A similar problem arises in the complex analytic and symplectic categories where some results are available for Kähler manifolds).

$7\frac{4}{5}$. Novikov in codimension one

The first homotopy invariance result concerns manifolds which are not aspherical but rather look like $V = W \times S^1$, where the relevant part of π_1 is just $\mathbb{Z} = \pi_1(S^1)$.

(Novikov 1965). *The class $L_k(V)$ of a $(4k+1)$-dimensional manifold V is a homotopy invariant of V.*

Proof. The class L_k is determined by its values on $H_{4k}(V)$ and so we must prove the invariance of $\langle L_k(V), h\rangle$ for all $h \in H_{4k}(V)$. Every homology class h of codimension one can be realized by a co-oriented submanifold $W \subset V$ appearing as the pull-back of a regular value of a smooth map $\beta_h : V \to S^1$ representing the Poincaré dual class $h^{\text{dual}} \in H^1(V)$. Now $\langle L_k, h\rangle$ acquires a meaning as it equals the signature of W. Indeed, by an obvious functoriality of L_k,

$$\langle L_k(T(V)), h\rangle = \langle L_k(T(V)|W), [W]\rangle$$

(where we may assume V, and hence W, oriented without loss of generality) and as the normal bundle of W in V is trivial, $T(V)|W$ is stably equivalent to $T(W)$. Thus

$$\langle L_k(V), h\rangle = \langle L_k(T(W), [W]\rangle = \sigma(W)$$

by the signature theorem. So, to prove the Novikov theorem, we must give a homotopy interpretation of $\sigma(W)$ in terms of the original manifold V. This is done below in the framework of the *proper homotopy type* of the cyclic covering $\widetilde{V}_h \to V$ induced from the covering $\mathbb{R} \to S^1$ by the map β_h. This $\widetilde{V}_h$ has a distinguished homology class $\widetilde{h} \in H_{4k}(\widetilde{V}_h)$ corresponding to h, which is realized by a lift of W to $\widetilde{V}_h$, say $W_0 \subset \widetilde{V}_h$. This $\widetilde{h}$ defines a cup product pairing on $H^{2k}(\widetilde{V}_h)$ by

$$(h_1, h_2) \mapsto \langle h_1 \smile h_2, \widetilde{h}\rangle$$

and *the signature of this pairing*, denoted *cup*$(\widetilde{V}_h|\widetilde{h})$ (which is a homotopy invariant of V being a proper homotopy invariant of $\widetilde{V}_h$) *equals the signature of* W_0 (which is diffeomorphic to W). Let us prove the equality we claim,

$$\sigma(cup(\widetilde{V}_h|\widetilde{h})) = \sigma(W), \qquad (+)$$

(which is a pretty homological formula for $\langle L_k(V), h\rangle = \sigma(W)$, not just "a homotopy invariance").

Proof of $(+)$. Let V^+ be a non-compact $4k+1$-manifold with compact boundary W_0 and show that the form $cup(V^+|[W_0])$ on $H^{2k}(V^+)$ has $\sigma = \sigma(W_0)$. In fact this $\sigma \underset{\text{def}}{=} \sigma(cup(V^+|[W_0]))$ equals the signature of the intersection form I_0 on W_0 restricted to the space $H \subset H_{2k}(W_0)$ corresponding to the cycles $W_0 \cap C$ in W_0 for all (possibly) infinite cycles C in V^+; see Fig. 10.

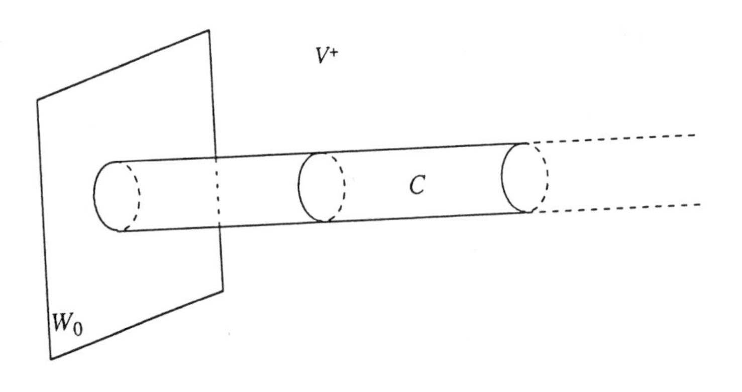

Figure 10

The $2k$-cycles in the I_0-orthogonal complement $H^\perp$ of H have zero intersection with all C's in V^+ and thus, by Poincaré duality in V^+, they bound in V^+, which implies vanishing of I_0 on $H^\perp$ (compare the proof of the cobordism invariance of σ in $7\frac{1}{4}$) and, by linear algebra, the desired equality $\sigma(I_0) = \sigma(I_0|H)$. Next, we take an open $4k+1$ manifold $\widetilde{V}$ (not necessarily anybody's covering) divided into two halves V^+ and V^- by some closed W_0 and conclude again that the form $\mathrm{cup}(\widetilde{V}|[W_0])$ has the same signature as the manifold W_0 moved deep into V^- without changing the signatures; see Fig. 11 below.

The signature of W_i is independent of i by the cobordism invariance while the form $\mathrm{cup}(\widetilde{V}|[W_i])$ is independent of i along with $[W_i] \in H_{4k}(\widetilde{V})$ and so

$$\sigma(W_0) = \sigma(W_i) = \sigma(\mathrm{cup}(V_i^+|[W_i])) = \sigma(\mathrm{cup}(V_i^+|[W_0]) \underset{i\to\infty}{\to} \sigma(\mathrm{cup}(\widetilde{V}|[W_0])).$$

Q.E.D.

Corollary to the proof. *The class L_k is a proper homotopy invariant of non-compact $(4k+1)$-manifolds.*

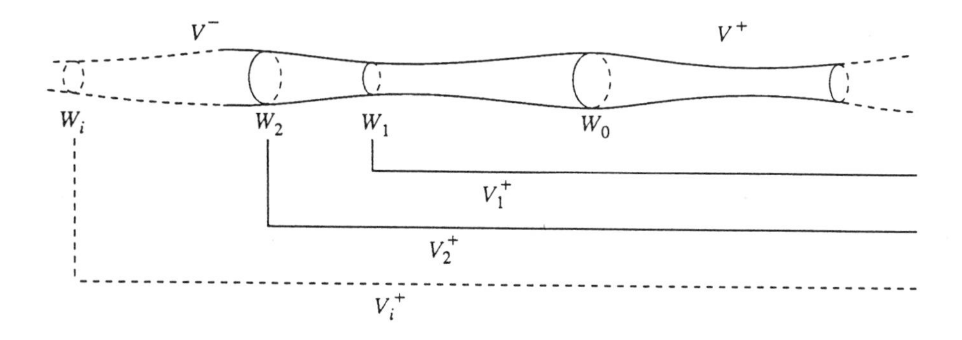

Figure 11

$7\frac{5}{6}$. Higher signatures σ_ρ.

Let us replace the circle S^1 in the above picture by an arbitrary closed aspherical manifold B with some fundamental group $\Pi = \pi_1(B)$ and look at a manifold V mapped to B. A homotopy class of such a map is determined solely by the homomorphism $\pi_1(V) \to \Pi$, and so our data actually consist of V and a homomorphism $\pi_1(V) \to \Pi$. We slightly perturb our map so it becomes smooth, say $\beta : V \to B$, and we take the pull-back $W = \beta^{-1}(b)$ of a regular value $b \in B$ of β. This W is a smooth submanifold in V of $\operatorname{codim} W = \dim B$ and we are keen on the signature of W in the case where $m = \dim W = 4k$.

We observe that the homology class $[W] \in H_m(V)$ can be described in more invariant terms as dual $\beta^*[B]^{\mathrm{co}}$, i.e. the Poincaré dual of the pull-back of the fundamental cohomology class $[B]^{\mathrm{co}} \in H^d(B)$, $d = \dim B$. Then we notice that the cobordism class of W is invariant under homotopies of β and movements of b. For example, if $b \in B$ is a regular value for a smooth homotopy $V \times [0,1] \to B$ between β_0 and β_1, then the pull-back of b in the cylinder $V \times [0,1]$ furnishes a cobordism between $W_0 = \beta_0^{-1}(b)$ and $W_1 = \beta_1^{-1}(b)$. Thus we see that the signature $\sigma(W)$ is a well-defined invariant of V with a given homomorphism $\pi_1(V) \to \Pi$. Another way to see it is by observing that $\sigma(W) = \langle L_k(V), [W] \rangle$, as in the case $B = S^1$. In fact, since W is the regular pull-back of a point, it has a trivial normal bundle in V (because it can be given by a nonsingular system of equations $\varphi_1(u) = 0, \ldots, \varphi_d(u) = 0$ in some neighbourhood $U \supset W$, where φ_i's come from local coordinates $\psi_1, \ldots, \psi_d$ in B at b for $d = \dim B$) and so all (stable!) characteristic classes $L_i(W) \underset{\mathbf{def}}{=} L_i(T(W)) = L_i(T(V)|W)$ are obtained by restricting $L_i(V)$ to W. In particular

$$\sigma(W) = L_k[W] = \langle L_k(V), [W] \rangle, \tag{σ_h}$$

or, cohomologically,

$$\sigma(W) = \langle L_k(V) \smile \beta^*[B]^{\mathrm{co}}, [V]\rangle, \tag{σ_{co}}$$

which is equivalent to σ_h) by the Poincaré duality. Finally, we generalize (σ_{co}) by introducing the *(higher) signature* for an arbitrary cohomology class $\rho \in H^*(B)$,

$$\sigma_\rho \underset{\mathrm{def}}{=} \langle L(V) \smile \beta^*(\rho), [V]\rangle. \tag{σ_ρ}$$

Here $L(V) = 1 + L_1(V) + \cdots \in H^*(V)$ and the evaluation of the cup-product $L(V) \smile \beta^*(\rho)$ refers to the degree n component for $n = \dim V$.

This definition of σ_ρ is quite general; it applies to an arbitrary aspherical space $B = B\Pi$ with $\pi_1(B) = \Pi$ and V with a (homotopy class of a) map $\beta : V \to B$. And the resulting σ_ρ is called the (higher) *ρ-signature* if V. Of course, this definition makes sense for non-aspherical spaces B as well, but aspherical B's are special as we shall see presently. (If B is an arbitrary closed oriented manifold of dimension d, and $\rho = [B]^{\mathrm{co}} \in H^d(B)$, then $\sigma_\rho = \sigma(W)$ for the pull-back W of a regular value in B, with the convention $\sigma(W) = 0$ for $\dim W$ not divisible by 4. This property, in fact, uniquely determines the class $L(V)$ if one uses maps to spheres).

Every ρ-signature of V can be visualized as the actual signature of some submanifold W in V. In fact, for every cohomology class $\gamma \in H^m(V)$ of codimension $4k (= n - m)$ there exists a closed *immersed* submanifold W in V with trivial normal bundle such that the fundamental class $[W]$ is Poincaré dual to some non-zero multiple $M\gamma$ of γ. (This is yet another consequence of Serre's finiteness. For example, if m is odd or if $n \geq 2m + 2$, then according to Serre, V admits a map $\alpha : V \to S^m$, such that $\alpha^*[S^m]^{\mathrm{co}} = M\gamma$ and thus $W = \alpha^{-1}(s)$, for a regular $s \in S^m$, is dual to $M\gamma$. In general, one should combine the above with the Hirsch immersion theorem). Then clearly,

$$\sigma(W) = \langle L_k(V), [W]\rangle = M\langle L_k(V) \smile \gamma, [V]\rangle.$$

Novikov conjecture for σ_ρ. Let $B = B\Pi$ be an aspherical space and $\rho \in H^*(B)$. *Then, for every smooth closed manifold V with a given (homotopy class of a) map $\beta : V \to B$, the ρ-signature σ_ρ is a homotopy invariant of V, i.e. for every homotopy equivalence $e : V' \to V$, the ρ-signature of V' for the composed map $\beta' = e \circ \beta$ equals σ_ρ. Equivalently, the β_*-image of the Poincaré dual of every rational Pontryagin class, $\beta_*(PDp_i) \in H_{4_i}(B\Pi; \mathbb{Q})$, is a homotopy invariant of V. (One can imagine Pontryagin classes of some singular spaces, in the spirit of Cheeger-Goresky-MacPherson, where the homological formulation will be preferable).*

We prefer to turn the conjecture to the following:

Question. For which Π and ρ is σ_ρ homotopy invariant for all (V, β) ?

Of course, it may happen that Novikov conjecture is universally true. But if not, our question only gains in validity.

Novikov proved the homotopy invariance of all σ_ρ for the free Abelian groups $\Pi = \mathbb{Z}^\ell$ which amounts to his codim 1 theorem for $\ell = 1$. In fact, Novikov was originally concerned with V homeomorphic to $V_0 \times T^\ell$, where $T^\ell = B(\mathbb{Z}^\ell)$ is the ℓ-torus, and to $V_0 \times T^{\ell-1} \times \mathbb{R}$, and general V's with $\Pi = \mathbb{Z}^\ell$ where handled later by Kasparov.

Lusztig reproved the Novikov-Kasparov theorem for $\Pi = \mathbb{Z}^\ell$ by generalizing the signature theorem to families of flat S^1-bundles, and he also extended this to some cohomology classes ρ in certain arithmetic groups (compare $8\frac{1}{2}$). Lusztig's argument, based on the index theorem for the signature operator (i.e. properly interpreted $d + d^*$ twisted with flat bundles), was generalized to certain infinite dimensional bundles by Miščenko who thus proved the Novikov conjecture for all ρ in closed Riemannian manifolds B with non-positive sectional curvature (eventually the conjecture was settled for all *complete* B with $K(B) \leq 0$). In fact, the validity of the Novikov conjecture seems to be intimately related to the macroscopic geometry of the universal covering $\widetilde{B}$ of B and/or of the group Π. Some of this is explained in $7\frac{6}{7}$ and §9. (Also see [Fa-Hs], [Fa-Jo], [NC+] and references therein).

$7\frac{6}{7}$. On topological invariance of L_i and Lipschitz geometry

Let us recall the original Novikov homotopy invariance theorem.

Let U be an oriented manifold diffeomorphic to $W \times T^\ell \times \mathbb{R}$, where W is a closed manifold of dimension $4k$. Then the value $\langle L(U), [W]\rangle$ is a proper homotopy invariant of U. Namely, if $\gamma : U' \to U$ is a proper homotopy equivalence, which happens to be smooth and transversal to $W = V \times t \times r$ for some $(t, r) \in T^\ell \times \mathbb{R}$, then

$$\sigma(\gamma^{-1}(W)) = \sigma(W).$$

Novikov proved that by (a seemingly circular surgery argument) constructing inductively a descending sequence of submanifolds in U', say $U' = W_0' \supset W_1' \supset W_2' \supset \cdots \supset W_{\ell-1}' \supset W_\ell'$, where each W_i' is homotopy equivalent to $W \times T^{\ell-i}$ with the inclusions $W_i' \subset W_{i-1}'$ homotopic to the standard ones, $W_i' \subset W_i' \times T^1 \approx W_{i-1}'$. The final manifold W_ℓ' is then homotopy equivalent to W and so has $\sigma(W_\ell') = \sigma(W)$. On the other hand, this W_ℓ' obviously has a trivial normal bundle in V' and so $\sigma(W_\ell') = \langle L_k(V'), [W_\ell']\rangle$.

Now, to prove the topological invariance of L_i (and hence, of p_i) for all manifolds V, we will show, following Novikov, that if some homology class

$h \in H_{4k}(V)$ is realized by an immersed submanifold W with trivial normal bundle and certain signature σ, then, in a homeomorphic manifold V', a similar realization W' of h has the same signature σ, i.e. $\sigma(W') = \sigma(W)$. Since the normal bundle of W is trivial, $T(V)|W = T(V) \oplus \mathrm{Triv}^{n-4k}$ and a tubular neighbourhood of W in V is diffeomorphic to $W \times \mathbb{R}^{n-4k}$ immersed (i.e. locally diffeomorphically mapped) into V. We take some embedded ℓ-torus $T^\ell \subset \mathbb{R}^{n-4k}$ for $\ell = n - 4k - 1$, with a tubular neighbourhood $T^\ell \times \mathbb{R} \subset \mathbb{R}^{n-4k}$ and form a (non-simply connected!) manifold $U = W \times T^\ell \times \mathbb{R}$ immersed in V. As we pass to a homeomorphic V', the corresponding U' remains homeomorphic to U and hence properly homotopy equivalent to U. Therefore, a smooth W' in U realizing the homology class $[W]$ in U' (and thus homologous to $W \subset U \to V$) has by the Novikov homotopy invariance theorem the same signature as W. Q.E.D.

(Notice that we used here the existence of W with trivial normal bundle realizing a non-zero multiple of a given homology class of V, which is a consequence of Serre's finiteness theorem).

A homotopy application of the topological invariance. It is an easy consequence of the above that the L-classes of vector bundles over an arbitrary base are invariant under fiberwise homeomorphisms between bundles, and the same is true for sphere bundles. We know that this is not true for proper fiberwise homotopy equivalences but it may be sometimes so for special homotopy equivalences. For instance one may speak of homotopy equivalence in the category of metric spaces and (proper) *Lipschitz* maps where the implied homotopies $X \times [0,1] \to Y$ must be Lipschitz for the product metric.

Basic example. Let V_1 and V_2 be compact homotopy equivalent Riemannian manifolds. Then, obviously, their universal coverings $\widetilde{V}_1$ and $\widetilde{V}_2$ are properly Lipschitz homotopy equivalent.

Question. Let X and Y be bundles over the same base with (smooth) Euclidean fibers and with fiberwise (not necessarily Euclidean) metrics. Suppose X and Y are fiberwise properly Lipschitz homotopy equivalent. Do they have equal L classes? Of course, the answer may depend heavily on the geometry of the fibers, and the most interesting case is where the fibers are properly Lipschitz homotopy equivalent to the universal covering of a compact manifold V.

Example: hyperbolic fibrations. Let the fibers of X and Y be complete simply connected Riemannian manifolds with negative curvatures $K \le -\kappa^2 < 0$. Then each fiber, say X_a of X, admits a compactification $\overline{X}_a$ homeomorphic to the closed n-hall, $n = \dim X_a$, where X_a sits in $\overline{X}_a$ as the interior of the ball. The *ideal boundary* $\partial X_a \underset{\text{def}}{=} \overline{X}_a - X_a$ is homeomorphic to S^{n-1}, and the S^{n-1}-bundle thus associated to X is fiberwise homeomorphic to the normal sphere bundle of a section $A \to X$. Furthermore, every fiberwise Lipschitz homotopy equivalence $X \leftrightarrow Y$ induces a fiberwise homeomorphism between the ideal boundary (spherical) bundles, and by Novikov's topology invariance, an equality between the L-classes of X and Y.

Hyperbolic manifolds. Let V be a closed manifold with $K(V) < 0$, and let $\widetilde{V} \to V$ be the universal covering viewed as a principal Π-bundle with $\Pi = \pi_1(V)$.

Take the associated fibration $X \to V$ with the fiber $\widetilde{V}$ for the Galois action of Π on $\widetilde{V}$ (X equals $\widetilde{V} \times \widetilde{V}$/diagonal action naturally fibered over $V = \widetilde{V}/M$), let V' be homotopy equivalent to V with the corresponding bundle $X' \to V'$ and bring this bundle to some $Y \to V$ via our homotopy equivalence $V \to V'$. The homotopy equivalence $V \leftrightarrow V'$ (obviously) induces a Lipschitz homotopy equivalence between X and Y over V (since V and V' are compact) and hence, in the case $K(V') < 0$, the equality of the L-classes, which are therefore invariant under homotopy equivalences between closed manifolds of negative curvature. And by the same token, L-classes are invariant under *Lipschitz* homotopy equivalences between *complete* manifolds of negative curvature.

Furthermore, by applying a Novikov-type argument on the large scale, one can drop the assumption $K(V') < 0$ (while keeping $K(V) < 0$) and eventually recover the full Novikov conjecture for V by topological means without using the index theorem (see [Fa-Hs], [Fa-Jo], [Pe-Ro-We], and references therein). In fact, this can be done quite elementarily using products of surfaces of genus ≥ 2 instead of tori (see $9\frac{1}{3}$).

The above L-equality problem for bundles may be preceded by the following.

Realization problems. Let Π be a finitely presented group which is S^{n-1} at infinity in the sense specified below. When does such a Π admit a discrete cocompact action on $\mathbb{R}^n$, or at least when does some Cartesian product $\Pi^k \times \mathbb{Z}^\ell$ admit such an action on $\mathbb{R}^{nk+\ell}$? Even if no such action exists, one may try to associate to each principle Π-bundle a "virtual Euclidean bundle" and define its L-classes.

On being S^{n-1} at infinity. There are several possible definitions. For example, if Π is a word hyperbolic group then one may speak of its ideal hyperbolic boundary $\partial\Pi$ and "$\partial\Pi$ homeomorphic to S^{n-1}" is one way to express the idea of "S^{n-1} at infinity". Here one knows for $n = 2$ that the realization problem has positive solution (without stabilization) but this, unexpectedly, is a difficult theorem (equivalent to the so-called Seifert conjecture recently solved by Gabai and by Casson with Jungreis). On the other hand, we do have $S^{n-1} = \partial\Gamma$ with a natural Γ action and so our spherical (and Euclidean) bundles automatically come along.

In general, for any finitely generated group, one can define its "homotopy type at infinity". So, for $n \geq 3$, we should require Π to be $(n-2)$-connected at infinity and have $H_{n-1} = \mathbb{Z}$ at infinity. This is, probably, equivalent (at least after some stabilization) to the existence of a complete Riemannian manifold V of dimension n, such that

1. V is *quasi-isometric to* Π with a word metric, i.e. V admits an ϵ-net Δ for some $\epsilon > 0$, which is bi-Lipschitz to Π.

2. V is *uniformly contractible*, i.e. there is a function $\rho_V(r)$, such that every r-ball in in V is contractible within the concentric ρ-ball for $\rho = \rho_V(r)$ (which is assumed $\geq r$).

In order to avoid possible complications, one may additionally require that this V is "large at infinity" in a suitable sense, e.g. admits a proper Lipschitz map $f : V \to \mathbb{R}^n$ of degree one, compare §4 and [Fe-We].

Finally, for an arbitrary complete Riemannian manifold (not necessarily homeomorphic to $\mathbb{R}^n$) we want to raise the question of (the existence and invariance of) characteristic classes for the "group" (H-space) of its Lipschitz homotopy equivalences. Again the main examples come from universal coverings of closed (not necessarily aspherical) manifolds where Lipschitz homotopy equivalences (individually and fiberwise in bundles) tend to preserve certain L-classes. (Compare [Pe-Ro-We] and §9).

$7\frac{7}{8}$. Wall-Witt groups of $R(M)$ and homomorphisms $WM : HBrd_* B\Pi \to Witt_*$ and $\alpha : H_*(B\Pi; \mathbb{Q}) \to HBrd_* \otimes \mathbb{Q}$.

Recall that the oriented bordism group of a topological space B, denoted $Brd_n B$, is formally generated by closed oriented n-dimensional manifolds V coming along with continuous maps $\beta : V \to B$, which are subject to the following:

Relations.

(1) Reversing the orientation of V reverses the sign of the bordism class,

$$[-V, \beta] = -[V, \beta].$$

(2) Disjoint union of manifolds (and maps) correspond to the addition in Brd_n,

$$[(V_1, \beta_1) \coprod (V_2, \beta_2)] = [V_1, \beta_1] + [V_2, \beta_2].$$

With this one sees that Brd_n is commutative.

(3) For every oriented $(n+1)$-dimensional manifold W with boundary $V = \partial W$ and a continuous map $\alpha : W \to B$,

$$[V, \alpha | V] = 0.$$

Actually, instead of generating a group by all (V,β) we may take the set $\{V,\beta\}$ itself with the semigroup structure for the disjoint union $\coprod$ and obtain

$$Brd_nB = \{V,\beta\}/(1) + (2) + (3).$$

For example, if B is a single point, then Brd_n is the usual Rochlin-Thom cobordism group of n-dimensional manifolds.

Next we add the following extra relation.

(4) If V_1 and V_2 are orientably homotopy equivalent and β_1 is homotopic to β_2 (or rather to $h \circ \beta_1$ for the implied homotopy equivalence $h : V_2 \to V_1$) then

$$[V_1, \beta_1] = [V_2, \beta_2].$$

Finally, we stabilize, by taking products with the complex projective plane, $V \rightsquigarrow V \times \mathbb{C}P^2$, where $\beta(v,c) = \beta(v)$, and by adding the corresponding relation

$$(5) \qquad [(V,\beta) \times \mathbb{C}P^2] = 0 \Rightarrow [V,\beta] = 0.$$

The essential property of $\mathbb{C}P^2$ here is the equality $\sigma(\mathbb{C}P^2) = 1$ which shows (with the Cartesian multiplicativity of σ, see (5) in $8\frac{1}{4}$) that this stabilization does not change the ρ-signature σ_ρ of (V,β) for every $\rho \in H^*(B)$ (compare $8\frac{5}{6}$). In fact, we could use any manifold W instead of $\mathbb{C}P^2$ of dimension $4k$ with $\sigma(W) = 1$ and arrive (after using the homotopy equivalence axiom (4)) at the same result (i.e. $HBrd_nB$ defined below).

Now we factorized the stabilized bordims by the homotopy equivalence relation and set

$$HBrd_nB = \{V,B\}/(1) + \cdots + (5) = Brd_nB/(4) + (5).$$

If B consists of a single point then $HBrd_nB$ is torsion for $n \neq 4k$ and $HBrd_{4k}$/torsion equals the ordinary *Witt group* of quadratic forms over $\mathbb{R}$. Recall that the Witt group of a field K is formally generated by the isomorphism classes of non-singular quadratic forms φ over K with the relations

(a) $[\varphi_1 \oplus \varphi_2] = [\varphi_1] + [\varphi_2]$, for the direct sum $\oplus$ of forms;

(b) $[-\varphi] = -[\varphi]$

(usually one takes instead of (b) the relation $[xy] = 0$ for the form xy on K^2 but this only has effect on the 2-torsion of the resulting group). In the case $K = \mathbb{R}$ everybody knows that Witt $\mathbb{R} = \mathbb{Z}$ with the isomorphism given by the signature $\varphi \mapsto \sigma(\varphi)$.

If B is simply connected, then again $HBrd_nB$ is torsion for $n \neq 4k$ (where one should assume B is a finite polyhedron to avoid irrelevant complications) and

$$HBrd_{4k}B/\text{torsion} = \mathbb{Z} = \text{Witt}\mathbb{R}(= \text{Witt}\mathbb{Z} = (\text{Witt}\mathbb{Q})/\text{torsion})$$

for the signature homomorphism $[V, \beta] \mapsto \sigma(V) \in \mathbb{Z}$, as follows from the Novikov-Browder theory.

The real story begins when we take a group Π and the classifying (aspherical) space $B = B\Pi$ (with $\pi_1 = \Pi$) where $HBrdB\Pi$ serves as a prototype for the definition of the Wall-Witt group of (yet unspecified group ring of) Π. For example, for the trivial group $\Pi = \{e\}$ this gives mod torsion (see the above), the Witt group ($= \mathbb{Z}$) of the (integral, rational (or) real group ring $\mathbb{R}(\{e\}) = \mathbb{R}$. This may still appear rather far-fetched but $HBrd_nB\Pi$ can be (essentially) recaptured in more algebraic terms of the (Wall) Witt group of a group ring of Π. This is defined for an arbitrary ring R with an involution denoted $r \mapsto \bar{r}$, where the relevant rings in topology are the following: the integers $\mathbb{Z}$, the ring $\mathbb{Z}[\frac{1}{2}]$ consisting of the fractions $n/2^k$, the rationals $\mathbb{Q}$, the reals $\mathbb{R}$, and finally all of $\mathbb{C}$. The involution is trivial for the first four of them which are subrings of $\mathbb{R}$ and it is the ordinary complex conjugation on $\mathbb{C}$.

The group ring $R(\Pi)$ consists, by definition, of the finite linear combinations $\Sigma_i r_i \pi_i$ (or, equivalently of functions $\Pi \to R$ with finite supports) with the obvious rules of addition and multiplication (which is called convolution on functions $\Pi \to R$). Besides, we have an involution on $R(\Pi)$ given by $\Sigma r_i \pi_i \mapsto \Sigma \bar{r}_i \pi_i^{-1}$ and denoted $s \mapsto s^*$. If $R \subset \mathbb{C}$ and we think of an $s = \Sigma r_i \pi_i$ as an operator acting on the complex Hilbert space $\ell^2(\Pi)$ of square summable functions $\Pi \to \mathbb{C}$ by convolution (group ring product) $\sigma \mapsto s\sigma$ for all $\sigma \in \ell^2(\Pi)$ (which is well defined being a finite linear combination of the π_i-translations on $\ell^2(\Pi)$, namely $\Sigma_i r_i \pi_i(\sigma)$, for $\pi_i(\sigma(\pi)) = \sigma(\pi_i^{-1}\pi))$, then s^* is the adjoint operator to s. The (Wall) Witt group $\text{Witt}_{2k}R$ is generated by the equivalence classes of non-singular bilinear $(-1)^k$-symmetric forms of finite rank over $R(\Pi)$. These are given by invertible square matrices $A = (a_{ij}), a_{ij} \in R(\Pi)$, with $A^* = (-1)^k A$ where A^* is defined as (a^*_{ji}). Two forms represented by matrices A_1 and A_2 of the same size are (called) *equivalent* if $A_1 = B^* A_2 B$ for an invertible B. The relations of the Witt group are two,

$$[A_1 \oplus A_2] = [A_1] + [A_2],$$

where we identify forms with matrices and denote by $\oplus$ the direct sum, and

$$\begin{bmatrix} 0 & 1 \\ (-1)^k & 0 \end{bmatrix} = 0,$$

which agrees with the usual Witt relation $[xy] = 0$ for the Witt group of quadratic forms over a field. (Wall also defined Witt_{odd} but we shall not go into this in our paper).

Examples. (a) If Π is trivial and $R(\Pi) = R$, then for k even

$$\mathrm{Witt}_{2k}R = \mathrm{Witt}R$$

if the involution on R is trivial. If $R = \mathbb{C}$ with complex conjugation, then the Witt group $\mathrm{Witt}_{2k}\mathbb{C}$ for k even is built of non-singular *Hermitian forms* A which, as real quadratic forms, are characterized in Witt by the signature. In fact, the inclusion $\mathbb{R} \subset \mathbb{C}$ (obviously induces an isomorphism $\mathrm{Witt}_{2k}\mathbb{R} \simeq \mathrm{Witt}_{2k}\mathbb{C} = \mathbb{Z}$ for k even where Witt_{2k} is isomorphically brought to $\mathbb{Z}$ by the signature, $[A] \mapsto \sigma(A) \in \mathbb{Z}$, and the same is true mod torsion for the inclusions $\mathbb{Z} \subset \mathbb{Z}(\frac{1}{2}) \subset \mathbb{Q} \subset \mathbb{R}$ but this is less obvious (see [Mi-Hu]). One likes $\mathbb{Z}(\frac{1}{2})$ because 2 is invertible in this ring and so there is no difference between quadratic and bilinear symmetric forms. If k is odd then $\mathrm{Witt}_{2k}\mathbb{R} = 0$ as all non-singular skew symmetric forms over $\mathbb{R}$ are equivalent to sums $\overset{m}{\underset{i}{\oplus}} x_i \wedge y_i$ (and the same is true mod torsion for the above subrings $\mathbb{Z}, \mathbb{Z}[\frac{1}{2}]$ and $\mathbb{Q}$ of $\mathbb{R}$). On the other hand, $\mathrm{Witt}_{2k}\mathbb{C}$ for odd k is isomorphic to that for k even by the correspondence $A(x,y) \to A(x, \sqrt{-1}y)$ turning skew-Hermitian forms into Hermitian ones. Notice that $\mathbb{C}$ with the trivial involution obviously has $\mathrm{Witt}_{2(\mathrm{even})} = \mathbb{Z}_2$ and $\mathrm{Witt}_{2(\mathrm{odd})} = 0$.

(b) Let $R = R_m$ be the (non-commutative) ring of complex matrices of order m with the Hermitian involution. Then $\mathrm{Witt}_{\mathrm{even}}R = \mathbb{Z}$, where the isomorphism is established by the signature. Namely every matrix $A = \{a_{ij}\}$ of order n with entries $a_{ij} \in R_m$ defines a (block) matrix, say $\widetilde{A}$, of order mn with complex entries and $\sigma(A) \underset{\mathrm{def}}{=} \sigma(\widetilde{A})$.

(c) Let R be the ring $\mathrm{Cont}(X)$ of continuous complex functions on a compact space X. Then a nonsingular Hermitian form of rank m over R amounts to a fiberwise non-singular Hermitian form A on the trivial bundle $\mathrm{Triv}^m = X \times \mathbb{C}^m \to X$. This bundle can be (homotopically uniquely) split into $T_+ \oplus T_-$ where A is positive on T_+ and negative on T_- and one defines the signature of A with the values in $K^0(X)$ by $\sigma(A) = [T_+] - [T_-]$. This $\sigma(A)$ is divisible by 2 in $K^0(X)$ as $[T_+] + [T_-] = 0 = [\mathrm{Triv}^m]$ (which would not happen if we had allowed non-trivial bundles to start with, i.e. forms on *projective* rather than free moduli over R. Conversely, for every vector bundle T over X one has the Hermitian form $\mathbb{1} \oplus -\mathbb{1}$ on the trivial bundle $T \oplus T^\perp$ which (easily) implies that $\mathrm{Witt}_{2(\mathrm{even})}R = 2K^0(X)$, and since $R \ni \sqrt{-1}$ we see as above (for $R = \mathbb{C}$) that $\mathrm{Witt}_{2(\mathrm{odd})}R = \mathrm{Witt}_{2(\mathrm{even})}R$.

Recall that for nice spaces (manifolds, cell complexes etc) $K^0(X)/\mathrm{torsion} \approx H^{\mathrm{even}}(X)/$ torsion, or better to say, $K^0(X) \otimes \mathbb{Q} = H^{\mathrm{even}}(X; \mathbb{Q})$, where the passage from the K-theory to the cohomology is given by the Chern character $[T] \mapsto \mathrm{ch}T$ (see $5\frac{3}{8}$) which is indeed an isomorphism over $\mathbb{Q}$ by the Serre finiteness theorem. Thus

$$(\mathrm{Witt}_{\mathrm{even}} R) \otimes \mathbb{Q} \underset{\mathrm{ch} \circ \sigma}{\simeq} H^{\mathrm{even}}(X; \mathbb{Q}).$$

(d) Let $\Pi = \mathbb{Z}^n$ and observe that the group ring $\mathbb{C}(\mathbb{Z}^n)$ (with our involution) is canonically isomorphic to the ring of complex valued functions (with the complex conjugation for *) on the torus $\mathbb{T}^n$ which are polynomials in the coordinates $z_i : \mathbb{T}^n \to S^1 \subset \mathbb{C}$ and $\overline{z}_i = z_i^{-1}$. For example, if $n = 1$ and $\mathbb{T}^1 = S^1 \subset \mathbb{C}$, then $s = \Sigma_i c_i i \in \mathbb{C}(\mathbb{Z})$ corresponds to the (Laurent) polynomial $p = \Sigma_i c_i z^i$ and $s^* \leftrightarrow \overline{p}$. One can see here an advantage of $\mathbb{C}$ over $\mathbb{R}$; the ring $\mathbb{R}(\mathbb{Z}^n)$ is harder to express in terms of functions on $\mathbb{T}^n$. Thus every Hermitian form A of rank m over $\mathbb{C}(\mathbb{Z}^n)$ defines a Hermitian form on the trivial bundle $\mathrm{Triv}^m \to \mathbb{T}^n$ and thus an element of $K^0(\mathbb{T}^n)$ denoted $\sigma(A) \in K^0(\mathbb{T}^n)$. Since Laurent polynomials are dense in the ring of complex valued continuous functions $\mathrm{Cont}(\mathbb{T}^n)$ and so every form on Triv^m can be perturbed to one with coefficients in $\mathbb{C}(\mathbb{Z}^n) \subset \mathrm{Cont}(\mathbb{T}^n)$; one might conclude that this inclusion induces an isomorphism on Witt_*. But this reasoning is faulty as a polynomial approximation to an invertible continuous function may be Laurent non-invertible. Yet (amazingly?) the conclusion is valid and the inclusion $\mathbb{C}(\mathbb{Z}^n) \subset \mathrm{Cont}(\mathbb{T}^n)$ does induce an isomorphism

$$\mathrm{Witt}_{2k}\mathbb{C}(\mathbb{Z}^n) \underset{\mathbb{Q}}{=} \mathrm{Witt}_{2k}\mathrm{Cont}(\mathbb{T}^n) = 2K^0(\mathbb{T}^n) \underset{\mathbb{Q}}{=} H^{\mathrm{even}}(\mathbb{T}^n).$$

(This is worth ≈ 30 % of the Novikov conjecture for $\Pi = \mathbb{Z}^n$ which claims here a specific geometrically defined homomorphism $WM\alpha : H_{\mathrm{even}}(\mathbb{T}^n) \to \mathrm{Witt}_{\mathrm{even}}\mathbb{C}(\mathbb{Z}^n)$ to be injective, compare below and $8\frac{1}{2}$). It is clear *now* that the ring $\mathbb{C}(\mathbb{Z}^n)$ has quite large Witt group (even if we complete this ring by the norm induced from the sup-norm on functions on $\mathbb{T}^n$), as large as $H^*(\mathbb{T}^n)$. To appreciate the hidden power of the above seemingly trivial formal discussion we suggest the reader would prove that $\mathrm{Witt}_2\mathbb{C}(\mathbb{Z}^2) \neq 0$ without resorting to the topology of $\mathbb{T}^2$ but by honestly exhibiting a skew-Hermitian form A over $\mathbb{C}(\mathbb{Z}^2)$ (see $7\frac{8}{9}$ for such an example) non-equivalence of A to zero in Witt_2 perceived by a direct algebraic reasoning.

From $HBrd_n$ to $Witt_n$. There is a natural (Wall-Miščenko) homomorphism $WM_{\mathbb{C}}$ from $HBrd_n B\Pi$ to $\mathrm{Witt}_n\mathbb{C}(\Pi)$ for all groups Π defined, roughly, as follows. Take a manifold V of dimension n (representing an element in $HBrd_n$) with some triangulation and observe that the chain complex of the Π-covering $\widetilde{V} \to V$ is a free $\mathbb{Z}(\Pi)$-module where one uses lifts of simplices from V to $\widetilde{V}$ for a basis (of cardinality equal the number of simplices in V). If c_1 and c_2 are two simplicial chains in $\widetilde{V}$ of complementary dimensions, one may define (sometimes ambiguously) their intersection index $c_1 \frown c_2 \in \mathbb{Z}$ which then gives us a (partially defined) pairing with values in $\mathbb{Z}(\Pi)$, i.e. in functions $\Pi \to \mathbb{Z}$ by $\pi \mapsto (\pi c_1) \frown c_2$.

A more careful look reveals that neither the ambiguity (localized at the boundaries of chains) nor degeneracy (tempered by the Poincaré duality on the chain level) of this pairing matters as one passes to Witt_n (see $8\frac{8}{9}$; we only speak of n even but this formalism can be actually used to define $\mathrm{Witt}_{\mathrm{odd}}$). Furthermore, surgeries of V essentially amount to adding direct hyperbolic summands $\begin{pmatrix} 0 & 1 \\ (-1)^k & 0 \end{pmatrix}$ (for $n = 2k$) and homotopy equivalences correspond to equivalences of forms. Thus we obtain a (natural homomorphism WM_R : $HBrd_n B\Pi \to \mathrm{Witt}_n R(\Pi)$ for $R = \mathbb{Z}[\frac{1}{2}]$ and hence for R equal $\mathbb{Q}, \mathbb{R}, \mathbb{C}$ as they contain $\mathbb{Z}[\frac{1}{2}]$.

There are certain additional points to settle if one works over $\mathbb{Z}$ where 2 is non-invertible which lie beyond the scope of the present paper and its author.

Now we return to (the Novikov conjecture on the homotopy invariance of) the ρ-signature σ_ρ for $\rho \in H^*(B\Pi)$ (see $7\frac{5}{6}$) which assigns, loosely speaking, to each V mapped to $B\Pi$ the signature of the pull-back of a suitable cycle in $B\Pi$ Poincaré dual to ρ. The relations (1), (2), (3) of the bordism group $Brd_n B\Pi$ are matched by the properties (1), (2), (3) of σ in $7\frac{1}{4}$ while the relation (5) for $HBrd_n$ goes along with the Cartesian multiplicativity property (5) in $7\frac{1}{4}$. Thus σ_ρ defines a homomorphism, also called $\sigma_\rho : Brd_* B\Pi \to \mathbb{Z}$ for each $\rho \in H^*(B\Pi)$. The Novikov conjecture for ρ claims that σ_ρ survives the homotopy invariance condition (see (4) above) for $HBrd_n$, which amounts to the existence of a homomorphism $Nov_\rho : HBrd_* B\Pi \to \mathbb{Z}$ making the following diagram commutative

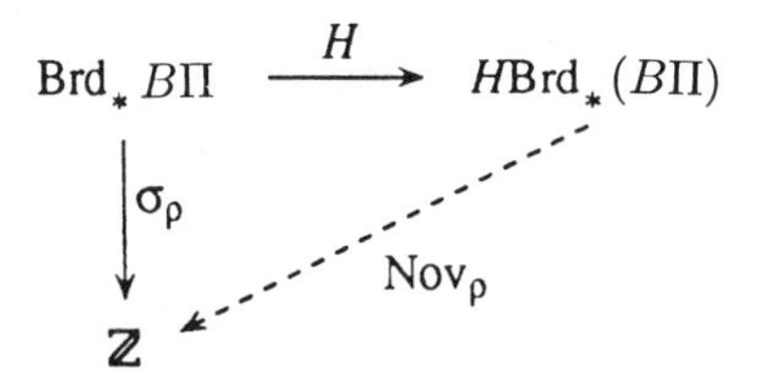

where H is the quotient map (for $HBrd_* = Brd_*/(4) + (5)$). In particular, it would suffice to construct homomorphism $\mathrm{Nov}_\rho^{\mathbb{Q}} : \mathrm{Witt}_*\mathbb{Q}(\Pi) \to \mathbb{Z}$ or even better $\mathrm{Nov}_\rho^{\mathbb{Q}} : \mathrm{Witt}\mathbb{C}(\Pi) \to \mathbb{Z}$ commutatively completing the diagram,
In fact, the analytic approach to the Novikov conjecture (see $8\frac{2}{3}$) delivers such a homomorphism from Witt of even a bigger ring, namely the *C^*-algebra* $C^*(\Pi)$, the completion of $\mathbb{C}\Pi$ in the operator topology for the natural embedding of $\mathbb{C}(\Pi)$ into bounded operators on $\ell^2(\Pi)$ (where, recall, $s \in \mathbb{C}(\Pi)$ acts on $\ell^2(\Pi)$ by the convolution, and observe that $C^*(\mathbb{Z}^n) = \mathrm{Cont}(\mathbb{T}^n)$).

Now write $\sigma_\rho(b) = \sigma_b(\rho)$, thus relating to each $b \in Brd_* B\Pi$ the homo-

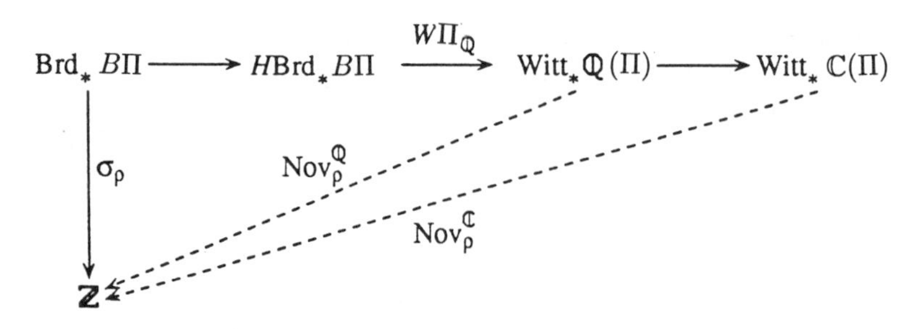

morphism $\sigma_b : H^*(B\Pi) \to \mathbb{Z}$. We tensor everything with $\mathbb{Q}$ and denote by $\sigma_b^{\mathbb{Q}} \in H_*(B\Pi;\mathbb{Q})$ the class corresponding to σ_b. The resulting homomorphism $h : (Brd_* B\Pi) \otimes \mathbb{Q} \to H_*(B\Pi;\mathbb{Q})$ for $h : b \mapsto \sigma_b^{\mathbb{Q}}$ can be described as follows. Recall that the bordism group of $B\Pi$ (as well as of any other space) tensored with $\mathbb{Q}$ equals the tensor product of $H_*(B\Pi;\mathbb{Q})$ with $Brd_*\{\text{point}\}$. In fact, a multiple of each $b \in Brd_* B\Pi$ can be represented by an integer combination of (bordism classes of) maps $\beta_\mu : V_\mu \times W_\mu \to B(\Pi)$ constant in $w \in W_\mu$ where V_μ are stably parallelizable (and hence cobordant to zero for $\dim V_\mu > 0$), W_μ are Cartesian products of complex projective spaces $\mathbb{C}P^{2k_\mu}$ and $\beta_\mu^*[V_\mu]$ form a basis in $H_*(B\Pi;\mathbb{Q})$.

The above homomorphism $h : Brd_* \to H_*$ assigns to such a β_μ the class $\sigma(W_\mu)(\beta_\mu)_*[V_\mu]$ and to combinations of β_μ's the corresponding combinations of these. This agrees with the homomorphism $H : Brd_* \to HBrd_*$ which sends $[V \times W, \beta]$ to $\sigma(W)H[V,\beta]$ (according to the stabilization axiom (5)) in the definition of $HBrd_*$). The relation $Brd_* B\Pi \underset{\mathbb{Q}}{=} H_*(B\Pi) \otimes Brd_*\{\text{point}\}$ (trivially) implies that $h : Brd_* B\Pi \to H_*(B\Pi)$ is surjective when tensored with $\mathbb{Q}$. In fact, a multiple of every homology class in $B\Pi$ (as well as in any space) is representable by $\beta^*[V]$ for a suitable (stably parallelizable) V and $\beta : V \to B\Pi$ as follows again from the Serre finiteness theorem. With this we obtain the homomorphism

$$\alpha : H_*(B\Pi;\mathbb{Q}) \to (HBrd_* B\Pi) \otimes \mathbb{Q})$$

as $H : Brd_* \to HBrd_*$ vanishes on the kernel of $h : Brd_* \to H_*$ by the above discussion. Here is the full diagram,
where ev_ρ for $\rho \in H^*(B\Pi)$ is the usual evaluation (pairing of cohomology on homology).

Conclusion. If α is injective then $\mathrm{Nov}_\rho \otimes \mathbb{Q}$ exists for all ρ which implies the Novikov conjecture for all ρ. Indeed $\mathrm{Nov}_r \otimes \mathbb{Q}$ may be obtained by just linearly extending ev_ρ from $H_*(B\Pi;\mathbb{Q})$ to $HBrd_* \otimes \mathbb{Q} \supset H_*(B\Pi;\mathbb{Q})$. In fact, a little extra thought shows that homotopy invariance of all σ_ρ is equivalent to the injectivity of our $\alpha : H_*(B\Pi;\mathbb{Q}) \to (HBrd_* B\Pi) \otimes \mathbb{Q}$. Furthermore, one may pass to the Witt groups and observe that the Novikov conjecture would follow from the injectivity of each of the homomorphisms obtained by composing

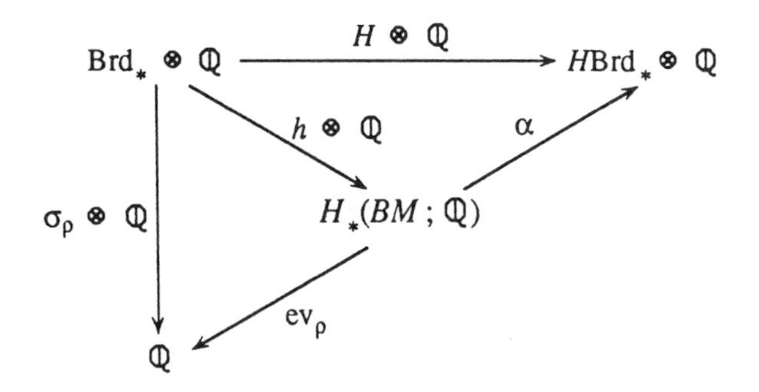

α with the (Wall-Miščenko) homomorphism WM from ($HBrd_*$ to the Witt groups of Π over $\mathbb{Z}[\frac{1}{2}], \mathbb{Q}, \mathbb{R}$ and $\mathbb{C}$. In fact, the Novikov conjecture is known to be equivalent to the injectivity of $WM_{\mathbb{Q}} \circ \alpha : H_*(B\Pi;\mathbb{Q}) \to (\mathrm{Witt}\,\mathbb{Q}(\Pi)) \otimes \mathbb{Q}$. Notice that both H_* and Witt_* are associated to Π by purely algebraic constructions while the homomorphism $WM_{\mathbb{Q}} \circ \alpha$ goes via cobordisms (and uses the Serre finiteness theorem at some stage). Yet the qualitative

$$\mathrm{rank}\,\mathrm{Witt}\,\mathbb{Q}(\Pi) \geq \mathrm{rank}\,H_*(B\Pi) \qquad (*)$$

is stated in purely algebraic terms and so one dreams of an algebraic proof of this for many groups Π. But the known proofs of the Novikov conjecture in the majority of cases use analysis (sometimes topology) and no direct algebraic approach to $(*)$ is available except for rather special groups Π.

According to Alain Connes' philosophy the difficulty of identifying Witt_* with H_* is due to the fact that the habitats of these groups are different. Witt_* is naturally defined on the operator norm completion $C^*(\Pi)$ of $\mathbb{C}(\Pi)$ where Witt_* identifies with K_* (as for rings of continuous functions) while the homology, or rather cohomology $H^*(\Pi)$, defined via cyclic cocycles, survives only much smaller extensions of $\mathbb{C}(\Pi)$. In some cases, e.g. for hyperbolic groups, the gap can be filled in but it remains wide open in general, compare [Co-Mo].

$7\frac{8}{9}$. Remarks and references concerning Serre, Witt and topological Pontryagin classes

(a) The business of topology is finding certain quantities, preferably numbers, attached to geometric objects which are smooth, homeomorphic, or best of all, homotopy invariants. These, when found, should be evaluated in specific cases to make sure they are non-zero for sufficiently many examples. The basic instance of this is the index of intersection between two cycles c_1 and c_2 of complementary dimensions, say i and $n-i$, in a manifold V. This is a homological invariant of the cycles and a (proper) homotopy invariant of V; if this index $\neq 0$ for some c_1 and c_2 we conclude that the homology groups

$H_i(V)$ and $H_{n-i}(V)$ do not vanish, and according to the Poincaré duality, the intersection of cycles yields 100% control over vanishing/non-vanishing of the (rational) homology. Another (essentially equivalent) test for non-triviality of a cycle c is provided by closed differential forms ω via the implication $\omega(c) \overset{\text{def}}{=} \int_c \omega \neq 0 \overset{\text{Stokes}}{\Rightarrow} c$ is non-homologous to zero which manifests the duality between homology and cohomology in the *same* dimension.

One proceeds similarly in the (Novikov) problem of detecting non-zero elements c in $HBrd_* B\Pi$ or in some Witt_* by constructing computable linear functions (signatures σ on these groups where the non-vanishing of $\sigma(c)$ is verifiable and where the pertinent invariance mechanism (see §$8\frac{2}{3}$) is K-theoretic rather than homological). Namely, intersections of cycles (and integrals of forms) are replaced (quantized?) by indices of Fredholm operators in Hilbert spaces where the invariance of the indices under homotopics of operators plays the pivotal role (compare $[\mathrm{At}]_{\mathrm{GAE}}$).

Illustration. Consider a vector bundle X over a closed manifold V and try to show X is non-trivial. This can be done (co)homologically by taking a characteristic cohomology class of X and evaluating it on a cycle in V, e.g. by integrating a suitable (Chern-Weil) curvature form of X over V. But instead one may take the Dirac (or signature operator twisted with X, say D_X) and derive the desired non-triviality of X from the non-equality $\operatorname{ind} D_X \neq \operatorname{ind} D_{\mathrm{Triv}}$. (Of course, the index theorem reduces the actual computation of the indices to cohomology but this is not our concern at the moment).

(b) ***Serre theorem and applications.*** This theorem comes in many disguises and says, in effect, that the rational (i.e. numerical) homotopy invariants of many *simply connected* spaces are essentially the (co)homological ones and so there is nothing new and unexpected down there hidden from our eyes in the depth of homotopies. Here are specific formulations.

I. ***The stable homotopy groups of spheres of positive codimension are finite.***

In fact $\mathrm{cardHomot}(S^n \to S^N) < \infty$, unless $n = N$ or $n = 2N - 1$ for N even. Thus the only numerical homotopy invariants for maps $S^n \to S^N$ are *the degree* (for $n = N$) and *the Hopf invariant* (for n even and $n = 2N - 1$) which are both obtained by integrating form over cycles.

II. *For every finite complex V the homotopy classes of maps $f : V \to S^N$ are classified modulo torsion (i.e. $\otimes\mathbb{Q}$) by the cohomology $H^N(V)$ via the correspondence $[f]_{\mathrm{homot}} \leftrightarrow f^*[S^N]^{\mathrm{co}}$, provided N is odd or $2N > \dim V + 1$.*

This means that a non-zero multiple of each class $c \in H^N(V)$ is representable by $f^*[S^N]^{\mathrm{co}}$ for some f and if two maps f_1 and f_2 have equal pull-backs of $[S^N]^{\mathrm{co}}$ to $H^N(V;\mathbb{Q})$, then some non-zero multiples Mf_1 and Mf_2 are homotopic, where Mf refers to composing f with a self-mapping $S^N \to S^N$ of degree M.

If V is a smooth closed n-dimensional manifold, then $H^N(V)$ is isomorphic to $H_{n-N}(V)$ where *spherical* classes $c \in H^N(V)$, i.e. of the form $f^*[S^N]^{\mathrm{co}}$ corresponds to $(n-N)$-dimensional homology classes in V representable by *submanifolds* $W \subset V$ *with trivial normal bundles* which appear as pull-backs of regular values of smooth maps $f : V \to S^N$. Thus, for N odd, a multiple of every class in $H_{n-N}(V)$ is representable by such a manifold. No direct geometric proof of this has been ever found!

Multiplicativity of signature. Let $\widetilde{V}$ be a finite Galois G-covering of V and prove the identity $\sigma(\widetilde{V}) = (\operatorname{card} G)\sigma(V)$ by showing that $M\widetilde{V}$ is cobordant to $M(\operatorname{card} G)V$ where kV denotes the disjoint union of k copies of V. Let $X_0 \to V$ be a vector bundle associated to the principle fibration $\widetilde{V} \to V$ via a representation $G \to GL_k$ where the action of G is free at a generic unit vector $x \in \mathbb{R}^k$ (e.g. G acts on the space $\mathbb{R}^{\operatorname{card} G} = \operatorname{maps}(G \to \mathbb{R})$ in the usual way). Then the orbit $G(x)$ defines a G-valued section of X_0, i.e. an embedding $\widetilde{V} \to X_0$ intersecting each fiber across a G-orbit. We add a complementary bundle, say X_1 to X_0, so that $X = X_0 \oplus X_1$ is trivial and the above embedding lands in the unit sphere bundle of X which is $V \times S^N$ for some N (as large as we want) and where the normal bundle of $\widetilde{V}$ there is trivial. (Actually, $\widetilde{V}$ has already trivial normal bundle in X_0 but X is slightly more convenient). This $\widetilde{V} \subset V \times S^N$ can be represented as the regular pull-back of some map $f : V \times S^N \to S^N$ which is, obviously, cohomologous to the $(\operatorname{card} G)$-multiple of the projection $f_0 : V \times S^N \to S^N$, and by Serre $M[f]_{\mathrm{homot}} = M(\operatorname{card} G)[f_0]_{\mathrm{homot}}$. The pull-back of a regular point of the implied (smooth) homotopy $V \times S^N \times [0,1] \to S^N$ provides the required cobordism realized by a submanifold in $V \times S^N \times [0,1]$ with a trivial normal bundle.

Conclude by observing that the range of Serre's theorem includes, besides spheres, all compact homogeneous spaces and among non-homogeneous ones such spaces as Kähler manifolds. But understand this needs Sullivan's theory of minimal (algebraic) models of rational homotopy types.

(c) ***Definition of*** Witt_* ***with algebraic Poincaré complexes.*** Let us indicate (following [Miš]) a unified definition bringing $HBrd_*B(\Pi)$ and $\mathrm{Witt}_* R(\Pi)$ to a common ground. Recall that $\mathrm{Witt}_{\mathrm{even}} \Rightarrow$ for an arbitrary ring $\Rightarrow$ with involution (e.g. for $\mathcal{R} = R(\Pi)$ or for the ring of continuous functions over some space) is built out of non singular (skew) Hermitian forms, or equivalently, isomorphisms $A : M \to M^*$, where M is a free module of finite rank over $\mathcal{R}$ and M^* is the (Hermitian) dual (also free) module. Now, we generalize by replacing an individual M by a complex of free moduli, of formal dimension n,

$$(M, \partial) = 0 \to C_n \overset{\partial_n}{\to} C_{n-1} \overset{\partial_{n-1}}{\to} \dots \overset{\partial_1}{\to} C_0 \to 0$$

where the basic examples are the chain complexes of Π-coverings $\widetilde{V}$ of triangulated n-dimensional manifolds V, and consider the Hermitian dual complex (of cochains)

$$(M^*, \delta = \partial^*) = 0 \to C^0 \overset{\partial_0}{\to} C^1 \overset{\partial_1}{\to} \dots \overset{\partial_{n-1}}{\to} C^n \to 0$$

(with a suitable Hermitian sign adjustment in the definition of δ). We work, instead of isomorphisms, with chain homotopy equivalences $A : M \to M^*$ where $A_i : C_i \to C^{n-i}$ (which embody the Poincaré duality for $\widetilde{V}$). We use $A_1 \oplus A_2$ for addition and the equivalence relation is made by emulating cobordism in the algebraic language. Namely, we carefully record the algebraic effect of an individual surgery of V (mapped to $B\Pi$) as adding and/or eliminating certain generators in M and declare M_1 and M_2 equivalent (algebraically cobordant) if they can be joined by a chain of such algebraic surgeries. More conceptually, we define algebraic Poincaré complexes with boundaries, thus introducing the algebraic counterpart of cobordisms. If $n = 2k$, one can kill all C_i and C^i for $i \neq k$ by algebraic surgery thus arriving at an isomorphism $A' : C_k \to C^k = C_k^*$ equivalent to the original A and equating the new $\mathrm{Witt}_{\mathrm{even}}$ built out of Poincaré complexes with the old $\mathrm{Witt}_{\mathrm{even}}$ made of Hermitian forms. (Similar simplification is possible for n odd where the algebraic surgeries bring the Miščenko definition down to the original one of Wall).

The algebraic cobordism relation is stronger than the geometric one as it includes homotopy equivalences and so the group $HBrd_* B\Pi$ happily maps into Witt_*. (See [Miš], [Kas], [Ran]$_{\mathrm{ALT}}$, [Ran]$_{\mathrm{LKLT}}$ and [Ran]$_{\mathrm{NC}}$ for details and further references).

Example. Let $\Pi = \mathbb{Z} \oplus \mathbb{Z}$ where $\mathbb{Q}(\Pi)$ equals the Laurent polynomial ring in the variables $t_i^{\pm 1}$, $i = 1, 2$. Then the (symplectic) form over $\mathbb{Q}(\Pi)$ corresponding to the 2-torus, (i.e. the Poincaré complex of this torus) can be given by the following invertible matrix A

$$A = \begin{pmatrix} ((t_2)^{-1} - t_2)/2 & (1 + (t_1)^{-1} - t_2 + (t_1)^{-1} t_2)/2 \\ (-1 - t_1 + (t_2)^{-1} - t_1 (t_2)^{-1})/2 & ((t_1)^{-1} - t_1)/2 \end{pmatrix}$$

kindly communicated to me by Andrew Ranicki. It is not at all obvious that the class of A does not vanish in $\mathrm{Witt}_2\, \mathbb{Q}(\Pi)$; but it is known to be non-zero even in $\mathbb{C}(\Pi) \supset \mathbb{Q}(\Pi)$ and in the C^*-algebra $C^*(\Pi) \supset \mathbb{C}(\Pi)$ as follows, for example, from Lusztig's theorem (see $8\frac{5}{8}$).

(d) ***Historical reminiscences.*** Everything presented in §7 belongs to history, 20 years back and more. The key idea of using the signatures of submanifolds for the invariance proofs of Pontryagin classes is due, independently, to Rochlin and Thom. Rochlin noticed in 1957 that the invariance of the signature under *topological* cobordisms (by Poincaré duality) implies the topological invariance of $L_{4k}(V^{4k+1})$. Thom and Rochlin-Švazc independently observed in 1957-1958 that the pull-backs under piecewise linear maps of generic points are manifolds in the combinatorial category and so the signatures of these pull-backs are combinatorial (co)bordism invariants. This allowed an extension of the Pontryagin classes to p.l. manifolds and, in particular, proved invariance of Pontryagin classes under p.l. homeomorphisms of smooth manifolds. (For several years afterwards Rochlin had been trying to prove the topological invariance of all L_i (and thus of Pontryagin classes) but was continuously sliding into the (proper) homotopy category where Serre's finiteness theorem predicted the lack of necessary structure for such a proof as surely as the laws of thermodynamics rule out the perpetual motion machine).

In 1965–66 Novikov realized that *non-simply connected* open subsets harbour sufficient homotopy information for the *topological* invariance and put forward his homotopy invariance conjecture for general groups Π. The analytic approach starts with the innocuous-looking 1969 paper by Gelfand and Miščenko (see [Ge-Mi]), where they compute $\mathrm{Witt}_* C^*(\mathbb{Z}^n)$ via $K^*(\mathbb{T}^n)$ (see (d) in $7\frac{7}{8}$), followed by Lusztig's 1972 artillery shell charged with the index theorem. We still live through the explosion in the atmosphere saturated with C^*-algebras, Fredholm representations, spectral flows, etc.

8. Signatures for flat and almost flat bundles and C^*-algebras

We approach the Novikov conjecture by systematically searching for homomorphisms $H\mathrm{Brd}_* B\Pi \to \mathbb{Z}$ which, by the very definition of $H\mathrm{Brd}_*$, are homotopy (as well as bordism) invariants of closed oriented manifolds V (mapped to $B\Pi$) representing the group $H\mathrm{Brd}_*$. As $H\mathrm{Brd}_*$ naturally goes to $\mathrm{Witt}_* \mathbb{C}(\Pi)$ we shall be quite content to have these homomorphisms extended to $\mathrm{Witt}_* \mathbb{C}(\Gamma) \to \mathbb{Z}$.

Non-example. Take a unitary representation $\rho : \Pi \to U(p)$ and extend it by linearity to an involutive homomorphism $\mathbb{C}(\Pi) \to \mathrm{Mat}_p \mathbb{C}$ where $\mathrm{Mat}_p \mathbb{C}$ is the ring of $p \times p$ matrices with the usual Hermitian involution. This induces a homomorphism $\mathrm{Witt}_{2k} \mathbb{C}(M) \to \mathrm{Witt}_{2k} \mathrm{Mat}_p \mathbb{C} = \mathbb{Z}$ (see $7\frac{7}{8}$) which on the level of $H\mathrm{Brd}_*$ can be described as follows. Let $(V, \beta : V \to B\Pi)$ represent some element in $H\mathrm{Brd}_* B\Pi$ and let $X_\rho \to B\Pi$ be the flat unitary bundle associated to ρ. We also denote by X_ρ the β-induced (flat unitary) bundle over V and we look at the cohomology of V with coefficients in X_ρ. If $\dim V = 2k$, then there is a $\mathbb{C}$-values pairing in the middle dimension on this cohomology, say $H^k(V; X_\rho) \oplus H^k(V; X_\rho) \to \mathbb{C}$ which is obtained by composing the following

1. The cup product

$$H^k(V;X_\rho)\otimes H^k(V;X_\rho)\to H^{2k}(V;X_\rho\otimes X_\rho)$$

(which is defined generally as $H^i(V;X_\rho)\otimes H^j(V;X_{\rho'})\to H^{i+j}(V;X_\rho\otimes X_{\rho'})$).

2. $H^{2k}(V;X_\rho\otimes X_\rho)\to H^{2k}(V;\mathbb{C})$ for the ($\mathbb{R}$-linear) map $X_\rho\otimes X_\rho\to\mathbb{C}$ given by the scalar product $(x\otimes x')\mapsto\langle x,x'\rangle$ in X_ρ.

3. Evaluation of $H^{2n}(V;\mathbb{C})$ on the fundamental class [V].

If we represent the cohomology by k-forms with coefficients in X_ρ, say by $\sum_i x_i\omega_i$ and $\sum_j x_j\omega_j$, then our pairing amounts to $\int_V\sum_{i,j}\langle x_i,x_j\rangle(\omega_i\wedge\omega_j)$). If k is even, this pairing is Hermitian and we may speak of its signature, denoted $\sigma_\rho(V)$ and in the odd case we pass from "skew-Hermitian" to "Hermitian with" the help of $\sqrt{-1}$ as earlier and define $\rho_\rho(V)$ just the same. This σ_ρ looks as good as the ordinary signature $\sigma(V)$ with the same charming properties (see (1)-(5) in $7\frac{1}{4}$) but ..., it just happens to be equal to $p\sigma(V)$ for all unitary representations ρ. (This follows from the index theorem and the vanishing of the Chern classes of flat unitary bundles. I wonder if there is a direct algebraic proof in the language of the homomorphism $\mathbb{C}(\Pi)\to\operatorname{Mat}_p\mathbb{C}$.

To help the problem, let us pass from the unitary group $U(p)$ to $U(p,q)$, the group of isometries of $\mathbb{C}^{p+q}$ endowed with the Hermitian (p,q)-form $\sum_{i=1}^p z_i\bar z_i-\sum_{j=p+1}^{p+q}z_j\bar z_j$. (The encouraging (p,q)-sign is a possible non-vanishing of $U(p,q)$ characteristic classes in agreement with the Chern-Weil theory). So we take a representation $\rho:\Pi\to U(p,q)$ which extends to an involutive homomorphism $\mathbb{C}(\Pi)\to\operatorname{Mat}_{pq}\mathbb{C}$, where $\operatorname{Mat}_{pq}\mathbb{C}$ is the ring of $(p+q)\times(p+q)$ matrices with the involution corresponding to our new (p,q)-Hermitian form. Namely A^* is defined, for all $A\in\operatorname{Mat}_{pq}\mathbb{C}$, by the rule $\langle x,A^*y\rangle_{pq}=\langle Ax,y\rangle_{pq}$, i.e. if we write A in (p,q)-blocks then the (p,q)-involution is expressed as usual by

$$\begin{pmatrix}A_{11}&A_{12}\\A_{21}&A_{22}\end{pmatrix}\overset{\mapsto}{{}_{*pq}}\begin{pmatrix}A_{11}^*&-A_{21}^*\\-A_{12}^*&A_{22}^*\end{pmatrix}.$$

One easily sees that $\operatorname{Witt}_{2k}M_{pq}=\mathbb{Z}$ and so each $\rho:\Pi\to U(p,q)$ defines a homomorphism $\sigma_\rho:\operatorname{Witt}_{2k}\mathbb{C}(\Pi)\to\mathbb{Z}$. If this σ_ρ is applied to a manifold V with a flat $U(p,q)$-bundle X_ρ induced by the implied map $\beta:V\to B\Pi$ from such an X_ρ over $B\Pi$, then the resulting $\sigma_\rho(V)=\sigma_\rho(V,\beta)$ can be easily identified with the signature $\sigma(V;X_\rho)$ of the cup-product pairing on $H^k(V;X_\rho)$. If k is even it is the true signature but for k odd the pairing is skew-Hermitian and, before taking the signature, "skew" must be compensated by $\sqrt{-1}$. This signature $\sigma(V;X_\rho)$ is as cute and pretty as our old $\sigma(V)$ (corresponding to the trivial representation) and it displays all the beautiful formal features (1)-(5) of σ indicated in $7\frac{1}{4}$. But, first of all, $\sigma(V;X_\rho)$, being a homological creature, is *homotopy invariant*, exactly like ordinary σ. And now come the ρ-counterparts of (1)-(5) where we start with (3_ρ) leaving out (1_ρ) and (2_ρ), which do not merit being written down more than once.

(3_ρ) ***Bordism invariance.*** If V equals the boundary of some compact W and X_ρ extends to a flat $U(p,q)$-bundle over $W \supset V$ (i.e. the implied homomorphism $\pi_1(V) \to U(p,q)$ extends to $\pi_1(W)$), then $\sigma(V;X_\rho) = 0$. In fact, this only relies on the Poincaré duality and so equally applies to p.l. and even to the topological category.

(4_ρ) ***Multiplicativity.*** If $\widetilde{V} \to V$ is a finite d-sheeted covering then

$$\sigma(\widetilde{V}, \widetilde{X}_p) = d\sigma(V;X_\rho),$$

for the $\widetilde{X}_\rho$ induced by this covering form $\widetilde{X}_\rho$. The above proof of multiplicativity of σ applies here as well.

(5_ρ) ***Cartesian multiplicativity.***

$$\sigma(V_1 \times V_2; X_{\rho_1} \otimes X_{\rho_2}) = \sigma(V_1;X_{\rho_1})\sigma(V_2;X_{\rho_2}).$$

(This is clear. In the important special case, where ρ_2 is trivial, this reduces to

$$\sigma(V \times W;X_\rho) = \sigma(W)\sigma(V;X_\rho).$$

(6_ρ) ***Additivity.*** Since the signature makes sense for singular forms, $\sigma(V;X_\rho)$ is defined for open manifolds V via the pairing on the cohomology with compact support.

The statement and the proof of additivity we leave to the reader.

(7_ρ) ***Codim 1-formula*** (compare $7\frac{4}{5}$). Let W be a closed hypersurface of dimension $2k$ in an open connected manifold V with X_ρ over it. Then the signature of the cup product pairing on $H^k(V;X_\rho)$ with evaluation on [W] equals $\sigma(W;X_\rho|W)$, provided V is divided by W into two halves as in Fig. 11. The proof is the same as in $7\frac{4}{5}$ by Poincaré duality.

Remark on real bundles. If X_ρ is a flat $O(p,q)$ bundle, i.e. with a quadratic (p,q)-form in the $\mathbb{R}^{p+q}$-fibers, then $\sigma(V;X_\rho)$ is defined whenever $\dim V = 4k$ and it is extended as zero for the dimensions not divisible by 4. And for $\dim V = 4k+2$ one may use flat sympletic bundles X_ρ corresponding to the representations ρ of Π into the group $\mathrm{Spl}\, 2p$, i.e. the automorphism group of $(\mathbb{R}^{2p}, \sum_{i=1}^{p} x_i \wedge y_i)$. In this case the cup pairing on $H^{2k+1}(V)$ with coefficients in X_ρ is symmetric and so the signature is defined. The above properties (1_ρ)-(7_ρ) obviously extend to the real case where one should remark that the tensor product of $(-1)^i$-symmetric and $(-1)^j$-symmetric forms is $(-1)^{i+j}$-symmetric (where symmetric $= (-1)^{2\ell}$-symmetric and skew symmetric $= (-1)^{2\ell+1}$-symmetric). In fact, one may reduce everyting to the complex case with the natural embeddings $O(p,q) \subset U(p,q)$ and $\mathrm{Spl}\,2p \subset U(2p,2p)$.

Examples. So far our discussion was void of actual content as we have not shown to the reader a single bundle X_ρ with $\sigma(V, X_\rho)$ not being a multiple of the ordinary signature. But these X_ρ do exist as was pointed out by Lusztig and Meyer (who brought in these conceptions). Namely, let Π be a torsion free discrete subgroup of a semisimple group G with no compact factor group and $B = \Pi\backslash G/\mathrm{maxcomp}$ be the locally symmetric space with $\pi_1(B) = \Pi$. Notice that $B = B\Pi$ as the universal covering of V, i.e. $G/\mathrm{maxcomp}$, has non-positive sectional curvature. Then each $U(p,q)$ representation of G gives us a representation ρ of $\Pi \subset G$ and thus a flat bundle X_ρ over B (and over each V mapped to B) among which one finds non-zero ρ-signatures, especially if $\Pi \subset G$ is an arithmetic (e.g. cocompact) subgroup.

A specific example is that of $G = \mathrm{Spl}\,2p$ and $\Pi = (\mathrm{Spl}\,2p)\,\mathrm{cup}\,GL_{2p}\mathbb{Z}$ (or rather a subgroup of finite index there without torsion) where many Riemann surfaces (as well as higher dimensional subvarieties) $V \subset B = \Pi\backslash\mathrm{Spl}\,2p/U(p)$ have non-zero σ_ρ-signatures for the Spl-bundle corresponding to the original representation $\rho : \Pi \hookrightarrow \mathrm{Spl}\,2p$. To get a perspective one should keep in mind that arithmetic groups are of exceptionally symmetric nature not dreamed of in the realm of general infinite groups. It is also likely, that apart from several exceptional cases such as π_1 (Riemann surface) the representations $\rho : \Pi \subset U(p,q)$ with sufficiently rich σ_ρ should be of an arithmetic nature.

Now we focus on a single flat Spl-bundle X-over a closed surface B with $\sigma(B, X) = s \neq 0$ (see $8\frac{2}{7}$) and derive from this.

$8\frac{1}{4}$. Quick proof of the topological invariance of Pontryagin clases

Our basic tools, besides the above flat symplectic bundle $X \to B$ with non-zero signature s of the *quadratic* form on $H^1(V; X)$ (dual to the intersection form on $H_1(V; X)$), will be the Rochlin-Thom expression for the value of the L_k-class (and thus of Pontryagin classes) at a homology class $h \in H_{4k}(V)$ by the signature of a $4k$-submanifold $W \subset V$ realizing h with the trivial normal bundle and Novikov's idea of using (non-tubular) neighbourhoods $U \subset U_{\mathrm{Tub}} = W \times \mathbb{R}^{n-4k} \subset V$, $n = \dim V$, of the form $U = W \times B^* \times \mathbb{R}$ for suitable closed hypersurfaces $B^* \subset \mathbb{R}^{n-4k}$ with $U_{\mathrm{Tub}} B^* = B^* \times \mathbb{R}$, where our B^* will be $B \times B \times B \times \cdots \times B$ instead of Novikov's T^ℓ.

Recall that a non-zero multiple Mh of every $h \in H_{4k}(V)$ for $\dim V - 4k$ *odd* can be represented by the fundamental class $[W]$ of some W with trivial normal bundle by the Serre finiteness theorem.

We first do the case $\dim V - 4k = 3$ as follows.

Basic Lemma. *Let H be a closed oriented $(4k+2)$-dimensional manifold* *and $\beta : H \to B$ a smooth map. Then*

$$\sigma(\beta^{-1}(b)) = s^{-1}\sigma(h; \beta^*(X)), \tag{$*$}$$

where $b \in B$ is a regular value of β and $\beta^*(X)$ denotes the pull-back of X to H.

Proof. Both signatures, on the left- and right-hand sides of $(*)$ are bordism invariants of (H, β), and since a non-zero multiple of every bordism class is a combination $\sum_i c_i \times W_i$, where c_i are cycles in B, i.e. points, circles, or copies of B, and where the implied maps β_i are projections $c_i \times W_i \to c_i \subset B$, one needs only to check $(*)$ for $H = c \times W$. If $c = B$, then $(*)$ follows for the Cartesian multiplicativity for $H = B \times W$ (see (5_ρ) above); otherwise, both signatures are zero. In fact, $\sigma(\beta^{-1}(b)) = 0$ since the $\beta^{-1}(b)$ is empty for (generic) $b \in B - c$ and $\sigma(H; \beta^*(X)) = \sigma(W)\sigma(c; X|c) = 0$ for the dimension reason if $\dim c = 0$ or 1 ($\sigma(W) = 0$ for $\dim W \neq 4k$ and $\sigma(c; X|c) = 0$ for $\dim c \neq 4k + 2$; so *both* factors vanish which is more than enough).

Corollary (A). *Let (H_1, β_1) and (H_2, β_2) be topologically bordant, i.e. there exists a compact topological manifold U with $\partial\overline{U} = W_1 - W_2$ such that β_1 and β_2 extend from $\partial\overline{U}$ to a continuous map $\overline{\beta} : \overline{U} \to B$ (where the minus sign refers to the reversed orientation). Then*

$$\sigma(\beta_1^{-1}(b)) = \sigma(\beta_2^{-1}(b)). \tag{$+$}$$

Notice that (H_i, β_i), $i = 1, 2$, are assumed smooth and so the regular pullbacks $\beta_i^{-1}(b)$ are manifolds. Also notice that $(+)$ does not directly involve X but this appears in the proof.

Proof. The cobordism invariance (see (3_ρ) of $\sigma(H; \beta^*(X))$ only uses the Poincaré duality and so allows topological manifolds $\overline{U}$. Hence $\sigma(\beta^{-1}(b)) = \sigma(H; \beta^*(X))$ is also a topological bordism invariant. Q.E.D.

Alternative corollary (B). *Take $H_1 = B \times W$, let U be properly homotopy equivalent to $H_1 \times \mathbb{R}$ and H_2 be a hypersurface in U separating the two ends of U (as $H_1 \times 0$ in $H_1 \times \mathbb{R}$). Then a smooth map $\beta_2 : H_2 \to B$, homotopic to the composition of the following three, the inclusion $H_2 \to U$, the homotopy equivalence $U \to H_1 \times \mathbb{R}$ and the projection $H_1 \times \mathbb{R} = B \times W \times \mathbb{R} \to B$ has $\sigma\beta_2^{-1}(b) = \sigma(W)$.*

Proof. Combine $(*)$ and (7_ρ).

Proof of the topological invariance of $\mathbf{L_k}$ for $\mathbf{\dim V - 4k = 3}$. We take $W \subset V$ with trivial normal bundle and a neighbourhood $U \subset V$ of W of the form $U = W \times B \times \mathbb{R} \subset U_{\mathrm{Tub}}W = w \times \mathbb{R}^3$, where $B \times \mathbb{R} \subset \mathbb{R}^3$ appears as the tubular neighbourhood of the surface B embedded to $\mathbb{R}^3$. Now, we change the smooth structure in V and thus in U, take a smooth hypersurface H_2 separating the ends of U for the new smooth structure and map $H_2 \to B$ by some smooth $\beta_2 : H_2 \to B$ in the homotopy class corresponding to $H_2 \hookrightarrow U = W \times B \times \mathbb{R} \to B$. What we have to show is the equality $\sigma(\beta_2^{-1}(b)) = \sigma(W)$ and this follows either from the above (A) or (B). Namely, H_2 does not interset $H_1 = W \times B \times r \in U = W \times B \times \mathbb{R}$ for some (say, sufficiently large) $r \in \mathbb{R}$, so H_1 and H_2 bound together $\overline{U} \subset U$ and A applies. Notice that $\overline{U}$ can not be made smoth to accommodate both H_1 and H_2 on its boundary. Alternatively, one may apply (B) in a similar obvious way. (Notice that (A)-argument mimics the Rochlin 1957 proof of the topological invariance of L_i for $\dim V - 4_i = 1$ and (B) imitates Novikov's codim 1-argument of 1965; see $7\frac{4}{5}$).

The proof for $\dim V - 4k = 2\ell + 1$. We use now $B^\ell = \underbrace{B \times B \times \cdots \times B}_{\ell}$ with $X^\ell = \underbrace{X \otimes X \otimes \cdots \otimes X}_{\ell}$ over it.

Basic lemma$_\ell$. *Let* $\dim H = 4k + 2\ell$ *and* $\beta : H \to B^\ell$ *be a smooth map with the following property concerning the projections* $p : B^\ell \to B^{\ell'}$, $\ell' < \ell$. *We require* $\sigma(H; (p \circ \beta)^* X^{\ell'}) = 0$ *for all* p *(there are* $\ell!/\ell'!(\ell - \ell')!$ *of them for each* ℓ'*) and all* $\ell' = 0, 1, \cdots, \ell - 1$. *Then*

$$\sigma(\beta^{-1}(b)) = s^{-\ell} \sigma(H; \beta^*(X^\ell)). \qquad (*_\ell)$$

Proof. Check $(*_\ell)$ as earlier for $H = c \times W$, where $c = c_1 \times c_2 \times \cdots \times c_\ell$ is a Cartesian product of our old cycles in B, i.e. points, circles or whole surfaces.

Observe that our requirement is satisfied for H itself (since product of surfaces have zero signature) and hence for $H = B^\ell \times W$ and that it is both topological bordism invariant by the (A)-argument as well the proper homotopy invariant of $H \times \mathbb{R}$ by (B). The rest of the proof for $\ell > 1$ is the same as for $\ell = 1$ with a negligible extra effort needed to embed $B^\ell \to \mathbb{R}^{2\ell+1}$. Finally, we take care of the remaining case $\dim V - 4k$ *even* by passing to $V \times S^1$. Q.E.D.

$8\frac{2}{7}$ Flat bundles over surfaces with non-zero signatures

Let B be a compact oriented surface possibly with connected boundary with strictly negative Euler characteristic $\chi(B)$ and $X \to B$ a flat vector bundle over B. A non-trivial example of this is some "square root" of the tangent bundle of B, call it $X_{sp} \to B$. This can be visualized topologically via the associated circle (unit sphere) bundle $UX_{sp} \to B$ which is obtained by taking some double covering of the unit tangent bundle $UT(B)$ non-trivial (i.e. connected) over each tangent circle. Such a covering is (essentially by definition) the same thing as a spin structure on B; this exists since the Euler class of $T(B)(=\chi(B))$ is even, but not unique. In fact spin structures are classified by $H^1(B,\mathbb{Z}_2)$. This is better seen if we view $T(B)$ and X_{sp} as complex line bundles so that $T(B)$ becomes the tensor square of X_{sp}. Then observe that the bundle $UT(B) \to B$ has a flat $PSL_2\mathbb{R}$-structure corresponding to the usual action of $PSL_2\mathbb{R} = SL_2\mathbb{R}/\{\pm 1\}$ on the unit disk identified with the universal covering of B which gives us a flat $SL_2\mathbb{R}$-structure for the bundle $X_{sp} \to B$.

This applies, strictly speaking if B has an empty boundary, if $\partial B \neq 0$, the universal covering is realized as a part of the unit disk. Notice that this bundle is symplectic as $Sp\ell 2p = SL_{2p}$ for $p = 1$. We shall see later that $|\sigma(B; X_{sp}| = 2\,|\chi(B)| \neq 0$ but now we want to show how to compute (co)homology of B with coefficients in X in general.

$\mathbf{H_2(B;X)}$. This is zero unless B is a closed surface and 2-cycles are exactly horizontal sections $B \to X$. In particular, if the underlying representation ρ of $\pi_1(B)$ to the group of the automorphism of the fiber (this is GL_m for $m = \operatorname{rank} X$) has no fixed vector $\neq 0$, then $H_2(B;X) = 0$. For example $H_2(B;X_{sp}) = 0$.

$\mathbf{H_0(B;X)}$. Here 0-cycles are just vectors in X_b, $b \in B$. If γ is a loop in B based at b with monodromy $A : X_b \hookleftarrow$, then $x - Ax$ is the boundary for each $x \in X_b$ and so $H_0(B;X) = 0$ unless ρ fixes a covector.

$\mathbf{H_1(B;X)}$. This is more interesting. Take a standard basis of loops $\gamma_1, \ldots, \gamma_m$, $m = b_1(B) = \operatorname{rank} H_1(B; \mathrm{Triv}^1)$, at some point $b \in B$ and let A_i, $i = 1, \ldots, m$ be the corresponding monodromies of the fiber X_b (i.e. $\rho(\gamma_i)$). Then the 1-cycles are m-tuples $(x_1, \cdots, x_m)$, $x_i \in X_b$, satisfying the equation $\sum_{i=1}^m x_i - A_i x_i = 0$. Notice, that if ρ has no invariant vector, the support of such a cycle cannot consist of a single loop γ_i and so this support is necessarily singular (not looking as a nice 1-cycle). Even without solving this equation we predict (Euler-Poincaré):

$$\operatorname{rank} H_1(B;X) = -\chi(B) \operatorname{rank} X + \operatorname{rank} H_0(B;X) + \operatorname{rank} H_2(B;X)$$

which gives us rank $H_1(B;X) = -\chi(B) \operatorname{rank} X$ for irreducible ρ. Furthermore, if we cut B into pieces B_j, $j = 1, \cdots, n$ along simple non-contractible curves, such that X has no parallel covector sections over these B_j, then

$$H_1(V;X) = \bigoplus_{i=1}^{n} H_1(B_j; X|B_j), \qquad (\oplus)$$

unless B has no boundary and X admits a parallel section over B (i.e. $H_2(X;B) \neq 0$).

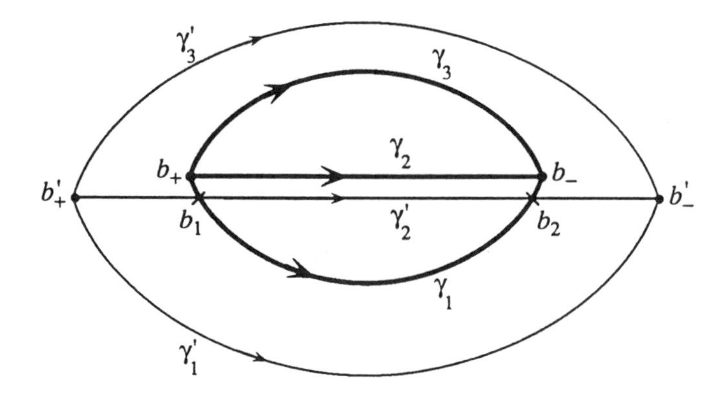

Figure 12

If, in addition, X comes along with a parallel skew-symmetric form ω which gives us a quadratic intersection form on $H_1(B;X)$, then the decompositon $(\oplus)$ is necessarily orthogonal for such a quadratic form since cycles with disjoint supports have zero indices of intersection.

Recall that this index is defined for pairs of 1-chains in general position $c = \sum_\mu x_\mu \gamma_\mu$ and $c' = \sum_\nu x_\nu \gamma_\nu$ where γ_μ and γ_ν are simple oriented arcs in B and x_μ and x_ν are sections of X over γ_μ and γ_ν. Whenever two arc transversally intersect, say γ_μ and γ_ν at some point $b_{\mu\nu}$, we take $\pm\omega(x_\mu, x_\nu)$ in the fiber $X_{b_{\mu\nu}}$ where the $\pm$ sign is the usual index of intersection between γ_μ and γ_ν, and $c \cap c' = \sum_{\mu,\nu} \pm\omega(x_\mu, x_\nu)$ which defines a quadratic form on $H_1(B;X)$ dual to the cup-product form on $H^1(B;X)$ mentioned earlier.

This implies (and refines in this special case) the additivity of the signature of the intersection form on $H_1(B;X)$ (see (σ_ρ) above) and reduces the computation of the intersection form to the case where B is a "pair of pants" i.e. S^2 minus three disks, where the homology and the intersection form can be computed with the chain complex on three arcs forming a 1-skeleton of B; see Figure 12 below.

Here the 1-chains are the sums $\sum_{i=1}^{3} x_i \gamma_i$ and the cycles are the solutions of the system

$$\sum_{i=1}^{3} x_i = 0 \quad , \quad \sum_{i=1}^{3} A_i x_i = 0$$

where $x_i \in X_{b_+}$, $i = 1,2,3$ and $A_i : X_{b_+} \to X_{b_-}$ are parallel transport operators along γ_i. The intersection form on the chains is $\omega_{b_+}(x_1, x_2') - \omega_{b_-}(A_1x_1, A_2x_2')$ (for the usual orientation on $\mathbb{R}^2$) which we write (as quadratic form) as $\omega(x_1, x_2) - \omega(A_1x_1, A_2x_2)$. To facilitate the computation of (the signature of) this form on cycles, we assume that the natural symmetry $s : B \to B$ of the third order (fixing (b_+, b_-) and permuting $\gamma_1 \mapsto \gamma_2 \mapsto \gamma_3 \mapsto \gamma_1$) extends to X. Let, moreover, X be real of rank 2 and s acts non-trivially on X_{b_+} and X_{b_-}. Then the s-invariant chains are (obviously) cycles and if $H_0(X;B) = 0$ all of H_1 is s-invariant. If, furthermore, (X_{b_+}, ω_{b_+}) is identified with the tangent plane $T_{b_+}(B)$ with the usual area form and the action $D_{b_+}s$, and (X_{b_-}, ω_{b_-}) is similarly identified with $T_{b_-}(B)$, then $\omega_{b_+}(x, sx) > 0$ for $0 \neq x \in X_{b_+} = T_{b_+}(B)$ and $\omega_{b_-}(x, sx) < 0$ for $0 \neq x \in X_{b_-} = T_{b_-}(B)$ since Ds rotates $T_{b_+}(B)$ counter-clockwise and $T_{b_-}(B)$ clockwise. Thus the *intersection form on $H_1(B; X)$ is positive definite.*

Let us explain why this applies to the above bundle $X_{sp} = \sqrt{T(B)}$. To see that some spin structure is s-invariant, we take the quotient $\underline{B} = B/\{1, s, s^2\}$ which is again an orientable Riemann surface whose threefold covering is B. We take some spin structure on $\underline{B}$, i.e. some $\underline{X}_{sp} = \sqrt{T(\underline{B})}$ and observe that the lift of this to B away from the (two) ramification points perfectly goes across these points so that the lifted structure is s-invariant and behaves at b and b' as required. Consequently

$$\sigma(B; X_{sp}) = 2 = -2\chi(B).$$

The sign here depends on how we orient X_{sp}. If we change the orientation by replacing $\omega \rightsquigarrow -\omega$, we thus change the sign of σ.

Conclusion. Each closed Riemann surface B has *positive definite* (quadratic) intersection form on $H_1(B; X_{sp})$ with $|\text{signature}| = |2\chi(B)|$. (This was pointed out to me by Bill Goldman who observes furthemore in [Gold] that for each $SL_2\mathbb{R}$-bundle X over a closed surface B,

$$\sigma(B; X) = 2e(X)$$

where $e(X)$ is the Euler class. Goldman also notices that his formula implies the *Milnor-Wood inequality* $|e(X)| \leq \frac{1}{2}|\chi(B)|$ for all flat $SL_2\mathbb{R}$-bundles over B. We elaborates this in $8\frac{1}{2}$. The proof follows by decomposing a general B into pairs of pants and by observing that the signature is invariant under homotopies of flat $SL_2\mathbb{R}$-bundles of closed surfaces (as the intersection form is non-singular by Poincaré duality) and so everything can be reduced to the symmetric case.

We suggest the reader extend our conclusion to surfaces with boundary where the statement, not the proof, needs an extra case and consult [Mey] for further study of the twisted signature.

$8\frac{1}{3}$. Pontryagin classes for topological manifolds

Let us indicate a modification of the above argument which allows an extension of the definition of L_k to the topological category (and at the same time reduces the role of bordisms to the Serre finitness theorem pure and simple). For this we need the following topological version of Novikov's formula (+) in $7\frac{4}{5}$ concerning signatures of cycles of codim = 1 in open manifolds. Here we shall be dealing with such a topological manifold of dimensions $2k+1$ and a distinguished $2k$-dimensional homology class k "dividing" U in the following sense. There is a proper function $\rho : U \to \mathbb{R}$ such that h is contained in the image of the inclusion homomorphism $H_{2k}(\varphi^{-1})[a,b]) \to H_{2k}(U)$ for some (and, hence for each) non-empty segment $[a,b]$, $-\infty \leq a \leq b \leq +\infty$. Given such an h and a $U(p,q)$-flat bundle X over W, we define the cup-pairing of $H^{2k}(U,X)$ on h in the obvious way and denote it by $\sigma(h;X)$.

Localization Lemma. *Let $U' \subset U$ be an open subset and h' be a $2k$-dimensional homology class of U' which goes to h under the inclusion homomorphism $H_{2k}(U') \to H_{2k}(U)$. Then*

$$\sigma(h';X|U') = \sigma(h;X). \qquad (+)'$$

Before going into the proof we indicate several examples.

(1) Suppose U is a closed manifold. Then, necessarily $h = 0$ and the lemma is vacuous. (It is not true, in general, that the signature of a cycle h in U does not change if we pass to some neighbourhood of this cycle; we do need the "dividing" property of h).

(2) Let h be a realized by a closed submanifold H of codimension one in U. Then small (say, tubular) neighbourhoods $U' \subset U$ of H obviously have $\sigma([H];X|U') = \sigma(H;X|H)$ and our lemma reduces to that in $7\frac{4}{5}$ with the additional X-twist.

(3) Let U be the interior of a compact manifold $\overline{U}$ with boundary $H = \partial\overline{U} = \overline{U} - U$ and $U' \subset U$ be a small (e.g. tubular) neighbourhood of infinity (i.e. of H in $\overline{U}$). Then $[H]$ vanishes in $H_{2k}(\overline{U})$ and so its signature in U' must be zero which amounts to the vanishing of the signature of H which agrees with the cobordism invariance of the signature.

The proof of $(+)'$. One can assume that $U' = \varphi^{-1}(]-1,1[) \subset U$ (for a suitable φ), write $U' = U_+ \cap U_-$ for $U_+ = \varphi^{-1}(]-1,\infty[)$ and $U_- = \varphi^{-1}(]-\infty,1[)$ and $(+)'$ would follow from the corresponding equalities for $U_- \subset U$ and $U' \subset U_-$. Or we may assume U' contains one of the two ends of U, i.e. either U_+ or U_-, say U_- and now we use the same argument as in $7\frac{4}{5}$. Namely, we first observe that $\sigma(h;X)$ equals the signature of the intersection form on the image Im of the restriction homomorphism of $H^{\text{infs}}_{2k+1}(U;X) \underset{PD}{=} H^{2k}(U;X)$ to $H^{\text{infs}}_{2k+1}(U_-;X) \underset{PD}{=} H^{2k+1}(U_-;X)$, where H^{infs}_* denotes the homology with infinite supports which equals the cohomology via the Poincaré duality and where the intersection form in question is

$$(h_1,h_2) \mapsto h_1 \cap h_2 \cap h \in \mathbb{R}.$$

This can be defined via the Poincaré duality or by combining the geometric intersection of the supports of cycles with the scalar product in X. Next, notice that if some $h^{\perp} \in H_{2k+1}(U_{-};X)$ satisfies $h^{\perp} \cap h_1 \cap h = 0$ for all $h_1 \in \mathrm{Im}$, then $h^{\perp} \cap h \in H_{2k}(U_{-};X)$ goes to zero under the inclusion homomorphism $H_{2k}(U_{-};X) \to H_{2k}(U;X)$ by the Poincaré duality in U and so $h^{\perp} \cap h$ belongs to the image of the boundary homomorphism $\partial : H_{2k+1}(U;X) \to H_{2k}(U;X)$. Hence, our intersection form vanishes on the space $\mathrm{Im}^{\perp}$ of all these $h^{\perp}$ (compare Figure 10 in $7\frac{4}{5}$) and the proof follows by linear algebra.

Remark. All we have actually used in this argument was the (local) Poincaré duality in U over $\mathbb{R}$.

Iterated (co)bordisms. Our objects are triples (U, h, X) with $\dim U = 2k+1$ as above and two such triples (U_i, h_i, X_i), $i = 1, 2$, are called *precobordant* if there exists a third such triple, say (U, h, X), and (equidimensional!) embeddings $U_i \to U$, $i = 1, 2$, such that h_i go to h, and X restricts to X_i. Our lemma says, in effect, that the signature $\sigma(U, h, X) \overset{\text{def}}{=} \sigma(h; X)$ is a pre-cobordism invariant in the topological category. Next we observe the "pre-cobordism" is not, a priori, an equivalence relation in the topological case (but clearly so in the smooth and p.l.-categories where h can be realized by a codim 1-submanifold); see Figure 13 below.

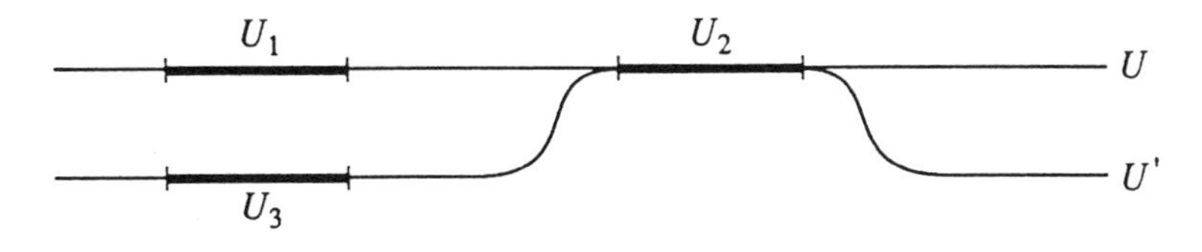

Figure 13

And we define *cobordism* as the equivalence relation spanned by pre-cobordisms. Of course, the signature is a cobordism invariant.

Homotopy invariance of cobordism. *Let S be a topological manifold, $\underline{U} \subset S$ be an open subset with compact closure, $\underline{X} \to \underline{U}$ a flat $U(p,q)$ bundle and $h \subset H_{\ell-1}(\underline{U})$ be a dividing cycle in $\underline{U}$ for $\ell = \dim U = \dim S$. Then for every topological manifold V and proper continuous map $f : V \to U$ the cobordism class of the pull-back $f^{-1}(\underline{U}, \underline{h}, \underline{X})$ is invariant under proper homotopies of f, where*

$$f^{-1}(\underline{U}, \underline{h}, \underline{X}) \underset{\text{def}}{=} \left(f^{-1}(U), f^{*}(\underline{h}), f^{*}(\underline{X})\right)$$

for $f^*(\underline{h}) = \mathcal{PD}f^*(\mathcal{PD}\underline{h})$.

Proof. The cobordism class of the pull-back does not change if we replace $\underline{U}$ by a slightly smaller open subset, set $\underline{U}' \subset U$ with compact closure in $\underline{U}$. Now if $f_{t'}$ is sufficiently close to f_t, we have the inclusion $f_{t'}^{-1}(\underline{U}') \subset f_t^{-1}(\underline{U})$ which provides a cobordism between the f_t and $f_{t'}$ -pull-backs of $(\underline{U}, \underline{h}, \underline{X})$. As every homotopy f_t can be divided into small steps, the invariance follows.

Definition of L_k ***for*** $n - 4k = 3$. Let U be a compact topological manifold and define $L_k \subset H^{4k}(V;\mathbb{Q})$ by prescribing it values on each homology class $g \in H_{n-3}(V)$, $n = \dim V$, as follows. Take $\underline{U} = B \times \mathbb{R} \subset S^3$ for a Riemann surface B (of genus ≥ 2), extend our flat Spl-bundle from B to $\underline{X}$ over $\underline{U}$ and take a map $f : V \to S^3$ so that $f^*[S^3]^{\text{co}} = \mathcal{PD}(Mg)$ for some $M \neq 0$. Then set $\langle L_k, g\rangle = (sM)^{-1}\sigma(f^{-1}(\underline{U}, \underline{h}, \underline{X}))$ for $\underline{h} = [B] \in H_2(\underline{U})$ and s being the signature of our basic bundle over B. Clearly, this defines a *linear* function on $H_{n-3}(V)$, i.e. a class in $H^{n-3}(V;\mathbb{Q})$ which we call L_k.

Notice that if V is smooth we can arrange the matters so that $U = f^{-1}(\underline{U}) = \underline{U} \times W$ for $W = f^{-1}(u)$ where the equality $\sigma(U;h;X) = s\sigma(W)$ follows from the Cartesian multiplicativity (and so we replace the cobordism theory in our earlier topological invariance proof by the elementary homotopy invariance of cobordisms).

The case $n - 4k = 2\ell \geq 3$. Proceed as above, but now with $\underline{U} = B^\ell \times \mathbb{R} \subset S = S^{2\ell+1}$.

Open manifolds V. Use *proper* maps to $\mathbb{R}^{2\ell+1}$ instead of maps to $S^{2\ell+1}$.

The case $n - 4k$ ***even.*** Stabilize to $V \times \mathbb{R}$ or $V \times S^1$.

On stabilization. Our L_k are *not*, a priori, stable for $V \rightsquigarrow V \times \mathbb{R}$ but they *are* stable for $V \times \mathbb{R} \rightsquigarrow V \times \mathbb{R}^3$. In fact $V \times B \times \mathbb{R}$ obviously *embeds* into $V \times \mathbb{R}^3$ and when we use some $U \subset V \times \mathbb{R}$ to define L_k, we take $U \times B \subset V \times \mathbb{R}^3$ for the composed embedding $U \times B \subset (V \times \mathbb{R}) \times B \subset V \times \mathbb{R}^3$, in order to define L_k of $V \times \mathbb{R}^3$ and apply the Cartesian multiplicativity, $\sigma(u \times B, \cdots) = s\sigma(U, \cdots)$. Now we may speak of the *stable* classes L_k which have an advantage of being functorial for equidimensional topological immersions $V_1 \to V_2$ as every such immersion can be turned into an embedding $V_1 \times \mathbb{R}^N \to V_2 \times \mathbb{R}^N$ for large N and functoriality for *equidimensional embeddings* is obvious with our definition of L_k (even before stabilization). To finish the story one should proof the (Cartesian) multiplicativity of L_k for $V = V_1 \times V_2$ (this is easy if V_1 is smooth or p.l. but I do not see how to do it for both V_1 and V_2 topological without dirting my hands in the topological topology) and/or the corresponding property for Whitney sums of topological bundles. (See [Ki-Si] and [Ran]$_{\text{Haup}}$ for the classical approach).

Multiplicativity for coverings. This follows directly from the homotopy invariance of cobordism and the Serre finitness theorem, as usual (while the original proof by J. Schafer appealed to the topological transversality theory of Kirby and Siebenmann).

$8\frac{1}{2}$. Lusztig signature theorem for flat (skew) Hermitian bundles and norms on Witt_{2k}

We are back to a general situation of a flat $U(p,q)$-bundle $X_\rho \to V$ and before stating Lusztig theorem we observe some additional properties of the signature $\sigma_\rho = \sigma(V; X_\rho)$.

Extendability of σ_ρ to $\mathrm{Witt}\,\mathbb{C}(\Pi)$. Since the bordism invariance of σ_ρ depends solely on the Poincaré duality, it remains valid for algebraic cobordisms of algebraic Poincaré complexes and thus defines a homomorphism $\sigma_\rho : \mathrm{Witt}_{2k}\,\mathbb{C}(\Pi) \to \mathbb{Z}$ compactable with the homomorphism $\mathrm{Brd}_{2k}\, B\Pi \to \mathrm{Witt}_{2k}\,\mathbb{C}(\Pi)$ (which factors through $H\,\mathrm{Brd}_{2k}\, B\Pi$). More algebraically, $\rho : \Pi \to U(p,q)$ defines an involutive homomorphism $\mathbb{C}(\Pi) \to \mathrm{Mat}_{pq}\,\mathbb{C}$ and hence a homomorphism

$$\mathrm{Witt}_{2k}\,\mathbb{C}(\Pi) \underset{w_\rho}{\to} \mathrm{Witt}_{2k}\,\mathrm{Mat}_{pq}\,\mathbb{C} \underset{\sigma}{\overset{\sim}{\to}} \mathbb{Z}$$

recapturing σ_ρ as $\sigma \circ w_\rho$, by an easy argument.

A cellular bound on σ_ρ. If V admits a cell decomposition (e.g. given by a Morse function) with at most b middle dimensional cells then, obviously, $|\sigma(V; X_\rho| \leq b \,\mathrm{rank}\, X_\rho$ for all ρ. Consequently, if $\sigma(V; X_\rho) \neq 0$ for some ρ, then a d-sheeted covering $\widetilde{V}_d$ of V needs at least $d/\,\mathrm{rank}\, X_\rho$ cells.

Examples.

(a) If V fibers over the circle then cyclic d-sheeted coverings of V can be decomposed into c cells with c independent of d. Thus $\sigma(V; X_\rho) = 0$ for all representations ρ. Recall that some hyperbolic 3-manifold V_0 fiber over S^1 and so the signature vanishes on $V = V_0 \times V_1$ for all V_1.

(b) Let V be a Cartesian product of closed surfaces of genera ≥ 2. Then, clearly, $\widetilde{V}_d$ needs at least $\left|\chi(\widetilde{V}_d)\right| = d\,|\chi(V)|$ cells for any decomposition which is $\geq d2^k$, $k = \dim V/2$. What is less obvious is that *every manifold V' which admits a map $\beta : V' \to V$ of degree d contains $\gtrsim d$ cells (of dimension $k = n/2$) in every of its cell decompositions.*

Proof for $\dim V = 4$. We know V admits a flat quadratic bundle $X \to V$ with $\sigma(V; X) = s \neq 0$. And every V' mapped to V with degree d satisfies $\sigma(V'; X') - \sigma(V') = ds$ where $X' \to V'$ denotes the bundle induced from X by the implied map $\beta : V' \to V$. In fact, all three quantities, $\sigma(v'; X')$, $\sigma(v')$ and $ds = s \deg \beta$ are linear functions on the oriented bordism group $\mathrm{Brd}_4 V$, where the elements are represented by pairs $(V', \beta : V' \to V)$. This group is spanned over $\mathbb{Q}$ by (compare $7\frac{7}{8}$).

I. $(V' = V, \beta = \mathrm{id})$,
II. $(V' = \mathbb{C}P^2\ , \beta = \mathrm{const})$.

Our formula is obviously valid for I and II, hence it is valid for (V', β). Since both signatures, $\sigma(V')$ and $\sigma(V'; X')$ are bounded in the absolute values by the number c of cells in V, we have $2c \geq d|s|$. Q.E.D. The proof for $\dim V \geq 4$ is similar and left to the reader.

Definition of the rank norm. Let $\mathrm{rank}(w)$, $w \in Witt_k$, be the minimum of ranks (which we assume to be well-defined) of quadratic moduli representing w. This applies to both even and odd k for $Witt_k$ of a ring $\mathcal{R}$ and a similar definition is valid for the K-groups of $\mathcal{R}$. Then we introduce the *rank norm* $\|w\|$ by

$$\|w\| = \lim_{i \to \infty} i^{-1} \mathrm{rank}(iw).$$

Examples.

(a) Let $\mathcal{R}$ be the ring of continuous functions on a compact connected topological space X. Then $K_*(\mathcal{R}) = K^*(X)$ and if X is finite dimensional then every element of the *reduced* group $K^*(X)$ (obtained from $K^*(X)$ by factoring away K^* {point}) can be represented by a (virtual) vector bundle of rank $\leq 2 \dim X$ and so the rank norm vanishes on the reduced $K_*(X)$.

(b) Let V be a closed oriented manifold admitting a sequence $(V_i', \beta_i : V_i' \to V)$ where V_i' are closed oriented manifolds of the same dimension as V and β_i are continuous maps such that

(i) $\deg \beta_i = d_i \underset{i \to \infty}{\longrightarrow} \infty$;

(ii) the induced tangent bundles $\beta_i^*(T(V))$ are $\mathbb{Q}$ equivalent to $T(V_i')$, i.e. $[\beta_i^* T(V)]$ equal $[T(V_i')]$ in $K^0(V_i') \oplus \mathbb{Q}$.

(iii) V_i' can be decomposed into c_i cells with $c_i/d_i \underset{i \to \infty}{\longrightarrow} 0$.

Then the class of $[V]_{\mathrm{Witt}}$ of V in $\mathrm{Witt}_n \mathbb{C}(\Pi)$, for $\Pi = \pi_1(V)$ defined by the Wall-Miščhenko homomorphism $WM : \mathrm{Brd}_n V \to \mathrm{Witt}_n \mathbb{C}(\Pi)$ has zero rank norm. In particular, if V fibers over S^1, then $\|[V]_{\mathrm{Witt}}\| = 0$.

Let V be a Cartesian product of Riemann surfaces of genera ≥ 2. Then $\|[V]_{\mathrm{Witt}}\| \neq 0$ as follows for the existence of a flat symplectic or quadratic bundle $X \to V$ with $\sigma(V; X) \neq 0$. In fact one can identify in this case the subspace $\Delta \subset \mathrm{Brd}_* V \oplus \mathbb{Q}$ on which the norm $[V', \beta] \mapsto \|WM[V', \beta]\|$ vanishes. This Δ is spanned by those $\left[V_\mu' \times W_\mu, \beta = \beta(v')\right]$ where the classes $\beta_\mu \left[V_\mu'\right] \subset$

$H_*(V)$ have 1-dimensional components in their Künneth decomposition. (Since V equals a product of surfaces, $H_*(V)$ is built of those cycles which are products of surfaces and circles ; the above condition requires a presence of circles in all $\mu_\mu\,[V'_\mu]$).

Signature theorem. We recall the classical argument identifying the ordinary signature of a closed oriented $4k$-manifold V with the index of the Hodge-de Rham signature operator. We take some Euclidean norms on the bundles of exterior forms $\Lambda^i(V)$ and some smooth measure dv on V. With this we have the L_2-norms on forms for $\|\lambda\|_{L_2} = \left(\int_V \|\lambda_v\|^2\,dv\right)^{\frac{1}{2}}$ and define the adjoint operator d^* to the exterior differential on the forms. This d^* maps smooth $(i+1)$-forms to i-forms according to the formula $\int_V \langle d^*\lambda, \lambda'\rangle_v\,dv = \int_V \langle \lambda, d\lambda'\rangle_v\,dv$ for all smooth $(i+1)$-forms λ and i-forms λ'. One checks that the operator $d+d^* : C^\infty\Lambda^*(V) \hookleftarrow$ is elliptic (notice that $d+d^*$ mixes degrees) and that the kernel of $d+d^*$ canonically identifies with $H^*(V;\mathbb{R})$. Then one observes that $d+d^*$ sends even forms to odd ones and vice versa and the index of $d+d^* : C^\infty\Lambda^{\text{even}}(V) \to C^\infty\Lambda^{\text{odd}}(V)$ equals the Euler characteristic of V. There is nothing specially "manifoldish" about it. One could start for example, with the boundary operator ∂ on a finite cell complex V and arrive at the same interpretation of $\chi(V)$ as $\operatorname{ind}\partial+\partial^*$.

Next, we want to split the bundle $\Lambda^*(V)$ into two pieces in a less trivial way, say into $\Lambda^* = \Lambda^*_+ \oplus \Lambda^*_-$, so that $d+d^*$ should map $C^\infty(\Lambda^*_+)$ into $C^\infty(\Lambda^*_-)$ with ind $=$ signature $\sigma(V)$. This becomes possible if we choose our norms in Λ^i with more care starting from a single norm in $T(V)$ (or in $\Lambda^1(V)$) i.e. with a Riemannian metric, say g, in V, which will also be used for the definition of the measure dv. Here is the relevant linear algebra.

Let $\mathbb{R}^n$ be the Euclidean space with the usual metric and embed the Grassmann manifold $\mathrm{Gr}_i\,\mathbb{R}^n$ of oriented i-subspaces into $\Lambda^i\mathbb{R}^n$ by assigning to each $L \subset \mathbb{R}^n$ the pull-back of the oriented volume form on L (of degree $i = \dim L$) under the orthogonal projection $\mathbb{R}^n \to L$. Then observe that the oriented orthogonal complement $L \mapsto L^\perp$ defines a map $\perp\ \mathrm{Gr}_i\,\mathbb{R}^n \to \mathrm{Gr}_{n-i}\,\mathbb{R}^n$ which uniquely extends to a *linear* map on forms, denoted $* : \Lambda^i(\mathbb{R}^n) \to \Lambda^{n-i}(\mathbb{R}^n)$. Linear extendability of $\perp$ follows, by a little thinking, from its $O(n)$-invariance. Now, using $*$ we define the scalar product on $\Lambda^i(\mathbb{R}^n)$ by $\langle\lambda,\lambda'\rangle = \langle\lambda\wedge *\lambda\rangle \in \Lambda^n\mathbb{R}^n = \mathbb{R}$, where Λ^n is identified with $\mathbb{R}$ via the oriented Euclidean volume form on $\mathbb{R}^n$ and observe that $* : \Lambda^i(\mathbb{R}^n) \to \Lambda^{n-i}(\mathbb{R}^n)$ is isometric for the norm $\|\lambda\| = \langle\lambda\wedge *\lambda\rangle^{\frac{1}{2}}$ as $*^2 = \pm 1$ (i.e. $*\lambda_1 \wedge *\lambda_2 = \lambda_1\wedge\lambda_2$ for forms of complementary degrees).

One checks with a minor effort that this is indeed a *symmetric* and *positive definite* scalar product, and also one sees that if n is even, then $*^2 = (-1)^i$. In particular, if $n = 4k$, then $*$ is an involution on $\Lambda^{2k}(\mathbb{R}^n)$, i.e. $*^2 = 1$, and one can modify $*$ to an involution on all of $\Lambda^*(\mathbb{R}^n)$ by taking $\widehat{*}_i = \pm *_i$ with a suitable $\pm$ sign, e.g. $\widehat{*}_i = *$ for $i \le 2k$ and $\widehat{*}_i = (*_{n-1})^{-1}$ for $i \ge 2k$.

Now we return to our manifold V, and we define the norms in $\Lambda^i(V)$ using some Riemannian metric g on V, i.e. a Euclidean structure on $T(V)$, and the corresponding $* = *_g : \Lambda^i(V) \to \Lambda^{n-i}(V)$. Thus

$$\|\lambda\| = \left(\int_V \lambda \wedge *\lambda\right)^{\frac{1}{2}} \quad , \quad \lambda \in C^\infty \Lambda^i(V), \tag{$*$}$$

and the only link of this norm with d is via the Leibniz and Stokes formulae

$$d(\lambda \wedge \mu) = d\lambda \wedge \mu + (-1)^i \lambda \wedge d\mu \tag{Lei}$$

$$\int_V d(\lambda \wedge \mu) = 0. \tag{Sto}$$

for arbitrary smooth forms of degrees i and $n-i-1$ respectively. Observe, that these three formulae do not mix $*$ and d, yet as a conclusion one has the following relation between these two operators.

$$d^* = - * d * \qquad \text{for} \quad \dim X \,\text{even}\,.$$

Indeed $*d * \lambda \wedge *\lambda = d * \lambda \wedge \lambda'$ and by (Lei)

$$\int *d * \lambda \wedge *\lambda' = \int d(*\lambda \wedge \lambda') - (1)^i \int *\lambda \wedge d\lambda',$$

for $i = \deg \lambda$, where $*\lambda \wedge d\lambda' = *^2 \lambda \wedge *d\lambda' = (-1)^i \lambda \wedge *d\lambda'$. Thus by (Sto)

$$\int *d * \lambda \wedge *\lambda' = - \int \lambda \wedge *d\lambda'$$

which makes $*d* = -d^*$ by the definition of d^* for our scalar product. Next we assume $n = 4k$, recall the involution $\widehat{*} = \pm *$, and observe that $d + d^*$ anticommute with $\widehat{*}$, as $(d+d^*)\widehat{*} = d\widehat{*} - *d*\widehat{*} = \widehat{*}^2 d\widehat{*} - \widehat{*}^2 * d * \widehat{*} = -\widehat{*}d^* - \widehat{*}d = -\widehat{*}(d^* + d)$. Thus $d+d^*$ interchanges the $+1$ and -1 eigenspaces of $\widehat{*}$, denoted $\Lambda^*_+(V)$ and $\Lambda^*_\pm(V)$. Then the index of the operator $\mathcal{L} \underset{\text{def}}{=} d + d^* : C^\infty\Lambda^*_+(V) \to C^\infty \Lambda^*_-(V)$ equals $\dim \mathcal{H}^*_+ - \dim \mathcal{H}^*_-$ where $\mathcal{H}^* = \bigoplus_{i=0}^n \mathcal{H}^i$ denotes the space of harmonic form, i.e. the kernel of $d + d^*$ and $\mathcal{H}^*_\pm = \mathcal{H}^* \cap C^\infty(\Lambda^*_\pm(V))$. Since $\mathcal{H}^*$ is invariant under the operator $*$ (which is obvious) and hence under $\widehat{*}$, which (as well as $*$) interchanges $\mathcal{H}^i$ and $\mathcal{H}^{n-i}$ for $i \neq n/2$, we conclude that $\dim \mathcal{H}^*_+ - \dim \mathcal{H}^*_- = \dim \mathcal{H}^{2k}_+ - \dim \mathcal{H}^{2k}_-$ where, recall $n = 4k$. Finally we observe that the product pairing $(\lambda_1, \lambda_2) \mapsto \int_V \lambda_1 \wedge \lambda_2$ is symmetric in the middle dimension and Λ^{2k}_+ consists of those λ where $(\lambda, \lambda) = \langle \lambda, \lambda \rangle$ while Λ^{2n}_-

consists of the form λ satisfying $(\lambda, \lambda) = -\langle \lambda, \lambda \rangle$. Thus the cup-product form on $H^{2k}(V;\mathbb{R}) = \mathcal{H}^{2k}$ is positive on $\mathcal{H}^{2k}_+$ and negative on $\mathcal{H}^{2k}_-$. Hence,

$$\operatorname{ind} \mathcal{L} = \sigma(V), \tag{σ}$$

where, recall $\mathcal{L}$ equals $d + d^*$ restricted to $C^\infty \Lambda^*_+ \to C^\infty \Lambda^*_-$. Finally, to make full use of (σ), we invoke the general Atiyah-Singer index theorem which expresses ind $\mathcal{D}$ in terms of characteristic classes and which specializes in this case to (compare $7\frac{1}{2}$)

$$\sigma(V) = \operatorname{ind} \mathcal{L} = L[V] \tag{$\sigma = \mathrm{L}$}$$

Remark. Recall that $L[V]$ is a characteristic number of V and so is multiplicative under finite coverings $\widetilde{V} \to V$ which pull-back $T(V)$ to $T(\widetilde{V})$. But the issuing multiplicativity of the signature $\sigma(V)$ does not need the full force of the identity $\sigma = L$, but only, (as was pointed out by Atiyah) the easy part, $\sigma = \operatorname{ind} \mathcal{L}$. In fact, the index of any elliptic operator $\mathcal{D}$ can be computed as the difference of traces $\operatorname{Tr} P_+ - \operatorname{Tr} P_-$, where P_+ and P_- are integral operators with smooth kernels canonically constructed out of $\mathcal{D}$, such that these kernels, say $K_+(v,v')$ and $K_-(v,v')$ are supported in a given (arbitrarily small) neighbourhood of the diagonal $\Delta_V \subset V \times V$. Thus the index appears as an integral of a local quantity, namely $K_+(v,v) - K_-(v,v)$ (or more precisely of $\operatorname{Tr}_v K_+(v,v) - \operatorname{Tr}_v K_-(v,v)$ as K_+ and K_- are matrix valued functions) and so is multiplicative for coverings.

Signature for flat bundles. We consider separately two cases.

1. The manifold V in question is $4k$-dimensional and our flat bundle $X \to V$ is (indefinite) orthogonal.

2. $\dim V = 4k + 2$ and $X \to V$ is a flat symplectic bundle.

Case 1. We denote by Q the implied non-singular quadratic form on X and by evaluating Q on the exterior product of X-valued forms on V we obtain a pairing $(\Lambda^i \otimes X) \otimes (\Lambda^j \otimes X) \to \Lambda^{i+j}$ denoted $\alpha \wedge_Q \beta$ which satisfies the Leibniz formula $d(\alpha \wedge_Q \beta) = (d_X \alpha) \wedge_Q \beta + (-1)^i \alpha \wedge_Q d_X \beta$, where d_X is the exterior differential twisted with X (i.e. $d(\lambda \otimes x) = (d\lambda) \otimes x$ for horizontal sections x of X) since locally $\Lambda^i \otimes X$ is just the Cartesian sum of several copies of Λ^i as (X, Q) is flat. Next, we fix a Riemannian metric on V and some positive definite scalar product $\langle \ , \ \rangle_0$ on X. Then there (obviously) exists a unique splitting $X = X_+ \oplus X_-$ which is both Q and $\langle \ , \ \rangle_0$ orthogonal and such that $Q|X_+ \geq 0$ and $Q|X_- \leq 0$. We denote by τ the involution on X equal $+1$ on X_+ and -1 on X_- and observe that the quadratic forms $\langle x, x' \rangle = Q\langle x, \tau x' \rangle$

is positive definite. Then we define the scalar product on X-valued forms with the pairing

$$(\lambda \otimes x, \lambda' \otimes x') \mapsto \int_V \langle x, x' \rangle \lambda \wedge *\lambda' = \int_V Q\langle x, \tau x' \rangle \lambda \wedge *\lambda$$

which bilineary extends to all of $(\Lambda^i \otimes X) \otimes (\Lambda^i \otimes X)$, $i = 0, 1, \ldots, n$, where it is clearly positive definite. Now we are in the same situation as earlier with the involution $\lambda \otimes \lambda \mapsto (*\lambda) \otimes \tau x$ on the middle dimensional forms which extends as earlier with an adjustment of $\pm$ sign to an involution on $\Lambda^* \otimes X$ and which is still called $\widehat{*}$. So again we have an elliptic operator, $\mathcal{L}_{X,Q}$ equal $d_X + d_X^*$ on X-valued forms which sends the $(+1)$-eigenspace of $\widehat{*}$, say $C^\infty(\Lambda^* \otimes X)_+$, to (-1)-eigenspace $C^\infty(\Lambda^* \otimes X)_-$ and the index of $\mathcal{L}_{X,Q}$ equals $\sigma(V; X)$ for the same reason as earlier (since the formal properties of $(d_X, \widehat{*})$ are the same here as in the case of $X = \mathrm{Triv}^1$, $Q = x^2$.

We see already at this stage that $\sigma(V; X)$ is multiplicative. Moreover, by the general index theorem

$$\sigma(V; X) = \operatorname{ind} \mathcal{L}_{X,Q} = L_V \operatorname{ch}(\mathbb{C}X_+ - \mathbb{C}X_-)[V], \qquad (\sigma = L)_Q$$

where $\mathbb{C}X_+$ and $\mathbb{C}X_-$ are the complexifications of the Q-positive and Q-negative parts of X. In fact, $\mathcal{L}_{X,Q}$ is homotopic to the operator $\mathcal{L}_{X_+} \oplus \mathcal{L}^*_{X_-}$, where $\mathcal{L}_{X_+}$ is $\mathcal{L}$ twisted with X_+ for some orthogonal (non-flat) connection on X_+ and $\mathcal{L}^*_{X_-}$ is the adjoint to the twist of $\mathcal{L}$ with X_-. Thus $\operatorname{ind} \mathcal{L}_{X,Q} = \operatorname{ind} \mathcal{L}_{X_+} - \operatorname{ind} \mathcal{L}_{X_-}$ where $\operatorname{ind} \mathcal{L}_{X_\pm} = L_V(\operatorname{ch} \mathbb{C}X_\pm)[V]$ by the index theorem.

Case 2. (Symplectic). If S is a (parallel) symplectic (i.e. non-singular skew-symmetric) form on X then one obtains, with an auxiliary scalar product $\langle\ ,\ \rangle_0$ on X, an anti-involution λ on X, i.e. $\tau^2 = -1$, which preserves both forms, ω and $\langle\ ,\ \rangle_0$, and for which the pairing $\langle x, x' \rangle_0 = S(x, zx')$ is *positive definite* (and symmetric as τ preserves ω). Here again $\lambda \otimes x \mapsto *\lambda \otimes \tau x$ is an involution on the middle dimensional X-valued forms (now, recall, $\dim V = 4k + 2$ and $*$ is an anti-involution on Λ^{2k+1}) which extends with a sign adjustment as earlier to an involution $\widehat{*}$ on all of $\Lambda^* \otimes X$. We split $\Lambda^* \otimes X$ as before according to $\pm$ sign of the eigenvalues of $\widehat{*}$ and identify the signature $\sigma(V; X)$ with the index of the resulting operator

$$\mathcal{L}_{X,S} = d_X + d_X^* : C^\infty(\Lambda^* \otimes X)_+ \to C^\infty(\Lambda^* \otimes X)_-.$$

Finally, in order to compute the index of $\mathcal{L}_{X,S}$, we complexify (X, τ), take the *involution* $\sqrt{-1}\tau$ on $\mathbb{C}X$ and split $\mathbb{C}X$ into $X_+^{\mathbb{C}} \oplus X_-^{\mathbb{C}}$ according to the ± 1-eigenvalues of $\sqrt{-1}\tau$. Then

$$\operatorname{ind} \mathcal{L}_{X,S} = L_V \operatorname{ch}(X_+^{\mathbb{C}} - X_-^{\mathbb{C}})[V],$$

by the index theorem.

Hermitian case. If we start with a flat Hermitian bundle X, for $\dim V = 4k$, we split it into $X_+ \oplus X_-$ where the implied form is positive definite on X_+ and negative on X_- and obtain Lusztig's formula

$$\sigma(V;X) = L_V \operatorname{ch}(X_+ - X_-)[V].$$

If X is skew-Hermitian (on V of dimension $4k+2$) we pass to a Hermitian form $H(x,\overline{y}) = S(x, \sqrt{-1}\,\overline{y})$ and get the same formula with X_+ and X_- referring to H.

Application to the Novikov conjecture. Let Π be a group and $H^{\text{ev}}_{\text{fl}} \subset H^{\text{ev}}(B\Pi; Q)$ be the subspace spanned by the Chern characters $\operatorname{ch}(X_+ - X_-) = \operatorname{ch} X_+ - \operatorname{ch} X_-$ for all flat Hermitian and skew-Hermitian flat bundles X over $B\Pi$. Then every $\rho \in H^{\text{ev}}_{\text{fl}}$ satisfies the Novikov conjecture, i.e. for every manifold V mapped to $B\Pi$ by a continuous map $\beta : V \to B\Pi$ the ρ-signature of V, i.e. $L_V \smile \beta^*(\rho)[V]$, is homotopy invariant being equal to the ordinary signature of V with coefficients in the flat bundle $\beta^*(X)$ by the Lusztig theorem.

Examples. (c) We saw earlier that if $B\Pi$ is a Cartesian product of Riemann surfaces, then the above applies to the fundamental class $B\Pi$ as well as for the classes multiplicatively generated by the 2-dimensional classes induced from the fundamental classes of surfaces.

(a$'$) Let Π be a discrete group freely acting on the Cartesian product $\widetilde{B}$ of k copies of the hyperbolic plane (Poincaré disk) and $B = B\Pi = \widetilde{B}/\Pi$. Each of these planes gives us a Kähler form, call them $\omega_1, \omega_2, \ldots, \omega_k$, and their cohomology classes as well as the products of these sit in $H^{\text{ev}}_{\text{fl}} \subset H^{\text{ev}}(B;Q)$. In particular, if B is compact, then its fundamental class $[B]^{\text{co}}$ is in $H^{\text{ev}}_{\text{fl}}$ and hence satisfies the Novikov conjecture. If Π splits, this reduces to (a) but not all groups Π split.

(b) (See [Lus]). Let G be the real symplectic group $\operatorname{Spl} 2p$ and $\Pi \to G$ a homomorphism. Then the image of $H^*(BG;\mathbb{Q})$ in $H^*(B\Pi;Q)$ is contained in $H^{\text{ev}}_{\text{fl}}$. Furthermore if $\Pi \subset G = \operatorname{Spl} 2p$ is a discrete torsion-free subgroup, $\widetilde{B} = G/(\text{maxcomp.})$, and $B = B\Pi = \Pi\backslash\widetilde{B}$, then the cohomology classes of G-invariant forms on $\widetilde{B}$ descended to B are in $H^{\text{ev}}_{\text{fl}}$. In particular, if Π is cocompact, then the fundamental class $[B]^{\text{co}}$ is in $H^{\text{ev}}_{\text{fl}}$. This generalizes (a) for surfaces.

(b$'$) The above probably generalizes to all semi-semisimple real algebraic groups G as follows. Let $K \subset G$ be the maximal compact subgroup and $\widetilde{B} = G/K$. Then the (G-invariant) K-characteristic (Chern-Weil) forms on $\widetilde{B}$, when they descend to $\Pi\backslash\widetilde{B}$, must have their classes in $H^{\text{ev}}_{\text{fl}}$ (where the relevant flat bundles must come from suitable representations $G \to U(p,q)$ and/or $G \to \mathrm{Spl}\,2p)$. In particular, if $B = \Pi\backslash\widetilde{B}$ is compact with $\chi(B) \neq 0$, then the fundamental class $[B]^{\mathrm{co}}$ should be in $H_{\mathrm{fl}}^{\mathrm{ev}}$ according to our conjecture (which, whether true or false, must be obvious to anyone with some experience in the representation theory).*

Our interest in $H_{\mathrm{fl}}^{\mathrm{ev}}$ is not so much motivated by the Novikov conjecture which is known to be true for subgroups in Lie groups by the work of Kasparov anyway; (see [Kas]) but by the following stronger property of the homomorphism $H_{\mathrm{even}}(B(\Pi;\mathbb{Q}) \to \mathrm{Witt}_{\mathrm{even}}\,\mathbb{C}(\Pi)$ (which assigns to a homology class represented by a map $\beta : V \to B(\Pi)$, for a stably parallelizable manifold V, the Witt class of the algebraic Poincaré complex associated to some triangulation of V). *The norm on* H_{even} *induced from the rank norm on* $\mathrm{Witt}_{\mathrm{even}}$ *does not vanish on those* h *for which* $\langle\rho, h\rangle \neq 0$ *for some* $\rho \in H_{\mathrm{fl}}^{\mathrm{ev}}$.

This (cellular in nature) norm on $H_*(B\Pi)$ is similar in spirit to the simplicial norm (see $[\mathrm{Gr}]_{\mathrm{VBC}}$) and we shall investigate the relation between the two somewhere else. Here we indicate several questions concerning the rank norm on $\mathrm{Witt}_*\,\mathbb{C}(\Pi)$ and the corresponding norm on $H_*(B\Pi)$ and $\mathrm{Brd}\,B\Pi$.

Are there any lower bounds on these norms apart for the above $H_{\mathrm{fl}}^{\mathrm{ev}}$? In particular, is this norm ever non-trivial on $\mathrm{Witt}_{\mathrm{odd}}$? Is this norm non-zero on the fundamental classes of even dimensional manifolds of negative curvature? ("Yes" for constant curvature follows from Lusztig's remark on $O(n,1)$). How does this norm extend from $\mathrm{Witt}_*\,\mathbb{C}(\Pi)$ to $\mathrm{Witt}_*\,C^*(\Pi) = K_*C^*(\Pi)$? May this norm be non-zero on $\mathrm{Witt}_*\,\mathbb{C}(\Pi)$ and vanish in $\mathrm{Witt}_*\,C^*(\Pi)$? Or is the rank norm always zero on $C^*(\Pi)$? What is a possible asymptotic behaviour of $\mathrm{rank}(iw)$ for $w \in \mathrm{Witt}_*$ and $i \to \infty$ in the case where the rank norm, i.e. $\lim\limits_{i\to\infty} i^{-1}\,\mathrm{rank}(iw)$, vanishes?

Dirac twisted with flat $U(p,q)$***-bundles and*** $Sc > 0$. If V is spin and $\rho \in H_{\mathrm{fl}}^{\mathrm{ev}}$, i.e. a combination of $\mathrm{ch}(X_+ - X_-)$ for flat $U(p,q)$ bundles X over V, then we naturally expect that $(\widehat{A}_V \smile \rho)[V] = 0$. Indeed this is true as one can show that the (virtual) bundle $\kappa = [X_+] - [X_-]$ is almost flat in the unitary sense, i.e., UAFl in the terminology of $8\frac{3}{4}$, and thus the Dirac operator twisted with κ has index zero. Moreover, forget about $Sc > 0$, and assume that $(\widehat{A}_V \smile \rho)[V] \neq 0$ for some $\rho \in H_{\mathrm{fl}}^{\mathrm{ev}}$. Then the spectrum of the Dirac operator $\widetilde{D}$ on the universal covering $\widetilde{V}$ of V has $0 \in \mathrm{spec}\,\widetilde{D}$. Furthermore, if $\rho[V] \neq 0$, then $\widetilde{V}$ has infinite K-$\mathrm{length}_{\mathrm{st}}$. All this is especially easy to see if the implied representation of $\Pi = \pi_1(V)$ in $U(p,q)$ is proper (discrete) where

* When I asked Lusztig, he instantaneously pointed out that the fundamental spin representation settles the matter for $G = O(n,1)$ and promised to look into the general case at his leisure.

(at least for torsionless Π) one has a (classifying) map from V to the manifold $\Pi \backslash U(p,q)/U(p+q)$ of *non-positive* curvature.

Next, every countable subgroup $\Pi \subset U(p,q)$ can be made act properly on a suitable product of Bruhat-Tits buildings associated to $U(p,q)$, which also have non-positive curvatures (in a generalized sense) and so the above claim extends to the general (non-proper) case.

But the use of Bruhat-Tits is definitely an overkill (which may be necessary for the Novikov conjecture for *all* cohomology of a subgroup $\Pi \subset U(p,q)$) as we are concerned with rather special cohomology classes in $H^*(\Pi)$, namely those coming from $BU(p,q)$ and one may use another, more functorial approach due to Alain Connes. Namely, the (possibly non-proper) action of Π on the symmetric space $Z = U(p,q)/U(p+q)$ gives rize to a class of *Fredholm representations* of Π defined, roughly as follows (compare $9\frac{2}{3}$).

Take some natural $U(p,q)$ invariant elliptic operator Δ over Z, e.g. the Dirac operator and let H be the Hilbert space $\operatorname{Ker}\Delta$ of Δ acting on the pertinent L_2-space of sections. Consider the covector field $d\mu$ for the distance function $\mu(z) = \operatorname{dist}_Z(z, z_0)$ regularized at zero (as in $8\frac{2}{3}$). Then the Clifford multiplication of spinors in H by $d\mu$ (or by $d\mu/\|d\mu\|$ composed with the orthogonal projection (L_2-space of section) $\to H$ is a Fredholm operator $F : H \to H$ (at least if zero is isolated in the spectrum of Δ). Now, given a flat $U(p,q)$-bundle over V, we take the associated Z-bundle and the corresponding Hilbert bundle $\mathcal{H}$ with the fibers $H_v = H(Z_v)$. Since the fibers $Z_v (= Z)$ are contractible, there is a section $v \mapsto z_v \in Z_v$ and we get with $\mu = \operatorname{dist}(z, z_v)$ in each fiber Z_V, the Fredholm endomorphism $\mathcal{F} = \mathcal{H} \to \mathcal{H}$, defining some K-class $\kappa \in K_0(V)$. Then the Dirac operator on V can be twisted with this κ and $\operatorname{ind}\mathcal{D}_\kappa$ can be expressed in terms of the "universal" index of $\mathcal{D}$ with values in $K_0(C^*(\Pi))$ so that

$$0 \notin \operatorname{spec}\widetilde{\mathcal{D}} \Rightarrow \operatorname{ind}\mathcal{D}_\kappa = 0.$$

This can be used in conjunction with the index formula

$$\operatorname{ind}\mathcal{D}_\kappa = (\widehat{A}_V \smile \operatorname{ch}\kappa)[V]$$

which is pertinent since one can arrange the matters with Δ so that $\operatorname{ch}\kappa$ is "sufficiently far" from zero being non-trivially connected to $\operatorname{ch}([X_+] - [X_-])$. Namely, there are sufficiently many Δ's (and one can, probably, gain extra mileage by using representations of $U(p,q)$ by isometries of symmetric spaces $Z' \neq Z$) to make the ring generated by $\operatorname{ch}\kappa$'s (at least) as large as $H^{\mathrm{ev}}_{\mathrm{fl}}$. (See the original paper [Con]$_{\mathrm{CCTC}}$ and also §III.7 in the book [Con]$_{\mathrm{NCG}}$ for a wealth of ideas yet awaiting their full commutative geometric implementation, compare the "non-proper" discussion in §III of [C-G-M]$_{\mathrm{CCLC}}$).

$8\frac{5}{8}$. Families of Hermitian bundles

Let $\rho_{\underline{b}} : \Pi \to U(p,q)$ be a family of representations parametrized by a space $\underline{B} \ni \underline{b}$ and $X = \left\{X_{\underline{b}} = X_{\rho_{\underline{b}}}\right\}$ be the corresponding family of flat Hermitian bundles over a (connected closed oriented) manifold V with $\pi_1(V) = \Pi$. We want to define, following Lusztig, a (homotopy invariant!) signature $\sigma(V;X) \in K_*(\underline{B})$ and then express it in term of the characteristic classes of V and X. To do this we interpret $\rho_{\underline{b}}$ as a homomorphism ρ from Π to the group of $(p+q)$-matrices over the ring $\underline{R} = \operatorname{Cont}\underline{B}$ of continuous functions $\underline{B} \to \mathbb{C}$, and then as earlier, we obtain a homomorphism $\operatorname{Witt}_* \mathbb{C}(\Pi) \to \operatorname{Witt}_* \operatorname{Mat}_{pq}\underline{R}$ induced by the involutive homomorphism $\mathbb{C}(M) \to \operatorname{Mat}_{pq}\underline{R}$ associated to ρ (where the involution in $\operatorname{Mat}_{pq} R = \operatorname{Mat}_{pq} \otimes \underline{R}$ comes from the complex conjugation in $\underline{R}$ and the $U(p,q)$-involution in Mat_{pq}). Now, to avoid irrelevant technicalities, we assume $\underline{B}$ is compact and use the natural homomorphism $\sigma : \operatorname{Witt}_* \operatorname{Mat}_{pq}\underline{R} \to K_*(\underline{R}) = K^*(\underline{B})$ as in example (c) of $7\frac{7}{8}$. In particular, we obtain with $\rho_{\underline{b}}$ a family of chain $\mathbb{C}(\Pi)$-complexes $C^*_{\underline{b}}$ of (some triangulation of) V with coefficients in $X_{\underline{b}}$, $\underline{b} \in \underline{B}$, which we view as a single $\underline{R}(\Pi)$-complex which then can be made "short" by algebraic surgeries reducing it, in the case $\dim V$ even, to a single non-singular (skew)-symmetric form over $\underline{R}$, i.e. such a form in some vector bundle $Y \to \underline{B}$, with $\sigma(V;X)$ becoming the difference $[Y_+] - [Y_-] \in K^{\mathrm{ev}}(\underline{B}) = K_{\mathrm{ev}}(\underline{R})$. It is (more or less) obvious that this $\sigma(V)$ is a homotopy invariant of V (compare $7\frac{7}{8}$).

Next, we consider the family of the (differential) signature operators $\mathcal{L}_{X_{\underline{b}}}$ over V, $\underline{b} \in \underline{B}$, and recall that the index

$$\operatorname{Ker}\mathcal{L}_{X_{\underline{b}}} - \operatorname{Ker}\mathcal{L}^*_{X_{\underline{b}}}, \quad \underline{b} \in \underline{B},$$

(despite the fact that the dimensions of these kernels may vary with $\underline{b}$) is defined as an element of $K^{\mathrm{ev}}(\underline{B})$ and denoted $\operatorname{ind}\mathcal{L}_X \in K^{\mathrm{ev}}(\underline{B}) = K_{\mathrm{ev}}(\underline{R})$. Now we may state the ***Lusztig signature theorem for flat families.***

$$\sigma(V;X) = \operatorname{ind}\mathcal{L}_X,$$

where, $\operatorname{ind}\mathcal{L}_X$ *can be expressed according to the index theorem for families as*

$$\operatorname{ch}\operatorname{ind}\mathcal{L}_X = \underline{\operatorname{Gys}}\left(L_V \smile \operatorname{ch}(X_+ - X_-)\right).$$

Here X is regarded as a bundle over $V \times \underline{B}$ and L_V refers to the pull-back of the L-class of V for the projection $V \times \underline{B} \to V$ while $\overline{\text{Gys}}$ denotes the Gysin push forward homomorphism $H^*(V \times \underline{B}) \to H^*(\underline{B})$ for the projection $V \times \underline{B} \to B$.

Remark. Since $\sigma(V;X)$ is a homotopy invariant of V, so is $\underline{\mathrm{Gys}}\,(L_V \smile \mathrm{ch}(X_+ - X_-))$ which, for interesting X, provides non-trivial homotopy invariance properties of L_V.

Example. Let $\Pi = \mathbb{Z}^n$ and $\underline{B}$ be the dual *n-torus*, i.e. $\underline{B} = \mathrm{Hom}(\mathbb{Z}^n \to \mathbb{T}^1)$. We recall that the group ring $\mathbb{C}(\mathbb{Z}^n)$ is canonically isomorphic to the (dense) subring $\underline{R}_0 \subset \underline{R} = \mathrm{Cont}(\mathbb{T}^n)$ consisting of polynomial functions in the variables t_i and t_i^{-1}, $i = 1, \dots, n$. We view homomorphisms $\mathbb{Z}^n \to \mathbb{T}^1$ (parametrized by $\underline{B}$) as one dimensional complex representations and thus obtain a representation of $\mathbb{Z}^n$ over the ring $\underline{R} = \mathrm{Cont}\,\underline{B}$ of rank 1, i.e. an $\underline{R}$-linear action of $\mathbb{Z}^n$ on $\underline{R}$. This action preserves $\underline{R}_0 = \mathbb{C}(\mathbb{Z}^n)$ where it coincides with the ring group product (convolution) in $\mathbb{C}(\mathbb{Z}^n) \supset \mathbb{Z}^n$ and so our homomorphism $\mathrm{Witt}_* \,\mathbb{C}(\Pi) \to \mathrm{Witt}_* \,\underline{R}$ for $\Pi = \mathbb{Z}^n$ and $\underline{R} = \mathrm{Cont}(\underline{B} = \mathbb{T}^n)$ coincides with the one by the inclusion $\mathbb{C}(\Pi) = \underline{R}_0 \subset \underline{R}$ (compare Example (d) in $7\frac{7}{8}$).

Now we are able to prove the Novikov conjecture for $\Pi = \mathbb{Z}^n$ by showing that the composed homomorphism

$$H_{\mathrm{ev}}(BM;\mathbb{Q}) \to (H\,\mathrm{Brd}_{\mathrm{ev}}\,BM) \otimes \mathbb{Q} \to (\mathrm{Witt}_{\mathrm{ev}}\,\mathbb{C}(M) \otimes \mathbb{Q} \to K^0(\underline{B})$$

$$\underline{\kappa}$$

is injective. If a class in $H_{2k}(B\Pi;\mathbb{Q})$ is realized by a $2k$-dimensional stably parallelizable manifold V mapped to $\underline{B} = \mathbb{T}^n$ then, by the above discussion, $\underline{\kappa}([V]) = \sigma(V;X)$ for our bundle $X \to V \times \underline{B}$ (which is here of $\mathbb{C}$-rank one) for $\underline{B} = \mathbb{T}^n$, and by the index theorem

$$\mathrm{ch}\,\sigma(V;X) = \underline{\mathrm{Gys}}(\mathrm{ch}\,X).$$

Evaluation of $\underline{\mathrm{Gys}}(\mathrm{ch}\,X)$. Recall that the bundle X is naturally associated to the family of representations $\pi_1(V) \to \pi_1(B\mathbb{Z}^n) = \mathbb{Z}^n \xrightarrow[\rho_{\underline{b}}]{u} \mathbb{T}^1$ parametrized by $\underline{B} = \mathbb{T}^n$ (where we do not actually have to assume $\pi_1(V) = \mathbb{Z}^n$, a homomorphism $\pi_1(V) \to \mathbb{Z}^n$ will do), such that $X|V \times \underline{b} = X_{\rho_{\underline{b}}}$. Let us determine the first Chern class $c_1(X) \in H^2(V \times \underline{B})$ by evaluating it on each 2-subtorus $S_V^1 \times S_{\underline{B}}^1 \subset V \times \underline{B}$ for oriented circles in S_V^1 in V and $S_{\underline{B}}^1$ in $\underline{B}$. As $\underline{b}$ (parametrizing the representation $\rho_{\underline{b}} : \mathbb{Z}^n \to \mathbb{T}^1$) turns $S_{\underline{B}}^1$ the image of $[S_V^1]$ in $\mathbb{T}^1$ under the composed map $[S_V^1] \in \pi_1(V) \to \mathbb{Z}^n \xrightarrow[\rho_{\underline{b}}]{u} \mathbb{T}^1$ turns around $\mathbb{T}^1$ an integer number of times and this integer (obviously) equals $\left\langle c_1(X), \left[S_V^1 \times S_{\underline{B}}^1\right]\right\rangle$. It follows that for $n = 1$ and $V = S^1$

$$\underline{\mathrm{Gys}}(\mathrm{ch}\,X) = 1 \in H^1(\underline{B} = \mathbb{T}^1). \qquad (*)$$

Since $1 \neq 0$ this is (essentially) equivalent to Novikov's codim 1-theorem.

Next we observe that $\underline{\text{Gys}}(\text{ch}\, X)$ can be computed in terms of the homology class $[V] \in H_*(B\mathbb{Z}^n)$ where $B\mathbb{Z}^n = (S^1)^n$ is the dual torus to $\mathbb{T}^n$. We project the product $(S^1)^n \times \mathbb{T}^n$ to the two factors, $(S^1)^n \underset{p}{\longleftarrow} (S^1)^n \times \mathbb{T}^n \underset{\underline{p}}{\longrightarrow} \mathbb{T}^n$, and recall that $\underline{\text{Gys}}$ is obtained by combining P_* on homology with the Poincaré duality. Thus

$$\text{ch}(\underline{\kappa}([V])) = \underline{\text{Gys}}\,\text{ch}\, X = \mathcal{PD}(\underline{p}_*(p^{-1}[V] \frown \mathcal{PD}(\text{ch}\, X_n))) \qquad (+)$$

where X_n is the (universal) line bundle on $(S^1)^n \times \mathbb{T}^n$ arising from our family $\underline{B} = \mathbb{T}^n$ of representations $\pi_1(S^1)^n = \mathbb{Z}^n \to \mathbb{T}^1$, and $p^{-1} = \mathcal{PD}\underline{\text{Gys}}$, i.e. $\mathcal{PD}p^*\mathcal{PD}$.

Notice that for every cohomology class $\underline{h} \in H^*(\mathbb{T}^n)$ the equality (+) implies

$$\begin{aligned}
&\langle \text{ch}(\underline{\kappa}[V]) \smile \underline{h}, [\mathbb{T}^n] \rangle = \langle p_*(p^{-1}[V] \frown \mathcal{PD}(\text{ch}\, X_n), \underline{h} \rangle \\
&= \langle p^{-1}[V] \frown \mathcal{PD}\,\text{ch}\, X_n, \underline{p}^*(\underline{h}) \rangle = \langle \mathcal{PD}(p^{-1}[V]) \smile \text{ch}\, X_n \smile \underline{p}^*(\underline{h}), [(S^1)^n \times \mathbb{T}^n] \rangle \\
&= \langle p^*(h) \smile \text{ch}\, X_n \smile \underline{p}^*(\underline{h}), [(S^1)^n \times \mathbb{T}^n] \rangle \in \mathbb{Q}
\end{aligned}$$

for $h = \mathcal{PD}[V] \in H^*((S^1)^n)$. The latter formula, applied to arbitrary h and $\underline{h}$, defines a pairing, denoted $\Phi_{X_n} : H^*((S^1)^n; \mathbb{Q}) \otimes H^*(\mathbb{T}^n; \mathbb{Q}) \to \mathbb{Q}$, which is (by the above computation) non-singular iff the homomorphism $\text{ch} \circ \underline{\kappa} : H_*((S^1)^n; \mathbb{Q}) \to H^*(\mathbb{T}^n; \mathbb{Q})$ is injective. Since the bundle X_n (obviously) equals the (Cartesian) tensor product of n copies of $X_1 \to S^1 \times \mathbb{T}^1$, so $\text{ch}\, X_n = \underset{n}{\smile} \text{ch}\, X_1$, and the pairing Φ_{X_n} equals the tensor product of n-copies of Φ_{X_1}, where

$$\begin{aligned}
\Phi_{X_1}((h_0, h_1) \otimes (\underline{h}_0, \underline{h}_1)) &= h_0\underline{h}_0 + h_1\underline{h}_1 \\
\text{for} \quad h_0 &\in H^0(S^1) = \mathbb{Z} = H^1(S^1) \ni h_1 \\
\text{and} \quad \underline{h}_0 &\in H^0(\mathbb{T}^1) = \mathbb{Z} = H^1(\mathbb{T}^1) \ni \underline{h}_1
\end{aligned}$$

as we saw earlier.

More precisely, we see by induction on n, that the pairing Φ_{X_n} between the exterior algebras $H^*((S^1)^n) = \Lambda(x_1, \ldots, x_n)$ and $H^*(\mathbb{T}^n) = \Lambda(y_1, \ldots, y_n)$ is given by

$$(x + x' \wedge x_n)c_{n-1} \wedge (1 + x_n \wedge y_n) \wedge (y + y_n \wedge y'),$$

for $x, x' \in \Lambda(x_1, \ldots, x_{n-1})$, $y, y' \in \Lambda(y_1, \ldots, y_{n-1})$ and $c_{n-1} = \text{ch}\, X_{n-1}$. This exterior product develops to $x \wedge y \wedge c_{n-1} \wedge x_n \wedge y_n + x' \wedge c_{n-1} \wedge x_n \wedge y_n \wedge y'_n = \Phi_{X_{n-1}}(x, y) + \Phi_{X_{n-1}}(x', y')$, (where, recall all components of c_{n-1} have even degrees and so commute with x_n), which makes non-singularity of Φ_{X_n} follow from that of $\Phi_{X_{n-1}}$. This proves Novikov's conjecture for $\Pi = \mathbb{Z}^n$ modulo Lusztig's signature theorem $\sigma(V;X) = \operatorname{ind}\mathcal{L}_X$.

Idea of the proof of Lusztig's theorem. First, we redefine the Wall-Miščenko class $WM([V]) \subset Witt_*$ in differential terms without referring to any triangulation of V by using, instead of chains, the de Rham complex of smooth forms on V with the pairing given by the exterior product, $(\omega_1, \omega_2) \to \int_V \omega_1 \wedge \omega_2$. This is, of course, an infinite dimensional complex, but it is Fredholm (or elliptic) which allows a reduction to a finite dimensional one. The Fredholm property can be seen, for example, with a smoothing operator on forms given by $\omega \mapsto \int_M \operatorname{Diff}_\mu^*(\omega) d\mu$ where M is a compact connected family of diffeomorphisms close to the identity with a probability measure $d\mu$ on M. This smoothing gives us a compact endomorphism of the de Rham complex homotopic to the identity and commuting up to (properly understood) homotopy with the above exterior product pairing, which is sufficient for a de Rham definition of $WM([V]) \in \operatorname{Witt}_* \mathbb{C}(\Pi)$.

Alternatively, one may use a Riemannian metric on V and restrict the de Rham complex to eigenforms of the Hodge-Laplace operator belonging to eigenvalues below a certain level. The former definition, being rather local, is better adjusted to infinite coverings $\widetilde{V} \to V$, while the latter is good enough for our families of compact manifolds. Then one identifies the de Rham version of WM with the combinatorial one by observing that the relevant algebraic Poincaré complexes are homotopy eiquivalent by proceeding as in the usual de Rham theorem. Both definitions perfectly work for families of compact manifolds V. Moreover, for such families, one can use the second smooth definition of WM, which provides a suitable context for bringing in the signature operator $\mathcal{L}$ and its index. Actually, Lusztig's proof (see [Lus]) of the identity $\sigma(V;X) = \operatorname{ind}\mathcal{L}_X$ consists in a construction of a fiberwise homotopy between relevant bundles of complexes over $\underline{B}$ built of eigenform in the fibers $V \times \underline{b} \subset V \times \underline{B}$ (rather than an individual isomorphism of the previously considered case where $\underline{B} = \{\underline{B}_0\}$).

Remarks.

(a) Lusztig's proof extends to families over $\underline{B}$ which are not products and/or where bundles X do not come from representations. All one needs is a smooth fiber bundle $A \to \underline{B}$ with smooth fibers V and smooth $U(p,q)$-bundle $X \to A$ with a flat structure along the fibers. Notice that the combinatorial definition of $WM \in \operatorname{Witt}_* \underline{R}$ and/or $\sigma \in K^*(\underline{B})$ becomes technically slightly more complicated since we must match Poincaré complexes over different points $\underline{b} \in B$ where the fibers $A_{\underline{b}}$, diffeomorphic to a fixed V, have non-isomorphic triangulations. This actually may lead to an interesting signature even for the trivial bundle $X \to A$; see $[\mathrm{At}]_{\mathrm{SFB}}$.

(b) The K-theoretic signature $\sigma(V;X) \in K^*(\underline{B}) = K_*(\underline{R} = \operatorname{Cont}\underline{B})$ can be brought to an equal footing with the ordinary one with values in $\mathbb{Z} = K_0(\mathbb{C})$

(see (c) below), but there is (at least for a casual eye) an essential difference between the two due to the fact that the K-valued signature is *not a homological* invariant. In particular, it is much harder to define it for *topological* manifolds (where there is no obvious class of associated Poincaré complexes) and it seems impossible (?) to make sense of "K-signature of a homology class" in V (but the K-signature for manifolds with boundary may stand a chance).

(c) ***C^*-algebras.*** Whenever one has a representation of a group Π in a free Hermitian module M of a finite rank over some involutive algebra R, one defines, for each $(V, \beta : V \to B\Pi)$, a flat M-fibered bundle X over V and $WM[V] \in \mathrm{Witt}_* R$. We dealt above with the cases of $R = \mathbb{C}$ and $R = \mathrm{Cont}\,\underline{B}$.

Another important class of examples is given by the group algebra $\mathbb{C}(\Pi)$ itself and its extensions, such as $C^*(\Pi) \supset \mathbb{C}(M)$ which is the completion of $\mathbb{C}(\Pi)$ in the operator norm topology. This $C^*(\Pi)$, as well as $\underline{R} = \mathrm{Cont}\,\underline{B}$, is identified with an involutive (for taking adjoints) subalgebra of operators on a Hilbert space closed in the operator norm topology (where continuous functions on $\underline{B}$ act by multiplication on $L_2(\underline{B})$ and the group ring acts on $\ell_2(\Pi)$ by convolution). Such algebras are called *C^*-algebras* and they, albeit non-commutative, share many common properties with algebra of continuous functions and may be thought of as algebras of continuous functions on certain *non-commutative* or *quantum* "spaces". For example, Hermitian forms are diagonalizable over such algebras (by the spectral theorem) and one has a natural homomorphism $\mathrm{Witt}_0(R) \to K_0(R)$ defined by $M \to [M_+] - [M_-]$ as in the case of $R = \mathrm{Cont}\,\underline{B}$ (see $[\mathrm{Ros}]_{\mathrm{ANFT}}$ for a definition of $\mathrm{Witt}_n \to K_n$ for all n). Furthermore, one can define the index of the signature operator $\mathcal{L}_X$ with values in $K_0(R)$ as well as of any other elliptic (pseudo)differential operator on V twisted with X, such as the Dirac operator.

This, actually, can be done for not necessarily flat C^*-algebra bundles over V; (see [Mi-Fo], [Kas]). Also non-trivial fibrations $A \to \underline{B}$ with fibers $A_{\underline{b}}$ diffeomorphic to V (see above (a)) fit into this context as the spaces of the fiberwise differential forms are $\underline{R}$-moduli and the fiberwise elliptic operators are Fredholm over $\underline{R}$ which allows the definition of the index $\in K_*(\underline{R}) = K^*(\underline{B})$.

(c$'$) ***The homotopy invariance of $\mathcal{L}_X$.*** Lusztig's argument generalizes to the non-commutative C^*-algebra context (see [Ka-Mi] and references therein) and shows that $\mathrm{ind}\,\mathcal{L}_X \in K_0(R)$ is a homotopy invariant of V. In fact, this index can be defined for quite general Hermitian Fredholm complexes over C^*-algebras where one can prove its invariance under chain homotopy equivalences (see [Ka-Mi]). However, this does not directly lead to the Novikov conjecture as, for all we know, the group $K_0(R)$ can be too small to contain sufficient information about the characteristic classes of V. But it suggests another version of the Novikov conjecture, called *strong*, or *C^*-Novikov*, which claims, essentially, that for $R = C^*(\Pi)$, the group $K_*(R) \otimes \mathbb{Q}$ is as big as $H_*(B\Pi; \mathbb{Q})$, which is manifested by injectivity of the composed map

$$\underbrace{H_*(B\Pi;\mathbb{Q}) \to (\mathrm{Witt}_*\,\mathbb{C}(\Pi))\otimes\mathbb{Q} \to (\mathrm{Witt}_*\,C^*(\Pi))\otimes\mathbb{Q} \to K_*(C^*(\Pi))}_{\underline{\kappa}}\otimes\mathbb{Q}$$

In fact, one can cast the construction of $\underline{\kappa}$ in a purely K-theoretic framework by replacing $H_*(B\Pi)$ by $K_*(B\Pi)$ and the defining corresponding homomorphism, call it $\kappa : K_*(B\Pi) \to K_*(C^*(\Pi))$, operator-theoretically without using $H_*(B\Pi)$ and $\mathrm{Brd}_*(B\Pi)$ (see below).

$8\frac{2}{3}$. Index homomorphism $\mathbf{K} : \mathbf{K_0(B\Pi) \to K_0(C^*(\Pi))}$ and the strong Novikov conjecture

Recall the definitions.

***Definition of** $K_0(R)$.* This is defined for an arbitrary ring R as the (Grothendieck) group of isomorphism classes of *projective* moduli M over R of finite rank. In other words, this is the Abelian group generated by these M's with the relations

(1) *if* M_1 *isomorphic to* M_2 *then* $[M_1] = [M_2]$, *where* $[M]$ *refers to the class of* M *in* K_0,

(2) *if* $M = M_1 \oplus M_2$ *then* $[M] = [M_1] + [M_2]$.

Definition of "projective". Here "projective of finite rank" signifies that M is a direct summund of free module of finite rank, i.e. M embeds into $R^N = \underbrace{R \oplus R \oplus \cdots \oplus R}$, where it admits a projection $P : R^N \to M \subset R^N$ fixing M. Thus every M is represented by an idempotent in the matrix ring $\mathrm{Mat}\,R$, i.e. an operator $P \subset \mathrm{Mat}\,R$ with $P^2 = P$. For example, if R is a field, our M are just finite dimensional vector spaces. Relation (a) and (b) turn them into the semigroup of positive integers but as we say "group" we complete it to the group of integers. Another example is $R = \mathrm{Cont}\,B$ for a compact metric space B. Here free moduli R^N correspond to trivial bundles $\mathrm{Triv}^N \to B$ and their projective submoduli correspond to subbundles since the indempotents $p : R^N \to R^N$ appear as bundle endomorphisms satisfying $P^2 = P$. Thus "projective moduli over R" translates to "vector bundles over B" and $K_0(\mathrm{Cont}\,B) = K^0(B)$. K_0 is a covariant function while K^0 is a contravariant, one which fits with $B \rightsquigarrow \mathrm{Cont}\,B$ being a contravariant functor.

The rings we care most about are group rings such as $\mathbb{C}(M)$ and various completions of $\mathbb{C}(M)$ for infinite groups Π. The K_0-groups of group rings without completion tend to be rather small. For example, $K_0\mathbb{C}(\mathbb{Z}^n) = 0$ and there is a conjecture that $K_0(\mathbb{C}(\Pi)) = 0$ for all torsionless finitely presented (finitely generated?) groups Π (related to the Kaplansky conjecture claiming that the relation $r_1 r_2 = 0$ in the group ring of Π without torsion implies that either r_1 or r_2 is zero).

Definition of $K_0(B)$. This is motivated by the following observation (due to Atiyah). Let D be a pseudo-differential operator of *order zero* over a compact manifold V. Such a $\mathcal{D}$ acts between the L_2-spaces of sections of the implied bundles, say $\mathcal{D} : H_+ \to H_-$, and it is a *bounded Fredholm* operator between these Hilbert spaces of sections. Furthermore, $\mathcal{D}$ *almost commutes* with multiplication by *continuous functions* f on V in the sense that the commutator $(\mathcal{D} \circ f - f \circ \mathcal{D}) : H_+ \to H_-$ is a *compact* operator for all $f \in \operatorname{Cont} V$. On the other hand, one can twist $\mathcal{D}$ with an arbitrary vector bundle X over V and define the index of the twisted operator, say $\operatorname{ind} \mathcal{D}_X \in \mathbb{Z}$, which gives one a homorphism $K^0(V) \to \mathbb{Z}$ for $[X] \mapsto \operatorname{ind} \mathcal{D}_X$. An appropriate general twisting procedure of $\mathcal{D}$ with X is as follows. First, for $X = \operatorname{Triv}^N$ we just take $\mathcal{D}^N = \underbrace{\mathcal{D} \oplus \mathcal{D} \oplus \cdots \mathcal{D}}_{N}$ and then we *compress* $\mathcal{D}^N$ to a given subbundle $X \subset \operatorname{Triv}^N$ by composing with a projection $P : \operatorname{Triv}^N \to X$, i.e. by setting $\mathcal{D}_X = P \circ \mathcal{D}^N$. (Recall that originally, $\mathcal{D}$ acts between sections of bundles, say $\mathcal{D} : H_+ = L_2(S_+) \to H_- = L_2(S_-)$. Then $\mathcal{D}^N$ acts between sections of the tensor products $S_+ \otimes \operatorname{Triv}^N$ and $S_- \otimes \operatorname{Triv}^N$ while $\mathcal{D}_X$ acts from sections of $S_+ \otimes X \subset S_+ \otimes \operatorname{Triv}^N$ to those of $S_- \otimes X \subset S_- \otimes \operatorname{Triv}^N$ by $\mathcal{D}_X(s_+ \otimes x) = P\mathcal{D}_N(s_* \otimes x)$ where P applies to the sections of $S_- \otimes \operatorname{Triv}^N$ via the second component. This agrees with the twist for *differential* operators $\mathcal{D}$ of the first order with (X, ∇) for the connection ∇ on X induced from the trivial one on Triv^N by the compression $\nabla^{\operatorname{Triv}}$ with P; compare $6\frac{1}{2}$).

Example. Suppose we start with a first order elliptic differential operator, say $\mathcal{D} : C^\infty(S_+) \to C^\infty(S_-)$, such as the Dirac or signature operator. This can be directly twisted with bundles X which defines the index homomorphism $\operatorname{ind} : K^0(V) \to \mathbb{Z}$ (for $\operatorname{ind}_{\mathcal{D}}[X] = \operatorname{ind} \mathcal{D}_X$). Alternatively, we may first modify $\mathcal{D}$ in order to make it L_2-bounded by taking its *polar part*, defined by

$PP\mathcal{D} = \mathcal{D}$ on $\ker \mathcal{D}$

$PP\mathcal{D} = \mathcal{D}(D^*\mathcal{D})^{-\mathbb{1}}$ away from $\ker \mathcal{D}$.

Or, if one does not want to bother with $\ker \mathcal{D}$, one may take $\widehat{\mathcal{D}} = \mathcal{D}(1 + \mathcal{D}^*\mathcal{D})^{-1}$ and observe that this is a zero order pseudo-differential operator with the property $\operatorname{ind} \widehat{\mathcal{D}}_X = \operatorname{ind} \mathcal{D}_X$ for all vector bundles X over V.

Now we are psychologically prepared for the definition of $K_0(B)$. This is done via $K^0(R = \operatorname{Cont} B)$ which, in fact, will be done now for all algebras R over $\mathbb{C}$ with involutions as follows. First we introduce K*-cycles* as *Fredholm representations of* R, i.e. pairs of actions of R on Hilbert spaces, say on H_+ and H_- (i.e. involutive homomorphisms of R into the algebras Bnd.oper (H_+) and Bnd.oper (H_+)) and a *bounded Fredholm* operator $\mathcal{D} : H_+ \to H_-$ which *almost* *commutes* with these actions in the sense that the commutator $\mathcal{D} \circ f - f \circ \mathcal{D}$ is a compact operator on H_- for all $f \in R$ (where "Fredholm" signifies the existence of an "appropriate inverse" bounded operator $\mathcal{D}' : H_- \to H_+$, such that $\mathcal{D}'\mathcal{D} - 1$ and $\mathcal{D}\mathcal{D}' - 1$ are compact operators). These cycles form a semigroup for the direct (Cartesian) sum of underlying Hilbert spaces and representations. Then we add the following (equivalence) relations between the K-cycles.

(1) ***Isomorphism.*** The existence of bounded linear isomorphisms $H_+ \leftrightarrow H'_+$ and $H_- \leftrightarrow H'_-$ which commute with the operators on both sides.

(2) ***Homotopy.*** This refers to homotopies $D_t : H_+ \to H_-$, $t \in [0,1]$, which are supposed to be norm continuous in t and almost commute, for all $t \in [0,1]$ with implied actions of R on H_+ and H where these actions stay still with t running over $[0,1]$. The resulting K-cycles, for $t = 0$ and $t = 1$ are declared equivalent (by this homotopy).

(3) ***Degeneration.*** A K-cycle is called *degenerate* if the corresponding operator $\mathcal{D}$ is a bounded linear *isomorphism* between H_+ and H_- which *commutes* with the actions of R on H_+ and H_-. And the degenerate cycles are declared zero in $K^0(R)$.

Now we divide the semigroup of K-cycles by (1) + (2) + (3) and obtain $K^0(R)$. Notice that taking inverse in this group corresponds to $H_+ \leftrightarrow H_-$ and $\mathcal{D} \leftrightarrow \mathcal{D}^*$.

Observe that $K^0(R)$ stands up to the notation being a *contravariant* functor from algebras to Abelian groups and thus $B \rightsquigarrow K_0(B) = K^0(\mathrm{Cont}\, B)$ is *covariant.* It takes some effort to prove that K_0 is a homology theory, e.g. it is a homotopy functor (which amounts to showing that $K_0(B \times [0,1]) = K_0(B)$. Also one has to prove that $K_0\,\{\text{point}\} = \mathbb{Z}$ for the homomorphism $[\mathcal{D}] \mapsto \mathrm{Ind}\,\mathcal{D}$. (This follows from Kuiper's theorem claiming connectivity (and even, contractibility) of the group of bounded linear automorphism of an infinite dimensional Hilbert space). But we shall not need all these properties of $K_0(B)$, but only the existence of a (index) pairing between K_0 and K^0 and of a homomorphism $\mathrm{ch}' : H_{\mathrm{ev}}(B; \mathbb{Q}) \to K_0(B) \otimes \mathbb{Q}$ (defined later on with the signature operator) injective with respect to this pairing, which means $\langle \mathrm{ch}'\, h_{\mathrm{ev}}, k^0 \rangle = 0$ for all $k^0 \in K^0(B)$ implies $h_{\mathrm{ev}} = 0$ and which yields the ordinary injectivity of ch'.

Index pairing between K_0 and K^0. We define $\langle [R], [\mathcal{D}] \rangle = \mathrm{ind}\,\mathcal{D}$, where $[R] \in K_0(R)$ is the (distinguished) element represented by the free 1-dimensional module over R identified with R and $[\mathcal{D}] \in K^0(R)$ the class of a cycle $\mathcal{D} : H_+ \to H_-$. Next, for a free module R^N we take $\mathcal{D}^N = \underbrace{\mathcal{D} \oplus \mathcal{D} \oplus \cdots \oplus \mathcal{D}}_{N} : H_+^N = H_+ \otimes_R R^N \to H_-^N = H_- \otimes_R R^N$ and set $\langle [R^N], [D] \rangle =$

$\operatorname{ind}\mathcal{D}^N (= N \cdot \operatorname{ind}\mathcal{D})$. Finally for a projective submodule $X \subset R^N$ we define $\mathcal{D}_X : H_+ \otimes_R X \to H_- \otimes_R X$ by using the embeddings $H_\pm \otimes_R X \subset H_\pm^N = H_\pm \otimes_R R^n$ and the projection $P_- : H_-^N \to H_- \otimes_R X$ corresponding to $P : R^N \to X$ implied by the definition of projectivity of X. Namely, $\mathcal{D}_X$ acts on $h = (h_1, \dots, h_N) \in H_+ \otimes_R X \subset H_+^N$ by $h^+ \mapsto P_- \mathcal{D}^N(h)$ where P_- projects $H_-^N = H_- \otimes_R R^N$ to $H_- \otimes_R X$ according to $h_- \otimes r \mapsto h_- \otimes Pr$. One checks easily that $\mathcal{D}_X$ is Fredholm (with the appropriate inverse $\mathcal{D}'_X = P_+(\mathcal{D}')^N$) and set

$$\langle [X], [D] \rangle \underset{\text{def}}{=} \operatorname{ind}\mathcal{D}_X.$$

Index pairing between $K^0(R)$ ***and*** $K_0(R \otimes \underline{R})$ ***with values in*** $K_0(\underline{R})$. The basic example is where we have a family of elliptic operators on a manifold V of the form $\mathcal{D}_{X_b}$ where $\mathcal{D}$ is a fixed operator over V and X_b is a variable bundle over V parametrized by $\underline{B} \ni b$. Or, we have a bundle X over $V \times \underline{B}$ and $X_b = X|V \times b$. The index of this family lies in $K^0(\underline{B})$. Now, for general C^*-algebras R and $\underline{R}$, we imitate the construction of $\mathcal{D}_{X_b}$ and $\operatorname{ind} \in K^0(\underline{B})$ as follows. Given a K-cycle $\mathcal{D} : H_+ \to H_-$ over R and a projective module $\mathcal{X} \subset (R \otimes \underline{R})^N$ with $P : (R \otimes \underline{R})^N \to \mathcal{X}$ we take the tensor products $\underline{H}_\pm = H_\pm \otimes \underline{R}$ which come along with the structures of *Hilbert moduli* over $\underline{R}$, which means they possess besides the actual, say right, $\underline{R}$ moduli structures, scalar products with values in $\underline{R}$ having the same formal properties as the usual scalar product and where the model example is a Hilbert vector bundle $\underline{X}$ over a space $\underline{B}$ with the Cont $\underline{B}$-valued scalar product on the space $\underline{H}$ of its continuous sections corresponding to the pointwise scalar product $\langle \underline{h}_1, \underline{h}_2 \rangle_b$, $b \in \underline{B}$ (see [Kas] for details). We tensor our $\mathcal{D}$ with $\underline{R}$ and obtain an $\underline{R}$-*Fredholm* operator $\underline{\mathcal{D}} : \underline{H}_+ \to \underline{H}_-$, which means that there exists an *approximate inverse over* $\underline{R}$ that is a bounded Hilbert module morphism $\underline{\mathcal{D}}' : \underline{H}_- \to \underline{H}_+$ such that $\underline{\mathcal{D}}\,\underline{\mathcal{D}}' - 1$ and $\underline{\mathcal{D}}'\,\underline{\mathcal{D}} - 1$ are *compact over* $\underline{R}$ i.e. lie in the operator norm closure of the span of the "rank-one operators", i.e. $\underline{R}$-morphism of the form $\pi_\pm : H_\pm \to \underline{R}h_\pm \subset H_\pm$ for $h_\pm :\subset \underline{H}_\pm$.

Finally, one twists $\mathcal{D}$ with a projective module $\mathcal{X}$ over $R \otimes \underline{R}$ as earlier (by composing with P) and gets an operator $\underline{\mathcal{D}}_\mathcal{X} : \underline{H}_+ \otimes \mathcal{X} \to \underline{H}_- \otimes \mathcal{X}$ where we tensor over $R \otimes \underline{R}$ as R acts on $\underline{H}_+$ on the left and $\underline{R}$ acts on the right. The operator $\underline{\mathcal{D}}_\mathcal{X}$ is $\underline{R}$-Fredholm and can be perturbed to another $\underline{R}$-Fredholm morphism $\underline{\mathcal{D}}_1$ having closed image and such that $\operatorname{Ker}\underline{\mathcal{D}}_1$ and $\operatorname{Coker}\underline{\mathcal{D}}_1 = \underline{H}_- \otimes \mathcal{X}/\operatorname{Im}\underline{\mathcal{D}}_1$ are projective moduli of finite rank over $\underline{R}$. Then one defines $\langle [\mathcal{X}], [\mathcal{D}] \rangle = [\operatorname{Ker}\underline{\mathcal{D}}_1] - [\operatorname{Coker}\underline{\mathcal{D}}_1] \in K_0(\underline{R})$; (see [Kas] and $[\text{Ros}]_{\text{KKK}}$ for details and references).

Construction of $\mathbf{K} : \mathbf{K_0(B\Pi)} \to \mathbf{K_0(C^*(\Pi))}$. Suppose $B\Pi$ is compact and let $R = \operatorname{Cont} B\Pi$ and $\underline{R} = C^*(\Pi)$. As the fundamental group Π of B, Π

(obviously) acts on $C^*(\Pi)$ we have a flat $\underline{R}$-fibered bundle X over $B\Pi$ associated to the universal covering of $B\Pi$. As the total space of X is acted upon by Π, the space $\mathcal{X}$ of continuous sections $B\Pi \to \mathcal{X}$ has an $\underline{R}$-module structure as well as the (obvious) R-module structure and thus an $R\otimes\underline{R}$-module structure. In fact this module is projective of finite rank over $R\otimes\underline{R}$ since X, being a locally trivial vector bundle, embeds into finite sum of trivial vector bundles X_i over $B\Pi$ where X_i equals X over some neighbourhood $U_i \subset B\Pi$ with U_i, $i=1,\dots,N$, covering $B\Pi$. Now we define our

$$\mathbf{K} : K_0(B\Pi) = K^0(R) \to K_0(\underline{R} = C^*(\Pi))$$

by pairing $K^0(R)$ with $[\mathcal{X}]$ as described above. One loosely can say that $\kappa([\mathcal{D}])$ equals $\operatorname{ind}\mathcal{D}_X \in K_0(R)$ for the operator $\mathcal{D}$ twisted with X. In fact, if the K-class of $\mathcal{D}$ comes from that of the signature operator $\mathcal{L}$ of a manifold V mapped to $B\Pi$, then $\mathbf{K}([\mathcal{D}]) = \operatorname{ind}\mathcal{L}_X$, for X' over V induced from X.

Example. Let $\Pi=\mathbb{Z}^n$ and $B\Pi$ be the torus $(S^1)^n$. Then the C^*-algebra $C^*(\Pi=\mathbb{Z}^n)$ is isomorphic to $\operatorname{Cont}\mathbb{T}^n$ (for the torus $\mathbb{T}^n$ dual to $(S^1)^n$) and there is a canonical complex line bundle X_n over $(S^1)^n\times\mathbb{T}^n$ (see the example following the Lusztig signature theorem in $8\frac{5}{8}$). The space of sections of X_n is a projective $R\otimes\underline{R}$-module for $R=\operatorname{Cont}(S^1)^n$ and $\underline{R}=\operatorname{Cont}\mathbb{T}^n = C^*(\mathbb{Z}^n)$, which can be easily identified with the above X and our $\mathbf{K}$ applied to the signature operator $\mathcal{L}$ on $(S^1)^n$ (or rather to $\widehat{\mathcal{L}} = \mathcal{L}(1+\mathcal{L}^*\mathcal{L})^{-\frac{1}{2}}$) is exactly the index of the family of the signature operators which we denoted earlier $\underline{\kappa}[(S^1)^n] \in K^0(\mathbb{T}^n)$. Moreover, for every V mapped to $(S^1)^n$ (according to a homomorphism $\pi_1(V)\to\mathbb{Z}^n$) the index $\underline{\kappa}[V]\in K^0(\mathbb{T}^n)$ of the induced family equals $\mathbf{K}(\widehat{\mathcal{L}}(V))$ where $\widehat{\mathcal{L}} = \mathcal{L}(1+\mathcal{L}^*\mathcal{L})^{-\frac{1}{2}}$ for the signature operator $\mathcal{L}=\mathcal{L}(V)$ and where $\mathbf{K}$ is defined via the induced line bundle over $V\times\mathbb{T}^n$. This follows from the Lusztig signature theorem as $\mathcal{L}$ and $\widehat{\mathcal{L}}$ have equal indices over $\mathbb{T}^n$.

Non-commutative generalization of the Lusztig theorem. (Compare Remark (c′) in $8\frac{5}{8}$). Observe that there is a natural homomorphism, say $\widehat{L} : \operatorname{Brd}_{\mathrm{ev}}(B\Pi) \to K_0(B\Pi)$ which assigns to each $(V, \beta : V \to B\Pi)$ the β-image of the class $[\widehat{\mathcal{L}}(V)] \in K_0(V)$. If V is stably parallelizable, then

$$\left\langle \beta_*([\widehat{\mathcal{L}}(V)]), [X] \right\rangle = \langle \beta_*([V]), \operatorname{ch} X\rangle$$

for all vector bundles over $B\Pi$ by the index theorem. Thus $\widehat{L}$ defines a *monomorphism* (in fact, an *isomorphism*) called ch', from $H_{\mathrm{ev}}(B\Pi;\mathbb{Q})$ to $K_0(B\Pi)\otimes\mathbb{Q})$ since $\mathrm{ch} : K^0 \to H^{\mathrm{ev}}$ is an epimorphism (in fact an isomorphism) over $\mathbb{Q}$.

Theorem (See [Kas]). *The homomorphism* $\mathbf{K}\circ\mathrm{ch}' : H_{\mathrm{ev}}(B\Pi;\mathbb{Q}) \to$

$K_0(C^*(\Pi)) \otimes \mathbb{Q}$ *equals our old* $\alpha : H_{\text{ev}}(B\Pi; \mathbb{Q}) \to (\text{Witt}_{\text{ev}} \mathbb{C}(\Pi)) \otimes \mathbb{Q}$ (see $7\frac{7}{8}$) *composed with* $\text{Witt}_{\text{ev}} \mathbb{C}(\Pi) \to \text{Witt}_{\text{ev}} C^*(\Pi) \to K_0(C^*(\Pi))$.

Corollary. *If* $\mathbf{K}$ *is injective then so is* α.

Thus the Novikov conjecture for $H_{\text{ev}}(B\Pi)$ would follow from the injectivity of $\mathbf{K}$ (and the odd case of the Novikov conjecture of Π would follow from the injectivity of $\mathbf{K}$ for $\Pi' = \Pi \times \mathbb{Z}$).

This motivates the following.

Strong Novikov conjecture (according to Rosenberg). *The homomorphism* κ *is injective for all countable groups* Π. In general, $B\Pi$ is not compact but it can be obtained as a union of compact polyhedra $P_0 \subset P_1 \subset P_2 \subset \cdots P_i \subset \cdots$ and $K_0(B\Pi)$ is defined as the direct limit of $K_0(P_i)$.

Groups where strong Novikov is proved. (1) Lusztig's argument proves strong Novikov for $\Pi = \mathbb{Z}^n$.

(2) If Π is the fundamental group of a complete manifold B of non-positive sectional curvature, then strong Novikov is valid for Π (see [Miš], [Kas]).

(3) Strong Novikov is valid for the subgroup Π of the linear group $GL(N, \mathbb{R})$ for all $N = 1, 2, \ldots$ (see [Kas]). Notice that this gives an alternative proof of the Lusztig theorem concerning flat Hermitian bundles, but the two proofs seem to provide somewhat different information. Namely, Lusztig's argument does not apparently say anything about *strong* Novikov, but it gives a non-trivial lower bound on the rank-norm on Witt_* which, in a way, is stronger than strong Novikov.

(4) If $B\Pi$ can be represented by a complete n-dimensional Riemannian manifold B whose universal covering $\widetilde{B}$ admits a proper (uniformly) Lipschitz map $\widetilde{B} \to \mathbb{R}^n$ of non-zero degree, then Π satisfies strong Novikov. In fact, for more general Π, strong Novikov is valid for the *Lipschitz* (hyper-Euclidean) part of the cohomology of Π (see [C-G-M]$_{\text{GCLC}}$).

How strong Novikov is proved. One has to show the non-vanishing of somebody in $K_0(C^*(\Pi)) \otimes \mathbb{Q}$, namely, of $\kappa = \mathbf{K}([\mathcal{L}(V)])$ for a suitable manifold V. This can be done by finding a K-cycle Δ over $C^*(\Pi)$ (representing an element in $K^0(C^*(\Pi))$ such that our "somebody" does not vanish on Δ. This Δ, according to the definition of $K^0(R)$ specialized to $R = C^*(\Pi)$, must be *a Fredholm representation of* Π, i.e. a Fredholm operator between two unitary representations of Π, say $\Delta : H_+ \to H_-$ where Π unitary acts on $H_\pm$ and, most importantly, Δ almost commutes with these actions, i.e. commutes, modulo compact operators.

Now let $\Pi = \pi_1(B)$ where B is a complete manifold with (non-strictly) negative sectional curvature and $\widetilde{B}$ is the universal covering of B acted upon by

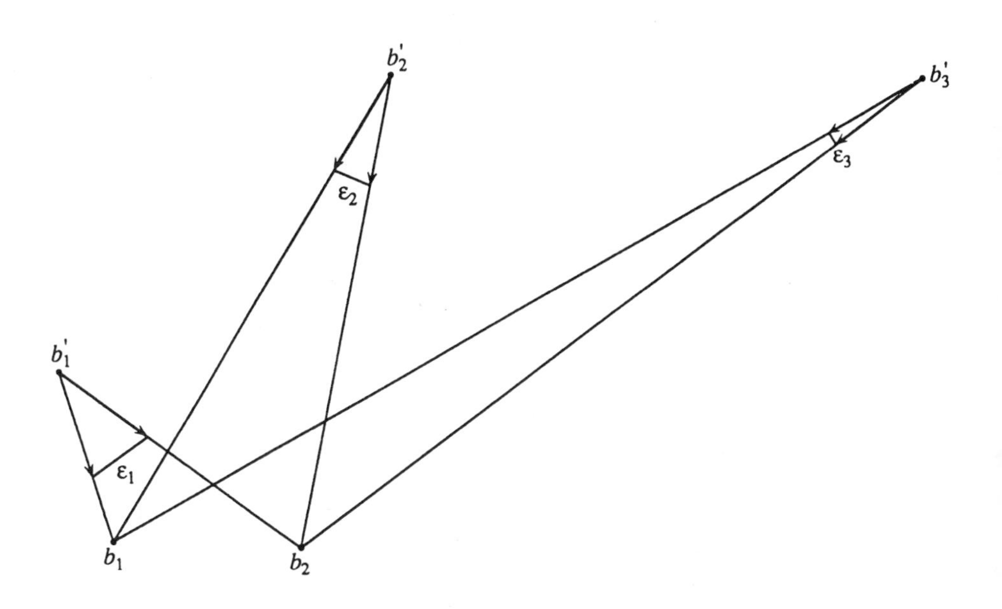

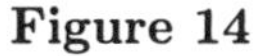
Figure 14

II. The distance function $\mu(b') = \mu_b(b') = \mathrm{dist}_B(b, b')$ is smooth for each $b \in \widetilde{B}$ and all $b' \neq b$, and its differential $d\mu(b')$ has $\|d\mu(b')\| = 1$ for all $b' \neq b$. The key property of $d\mu$, where the negative curvature enters, is a weak dependence of μ_b on b for $b' \to \infty$. Namely

$$\|d\mu_{b_1}(b') - d\mu_{b_2}(b')\| \to 0$$

for every fixed pair (b_1, b_2) and $b' \to \infty$. In fact $\|d\mu_{b_1}(b') - d\mu_{b_2}(b')\| \leq 2\,\mathrm{dist}(b_1, b_2)/\,\mathrm{dist}(b_1, b')$.

See Figure 14 below, where ε_i denotes $\|d\mu_{b_1}(b_i') - d\mu_{b_2}(b_i')\|$.

It follows that

$$\|d\mu(b') - d\mu(\pi b')\| \to 0 \quad \text{for} \quad b' \to \infty$$

for each $\mu = \mu_b$ and every $\pi \in \Pi$. Thus the operator $\delta = \delta_b : \Lambda^*(\widetilde{B}) \to \Lambda^*(\widetilde{B})$ defined by $\lambda \mapsto \lambda \wedge d\mu$ for $\mu = \mu_b(b')$, asmost commutes in the L_2-sense with the action of Π if we ignore what happens near b. To make it cleaner, we take a single Π-orbit, say $\Pi\widetilde{b} \in \widetilde{B}$, $\widetilde{b} \in \widetilde{B}$, missing a given point $b \in \widetilde{B}$, and consider the Hilbert space $H_{\widetilde{b}}$ of square summable forms on the tangent spaces of $\widetilde{B}$ along this orbit, i.e. $H_{\widetilde{b}} = \bigoplus_\Pi \Lambda^*_{\pi(\widetilde{b})}(\widetilde{B})$. Now, clearly, the above operator

$\delta = \delta_b$ restricted to $H_{\widetilde{b}}$ does have the almost commutation property (for the same reason as the multiplication operator on $\ell_2(\Pi)$ for $\varphi(\pi) \mapsto \delta(\pi)\varphi(\pi)$ where $\delta(\pi)$ is a function on Π which converges to a constant for $\pi \to \infty$); yet it is not Fredholm. But $\Delta_b = \delta_b + \delta_b^*$ is Fredholm since $\delta_b + \delta_b^*$ at each space $\Lambda^*_{\pi(\widetilde{b})}(\widetilde{B}) = \Lambda^*\mathbb{R}^n$, $\pi \in \Pi$, is an invertible (self-adjoint) operator. The required Δ can be eventually built of these Δ_b (by suitably "integrating" over $b \in \widetilde{B}$) and then non-vanishing of $\langle \kappa, [\Delta] \rangle$ is obtained by a cohomological computation (similar to the one for $B = (S^1)^n$ in $8\frac{1}{2}$) rendered possible by a suitable index theorem.

This unexpected intervention of negative curvature in the infinite dimensional realm was brought about by Misščenko in 1974, who also proved the relevant index theorem for the signature operator twisted with Fredholm representations of Π, replacing ordinary representations used by Lusztig. Namely, the representation of Π on $H_\pm$ define flat Hilbert bundles over $B\Pi$ associated to the universal covering and $\Delta : H_+ \to H_-$ gives rise to a fiberwise Fredholm homomorphism say, Δ_B between these bundles (at least for compact $B\Gamma$). Then $X' = \operatorname{Ker} \Delta_B - \operatorname{Coker} \Delta_B$ defines a virtual bundle over $B\Pi$ and Miščenko commutes the index of the signature operator twisted with X'. Thus Miščenko shows that every cohomology class $\rho \in H^*(B\Pi)$ of the form $\rho = \operatorname{ch} X'$ satisfies Novikov's conjecture (i.e. σ_ρ is homotopy invariant).

Finally we recall that the universal covering $\widetilde{B}$ of a complete manifold B with non-positive curvature is *Hyper-Euclidean* (see §4) i.e. it admits a proper Lipschitz map onto $\mathbb{R}^n$, $n = \dim B$, of non-zero degree, say $A : \widetilde{B} \to \mathbb{R}^n$ (for which we may take $\exp_b^{-1} : \widetilde{B} \to T_b(\widetilde{B}) = \mathbb{R}^n$) and such A suffices for the strong Novikov for $\Pi = \pi_1(B)$. Here one builds up the relevant Fredholm representation of Π out of (the Hilbert space of) maps $\lambda : \widetilde{B} \to \Lambda^*\mathbb{R}^n$ and takes $\delta(\lambda) = \lambda \wedge a(\widetilde{b})$ for $a(\widetilde{b}) = A(\widetilde{b})/(1 + \left\| A(\widetilde{b}) \right\|)$. (It is slightly more convenient to use spinors on $\mathbb{R}^n$ with the Clifford multiplication by $a(\widetilde{b})$ rather than Λ^* with the exterior product; see [C-G-M]$_{\mathrm{GCLC}}$. The Lipschitz property of A guarantees the almost commuting of this δ (and hence of $\Delta = \delta + \delta^*$) with the group action while "proper of positive degree" make the resulting K-cycle sufficiently nontrivial to detect non vanishing of relevant $\kappa \in K_0(C^*(\Pi))$. Cohomological sufficiency of this construction is explained in a slightly different situation in $9\frac{2}{7}$.

Spectral consequences of strong Novikov. The class $\mathbf{K}\beta_*([\widehat{\mathcal{L}}(V)]) \in K_0(C^*(\Pi))$ can be defined, for each closed oriented manifold V with a continuous map $\beta : V \to BM$, more directly via the Π-covering $\widetilde{V} \to V$ (see [Ros]$_{\mathrm{C^*APS}}$, [Roe]$_{\mathrm{CCIT}}$) and then non-vanishing $\mathbf{K}\beta_*([\widehat{\mathcal{L}}(V)]) \neq 0$ implies that the spectrum of the Hodge operator $\widetilde{d} + \widetilde{d}_*$ on L_2-forms on $\widetilde{V}$ contains zero. Thus $\operatorname{Spec}\widetilde{d} + \widetilde{d}_* \ni 0$ whenever Π satisfies the strong Novikov and the map β is not $\mathbb{Q}$-homologous to zero (i.e. $\beta_*[V] \in H_n(B\Pi)$ is not a torsion class).

If, furthermore, $\dim V$ is odd and $\kappa\beta_*([\widehat{\mathcal{L}}(V)])$ does not vanish in $K_1(C^*(\Pi))$ (compare $6\frac{8}{9}$) then the gaps in $\operatorname{Spec}\widetilde{d} + \widetilde{d}_*$ are bounded. But in most (all?) cases the relevant part of the *proof* of strong Novikov reduces to the Vafa-Witten argument, which is certainly easier than the full strong Novikov (see $6\frac{1}{2}$).

$8\frac{3}{4}$. Twisting the signature operator with almost flat bundles

Let us slightly change our view about the Novikov conjecture concerning the homotopy invariance of concerning $\sigma_\rho(V)$, $\rho \in H^*(B\Pi)$, by passing from cohomology of the (aspherical) classifying space $B\Pi$ to $K^0(B\Pi)$. Namely we take a vector bundle X over our manifold V induced from some bundle over $B(\Pi)$ by a map $\beta : V \to B\Pi$ and ask ourselves when the index of the signature operator on V twisted with X is a homotopy invariant. As this index is given by the Atiyah-Singer formula

$$\sigma'_X(V) \overset{\text{def}}{=} \operatorname{ind}\mathcal{L}_X = (L_V \operatorname{ch} X)[V] \overset{\text{def}}{=} \sigma_{\rho=\operatorname{ch} X}, \tag{$*$}$$

and $\operatorname{ch} : K^0(B\Pi) \to H^{\text{ev}}(B\Pi)$ is an isomorphism over $\mathbb{Q}$ we do not lose or gain in generality by shifting from the cohomology to the K-theory but change the language for the expected answer.

Model Example. (Lusztig's theorem, see $8\frac{1}{2}$). Let X be a flat Hermitian bundle over V split into $X_+ \oplus X_-$ (where the splitting does not have to agree with the flat connection) so that the implied Hermitian form is positive definite on X_+ and negative definite on X_-. Then the index of $\mathcal{L}_{X'}$ for $X' = [X_+]-[X_-]$ is a homotopy invariant of V being equal to the signature of V with coefficients in X. Here $[X_+] - [X_-]$ is the virtual difference, i.e. $[X_+] - [X_-] \in K^0(V)$ and $\operatorname{ind}\mathcal{L}_{X'} \underset{\text{def}}{=} \operatorname{ind}\mathcal{L}_{X_+} - \operatorname{ind}\mathcal{L}_{X_-}$ which, by additivity of the Chern character, satisfies ($*$) with X' in place of X.

UAFl bundles. Let us explain how to extend Lusztig's theorem from flat to *almost flat* bundles.

Let V be a compact Riemannian manifold. If X is a unitary bundle with a unitary connection ∇ over V, we denote (as in §4) by $\|\mathcal{R}(X)\|$ the operator norm of its curvature $\mathcal{R}(X) = \mathcal{R}(X, \nabla)$ thought of as an operator valued 2-form on V. Thus, the inequality $\|\mathcal{R}(X)\| \leq \varepsilon$ says in effect that the holonomy (or monodromy) transform, say M in X, around each loop in V which bounds a disk of area $\leq \delta$ satisfies $\|Ax - x\| \leq \varepsilon\delta\,\|x\|$ for all vectors $x \in X$. Then we extend this curvature norm to the K-theory of V, namely to $K_0(V)\otimes\mathbb{Q}$, by representing each $\kappa \in K_0(V) \otimes \mathbb{Q}$ by a (formal) rational combination $\mathcal{X} = \sum_i r_i X_i$ of unitarily bundles $X_i = (X_i, \nabla_i)$, and by setting

$$\|\mathcal{R}(\mathcal{X})\| \underset{\text{def}}{=} \max_i \|\mathcal{R}(X_i)\| \quad \text{and} \quad \mathcal{R}_{\text{un}}(\kappa) = \underset{\mathcal{X}}{\infty}\, \|\mathcal{R}(\mathcal{X})\|$$

for all representations of κ by $\mathcal{X}$. Then κ is called *unitarily almost flat* if $\mathcal{R}_{\text{un}}(\kappa) = 0$ and the subgroup of these κ is denoted $K^0_{\text{uafl}}(V) \subset K^0(V) \otimes \mathbb{Q}$. Clearly, this subgroup does not depend on the Riemannian metric in V and is, moreover, a homotopy invariant of V. In fact, it can be easily defined for every finite polyhedron (and with a minor extra effort, for an arbitrary compact metric space).

If V is connected and simply connected, then K^0_{uafl} equals the (infinite cyclic) group generated by the trivial line bundle as was essentially explained in $4\frac{1}{4}$. Furthermore, for every V, the group K^0_{uafl} dies on the universal covering $p : \widetilde{V} \to V$, i.e. $p^*(\kappa)$ vanishes in $K^0(\widetilde{V})$ for all $\kappa \in K^0_{\text{uafl}}(V)$. This means, for each arbitrarily large compact subset $\widetilde{U} \subset \widetilde{V}$, there exists a representation of $p^*(\kappa)$ by $\sum_i r_i \widetilde{X}_i$ where the bundles $\widetilde{X}_i | \widetilde{U}$ are all trivial. In fact, if the fundamental group $\Pi = \pi_1(V)$ is of *finite type up to dimension* $n = \dim V$, i.e. it admits a classifying space $B\Pi$ which is a cell complex with finitely many cells of dimension $\leq n = \dim V$, then there exists an $\varepsilon_0 = \varepsilon_0(V) > 0$ such that every ε_0-flat bundle X over V lifts to a *trivial* bundle $\widetilde{X}$ over $\widetilde{V}$ since triviality of $\widetilde{X} | \widetilde{U}$ for a large (but fixed) compact $\widetilde{U}$ makes $\widetilde{X}$ trivial on all of $\widetilde{V}$ for such a group Π. Furthermore, if there is a realization of $B\Pi$ by a *finite* cell complex, then every ε_0-flat bundle over V is induced by the classifying (i.e. isomorphic on π_1) map $V \to B \to B\Pi$ from some bundle over $B\Pi$ for some $\varepsilon_0 = \varepsilon_0(V) > 0$.

Novikov for UAFl. *The index of the signature operator with coefficients in a UAFl bundle, i.e. in κ in K^0_{uafl}, is a homotopy invariant.*

The proof will be explained later. Now we want to indicate some corollaries and generalizations. First, the homotopy invariance here signifies that if V and V' are homotopy equivalent, κ lies in $K^0_{\text{uafl}}(V)$ and $\kappa' \in K^0(V')$ corresponds to κ under the implied homotopy equivalence, then $\operatorname{ind} \mathcal{L}_\kappa = \operatorname{ind} \mathcal{L}_{\kappa'}$ which is equivalent to $(L_V \operatorname{ch} \kappa)[V] = (L_{V'} \operatorname{ch} \kappa')[V']$ (where, recall, almost flatness of κ implies that of κ'). This yields the homotopy invariance of the ρ-signature σ_ρ for $\rho = \operatorname{ch} \kappa$ where $\kappa \in K^0_{\text{uafl}}(B)$ for an arbitrary finite polyhedron B. The ρ-signature refers to manifolds V mapped to B, say by $\beta : V \to B$, and $\sigma_\rho \overset{def}{=} (L_V \beta^*(\rho)[V])$.

Example with infinite K-area. (Compare §4). Let B be a closed connected oriented manifold of infinite K-area and ρ be the fundamental cohomology class $[B]^{\text{co}} \in H^n(B)$, $n = \dim B$. Then σ_ρ is a homotopy invariant for all $(V, \beta : V \to B)$ (where, recall, $\sigma_\rho = \text{signature}\,(\beta^{-1}(b))$ for a regular value $b \in B$ of β). In fact, we know in this case that $M[B]^{\text{co}} = \operatorname{ch} \rho'$ for some $\rho' \in K^0_{\text{uafl}}(B)$ and $M \neq 0$.

Example with H^2. Let $\rho \in H^2(B)$ be an integer cohomology class which admits for every integer $d_0 \geq 1$ a finite covering $p : \widetilde{B} \to B$ for which the pull-back $p^*(\rho)$ is divisible by $d \geq d_0$ in $H^2(\widetilde{B};\mathbb{Z})$. Then σ_ρ is homotopy invariant for all $(V, \beta : V \to B)$. In fact, $\rho = c_1(X)$ for some complex line bundle, and the pull-back $\widetilde{X}$ of X to $\widetilde{B}$ admits a dth-root $(\widetilde{X})^{\frac{1}{d}}$ (whose d^{th} tensor power equals $\widetilde{X}$). Then the push-forward of $\widetilde{X}^{\frac{1}{d}}$ to X, say X_d over V, is approximately $1/d$-flat (this is a complex vector bundle of rank $= \deg p$) with $\operatorname{ch} X_d = \deg p (\operatorname{ch} X)^{\frac{1}{d}}$ (where $\deg p$ refers to the number of the sheets of $\widetilde{B} \to B$ in the case where B is non-orientable). It follows that $\operatorname{ch} K^0_{\text{uafl}}(B) \subset H^*(B)$ contains ρ (as well as ρ^i, $i = 1, 2, \ldots$) and so σ_ρ (as well as σ_{ρ^i} for $i \geq 2$) is a homotopy invariant.

Generalization of the K-area example. Recall that the K-area is invariant under *finite* Galois covering of B but not, with our present definition, under infinite ones. Thus the following is a genuine generalization. *If B admits a (possibly) infinite Galois covering of infinite* K-area, *then σ_ρ for $\rho = [B]^{\text{co}}$, is a homotopy invariant for all V mapped to B.*

Generalization of the H^2-example. *Let B be a (finite or infinite) polyhedron with the universal covering $\widetilde{B}$ having $H^2(\widetilde{B};\mathbb{Q}) = 0$ (e.g. $B = B\Pi$ for $\Pi = \pi_1(B)$). Then σ_ρ is a homotopy invariant for all $\rho \in H^*(B;\mathbb{Q})$ in the subring generated by $H^2(B;\mathbb{Q})$. In fact this remains true for the subring generated by $H^1(B;\mathbb{Q})$ and $H^2(B;\mathbb{Q})$.* Thus the Novikov conjecture is valid for the cohomology multiplicatively generated by the 1- and 2-dimensional classes in $H^*(B\Pi;\mathbb{Q})$.

Idea of the proof. The lifted line bundles $\widetilde{X}$ over $\widetilde{B}$ admit roots of all degrees d which makes them arbitrarily ε-flat with a possible descent back to B (see $9\frac{1}{7}$ for details).

$8\frac{8}{9}$. On the proof of Novikov for UAFl

Let us look at the homotopy invariance of the index of the untwisted signature operator $\mathcal{L}$ (see $8\frac{1}{2}$). The operator $\mathcal{L}$ is built out of the exterior differential d (which behaves well under homotopies) and the Hodge operator $*$ which apparently badly needs the Riemannian (and hence smooth) structure. But this $*$ is linked to the exterior product of form (which is a homotopy stable operation) by $\|\lambda\|^2_{L_2} = \int \omega \wedge *\omega$. This can be translated to a purely linearly algebraic (or operator theoretic) language along with other essential properties of d and $*$ (such as (Lei) and (Sto) of $8\frac{1}{2}$) where the homotopy equivalence between manifolds manifests itself by chain homotopy equivalence of de Rham complexes, and the homotopy invariance of $\operatorname{ind} \mathcal{L}$ follows by chasing a few diagrams. This equally works for $\mathcal{L}$ twisted with a flat bundle which may be Hermitian as in the Lusztig theorem or a bundle of finite projective moduli over a C^*-algebra R with $\operatorname{ind} \mathcal{L}_X \in K_0(R)$ (see Remark (c$'$) in $8\frac{5}{8}$). Since the index is a rather robust invariant, stable under small perturbations of our data, its homotopy invariance survives the passage from "flat" to "ε-flat" which analytically speaking corresponds to the condition $\|d^2\| \le \varepsilon$ instead of $d^2 = 0$ (see [Hi-Sc]). The situation is quite similar here to the homotopy invariance of the Novikov-Shubin invariants concerning the spectrum of $\widetilde{d} + \widetilde{d}^*$ near zero on an infinite Π-covering $\widetilde{V}$ of a compact manifold V (see [Gr-Sh] and $6\frac{10}{11}$).

Direct definition of the signature of $\sigma(V;X)$ for almost flat bundles X via almost homomorphism $\pi_1(V) \to U(p)$. Let $X = (X, \nabla)$ be an ε-flat unitary bundle of rank p over V, fix a base point $v_0 \in V$ and take a smooth loop γ at v_0 in each homotopy class. The parallel transport in X along this loop γ defines a map $\rho : \pi_1(V) \to U(p) = \mathrm{Aut} X_{v_0}$ which is close to being a homomorphism if ε is small. Namely if $[\gamma_1][\gamma_2] = [\gamma_3]$ in $\pi_1(V)$ then $\|\rho(\gamma_1)\gamma(\rho(\gamma_2) - \gamma(\gamma_3)\| \le A\varepsilon$, where A the area of the minimal disk in V spanning the loop $\gamma_1\gamma_2\gamma_3^{-1}$ and $\| \quad \|$ refers to the operator norm in $U(p)$ (compare Figure 4 in $4\frac{1}{4}$). For example, if X is flat, this ρ is a homomorphism. It is convenient always to use the same loop with the opposite orientation in the reciprocal homotopy class of γ, i.e. such that $[\gamma]^{-1} = [\gamma^{-1}]$. Then our ρ is symmetric, $\rho(g^{-1}) = (\rho(g))^{-1} = (\rho(g))^*$ for all $g \in \pi_1(V)$. In this case the linear extension of ρ to a linear map of the group ring $\mathbb{C}(\Pi)$ to the matrix ring Mat_p is involutive and thus sends (skew) Hermitian forms over $\mathbb{C}(\Pi)$ to those over Mat_p. In particular, let w be such a (non singular!) form (matrix) over $\mathbb{C}(\Pi)$ of rank r corresponding to the class $[V] \in \mathrm{Brd}_* B\Pi$, for $\Pi = \pi_1(V)$ under the Wall-Miščhenko homomorphism $\mathrm{Brd}_* B\Pi \to \mathrm{Witt}\,\mathbb{C}(\Pi)$ where we assume $\dim V$ is even, and denote by $\rho(w) \in \mathrm{Mat}_{rp}$ the image of w under (the linear extension of) ρ. This $\rho(w)$ gives us a (skew) Hermitian form over $\mathbb{C}$ whose signature we denote by $\sigma(V;X)$ or by $\sigma_\rho(V)$ (compare §8) which may depend, in general, upon our choice of w representing $WM[V] \in \mathrm{Witt}_*\,\mathbb{C}(\Pi)$ but for small $\varepsilon \to 0$ this dependence disappears. Namely, *for every pair of forms w_1 and w_2 representing the same class in* $\mathrm{Witt}_{\mathrm{ev}}\,\mathbb{C}\Pi$ *there exists an $\varepsilon_0 = \varepsilon_0(w_1, w_2) > 0$ such that for $\varepsilon \le \varepsilon_0$ the Hermitian matrices $\rho(w_1)$ and $\rho(w_2)$ have equal signatures.*

Proof. Let us isolate the relevant property of our linear map of $\mathbb{C}(\Pi)$ to Mat_p.

(F,δ)-homomorphisms. Let ρ be a linear map of an algebra R over $\mathbb{C}$ to a Banach algebra M and F be a subset in R. Then ρ is called an (F,δ)-*homomorphism* if $\|\rho(r_1)\rho(r_2) - \rho(r_1r_2)\| \le \delta$ for all r_1 and r_2 in F.

We observe that if r is invertible in R and both r and r^{-1} lie in F for some (F,δ)-homomorphisms ρ with $\delta < 1$, then $\rho(r)\rho(r^{-1})$ and $\rho(r^{-1})\rho(r)$ are invertible in M (being δ-close to 1). This implies invertibility of $\rho(r)$ if, for example, $M = \mathrm{Mat}_p$ or if M is a C^*-algebra and $\rho(r)$ and $\rho(r^{-1})$ are self-adjoint.

Coming back to our $\rho : \mathbb{C}(\Pi) \to M_p$ (which extends the map $\Pi \to U(p)$ called by the same name ρ) associated to an ε-flat bundle, we notice that it becomes (F,δ)-homomorphism for arbitrary (large) finite set F and (small) $\delta > 0$ if $\varepsilon > 0$ is small enough (i.e. $\forall F, \delta \exists \varepsilon \cdots$). This implies invertibility of $\rho(w)$ for a fixed w and small ε. Moreover, every finite chain $\mathcal{C}$ of mutually equivalent non-singular matrices over $\mathbb{C}(\Pi)$ transforms under ρ to such a chain in Mat_p with "equivalence" replaced by "δ-equivalence" which is good enough for preservation of the signature. Thus the equality $[w_1] = [w_2]$ in $\mathrm{Witt}_* \mathbb{C}(\Pi)$ implies $\sigma(\rho(w_1)) = \sigma(\rho(w_2))$, provided $\varepsilon \le \varepsilon_0(\mathcal{C}) = \varepsilon_0(w_1, w_2) > 0$. Q.E.D.

Corollary. *If X is an ε-flat bundle over V with a sufficiently small $\varepsilon > 0$, then $\sigma(V, X)$ is non-ambiguously defined and is a homotopy invariant of V where the smallness of ε depends on the implied homomotopy equivalence.* This means that, given a homotopy equivalence $f' : V' \to V$, for every sufficiently small $\varepsilon > 0$, the signature $\sigma(V'; f^*(X))$ equals $\sigma(V; X)$ for all ε-flat bundles X over V.

Actually, one can formalize the above by defining $\sigma(V, X_\varepsilon)$ in the limit for $\varepsilon \to 0$. Namely, let $\rho_i : \mathbb{C}(\Pi) \to \mathrm{Mat}_{\rho_i}$ be a sequence of involutive linear maps, such that, for every finite $F \subset \mathbb{C}(\Pi)$ and $\varepsilon > 0$, the maps ρ_i are (F,ε)-homomorphisms for all $i \ge i_0 = i_0(F, \varepsilon)$.

Of course, the sequence $\sigma(\rho_i(w))$ for a fixed form w over $\mathbb{C}(\Pi)$ does not necessarily stabilize for $i \to \infty$. But we may fix a non-principal ultrafilter in $\mathbb{N} \ni i$ and take the limit (eventual value) of $\sigma(\rho_i(w))$ over this ultrafilter, which may happen to be $\pm\infty$. Thus we define a homomorphism from $\mathrm{Witt}_{\mathrm{ev}} \mathbb{C}M$ to $\mathbb{Z}\,\mathrm{cup}\{\pm\infty\}$ (or, better, to the non-standard integers). Furthermore, we may extend this definition to *virtual* almost homomorphism which are formal rational combinations $\rho = \left\{\sum_j r_j \rho_{ij}\right\}_{i\in\mathbb{N}}$ by $\sigma(\rho(w)) = \lim\limits_{i\to\infty} r_j \sigma(\rho_{ij}(w))$ where "lim" refers to our ultrafilter. This allows us to define $\sigma(V, \kappa)$ for each $\kappa \in K^0_{\mathrm{uafl}}$, which is a well defined homotopy (!) invariant of V. Notice that a priori, this signature $\sigma(V; \kappa)$ depends on how κ is represented by a sequence of virtual bundles as well as on our ultrafilter, but in fact, it depends only on κ itself thanks to the following.

UAFl signature formula. (See [C-G-M]$_{\mathrm{PPl}}$ and [Ska]).

$$\sigma(V; \kappa) = \mathrm{ind}\, \mathcal{L}_\kappa \qquad (\star)$$

where, recall, $\mathrm{ind}\, \mathcal{L}_\kappa = (L_V \,\mathrm{ch}\, \kappa)[V]$ by the index theorem for the signature operator $\mathcal{L}$ on V twisted with κ.

This formula $(\star)$ relates two rather different quantities, where the first, the signature $\sigma(V; \kappa)$, heavily (in fact *too* heavily) depends on $\pi_1(V)$ (in the way it was defined) while the essential ingradient of $\mathcal{L}$ is the $*$ operator (as

$\mathcal{L} = d + d^*$ restricted to $\Lambda_+ \subset \Lambda^*(V)$). In fact $(\star)$ can be used to obtain non-trivial information about both σ and $\mathcal{L}$, albeit our primary purpose is the homotopy invariance of $\operatorname{ind} \mathcal{L}_\kappa$. As for $\sigma(V;\kappa)$, we can see with $(\star)$, for example, that it is multiplicative under finite coverings of V, but this seems to follow from the bordism theory (the Serre finiteness theorem) as well. What is more interesting is the behaviour of $\sigma(V;\kappa)$ under *infinite* Galois coverings, as we shall explain later on.

π_1 - ***Free definition of*** $\sigma(V;X)$. If X is a flat bundle then the definition of σ via the almost representations $\rho : \pi_1(V) \to \mathrm{Mat}_p$ is equivalent to the cohomological one, i.e. that of the ordinary signature of V with coefficients in X. Let us indicate a similar definition of $\sigma(V;X)$ where X is ε-almost flat. We fix a triangulation of V and trivialize X over each simplex Δ by using sections $\Delta \to X$ which are parallel along each straight segment issuing from the baricenter of Δ. With this we have natural boundary operators sections $(\Delta) \to$ sections (face of Δ) for all faces of Δ and thus operators $\partial_i : C_i = C_i(X) \to C_{i-1}(X)$ for $C_*(X) = \bigoplus_\Delta$-sections (Δ) satisfying $\|\partial_i\partial_{i-1}\| \le c\varepsilon$ where $c = c(V)$ equals, up to a universal constant const_n, the maximal number of neighbours a simplex $\Delta \subset V$ may have. We shall assume c is bounded in the course of our discussion and to save notations, pretend it equals one. Now, our main object is an *ε-complex* $(C_* = \bigoplus_{i=0}^n C_i(V;X), \partial_* = \bigoplus_{i=0}^n \partial_i)$ where C_i are finite dimensional Hilbert spaces (over $\mathbb{C}$) and ε refers to the bound $\|\partial^2\| \le \varepsilon$ for the operator norm $\| \;\; \|$. The signature of (V, X) is defined in terms of the intersection of chains in C_* imitating the Miščenko definition for $\varepsilon = 0$, where the intersection enters via the Poincaré duality given by an ε-homotopy equivalence between C_* and C^*. Here is the full diagram.

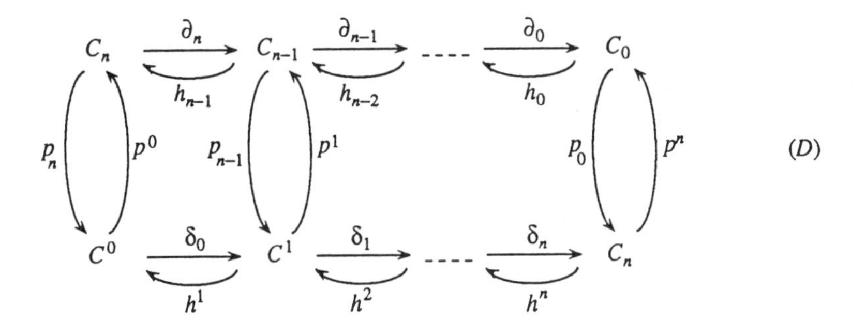

Here (C^*, δ) is the cochain ε-complex corresponding to (C_*, δ), where all square diagrams $(-1)^i$-commute up to ε, where $\overline{p}_i = (-1)^i p_{n-i}$ and $\overline{p}^i = (-1)^i p^{n-i}$ with $\overline{p}$ denoting the Hermitian conjugate for our Hilbert structure (and where, recall, we stick to n even). Furthermore, p_* and p^* are mutually inverse ε-homotopy equivalences with h^* and h_* serving as the corresponding ε-homotopies. This means $\|p^*p_* - 1 - h^*\delta - \delta h^*\| \leq \varepsilon$ and $\|p_*p^* - 1 - h_*\partial - \partial h_*\| \leq \varepsilon$. For every (small) $\varepsilon > 0$ and (big) $b > 0$ such ε-diagrams with all operators involved, i.e. ∂, δ, h, p bounded in the operator norm by b, form a semigroup under Cartesian sums, and one passes to the corresponding Grothendieck group, say $\mathcal{D}_n(\varepsilon, b)$. Then, again mimicking the case $\varepsilon = 0$, one introduces a subgroup of ε'-trivial ε-diagrams, $\mathrm{Tr}_n(\varepsilon', b) \subset \mathcal{D}_n(\varepsilon, b)$ takes the quotient group $\mathcal{D}_n(\varepsilon, b)/\mathrm{Tr}_n(\varepsilon', b)$ goes to the limit in the following order $\varepsilon \to 0$, $\varepsilon' \to 0$, $b \to \infty$, and checks (if one is able to unravel this mess of linear algebra, I hardly can do this myself) that the resulting group is isomorphic to $\mathrm{Witt}_{\mathrm{ev}}\, \mathbb{C} = K_0(\mathbb{C}) = \mathbb{Z}$ and so there is a well defined integer, called the *signature* $\sigma(D)$, assigned to each diagram D where ε is (very) small compared to b^{-1} which gives our $\sigma(V; X)$ for $C_* = C_*(V; X)$.

Example. Imagine we start with a flat bundle X_0 and then perturb it to an ε-flat X_ε. Suppose all non-zero eigenvalues of the operator $(\partial_0 + \partial_0^*)^2$ in the middle dimension, i.e. on C_m for $m = n/2$ are far away from zero, i.e. outside a fixed interval $[0, \delta]$ for δ much greater than ε. Then the spectrum of $(\partial_\varepsilon + \partial_\varepsilon^*)^2$ in the middle dimension has a well-defined part localized ε-close to zero, and the span H_ε of the corresponding eigenspaces is isomorphic to $H_m(C_*|_{\varepsilon=0})$. Furthermore, the intersection form on H_ε is isomorphic to that on $H_m(C_*|_{\varepsilon=0})$ being a small perturbation of the former and one can actually show that the signature of $C_*|_{\varepsilon\neq 0}$ equals the signature of the intersection form on H_ε, whenever there is such a well localized subspace $H_\varepsilon \subset C_m$. This is always the case if we keep the rank of C_* fixed for $\varepsilon \to 0$, but in general, the spectrum of $\partial + \partial^*$ may rather uniformly spread over the interval $[0, \delta]$ for $\delta >> \varepsilon$ and then the signature can not be recaptured without looking on all of C_* or at least on the eigenspaces belonging to the spectrum of $(\partial/\partial^*)^2$ on C_* close to zero. We shall indicate later specific examples where the localization of the spectrum does take place.

The π_1-free definition of $\sigma(V; X)$ has a de Rham counterpart where the equivalence with $\mathrm{ind}\, \mathcal{L}_X$ becomes a matter of simple (and painful) diagram chasing. Furthermore, one can generalize all this to cover the following.

(Non-unitary) Hermitian almost bundles X over V. These HAFl bundles are meant to generalize Lusztig's flat bundles as well as UAFl bundles. They come along with a connection ∇ preserving a Hermitian structure, i.e. a Hermitian form h as well as a unitary structure, denoted $\langle\ ,\ \rangle$ such that

1. The spectrum of h with respect to $\langle\ ,\ \rangle$ lies in an interval $[b^{-1}, b]$ for some (eventually large) constant $b > 0$.

2. The covariant derivative of $\langle\ ,\ \rangle$ is $\leq b$. This means that the monodromy operator $X_v \to X_{v'}$ along every path of length ≤ 1 has norm $\leq b$.

3. The curvature of ∇, as an operator valued 2-form on V, has norm $\leq \varepsilon$ with respect to $\langle\ ,\ \rangle$ (and a fixed Riemannian metric on V).

Then "HAFl" means a sequence of such bundles $X = X_{b,\varepsilon}$, where first $\varepsilon \to 0$ and then $b \to \infty$ (in fact, one could be more generous to b by just bounding it by something like $b \leq \exp \varepsilon^{-1}$), and K^0_{HAFl} is made of classes of rational linear combinations of $[X_{b,\varepsilon}]_+ - [X_{b,\varepsilon}]_-$. Everything we have said about UAFl extends to HAFl (but I admit I did not check it line by line), thus incorporating Lusztig into the AFl framework. In fact Connes's construction in $[\mathrm{Con}]_{\mathrm{CCTF}}$ allows a reduction of HAFl to UAFl with the following application.

Dirac twisted with HAFl bundles and $Sc > 0$. Every HAFl class $\kappa \in K^0(V)$ can be made UAFl if one allows infinite dimensional unitary bundles as in $9\frac{1}{4}$ by applying Connes' construction indicated in the end of $8\frac{1}{2}$. It follows that if V is a complete spin manifold with $ScV \leq \varepsilon^2 > 0$, then the twisted Dirac operator has $\operatorname{ind} \mathcal{D}_\kappa = 0$, which implies, as usual, that $(\operatorname{ch} \kappa \smile \widehat{A}_V)[V] = 0$. It would be interesting to find a geometric approach similar to the use of the Bruhat-Tits building indicated in $9\frac{1}{2}$.

$8\frac{14}{15}$. Families of UAFl and HAFl bundles parametrized by a space B

In this case the signature ranges in $K^0(B)$ as well as the index of $\mathcal{L}_X$ (compare Lusztig's theorem for families in $8\frac{5}{8}$) and the two are equal which implies the homotopy invariance of $\operatorname{ind} \mathcal{L}_X \in K^0(B)$ as the signature is homotopy invariant almost by definition. We shall say more about it in the end of this section and now turn to the basic example where our group Π is realized by the fundamental group $\pi_1(B)$ where B is a complete Riemannian manifold with the following property stronger (at least in spirit) than K-area $= \infty$.

Δ***-area*** $(B) = \infty$ ***with*** Δ ***for "diagonal".*** This means, by definition, the existence of a real vector bundle $Y \to B$ of rank $Y = \dim B$ and a *fiberwise proper* map $\widetilde{E}$ of the bundle $\widetilde{B}_\Delta \overset{def}{=} \widetilde{B} \times \widetilde{B}/\Pi \to B$ to Y (for the diagonal action of $\Pi = \pi_1(B)$ on the universal covering $\widetilde{B}$ of B so that the fibers of $\widetilde{B}_\Delta$ are copies of $\widetilde{B}$), such that $\widetilde{E}$ is of *non-zero degree* (where both B and Y are assumed oriented and where the basic example is $Y = T(B)$) *and area contracting on each fiber* of $\widetilde{B}_\Delta$, i.e. diminishing the areas of all smooth surfaces in these fibers.

Similar property with "Lipschitz" instead of "area contracting" appears in $[\text{C-M}]_{GCLC}$ under the heading "Families with avariable target" (and implicitly in the first paper by Miščenko [Miš]) where it is shown to imply the strong Novikov conjecture for $\Pi = \pi_1(B)$. (The basic example, already present in [Miš], is $\widetilde{\exp}^{-1} : \widetilde{B}_\Delta \to T(B)$ for manifolds B with negative curvature where $\widetilde{\exp}$ at each point $b \in B$ exponentiates $T_b(B)$ to the fiber $(\widetilde{B}_\Delta)_b = \widetilde{B}$, which can be thought of as the space $T_b(B) \to B$). "Area contracting" is, a priori, less demanding than "Lipschitz" but no actual group Π is known admitting B with Δ-area $= \infty$ but with no similar Lipschitz map. Also one should notice that area contracting maps (albeit more general) are more capricious characters than their Lipschitz counterparts (e.g. they do not stand convex combinations of maps and are harder to express in the discrete language of nets) and the formalism developed in $[\text{C-G-M}]_{\text{GCLC}}$ for Lipschitz map, does not extend (at least not directly) to the area contracting ones.

With these reservations in mind we state

Novikov conjecture for Δ-area $= \infty$. *If* $\Pi = \pi_1(B)$ *for a complete Riemannian manifold* B *with* Δ*-area* $(B) = \infty$ *then* Π *satisfies Novikov conjecture.*

This is one of the main applications of the general UAFl-theorem in $[\text{C-G-M}]_{\text{PPl}}$ and we now indicate the proof of it under the simplifying assumption (removed in $9\frac{1}{6}$) of Π being a *residually finite group.* So we take a cohomology class $\rho \in H^*(B)$ a map $\beta : V \to B$ and look at the value $(L_V\beta^*(\rho))[V]$ which we want to express in terms of the (homotopy invariant!) index of $\mathcal{L}$ twisted with a family of UAFl bundles over V. This family will be induced from B where it is constructed with the map $\widetilde{E} : \widetilde{B}_\Delta \to Y$. To simplify the picture we assume the bundle Y is trivial, which can always be achieved by replacing B by the total space of a bundle $Y^\perp \to B$ where $Y^\perp \oplus Y$ is trivial (compare "fixing the target" in $[\text{C-G-M}]_{\text{GGLC}}$, and then $\widetilde{E}$ reduces to a family of area contracting maps $\widetilde{E}_b : \widetilde{B}_b \to \mathbb{R}^n, b \in B$, where $\widetilde{B}_b = (\widetilde{B}_\Delta)_b$ is the universal covering of B with a marking $\widetilde{b} \in \widetilde{B}$ over $b \in B$, and where each map $\widetilde{E}_b$ is proper of positive degree and area contracting.

We compose these $\widetilde{E}_b$ with an ε-contracting map $\mathbb{R}^n \to S^n$ sending a neighbourhood of the infinity in $\mathbb{R}^n$ to a fixed point $s_0 \in S^n$ and thus pass to the ε-area contracting family $\widetilde{\Sigma}_b : \widetilde{B}_b \to S^n$. Now, if $\Pi = \pi_1(B)$ is residually finite, one may limit these $\widetilde{\Sigma}_b$ to certain maps $\widetilde{\Sigma}_b^N : \widetilde{B}_b^N \to S^n$ where $\widetilde{B}_b^N \to B$ is some (marked) N-sheeted covering (with $N < \infty$) approximating $\widetilde{B}$ such that the supports of the maps $\widetilde{\Sigma}_b$ (where supp $\widetilde{\Sigma}_b \overset{def}{=} C\ell\,\widetilde{\Sigma}_b^{-1}(s_0))$ *inject* under the (covering) maps $\widetilde{B}_b \to \widetilde{B}_b^N$. If B is non-compact this may be impossible for *all* $b \in B$ simultaneously but we shall need it only for $b \in \beta(V) \subset B$ for *compact* manifolds V.

Next we construct a family X of bundles $X(b)$ over B parametrized by B itself in three steps.

1. Take a unitary vector bundle $X_0 = (X_0, \nabla_0)$ over the sphere S^n for $n = 2m = \dim B$ with $c_m(X_0) \neq 0$ (where for odd n we just stabilize to $B \times \mathbb{R}$ as we often did).

2. Pull-back X_0 to $\widetilde{B}_b^N$ by the map $\widetilde{E}_b^N$.

3. Take the push-forward of the above $(\widetilde{E}_b^N)^*(X_0)$ under the covering map $\widetilde{B}_b^N \to B$ and call it $X(b) \to B$.

The bundles $X(b)$, for all $b \in B$, clearly are ε'-flat with $\varepsilon' \to 0$ for $\varepsilon \to 0$ and we may as well call them ε-flat to save notations. What remains to do is to compute the index of the signature operator $\mathcal{L}$ on V twisted with $\beta^*(X)$. To make it visual let us pretend that the maps $\widetilde{\Sigma}_b$ were actually defined over B itself. Namely, suppose we have maps $\Sigma_b : B \to S^n$ each of which sends the complement of a small neighbourhood $U_b \subset B$ of b to $s_0 \in S^n$ and $X(b) = \Sigma_b^*(X_0)$. The trouble with these Σ_b is that they can not be (area) contracting for small U_b and if, for example, B is compact, U_b can not be (arbitrarily) large. To help this we enlarge U_b, not in B itself but in the universal covering $\widetilde{B}$ or in some large but yet finite covering $\widetilde{B}^N$. So, as U_b grows, it becomes a "multivalued" set in B, i.e. a subset in $\widetilde{B}$ mapped to B. (The simplest example is that of a flat n-torus $\mathbb{T}^n$ where one starts with a small metric ball $U_b \subset \mathbb{T}^n$ which grows to a large ball in $\mathbb{R}^n$ and then is mapped finite-to-one back to $\mathbb{T}^n$ by the covering map $\mathbb{R}^n \to \mathbb{T}^n$). The bundle $X(b)$, defined via a proper map $U_b \to S^n$, extends from U_b to all of B by declaring it trivial ouside U_b in so far as U_b *injects* into B. But when U_b outgrows B we have to take the push-forward bundle which becomes infinite dimensional if we use the original (infinite) covering map $\widetilde{B} \to B$ and which has finite rank $N \operatorname{rank} X_0$ for $\widetilde{B}^N \to B$. Notice that for small U_b (injected into B) the push-forward bundle on B for the map $\widetilde{B}_b^N \to B$ is the same thing as the extended bundle plus $N-1$ copies of the trivial bundle of rank $= \operatorname{rank} X_0$ and as we enlarge U_b in $\widetilde{B}_b^N$ the isomorphism class of this bundle does not change. Thus we can compute the index of $\mathcal{L}_X$ using the family X coming from $\Sigma = \{\Sigma_b\}_{b \in B}$ as the error equals the index of $\mathcal{L}$ twisted with the above trivial bundle. So, let us compute this index of $\mathcal{L}_X$, $X = \{X(b)\}_{b \in B}$ which is an element $\kappa \in K_0(B)$ (where for non-compact B this K_0 is made of bundles with compact supports) by invoking the index theorem for families. Namely, we denote by $X^\beta \to V \times B$ the vector bundle corresponding to the family $\beta^* \left(X = \{X(b)\}_{b \in B}\right)$ over V and observe that the Poincaré dual of $\operatorname{ch} X^\beta$ equals a non-zero multiple of the image of the fundamental class [V] under the *graph* of β, i.e. the map $v \mapsto (v, \beta(v)) \in V \times B$. Namely,

$$\mathcal{PD} \operatorname{ch} X^\beta = M(\Gamma_\beta)_*[V] \qquad (*_\beta)$$

for $M = d \operatorname{ch} X_0[S^n]$ where d is the degree of $\widetilde{E} : \widetilde{B}_\Delta \to Y$ or equivalently, the degree of the maps $\Sigma_b : B \to S^n$. In fact, our family of bundles over B, viewed as a bundle $X^\Delta \to B \times B$ induced by the map $\Sigma : B \times B \to S^n$ for $\Sigma = \{\Sigma_b\}_{b \in B}$ has

$$\mathcal{PD}(\operatorname{ch} X^\Delta) = M[B_\Delta] \qquad (*_\Delta)$$

for $B_\Delta = B$ diagonally imbedded into $B \times B$ since $\Sigma^*([S^n]^{co})$ clearly equals $\mathcal{PD}(d[B_\Delta])$ and $(*_\Delta)$ implies $(*_\beta)$ by functoriality. Here we assume $\operatorname{ch} X_0$ equals a multiple of $[S^n]^{co}$ which is only possible for virtual bundles and in truth, we must use $[X_0] - [\mathrm{Triv}_0]$ where $\operatorname{rank} \mathrm{Triv}_0 = \operatorname{rank} X_0$. Now we see that the Chern character of $\kappa = \operatorname{ind} \mathcal{L}_{X^\beta} \in K_0(B)$, can be computed with the index theorem for the projection (family) $p : V \times B \to B$ and $(*)$ as follows:

$$\begin{aligned}\operatorname{ch} \kappa = \mathrm{Cys}_p(L_V \operatorname{ch} X^\beta) &= M \cdot \mathcal{PD}p_*(\mathcal{PD}L_V \frown (\Gamma_\beta)_*[V])) \\ &= M \cdot \mathcal{PD}(\beta_*(\mathcal{PD}L_V)).\end{aligned}$$

Therefore, the homotopy invariance of $\kappa = \operatorname{ind} \mathcal{L}_{X^\beta}$ implies that of $\beta_*(\mathcal{PD}L_V)$ and hence of the higher signatures σ_ρ as $\sigma_\rho \overset{def}{=} (L_V \vee \beta^*(\rho))[V]) = \rho(\mathcal{PD}L_V)$ for all $\rho \in H^*(B)$. Q.E.D.

Recall that the Novikov conjecture can be stated homologically as the homotopy invariance of $\beta_*(\mathcal{PD}L_V) \in H_*(B\pi; \mathbb{Q})$ for the classifying map $\beta : V \to B\Pi$ for $\Pi = \pi_1(V)$.

On the proof of the homotopy invariance of $\operatorname{ind} \mathcal{L}_X$ ***for UAFl (and HAFl) families*** X. A family of bundles parametrized by $b \in B$ should be viewed as a single bundle over the C^*-algebra Cont$B$, and everything we said about UAFl (and HAFl) bundles generalizes to the C^*-algebra framework (compare [Hi-Sc]). Notice that despite the appearance of C^*-algebras, the *strong* Novikov conjecture remains problematic for Δ-area $= \infty$. The difficulty stems from the fact that our almost representation $\Pi \to U(p)$ does not extend to $C^*(\Pi)$. Yet it extends to $\ell_1(\Pi)$ and so the gap lies between $K_0C^*(\Pi)$ and $K_0\ell_1(\Pi)$. (This circle of ideas was patiently explained to me by Henri Moscovici).

$8\frac{15}{16}$. On the classification of AFl bundles

What we want to know is the homotopy type of the space of UAFl (and HAFl) bundles, not their bare existence.

Example. Let V be an aspherical 4-dimensional manifold such that

(1) The universal covering is hyper-Eucliean;

(2) $H^2(V; \mathbb{Q}) \neq 0$;

(3) the fundamental group $\Pi = \pi_1(V)$ is residually finite and, moreover there is a class $h \in H^2(V)$ with $h^2 \neq 0$ and a sequence of finite coverings $p_i : \widetilde{V}_i \to V$, $i = 1, 2, \ldots$, such that $p_i^*(h) \in H^2(\widetilde{V}_i)$ is divisible by i.

Then one can form two kinds of UAFl bundles over V. The first group comes from mapping large finite coverings $\widetilde{V}_N$ of V (unrelated to the above $\widetilde{V}_i$) to S^n by (area) ε-contacting maps of non-zero degree, pulling back a suitable $X_0 \to S^n$ to $\widetilde{V}$ and pushing it down to X_ε on V.

Another construction consists of taking a complex line bundle Y on V with $\mathrm{ch}_1 Y = h$, pulling it to $\widetilde{U}_i \to \widetilde{V}_i$, taking the i-th root $\left(\widetilde{Y}_i\right)^{\frac{1}{i}}$ and pushing it down to Y_ε with $\varepsilon \approx i^{-1}$ on V. Then one can combine tensor products and exterior powers of X_ε and Y_ε and find among these some representing equal elements in $K_0(V)$. The question is when these UAFl bundles, say X'_ε and Y'_ε with $[X'_\varepsilon] = [Y'_\varepsilon]$ can be joined by a homotopy of UAFl bundles after a suitable stabilization. One also can throw into the game HAFl bundles, e.g. *flat* Hermitian bundles, (for example for V = surface $\times$ surface) and ask the same question in the HAFl category.

Global almost homomorphisms. The above question can be reformulated in terms of, say unitary, almost representations of an abstract group Π, and one may strengthen the notion of an ε-homomorphism $\rho : \Pi \to U(p)$ by requiring the inequality

$$\|\rho(\pi_1)\rho(\pi_2) - \rho(\pi_1\pi_2)\| \leq \varepsilon$$

to hold for *all* π_1 and π_2 in Π. This is, in the terminology of $8\frac{3}{4}$, an (F,ε)-homomorphism with $F = \Pi$ which suggests intermediate classes where F is infinite but smaller than all of Π. It may seem that for many groups Π such a *global unitary ε-representation* with small ε must be a small perturbation of an actual representation. In fact this is known, thanks to D. Kazhdan, for the amenable groups by a non-linear overaging argument (used earlier by Grove, Karcher and Ruh for compact groups).

On the other hand, if V is a compact manifold with *strictly negative sectional curvature* $K(V) \leq -\delta^2$ and X is an ε-flat bundle over V of rank p, then the holonomy around *geodesic* loops at a given point $v_0 \in V$ is such a global ε'-representation of the fundamental group Π of V to $U(p)$ with $\varepsilon' \leq C\varepsilon$ for some constant $C = C(V) \leq \pi\delta^{-2}$. This follows from the fact that all geodesic triangles in the universal covering $\widetilde{V}$ of V bound disks of area $\leq \pi\delta^{-2}$. Then this generalizes to any compact V with a *hyperbolic* fundamental group (where one can define a suitable substitute for geodesics; see $[\mathrm{Gro}]_{\mathrm{HG}}$) and one sees furthermore that every unitary (F,δ)-representation of a hyperbolic group Π with F generating Π gives rise to a global ε-representation with $\varepsilon \leq C\delta$ for $C = C(\Pi; F)$.

Questions. Are there non-hyperbolic groups with this globalization property ? Are there non-amenable groups where every global ε-representation can be perturbed to an actual representation? Here one may suspect irreducible lattices in semisimple groups of $\mathbb{R}$-rank ≥ 2 as they are full of Abelian (actual and virtual) subgroups where such perturbations do exist by Kazhdan's theorem.

A closely related globalization property is the existence of an ε'-parallel frame in the lift of an ε-flat bundle X from V to the universal covering $\widetilde{V}$ of V. This is always possible if $\pi_1(V)$ is hyperbolic (e.g. for $K(V) < -\delta^2$) and also for some bundles over certain non-hyperbolic manifolds, e.g. pushforwards of suitable line bundles over (finite covers of) Kähler hyperbolic manifolds; (see [Gro]$_{\mathrm{KH}}$).

9. Open manifolds and foliations

We have been avoiding so far a direct encounter with non-compact manifolds and now time has come to meet them face to face. We start by recalling

$9\frac{1}{9}$. L_2-index theorem for infinite coverings

Let $\widetilde{V}$ be a Galois Π-covering of a compact Riemannian manifold V, let D be an elliptic operator over V (e.g. Dirac or the signature operator $\mathcal{L}$) and $\widetilde{D}$ the lift of D to $\widetilde{V}$. (Notice that one may lift to $\widetilde{V}$ also the pseudo-differential operator whose Schwartzian kernels are supported *close to the diagonal* in $V \times V$ but, in general, such lift is impossible. For example, the projection operator on $\operatorname{Ker} D$ admits no geometric lift to $\widetilde{V}$.) The lifted operator $\widetilde{D}$ acts on the L_2-sections of the relevant Π-invariant bundles (lifted from V), say $\widetilde{D} : L_2\, \widetilde{S}_+ \to L_2\, \widetilde{S}_-$ ($\widetilde{D}$ is unbounded and defined on a dense subspace in $L_2\, \widetilde{S}_+$ but we write it as if it were globally defined), and its kernel $\operatorname{Ker} \widetilde{D} \subset L_2\, \widetilde{S}_+$ and cokernel $L_2\, \widetilde{S}_- / C\ell \operatorname{Im} \widetilde{D}$ are moduli *over the von Neumann algebra* $\mathcal{N}(\Pi)$, which is the algebra of bounded operators on $\ell_2(\Pi)$ commuting with the (say right) action of Π. Every such operator is given by a function ν on Π which acts on ℓ_2-functions φ by convolution $\varphi \mapsto \nu * \varphi$ and, in fact, $\mathcal{N}(\Pi)$ equals the weak operator closure of the group ring $\mathbb{C}(\Pi)$.

The *von Neumann dimension* of $\operatorname{Ker} \widetilde{D}$, denoted $\dim_\Pi \operatorname{Ker} \widetilde{D}$, can be intuitively thought of as the ordinary dimension divided by the cardinality of the group Π and this is the true definition for finite groups Π. In the general case where Π is infinite, one computes $\dim_\Pi \operatorname{Ker} \widetilde{D}$ as the Π-*trace* of the (orthogonal) projection operator $\widetilde{P} : L_2\, \widetilde{S}_+ \to \operatorname{Ker} \widetilde{D}$. This $\widetilde{P}$ has a C^∞-smooth Π-invariant (Schwartzian) kernel on $\widetilde{V}$, denoted $\widetilde{P}(\widetilde{v}_1, \widetilde{v}_2)$, where each value $\widetilde{P}(\widetilde{v}_1, \widetilde{v}_2)$ is an operator from the fiber of the implied vector bundle $\widetilde{S}_+$ at $\widetilde{v}_1$ to that at $\widetilde{v}_2$. (Strictly speaking $\widetilde{P}(\widetilde{v}_1, \widetilde{v}_2)$ is (operator) $\otimes$ (volume density) but as we assume $\widetilde{V}$ oriented with a Π-invariant Riemannian metric, densities reduce to functions on $\widetilde{V}$.) We consider trace $\widetilde{P}(\widetilde{v}, \widetilde{v})$, a smooth function on $\widetilde{V}$ which is Π-invariant and thus descends to a function on V denoted by trace $\widetilde{P}(v, v)$. Then we can easily prove that

$$\dim_\Pi \operatorname{Ker} \widetilde{D} = \int_V \operatorname{trace} \widetilde{P}(v, v)\, dv \qquad (*)$$

or (if one resents the abstract definition of $\dim_\Pi$) take (*) for the *definition* of $\dim_\Pi$ (and pay the price of checking that $\dim_\Pi \operatorname{Ker} \widetilde{D}$ depends only on the *Hilbert space stucture of* $\ker \widetilde{D}$ *with the unitary* Π*-action* and not on the specific geometry attached to this space). Next one introduces $\dim_\Pi \operatorname{coker} \widetilde{D}$, which can be defined as $\dim_\Pi \ker \widetilde{D}^*$ for the adjoint operator $\widetilde{D}^* : L_2 \widetilde{S}_- \to L_2 \widetilde{S}_+$ and arrives at the notion of the L_2*-index*,

$$\operatorname{ind}_\Pi \widetilde{D} = \dim_\Pi \ker \widetilde{D} - \dim_\Pi \operatorname{coker} \widetilde{D}\,.$$

Atiyah L_2-index theorem. *The above L_2-index of $\widetilde{D}$ equals the ordinary index of the underlying operator D on the compact manifold V, i.e.*

$$\operatorname{ind}_\Pi \widetilde{D} = \operatorname{ind} D\,. \tag{$\star$}$$

Corollary. *If* $\operatorname{ind} D > 0$ *(which is a topological condition on* (V, D) *by the Atiyah-Singer theorem) then for every Galois covering* $\widetilde{V} \to V$ *the space of* $\widetilde{D}$*-harmonic L_2-sections on* $\widetilde{V}$ *(i.e. the space* $\ker \widetilde{D}$*) is non-empty and therefore infinite dimensional for infinite Galois coverings.*

Examples. (a) Let V be a Riemann surface of genus ≥ 2 and $D = d + d^*$ acting from $\Lambda^1(V)$ to $\Lambda^0(V) \oplus \Lambda^2(V)$. Here $\operatorname{ind} D = -\chi(V) > 0$ and so the universal covering $\widetilde{V}$ supports an infinite dimensional space of harmonic 1-forms. Similarly, if V is a closed 4-dimensional manifold with strictly *positive* Euler characteristic, and Π is infinite, then $\widetilde{V}$ supports non-trivial harmonic 2-forms. (There is no contribution to the index from the dimensions 0 and 4 as every harmonic L_2-function or 4-form vanish on *infinite* $\widetilde{V}$.) Conversely, if $\chi(V) > 0$, then $\widetilde{V}$ supports harmonic 1-forms as well as 3-forms, since the Hodge operator $\widetilde{*} : \widetilde{\Lambda}^1 \to \widetilde{\Lambda}^3$ establishes a Π-equivariant isomorphism between harmonic 1- and 3-forms on $\widetilde{V}$.

(b) If V is a $4k$-dimensional manifold with non-zero signature, then every Π-covering of V supports a non-zero harmonic L_2-form of degree $2k$, as follows from Atiyah's theorem for the lifted signature operator $\widetilde{\mathcal{L}}$ on $\widetilde{V}$.

The L_2-signature of $\widetilde{V}$ can be defined combinatorially with some triangulation of V lifted to $\widetilde{V}$. Here one has $\widetilde{\delta}$ (and $\widetilde{\delta}^* = \widetilde{\partial}$) operating on ℓ_2-cochains on $\widetilde{V}$ and the Π-invariant quadratic form on *the L_2-cohomology* $L_2\, H^{2k}(\widetilde{V})$ whose L_2-*signature* is well defined over $\mathcal{N}(\Pi)$. In fact the L_2-signature makes sense for the pairing $\widetilde{h} : L_2\, H^{2i} \otimes L_2\, H^{2i} \to \mathbb{R}$ defined for each $h \in H_{4i}(V)$ by composing the cup-product $L_2\, H^{2i} \otimes L_2\, H^{2i} \to L_1\, H^{4i}$ with the evaluation of L_1-cohomology of $\widetilde{V}$ on the homology of V. And also one can define the L_2-signature of $\widetilde{V}$ where V is a compact $4k$-dimensional manifold with boundary. These signatures enjoy the same formal properties as their compact counterparts (see $7\frac{1}{4}$)

and we invite the reader to look at this. But the *L_2-multiplicativity* formula $\sigma_\Pi(\widetilde{V}) = \sigma(V)$, immediate with the Atiyah theorem applied to $\operatorname{ind}_\Pi \widetilde{\mathcal{L}}$, seems hard to prove by a cobordism argument. Furthermore, the L_2-signature $\sigma_\Pi(\widetilde{V})$ can be easily defined for combinatorial and topological manifolds V but the proof of the equality $\sigma_\Pi = \sigma$ becomes more complicated (unless I am missing something obvious).

(c) Let V be a closed surface of genus ≥ 2 and $X \to V$ be a symplectic bundle with $\sigma(V;X) \neq 0$ as in $8\frac{2}{7}$. Then $\sigma_\Pi(\widetilde{V};\widetilde{X}) = \sigma(V;X) \neq 0$ for the universal covering $\widetilde{V} \to V$ (where $\Pi = \pi_1(V)$) and consequently $\widetilde{V}$ supports nontrivial harmonic forms with coefficients in $\widetilde{X}$. (Notice the $\widetilde{X}$ over $\widetilde{V}$ is a *trivial* bundle but it admits no Π-invariant unitary structure. Yet the L_2-Betti number $L_2\, b_1(\widetilde{V};\widetilde{X})$ is non-ambiguously defined and equals $-\chi(V)$ rank X as the corresponding $L_2\, b_0$ and $L_2\, b_2$ vanish.) In fact this $\sigma_\Pi(\widetilde{V};\widetilde{X})$ can be defined for all V in the purely topological category similarly to $\sigma_\Pi(\widetilde{V};\widetilde{\mathrm{Triv}})$ considered in (b) but these signatures, and non-unitarily twisted L_2-Betti numbers in general, have not yet been looked upon with due attention.

L_2-index from the point of view of $K_0\mathcal{N}(\Pi)$. Let X be the (flat) $\ell_2(\Pi)$-fibered bundle over V associated to the Π-covering (principal Π-bundle) $\widetilde{V} \to V$ for the left action of Π on $\ell_2(\Pi)$ and observe that the spaces $L_2\, \widetilde{S}_\pm$ are the same as L_2-sections of the underlying bundles $S_\pm \to V$ with coefficients in X and the lift of D from V to $\widetilde{D}$ on $\widetilde{V}$ amounts to twisting D with X on V. The action of D on the twisted bundle is (obviously) compatible with the (right) action of Π, and hence of $\mathcal{N}(\Pi)$, on the fibers. So one may speak of the index of D_X with values in $K_0\mathcal{N}(\Pi)$. Observe that $\mathcal{N}(\Pi)$ is a C^*-algebra (for example, if $\Pi = \mathbb{Z}^n$ it is canonically isomorphic to the algebra of bounded measurable functions on the torus $\mathbb{T}^n = \operatorname{Hom}(\mathbb{Z}^n \to \mathbb{T}^1)$) and one knows that $K_0\mathcal{N}(\Pi)$ is isomorphic to $\mathbb{R}$ where the K_0-class of each projective module M over $\mathcal{N}(\Pi)$ is determined by the von Neumann dimension $\dim_\Pi M$. Of course, the K-theoretic index $\operatorname{ind} D_X \in K_0\mathcal{N}(\Pi)$ is the same thing as the L_2-index of $\widetilde{D}$ on $\widetilde{V}$ for the isomorphism $K_0\mathcal{N}(\Pi) \underset{\sim}{\to} \mathbb{R}$ given by $\kappa \mapsto \dim_\Pi \kappa$, that is,

$$\operatorname{ind}_\Pi \widetilde{D} = \dim_\Pi \operatorname{ind} D_X\,.$$

Generalization of Atiyah's theorem. Let $\widetilde{V}$ be as earlier a complete Riemannian manifold acted upon isometrically by a group Π but now we somewhat relax our assumptions on the action. We require that

1. The action is discrete (but not necessarily free, not even faithful. One could even allow non-discrete Lie groups Π properly acting on $\widetilde{V}$ but this is not needed for the applications we have in mind).

2. The manifold $\widetilde{V}$ has *locally bounded geometry*, i.e. the sectional curvatures of V are bounded by $|K(V)| \leq \text{const}$ and V has no geodesic loops shorter than $(\text{const})^{-1}$ which amounts to $2\,\text{Inj Rad}\, V \geq \text{const}^{-1}$.

3. The quotient space V/Π has finite volume (but is not necessarily compact).

4. The action of Π lifts to the actions on our bundles $\widetilde{S}_\pm$ over $\widetilde{V}$ and these commute with $\widetilde{D}$ which is supposed to be here a *geometric* differential operator (e.g. Dirac or the signature operator).

One can show that the L_2-index of $\widetilde{D}$ is well defined and finite in this case and can be computed in terms of the curvature of $\widetilde{V}$ as follows. Denote by $\widetilde{\Omega}_D$ the Chern-Weil form built out of the curvature of $\widetilde{V}$ and the bundles $\widetilde{S}_\pm$ which is the case of *compact* $\widetilde{V}$ and trivial Π gives us the index of $\widetilde{D} = D$ by

$$\operatorname{ind} D = \int_{\widetilde{V}} \Omega_D\,.$$

Observe that in the presence of the Π-action this form is Π-invariant, and the integral $\int_V \Omega\, dv$ absolutely converges where $V = \widetilde{V}/\Pi$ and where we assume the action of Π on $\widetilde{V}$ is orientation preserving.

Integral formula for L_2-index.

$$\operatorname{ind}_\Pi \widetilde{D} = \int_V \Omega_D\,. \qquad (\star\star)$$

(This implies $(\star)$ as $\int_V \Omega_D = \operatorname{ind} D$, whenever D is defined, by the very definition of Ω_D.)

Corollary. *If $\int_V \Omega_D > 0$ then there is a non-zero $\widetilde{D}$-harmonic L_2-section of $\widetilde{S}_+$.*

References. Everything started with [At]$_{\text{EPDG}}$. The formula $(\star\star)$ is proven (in a slightly different setting) in [Ch-Gr]$_{\text{CN}}$. Geometric and algebraic applications of $(\star)$ and $(\star\star)$ appear in the next section and in [Ch-Gr]$_{\text{BVND}}$, [Ch-Gr]$_{\text{L}_2}$ and [Gro]$_{\text{KH}}$.

$9\frac{1}{8}$. L_2-obstructions to positive scalar curvature

Our objective is the following theorem.

Let V be a closed oriented n-dimensional manifold such that the fundamental class $[V]^{\text{Co}} \in H^n(V, \mathbb{Q})$ is contained in the subring generated by $H^1(V;\mathbb{Q})$ and the kernel of the homomorphism $p^* : H^2(V;\mathbb{Q}) \to H^2(\widetilde{V};\mathbb{Q})$ for the universal covering $p : \widetilde{V} \to V$. Then V admits no metric with Sc > 0, *provided*

$\widetilde{V}$ is spin. Moreover if $\mathrm{Sc} > 0$, *then* $(\widehat{A}_v \smile \rho)[V] = 0$ *for all ρ in the subring generated by H^1 and* $\ker p^* \mid H^2$.

We have already proved a similar result with an assumption concerning finite coverings of V implying the relation K-area $V = \infty$; now we are ready to do the same for infinite coverings (without working out at this stage the corresponding notion of the K-area with infinite dimensional bundles).

Idea of the proof. (Compare Example with H^2 in $8\frac{3}{4}$.) Every $c \in H^2(V)$ can be realized as the first Chern class $c_1(X)$ of a complete line bundle $X \to V$ and if $p^*(c) = 0$ the lift $\widetilde{X} \to V$ is topologically trivial. Hence, one may take the d-th root $\widetilde{X}_d = (\widetilde{X})^{\frac{1}{d}}$ for all $d = 1, \ldots,$ and the curvatures of these are bounded by d^{-1} (compare $[\mathrm{Gr}]_{\mathrm{KH}}$). The fundamental group $\pi_1(V)$ does not naturally act on $\widetilde{X}_d$ but the obvious $\mathbb{Z}/d\mathbb{Z}$-extension, say Π_d of $\pi_1(V)$ does act there, and one can twist the lifted Dirac operator $\widetilde{D}$ on $\widetilde{V}$ with $(\widetilde{X})^{\frac{1}{i}}$. Now we are in a position to apply $(\star\star)$ to $\widetilde{D}_{\widetilde{X}_d}$ and bring it to a contradiction with the assumption $\mathrm{Sc} \geq \delta > 0$ via the twisted BL-formula (see §5). (One takes care of non-strictly positive Sc as on page 140/352 of $[\text{G-L}]_{\mathrm{PSC}}$ by referring to a theorem of Kazdan.) We leave details to the reader.

Example. Let (V, ω) be a closed symplectic (e.g. Kähler) manifold where the universal covering is contractible. Then V admits no metric with $\mathrm{Sc} > 0$.

Generalization. Our formula $(\star\star)$ allows an application to non-compact complete manifolds V with $\mathrm{Sc} \geq \delta > 0$ provided the universal covering is spin and has locally bounded geometry. For example every such V necessarily has $\int_V \Omega_D = 0$ where Ω_D is the n-form representing the $\widehat{A}$-genus of V. We suggest the reader would similarly extend the above theorem to the general framework of the formula $(\star\star)$.

$9\frac{1}{7}$. Novikov conjecture for $H^2(\Pi)$

Let V be a closed manifold and let $\rho \in H^*(V; \mathbb{Q})$ lie in the subring generated by $H^1(V; \mathbb{Q})$ and the kernel of $p^* : H^2(V; \mathbb{Q}) \to H^2(\widetilde{V}; \mathbb{Q})$ for the universal covering $p : \widetilde{V} \to V$. *Then the ρ-signature* $\sigma_\rho = (\rho \smile L_V)[V]$ *is homotopy invariant.*

Idea of the proof. (Compare Example with H^2 in $8\frac{3}{4}$.) Everything boils down to showing that the L_2- index of the signature operator $\widetilde{\mathcal{L}}$ on $\widetilde{V}$ twisted with the above (d^{-1}-flat!) bundle $\widetilde{X}_d = (\widetilde{X})^{\frac{1}{d}}$ (or a bundle built out of these) is homotopy invariant for large $d \geq d_0$ (where d_0 depends on the homotopy in question) as this L_2-index equals $L_V \smile (\mathrm{ch}\, X)^{\frac{1}{d}}\, [V]$ according to $(\star\star)$ (where $(\mathrm{ch}\, X)^{\frac{1}{d}} = \exp d^{-1} c_1(X)$). To prove this we denote by $\mathbf{X}_d \to V$ the push-forward of the bundle $\widetilde{X}_d$ under the covering map $p : \widetilde{V} \to V$. This $\mathbf{X}_d$ is an infinite dimensional (roughly) d^{-1}-flat bundle over V of which every fiber is naturally acted upon by the group Π_d of the previous section. To see it clearly, take the covering $\widetilde{S}_d$ of the unit circle bundle S associated to X, such that $\widetilde{S}_d$ completely uncovers V and covers each (circle) fiber of S exactly d times. This $\widetilde{S}_d$ equals the unit circle bundle of $\widetilde{X}_d$ and the group Π_d is the Galois group of the covering $\widetilde{S}_d \to S$. Thus the group Π_d acts on $\widetilde{X}_d$ and consequently on the space of sections of $\widetilde{X}_d \mid p^{-1}(v)$ for each $v \in V$. But this space of sections is exactly the fiber $(\mathbf{X}_d)_v$ by the definition of $\widetilde{X}_d$.

This action of Π_d gives to $\mathbf{X}_d$ the structure of an $\mathcal{N}(\Pi_d)$ bundle and therefore $\operatorname{ind} \mathcal{L}_{\mathbf{X}_d} \in K_0 \mathcal{N}(\Pi_d)$ is a homotopy invariant for large d by the discussion in $8\frac{5}{8}$. The (sketch of the) proof is concluded by observing that the desired L_2-index of $\widetilde{\mathcal{L}}$ twisted with $\widetilde{X}_d$ satisfies

$$\operatorname{ind}_{\Pi_d} \widetilde{\mathcal{L}}_{\widetilde{X}_d} = \dim_{\Pi_d} \operatorname{ind} \mathcal{L}_{\mathbf{X}_d}$$

similarly to the identity $\operatorname{ind}_\Pi \widetilde{D} = \dim_\Pi \operatorname{ind} D_X$ in $9\frac{1}{9}$. (The above expands the last claim in section 6 of [C-G-M]$_{\mathrm{PP}\ell}$. More details and applications will eventually appear (I hope) in our continuation of [C-G-M]$_{\mathrm{PP}\ell}$.)

$9\frac{1}{6}$. Novikov conjecture for Δ-area $= \infty$ revised and Fredholm K-area

We want to remove the residual finiteness assumption on Π (see $8\frac{4}{5}$). To warm up we start with the case where our V, closed connected oriented Riemannian manifold, admits a (possibly infinite and non-Galois) covering $p : \widetilde{V} \to V$ with an ε-flat K^0-class $\widetilde{\kappa}$ on $\widetilde{V}$ with compact support, and show that the ρ-signature σ_ρ for the push-forward $\rho = \operatorname{Gys}(\operatorname{ch} \widetilde{\kappa}) \in H^*(V)$ is a homotopy invariant of V for small $\varepsilon \le \varepsilon_0$ where $\varepsilon_0 > 0$ depends on the implied homotopy equivalence. Here our $\widetilde{\kappa}$ is given by a pair of unitary ε-flat bundles, $\widetilde{\kappa} = [\widetilde{X}^+] - [\widetilde{X}^-]$, where these bundles are connected by a homomorphism $\widetilde{F} : \widetilde{X}^+ \to \widetilde{X}^-$ which is a unitary connection preserving isomorphism outside a compact subset in $\widetilde{V}$. We push forward $\widetilde{X}^\pm$ to Hilbert bundles $\mathbf{X}^\pm \to V$ where the fiber of $(\mathbf{X}^\pm)_v$ equals the space of ℓ_2-sections of $\widetilde{X}^\pm$ on $p^{-1}(v) \in \widetilde{V}$. These $\mathbf{X}^\pm$ are as flat as $\widetilde{X}^\pm$ and $\widetilde{F}$ descends to a *Fredholm* homomorphism $\mathbf{F} : \mathbf{X}^+ \to \mathbf{X}^-$. In fact $\mathbf{F}_v$ is a unitary isometry between subspaces of finite codimensions, say $Y_v^+ \subset \mathbf{X}_v^+$ and $Y_v^- \subset \mathbf{X}_v^-$, for each $v \in V$. This defines a K^0-class on V, namely $\kappa = \operatorname{ind} \mathbf{F} = [\ker \mathbf{F} - \operatorname{coker} \mathbf{F}]$ as usual, with $\operatorname{ch} \kappa = \rho$. We fix some loops at a point $v_0 \in V$ representing the elements of $\Pi = \pi_1(V)$ and thus obtain two unitary ε-representations of Π, say $r^\pm$, in the Hilbert spaces $H^\pm = \mathbf{X}_{v_0}^\pm$. Clearly (and most importantly) the homomorphism $\mathbf{F}$ commutes with $r_\pm$ modulo compact operators, i.e. $\mathbf{F}(r^+(\pi)) - r^-(\pi)$ is a compact operator in H^- for every $\pi \in \Pi$.

One can equivalently express these properties in a Π-free language by using two diagrams D^+ and D^- of chain-cochain complexes of V with coefficients in $\mathbf{X}^\pm$ as (D) in $8\frac{8}{9}$, with a connecting homomorphism $D^+ \to D^-$ which commute with the homomorphisms in D^+ and D^- modulo compact operators, where our homomorphism $D^+ \to D^-$ is naturally associated to $\mathbf{F}$ and consists of $F_i : C_i^+ \to C_i^-$ for the chain spaces $C_i^\pm$ in $D^\pm$ and $F^i : C_-^i \to C_+^i$ dual to F_i. Here is the schematic picture of the resulting mess (compare (D) in $8\frac{8}{9}$).

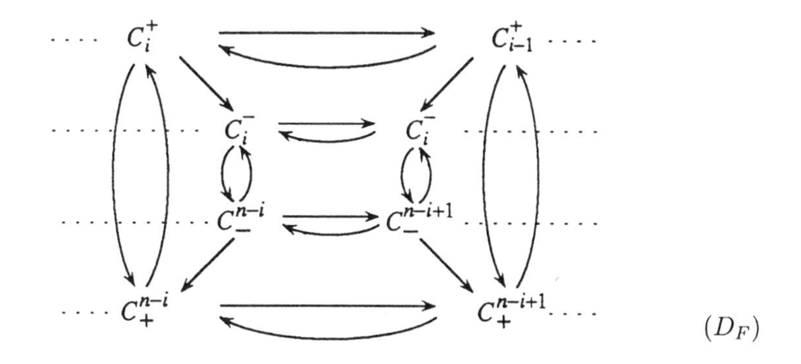

(D_F)

If $\varepsilon > 0$ is sufficiently small while the norms of the arrows in this diagram are not too large, one can extract a numerical invariant, called $\sigma(D_F) = \sigma(V;\kappa)$ as in $8\frac{2}{9}$ (for flat bundles this is done by Miščenko; see [Miš]) which is a homotopy invariant more or less by definition.

Example. Suppose all C_i and C^i are zero for $i \neq m$ (for $n = 2m$) and our diagram reduces to

$$\begin{array}{ccc} C_m^+ & \xrightarrow{F_m} & C_m^- \\ {\scriptstyle A^+}\downarrow & & {\scriptstyle A^-}\downarrow \\ C_+^m & \xleftarrow[F^m]{} & C_-^m \end{array}$$

where $A^\pm$ are invertible (by $A_\pm$ which are not notationally needed and which appear as $p_\pm^m$ in the diagram (D) in $8\frac{2}{9}$) and the diagram commutes modulo compact operators. To be specific, we assume the bilinear forms corresponding to $A^\pm$ are Hermitian (the skew-Hermitian case is similar) and identify $C^+ = C_m^+$ with (its dual) C_+^m as well as $C^- = C_m^-$ with C_-^m using the Hilbert structures in $C_m^\pm$. This simplifies our diagram to

$$A^+ \hookrightarrow C^+ \xrightarrow{F} C^1 \hookleftarrow A^-$$

where $A^{\pm}$ are bounded invertible Hermitian operators and F is a Fredholm operator almost (i.e. modulo compact operators) commuting with $A^{\pm}$. We orthogonally split C^+ and C^- according to the sign of the spectrum of $A^{\pm}$,

$$C^+ = C^+_+ \oplus C^+_- \quad \text{and} \quad C^- = C^-_+ \oplus C^-_- .$$

If C^+ and C^- were finite dimensional we would define the signature of this diagram as

$$\sigma = \sigma(A^+) - \sigma(A^-) = \dim C^+_+ - \dim C^+_- - \dim C^-_+ + \dim C^-_-$$

which is the same as

$$(\dim C^+_+ - \dim C^-_+) - (\dim C^+_- - C^-_-) \qquad (-)$$

where the latter makes sense in the infinite dimensional case as well since $F : C^+ \to C^-$ (obviously) provides a *Fredholm relation* between the negative and positive subspaces, $C^+_+ \sim\!\to C^-_+$ and $C^+_- \sim\!\to C^-_-$. That is, the composition of $\varphi \mid C^+_+$ with the orthogonal projection $C^- \to C^-_+$ is Fredholm, say $F_+ : C^+_+ \to C^-_+$, and similarly $F_- : C^+_- \to C^-_-$ is also Fredholm. Now $(-)$ makes sense and we define

$$\sigma = \operatorname{ind} F_+ - \operatorname{ind} F_- .$$

Next we define

Signature operator on V twisted with κ. We twist $\mathcal{L}$ on V with the Hilbert bundles $\mathbf{X}^{\pm}$ and consider the diagram

$$\begin{array}{ccc} C^\infty(\Lambda^*_+ \otimes \mathbf{X}^+) & \xrightarrow{\mathcal{L}_{\mathbf{X}^+}} & C^\infty(\Lambda^*_- \otimes \mathbf{X}^+) \\ \downarrow \varphi_+ & & \downarrow \varphi_- \\ C^\infty(\Lambda^*_+ \otimes \mathbf{X}^-) & \xrightarrow{\mathcal{L}_{\mathbf{X}^-}} & C^\infty(\Lambda^*_- \otimes \mathbf{X}^-) \end{array}$$

where the vertical arrows $\varphi_{\pm}$ are given by $\mathbf{F}$ and where $\Lambda^*_{\pm} = \Lambda^*_{\pm}(V)$ are the bundles splitting $\Lambda^*(V)$ by $\Lambda^* = \Lambda^*_+ \oplus \Lambda^*_-$ making $\mathcal{L}$ out of $d + d^*$ (see $8\frac{1}{2}$). Actually, it is slightly more convenient to use bounded (pseudo-differential of zero order operators) $\widehat{\mathcal{L}}^{\pm}$ instead of (differential operators) $\mathcal{L}_{\mathbf{X}^{\pm}}$ defined by $\widehat{\mathcal{L}}^{\pm} = (\mathcal{L}_{\mathbf{X}^{\pm}}) \circ (1 + \Delta^{\pm})^{-1/2}$ where $\Delta^{\pm}$ denotes the composition of $\mathcal{L}_{\mathbf{X}^{\pm}}$ with the adjoint operator (compare $8\frac{2}{3}$). Now we can use (Hilbert spaces) L_2 instead of C^∞ and rewrite our diagram as

$$\begin{array}{ccc} H_+^+ & \xrightarrow{\widehat{\mathcal{L}}^+} & H_-^+ \\ \varphi_+ \downarrow & & \downarrow \varphi_- \\ H_+^- & \xrightarrow{\widehat{\mathcal{L}}^-} & H_-^- . \end{array} \tag{$D_{\mathcal{L}}$}$$

If the above operators were Fredholm we would define the index of $(D_{\mathcal{L}})$ by

$$\text{index} = \text{ind}\,\widehat{\mathcal{L}}^+ - \text{ind}\,\widehat{\mathcal{L}}^- = \text{ind}\,\varphi_+ - \text{ind}\,\varphi_- .$$

Although neither $\widehat{\mathcal{L}}^\pm$ nor $\varphi_\pm$ are Fredholm, they are "Fredholm modulo each other". In particular, $(D_{\mathcal{L}})$ is commutative modulo compact operators and φ_+ nearly establishes a Fredholm relation between harmonic sections in H_+^+ and H_+^-, i.e. between $\ker \widehat{\mathcal{L}}^+ = \operatorname{Ker} \mathcal{L}_{\mathbf{X}^+}$ and $\ker \widehat{\mathcal{L}}^- = \ker \mathcal{L}_{\mathbf{X}^-}$. In fact the situation here can be reduced to the relative framework of $6\frac{4}{5}$ since (spaces of) sections of the Hilbert bundles $\mathbf{X}^\pm$ over V can be identified with (the spaces of) L_2-sections of $\widetilde{X}^\pm$ over the Π-covering $\widetilde{V}$ of V and $\text{ind}\,\mathcal{L}_\kappa$, i.e. the index of the diagram $(D_{\mathcal{L}})$ can be defined as in the Excision proposition of $6\frac{4}{5}$, with the twisted signature operators on $\widetilde{V}$, namely $\widetilde{\mathcal{L}}_{\widetilde{X}^+}$ substituting for D_+ in $6\frac{4}{5}$ and $\widetilde{\mathcal{L}}_{\widetilde{X}^-}$ for D'_+.

What remains to be done (which makes the bulk of work) is the identification

$$\sigma(V;\kappa) \overset{\text{def}}{=} \sigma(D_\varphi) = \text{ind}(D_{\mathcal{L}})$$

(which is done in the spirit of the discussion in $8\frac{5}{8}$) and then expressing $\text{ind}(D_{\mathcal{L}})$ by the Atiyah-Singer formula,

$$\text{ind}(D_{\mathcal{L}}) = (L_V \text{ ch } \kappa)\,[V] .$$

Remark. The logic of the K-theory has inevitably brought us into this tangle of "not quite Fredholm" diagrams where I can hardly grope my way. Fortunately, there is a simpler and more general approach to this case of the Novikov conjecture indicated in $9\frac{1}{4}$. On the other hand, the above discussion leads us to the promised land of

K-area via infinite dimensional bundles. This is defined for every Riemannian manifold V with a pair of ε-flat Hilbert bundles X^+ and X^- over V connected by a *Fredholm* homomorphism $F : X^+ \to X^-$ such that

(a) F almost commutes (i.e. commutes modulo compact operators) with the parallel transport in X^+ and X^- along each smooth path (e.g. loop) in V;

(b) F is a connection preserving unitary isomorphism outside a compact subset in V.

The minimal ε for which a compactly supported $\kappa \in K^0_{\mathrm{comp}}(V)$ can be represented as $\operatorname{ind} F$ for the above $X^\pm$ and F is denoted $\|F\mathcal{R}(\kappa)\|$. Then the corresponding *Fredholm K-area* of V is defined as $\sup \|F\mathcal{R}(\kappa)\|^{-1}$ over all κ with a non-zero Chern number. Now, clearly, this K-area is monotone increasing under all (finite or infinite) coverings of V trivial at infinity as the push-forward inequality from $4\frac{3}{5}$ applies in the present Fredholm framework to infinite coverings.

Exercise. We invite the reader to check the basic properties of the Fredholm K-area similar to what is done in §4 (e.g. finiteness for simply connected manifolds, compare $4\frac{1}{4}$) and also in §5 (e.g. the K-area inequality for $\operatorname{Sc} V \geq \varepsilon^2$, see $5\frac{1}{4}$).

Remark. The above notion of the ε-flatness for $\kappa \in K^0(V)$ and the corresponding almost flatness (for $\varepsilon \to 0$) appears in [Co-Hi] under the name of "nearly flat", where the authors raise the problem of finding examples of κ which are nearly flat but not representatble almost flatly by finite dimensional bundles. One can generalize further by admitting Hermitian rather than unitary flat Hilbert bundles $X^\pm$ in the spirit of K^0_{HAFE} and try to extend Connes' construction indicated in the end of $8\frac{1}{2}$ to some infinite dimensional symmetric spaces Z.

Idea of the proof of the Novikov conjecture for Δ-area $B = \infty$ and non-residually finite groups Π. We proceed essentially as in $8\frac{4}{5}$ but now our families are built of $\kappa(b)$, $b \in B$, for the above $\kappa(b) = \operatorname{ind}(X^+(b) \to X^-(b).)$ Here again it is useful to work in the language of C^*-algebras to avoid an explicit mentioning of B (encoded into the relevant C^*-algebra, i.e. $\operatorname{Cont} B$), as we may consider almost flat Hilbertian R-bundles $X^\pm \to V$ (for any C^*-algebra R, not only $R = \operatorname{Cont} B$), with an R-Fredholm homomorphism $F : X^+ \to X^-$, twist $\mathcal{L}$ with $\kappa = \operatorname{Ind}_R F$ and define $\operatorname{ind} \mathcal{L}_\kappa \in K_0(R)$. We claim that this index of $\mathcal{L}_\kappa$ is a homotopy invariant and equal to the $K_0(R)$-valued signature $\sigma(V;\kappa)$ (in accordance with section 6 in $[\text{C-G-M}]_{\text{FPP}}$ where we had more infinite dimensional aspirations) and we indicate the possibility of a HAFl-version of this claim. Unfortunately, it is unclear if we significantly (if at all) enlarge the class of group Π to which these more and more general homotopy invariance theorems apply.

Exercise. Define Fredholm K-area with the above bundles $X^{\pm}$ over C^*-algebras R and $\kappa = \operatorname{ind} F$ where the non-triviality condition on κ (replacing non-vanishing of a Chern number) is expressed in terms of the index pairing (with values in $K_0(R)$) of the fundamental K-homology class of V with the λ-subring generated by κ. Then extend the results of §§4 and 5 to this Fredholm C^*-K-area.

$9\frac{1}{5}$. Novikov conjecture for open Riemannian manifolds

Let $f : V' \to V$ be a proper homotopy equivalence between such manifolds and take a pair of cohomology classes $\rho \in H^*_{\mathrm{comp}}(V;\mathbb{Q})$ and $\rho' = f^*(\rho)$ with compact supports. We seek geometric conditions on V, V', f and ρ which would imply the equality

$$(L_V \smile \rho)[V] = L_{V'} \smile \rho')[V'] . \tag{NC}$$

This "open" NC sometimes implies the "closed" one, namely when V and V' are freely acted upon by Π and then NC for the push-forward Gys $\rho \in H^*(V/\Pi)$ (obviously) follows from that for ρ. In particular, NC for the fundamental cohomology class of a closed aspherical manifold B (pulled back to V by a map $\beta : V \to B$) follows from a suitable "open" NC, but for the rest of $H^*(B)$ one needs the "open" framework of a differential kind (see $9\frac{2}{7}$ and $9\frac{1}{2}$).

Example. Let $V = W \times \mathbb{R}^m$ for a closed manifold W of dimension $4k$ and $\rho \in H^{n-4k}_{\mathrm{comp}}(V)$ be the Poincaré dual to $[W] \in H_{4k}(V)$. Then

$$(L_V \smile \rho)[V] = \text{signature}\,(W)$$

and we ask whether a submanifold $W' \subset V'$ with trivial normal bundle which is homologous to $f^{-1}(W)$ has the same signature as W. We know it is false in general by the Serre finiteness theorem but we shall prove this below under the following three assumptions.

1. V is the *Riemannian* product, of W with $\mathbb{R}^m$.

2. The map f and the implied homotopy inverse, say $g : V \to V'$ are Lipschitz.

3. The implied homotopies $V \times [0,1] \to V$ and $V' \times [0,1] \to V'$ joining $f \circ g : V \to V$ and $g \circ f : V' \to V'$ with the identity maps are Lipschitz.

Notice that these assumptions are satisfied if f covers a (smooth) homotopy equivalence between $\mathbf{V} = W \times \mathbb{T}^m$ and some $\mathbf{V}'$, and thus the "open geometric" NC for $V = W \times \mathbb{R}^m$ implies Novikov's original homotopy equivalence theorem.

Our proof of NC under the assumptions 1, 2, 3 will follow the "quick proof" in $8\frac{1}{4}$ and $8\frac{1}{3}$. We assume $m = 2\ell + 1$ and take a (non-tubular) neighbourhood

U of $W = W \times 0 \subset V$ of the form $U = W \times \mathbf{U}$ where $\mathbf{U} \subset \mathbb{R}^m$ is a tubular neighbourhood of the Cartesian product B^ℓ of ℓ copies of a closed surface of genus ≥ 2 imbedded into $\mathbb{R}^m$. We know this $\mathbf{U} = B^\ell \times \mathbb{R}$ comes with a certain flat (symplectic or orthogonal) bundle $\mathbf{X} \to \mathbf{U}$ such that the lift of this bundle to U, call it $X \to U$, satisfies

$$\sigma_U \overset{\text{def}}{=} \sigma([W \times B^\ell]; X) = s\,\sigma(W) \qquad (*)$$

for some $s \neq 0$, where $\sigma([W \times B^\ell]; X)$ denotes the signature of the cup-product on $H^{2k+\ell}(U; X)$ evaluated on the class $[W \times B^\ell] \in H_{n-1}(U)$ for $n = 4k + 2\ell + 1 = \dim V$. We know this σ_U is (obviously) a proper homotopy invariant of U. Furthermore, if we perturb U to some $U_1 \subset V$ such that the intersection of the two contains the support (of some realization of) the homology class $h = [W \times B^\ell]$, then $\sigma_{U_1} \underset{\text{def}}{=} \sigma(h; X) = \sigma_U$. Furthermore, if we scale $\mathbf{U} \subset \mathbb{R}^m$ by a large λ, and take (large) $U = W \times \lambda \mathbf{U}$ then the pull-back $U' = f^{-1}(U) \subset V'$ satisfies

$$\sigma_{U'} = \sigma(h'; X') = \sigma_U$$

where the class h' is the image of $h = [W \times \lambda B^\ell]$ under $g_* = H_*(U) \to H_*(U')$. In fact if λ is sufficiently large compared to the implied Lipschitz constants of the maps and homotopies in question, then U' is homotopy equivalent to U modulo small (relative to λ) wiggling near the boundary which does not affect $\sigma(h; X)$ and $\sigma(h'; X')$ for homology classes h and h' having their supports λ-far from the boundaries of U and U' correspondingly. But we know on the other hand that $\sigma(h'; X') = \sigma(W')$ for a suitable W' imbedded into U' with trivial normal bundle (see $8\frac{1}{4}$, $8\frac{1}{3}$) which implies the desired equality $\sigma(W') = \sigma(W)$.
Q.E.D.

$9\frac{2}{9}$. A macroscopic criterion for vanishing of Pontryagin classes

Let now V be (the total space of) a vector bundle of rank m over a closed manifold W with a complete Riemannian metric g on V such that

(1) The restriction of g to each fiber V_w of $V \to W$ is flat.

(2) The distance function of $g \mid V_w$ is equivalent to $\mathrm{dist}_V \mid V_w$ for all fibers $V_w \subset V$. This means that the minimal path in V_w between a pair of points has

$$\text{length} \leq \text{const} \cdot \text{length (the minimal path in } V \supset V_w) ,$$

for some const ≥ 0 independent of the points.

(3) The fibers diverge at most sublinearly, ile.

$$\mathrm{dist}(v, V_w)/\mathrm{dist}(v, v_0) \to 0$$

for each fiber V_w, a fixed point $v_0 \in V$ and $V \to \infty$. Then, we claim, *the rational Pontryagin classes of the bundle $V \to W$ vanish.*

Proof. There obviously exists a (proper) retraction f_0 of V on a fiber, say on V_{w_0} which moves each point $v \in S(r) \subset V$ by at most $\varphi(r)$ for some function $\varphi(r)$ satisfying $r^{-1}\varphi(r) \to 0$ for $r \to \infty$. This retraction maps all fibers of $V \to W$ onto V_{w_0} properly with degree one and hence onto. It follows that there is a fiberwise map f' of the trivial bundle $V' = W \times \mathbb{R}^m$ to V, where $\mathbb{R}^m$ is identified with the fiber V_{w_0}, mapping $w \times \mathbb{R}^m$ onto V_w properly with degree one such that this map is roughly inverse to f_0, i.e.

$$\max\left(\mathrm{dist}_{\mathbb{R}^m}(x, f_0 \circ f'(x))\ ,\ \mathrm{dist}_V(v, f' \circ f_0(v))\right) \leq \varphi(r)$$

for all $x \in \mathbb{R}^m$ and $v \in V$ with r denoting the distance from x to the origin or from v to v_0, and where φ stands again for a sublinear function in r. Now, the *argument* of the previous section shows that if W is stably parallelizable of dimension $4k$, then $L_k(V) = 0$. Since every $4i$-dimensional homology class in W can be realized by a stably parallelizable manifold $W^{4i} \to V$, we conclude, by looking at the induced bundle over W^{4i}, that $L_i(V)$ vanish for every $i = 1, 2, \ldots$. Q.E.D.

Exercises. Fill in the details in the above proof. Show that the Euler class of the above V also vanishes. Relax (1) by allowing fibers with non-positive curvature. Find a metric on an arbitrary V satisfying (1) and (2) but not (3). See what happens to metrics on V of the form $g_{\mathrm{vert}} \oplus g_{\mathrm{hor}}$ associated to a Euclidean metric g_{vert} in the fibers and a metric g on W lifted to g_{hor} with some Euclidean connection.

$9\frac{1}{4}$. NC for bounded homotopies of multiply large manifolds

Let $\mathbf{B}$ be a locally compact metric space where all closed bounded subsets are compact. A homotopy $h : V \times [0,1] \to \mathbf{B}$ is called *bounded*, or just B, if $\mathrm{length}_{\mathbf{B}}\, h(v \times [0,1]) \leq \mathrm{const}$ for some const $= \mathrm{const}(h)$ independent of $v \in V$. This is essentially the same as h being *Lipschitz* for *some product* metric in $V \times [0,1]$. Then two manifolds *over* $\mathbf{B}$ i.e. V and V', coming along with *proper* maps β and β' into $\mathbf{B}$ are called *B-homotopy* equivalent if there exist maps $f : V' \to V$ and $g : V \to V'$ such that $f \circ g$ and $g \circ f$ are both B-homotopic to the identities in the above sense. If V and V' are compact then B-homotpy equivalence is the same as the ordinary homotopy equivalence, provided one has sufficiently many curves of finite length, e.g. if V and V' are Riemannian manifolds.

BN-***Problem.*** Take a cohomology class ρ on $\mathbf{B}$ with compact support. When is $\sigma_\rho = (L_V \smile \beta^*(\rho))[V]$ B-homotopy invariant?

The positive answer is given in [Pe-Ro-We] for $\mathbf{B} = \mathbb{R}^m$ and $\rho \in H^m_{\text{comp}}(\mathbb{R}^n)$ the fundamental class. This (obviously) implies the positive answer for all m-dimensional hyper-Euclidean Riemannian manifolds $\mathbf{B}$ (i.e. admitting proper Lipschitz maps $\mathbf{B} \to \mathbb{R}^m$ of positive degrees). On the other hand, the above argument positively solves BNP for a somewhat more general class of manifolds $\mathbf{B}$ called *multiply large* and defined as follows. For every $\varepsilon > 0$ there exists a *multi-domain* $\widetilde{U}$ over $\mathbf{B}$ i.e. a manifold $\widetilde{U}$ equidimensionally immersed into $\mathbf{B}$ and an ε-contracting proper map of positive degree of $\widetilde{U}$ onto the open unit Euclidean m-ball. For example, every $\mathbf{B}$ admitting a hyper-Euclidean covering is multiply large. A less obvious example comes from a metric g_ε on S^3 which has $K(g_\varepsilon) \leq \varepsilon$ and $\text{Diam}(S^3, g_\varepsilon) \leq 1$ (see $[\text{Gro}]_{\text{AFM}}$, [Bu-Gr] and [Bav]). Such a (S^3, g_ε) admits a λ-large multi-domain $\widetilde{U}$ for $\lambda = \frac{1}{2}\varepsilon^{-1}$ (namely, the exponentiated ε^{-1}-ball from $T_s(S^3)$) and a geometric connected sum (homeomorphic to $\mathbb{R}^3$) of these spheres (S^3, g_{ε_i}) with $\varepsilon_i \to 0$, $i = 1, 2, \ldots$, is multiply large, albeit it is very far from being hyper-Euclidean. A similar geometric phenomenon where a simply connected manifold has a large "partial covering" (which is not a part of an actual covering) may be observed in the universal coverings $\widetilde{V} \to V$ whenever the fundamental group $\pi_1(V)$ is logically complicated and so $\widetilde{V}$ contains many relatively short loops which must be stretched a lot in the process of contraction (see $[\text{Gro}]_{\text{AI}}$ and references therein).

BN ***for the fundamental classes*** $\rho \in H^m_{\text{comp}}(\mathbf{B})$ ***of multiply large manifolds*** $\mathbf{B}$. To prove the B-homotopy invariance of σ_ρ for $\beta : V \to \mathbf{B}$ we must express $\sigma_\rho = \sigma(\beta^{-1}(b))$ in B-stable terms. Here it is. Assume $m = \dim \mathbf{B}$ odd (if even, multiply $\mathbf{B}$ and V by $\mathbb{R}$) and take the tubular neighbourhood of the product of surfaces in $\mathbb{R}^m$ as earlier contained in the unit ball. This $\mathbf{U}$ is pulled back to $\widetilde{\mathbf{U}} \subset \widetilde{U} = \widetilde{U}_\varepsilon$ by our ε-contracting proper map $\widetilde{U} \to$ (unit ball in $\mathbb{R}^m$) and is $\approx \varepsilon^{-1}$-large in size. The $\widetilde{\mathbf{U}}$ is pulled back by β (via the fiber product construction) to a multidomain, say $\widetilde{V}$ over V. Our flat bundle also lifts to $\widetilde{V}$, say to $\widetilde{X} \to \widetilde{V}$, and we see as earlier that $\sigma_\rho = \sigma(\widetilde{h}; \widetilde{X})$ where $\widehat{h} \in H_{n-1}(\widetilde{V})$ is the homology class corresponding to the hypersurface in $\widetilde{V}$ obtained by pulling back the product of surfaces in $\mathbf{U}$ by the composed map $\widetilde{V} \to \widetilde{\mathbf{U}} \to \mathbf{U}$. If ε^{-1} is large compared to the (bounded) size of implied B-homotopies, then $\sigma(h; \widetilde{X})$ is invariant under the B-homotopy equivalence and so is σ_ρ. Q.E.D.

Terminological remark. If $\mathbf{B}$ is uniformly contractible then our B-inequality $\text{length}_B\, h(x \times [0,1]) \leq \text{const}$ follows from a weaker condition, namely $\text{Diam}_{\mathbf{B}}\, h(v \times [0,1]) \leq \text{const}$ which truly expresses boundedness rather than shortness of the paths $h(v \times [0,1]) \subset \mathbf{B}$. The shortness is, in general, stronger than the boundedness, as seen in our example of the connected sum of the spheres (S^3, ε). Our "length" really serves as shorthand for "the supremum of the diameters of the lifts of our paths to all possible multidomains over $\mathbf{B}$".

$9\frac{3}{11}$. Multiply large examples

Start with dim = 2 and observe that

A surface **B** *with a complete Riemannian metric is multiply large iff the universal covering of* **B** *is infinite.*

In fact we may pass to the universal covering and assume **B** is homeomorphic to $\mathbb{R}^2$. Take away a small topological disk D from **B** and observe that the universal covering $\widetilde{U}$ of the complement $U = \mathbf{B} - D$ is large; it admits ε-contracting maps of degree one onto the unit ball in $\mathbb{R}^2$ for all $\varepsilon > 0$. Q.E.D.

Dim = 3. Let **B** be a complete non-compact Riemannian manifold of dimension 3. Say that **B** is *uniformly connected at infinity* if for each $r > 0$ there exists $R = R(r) > 0$, such that every two points in **B** R-far from a metric r-ball in **B** can be joined by a path missing this ball.

If $H_2(\mathbf{B}) = 0$ *then "uniformly connected at infinity"* implies *"multiply large" for* $\dim \mathbf{B} = 3$.

Proof. Take a minimizing geometric segment γ in **B** of length $3R$, i.e. an isometric copy of $[0, 3R]$ and the ball D_r around the center of this segment. Then a short loop ℓ around γ near the center of D_r remains non-homologous to zero in $D_r - \gamma$ since the ends of γ can be joined by a path in $\mathbf{B} - D_r$ and so $D_r - \gamma$ admits an infinite cyclic covering $\widetilde{U} \to D_r - \gamma$ delooping ℓ. This $\widetilde{U}$ is roughly r-large where the relevant map to the (unit ball in) $\mathbb{R}^3$ is made out of the following three functions, distance (function) to the one of the ends of γ, distance to γ, the cyclic parameter of the covering (i.e. we use here a continuous map $D_r - \gamma \to S^1$ non-contractible on ℓ and the corresponding function from $\widetilde{U}$ to $\mathbb{R}$ covering S^1).

Corollary. *Let a closed* 3*-manifold* **B** *admits an infinite Galois covering with* $H_2 = 0$. *Then the fundamental class* $\rho = [\mathbf{B}]^{\mathrm{co}} \in H^3(\mathbf{B})$ *satisfies* NC, *i.e. for every* $V \to \mathbf{B}$ *the* ρ*-signature* $\sigma_\rho(V)$ *is a homotopy invariant of* V.

This is equivalent, by the 3-manifold theory, to NC for the fundamental group of every closed aspherical 3-manifold. On the other hand, the universal coverings of these have infinite stable K-areas. In fact, every uniformly contractible 3-manifold **B** of bounded local geometry has $K\text{-area}_{\mathrm{st}}\, \mathbf{B} = \infty$ (by an easy argument) and it is not impossible that these **B** are hyper-Euclidean. This would follow if for every metric on S^1 with filling radius $\geq R$ this S^1 had an ε-contracting map of degree $\neq 0$ to the unit circle with $\varepsilon \to 0$ for $R \to \infty$.

Codim 1-reduction. Let **B** and **B′** be complete uniformly contractible manifolds of dimensions n and let $n+1$ and $\varphi : \mathbf{B} \to \mathbf{B}'$ be a *quasi-isometric embedding* i.e. a Lipschitz map such that $\mathrm{dist}(\varphi(b_1), \varphi(b_2)) \geq R(\mathrm{dist}(b_1, b_2))$

for some function $R(d)$ satisfying $R(d) \to \infty$ for $d \to \infty$. Then if $\mathbf{B}$ is large in some sense then $\mathbf{B}'$ is comparably large in the same sense. For example *if* $\mathbf{B}$ *is multiply large then so is* $\mathbf{B}'$ *and the same is true for "hyper-Euclidean" in the place of "multiply large".*

Idea of the proof. We may pretend φ is a topological embedding and $\varphi(\mathbf{B}) \subset \mathbf{B}'$ divides $\mathbf{B}'$ into two halves, say $\mathbf{B}'_+$ and $\mathbf{B}'_-$; and we denote by $\delta : \mathbf{B}' \to \mathbb{R}$ the function $\operatorname{dist}(b', \varphi(\mathbf{B}))$ on $\mathbf{B}'_+$ and $-\operatorname{dist}(b', \mathbf{B})$ on $\mathbf{B}'_-$. Next, every map from $\mathbf{B}$ or a domain $\widetilde{U}$ over $\mathbf{B}$ into $\mathbb{R}^n$ which is ε-Lipschitz for the Riemannian metric in $\mathbf{B}$ can be modified to an ε'-Lipschitz map for the (non-Riemannian) metric induced from $\mathbf{B}'$ with $\varepsilon' \to 0$ for $\varepsilon \to 0$. Such a map can be ε''-Lipschitz extended to $\mathbf{B}'$ with $\varepsilon'' \leq n\varepsilon'$ and together with δ (scaled by a small ε) we obtain the required map $\mathbf{B}' \to \mathbb{R}^{n+1}$.

Example. If the fundamental group of an $(n+1)$-dimensional aspherical manifold contains $\mathbb{Z}^n$ as a subgroup, then the universal covering of this manifold is hyper-Euclidean.

codim 2-reduction. Now let $\dim \mathbf{B}' = \dim \mathbf{B} + 2$. We claim that *if* $\mathbf{B}$ *is multiply large then so is* $\mathbf{B}'$.

Idea of the proof. Use the infinite cyclic covering $\widetilde{U}$ of $\mathbf{B}' - \varphi(\mathbf{B})$ and the cyclic parameter there besides $\delta = \operatorname{dist}(b', \varphi(\mathbf{B}))$, as in the 3-dimensional case where we leave the actual proof to the (justifiably dissatisfied) reader who may consult §§7-12 in $[\text{G-L}]_{\text{PSC}}$ and [Yau] for similar results in the framework of $\text{Sc} > 0$.

Example. If the fundamental group of a closed $(n+2)$-dimensional aspherical manifold V contains $\mathbb{Z}^n$ as a subgroup, then this manifold is multiply large and, hence, its fundamental class satisfies the Novikov conjecture (and V admits no metric with $\text{Sc} > 0$).

The above makes plausible some largeness of uniformly contractible 4-manifolds (i.e. universal coverings of aspherical manifolds) as they may contain suitable surfaces (in agreement with the non-existence of metrics with $\text{Sc} > 0$ on closed aspherical 4-manifolds announced in [Sch]). On the other hand, there are examples of non-hyper-Euclidean uniformly contractible manifolds (see [Fe-We]), but these examples need non-bounded local geometry.

$9\frac{2}{7}$. BN for multiply large families

We want to extend the above to more general (non-fundamental) classes ρ with compact supports, which is done by using families as in $[\text{C-G-M}]_{\text{GCLC}}$ (also see $9\frac{1}{6}$). Namely let $p : \underline{\mathcal{C}} \to B$ be a topological submersion with locally compact (sometimes smooth oriented) m-dimensional fibers (i.e. each point

$c \in \underline{C}$ admits a split neighbourhood $U_\sigma \times \mathbb{R}^m \subset \underline{C}$ for some neighbourhood $U_\sigma \subset B$ of $b = p(c) \in B$, such that the coordinate changes are smooth and orientation preserving in the fiber direction with the derivatives continuously depending on $b \in B$ if $\underline{C}$ is assumed fiberwise smooth and oriented).

Examples. (a) A vector bundle $p : Y \to B$ is an essential example. This carries a distinguished cohomology class $\operatorname{Tom} Y \in H^m_{\mathrm{vc}}(B)$, where $m = \operatorname{rank} Y$ and "vc" means "with *vertically compact* supports", i.e. $(\operatorname{supp} \operatorname{Tom}) \cap Y_b$ is compact for all fibers $Y_b \subset Y$. If B is a manifold, then $\operatorname{Tom} Y$ is the Poincaré dual of the zero section $Y \underset{0}{\hookrightarrow} B$; if Y admits a *fiberwise proper* map $Y \to \mathbb{R}^m$ with degree d on the fibers, then $\operatorname{Tom} Y = d^{-1}$ (pull-back of the fundamental class $\rho \in H^m_{\mathrm{comp}}(\mathbb{R}^m)$). Notice, that if B is finite dimentional then Y often admits such maps to $\mathbb{R}^m$ with degree $d \neq 0$ by Serre's finiteness theorem. For example such a map exists if m is odd, or if $\dim B < \operatorname{rank} Y$.

Now we recall the *Gysin* push-forward homomorphism $H^i_{\mathrm{cv}}(Y) \to H^{i-m}(B)$ which is, in fact, defined for all fiberwise smooth submersions $\underline{C} \to B$, and observe that

$$\operatorname{Gys}(p^*(c) \smile \operatorname{Tom} Y) = c \,, \text{for all } c \in H^*(B)\,.$$

This agrees with the wrong way fonctoriality of the Thom class: if $Y' \to Y$ is a surjective homomorphism, then $\operatorname{Tom} Y = \operatorname{Gys} \operatorname{Tom} Y'$.

(a′) All the above applies to an arbitrary submersion $\underline{C} \to B$ with *contractible* m-manifold fibers.

(b) Let B be a manifold and $\underline{C} = B_\Delta \to B$ i.e. $B_\Delta = B \times B$ projected to the second component. If we embed B to B_Δ by the diagonal (section) $\Delta : B \to B_\Delta$ then the Tom class of the normal bundle of the so embedded $B \subset B_\Delta$, realized by a tubular neighbourhood $U_\Delta \subset B_\Delta$, equals the Poincaré dual of the homology class of the diagonal in $B \times B$. That is,

$$\text{Push-forward}\,((\operatorname{Tom} U_\Delta) \mapsto H^*(B \times B)) = \operatorname{PD}\,(\Delta[B])\,.$$

(b′) Now suppose the universal covering of $\widetilde{B}$ is contractible and let $\widetilde{B}_\Delta$ be obtained from B_Δ by taking the universal coverings of the fibers such that $\widetilde{B}_\Delta = (\widetilde{B} \times \widetilde{B})/\Pi$ for the diagonal action of $\Pi = \pi_1(B)$ on $\widetilde{B} \times \widetilde{B}$. We denote by $\widetilde{p} : \widetilde{B}_\Delta \to B$ the projection and by $\widetilde{\Delta} : B \to \widetilde{B}_\Delta$ the diagonal section and observe that $\operatorname{Tom} \widetilde{B}_\Delta$ is Poincaré dual to $\widetilde{\Delta}[B]$ and the push-forward of $\operatorname{Tom} \widetilde{B}_\Delta$ equals $\operatorname{PD} \Delta[B]$, as earlier.

Next, let V be a compact manifold, $\beta : V \to B$ a continuous map, and let $\widetilde{V} \to V$ denote the covering induced by the universal covering $\widetilde{B} \to B$. If we take $\widetilde{V}$ in each fiber of the (trivial) filtration $V_B = V \times B \to B$ mapped to $B_\Delta = B \times B$ by $\beta \times \mathrm{id}$, we obtain a $\widetilde{V}$-fibered bundle, say $\widetilde{q} : \widetilde{V}_B \to B$, naturally fiberwise mapped to $\widetilde{B}_\Delta$. Denote by $\mathrm{Tom}_{\widetilde{V}}\,\widetilde{B}_\Delta$ the pull-back of $\mathrm{Tom}\,\widetilde{B}_\Delta$ to $\widetilde{V}_B$ and take the cup-product of this $\mathrm{Tom}_{\widetilde{V}}$ with some characteristic (cohomology) class $\widetilde{\chi}$ of the vertical tangent bundle $T_{\mathrm{vert}}(\widetilde{V}_B) \to \widetilde{V}_B$.

Push-forward formula.

$$\mathrm{Gys}_{\widetilde{q}}(\widetilde{\chi} \smile \mathrm{Tom}_{\widetilde{V}}\,\widetilde{B}_\Delta) = \mathrm{Gys}_\beta\,\chi \in H^*_{\mathrm{comp}}(B)$$

for the corresponding class $\chi = \chi(T(V)) \in H^*(V)$.

Proof. The class $\mathrm{Tom}\,\widetilde{B}_\Delta$ is supported near the diagonal section $\widetilde{\Delta}(B) \subset \widetilde{B}_\Delta$ and so $\mathrm{Tom}_{\widetilde{V}}\,\widetilde{B}_\Delta$ sits near the diagonal $\widetilde{\Delta}(V) \subset \widetilde{V}_\Delta$ mapped to $\widetilde{B}_V = \widetilde{V} \times B$ where we picture V imbedded to B by β, see Fig. 15 below.

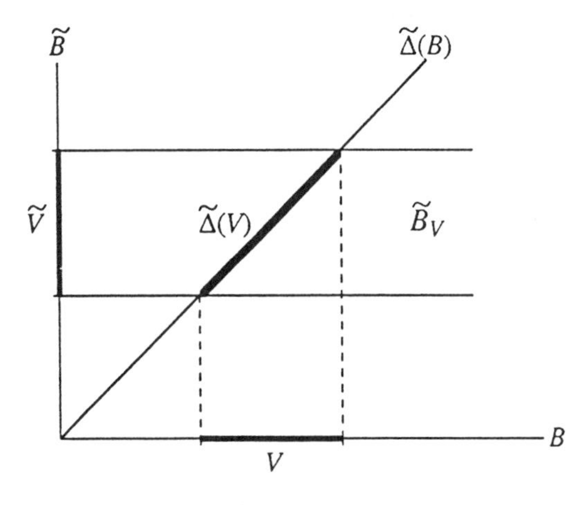

Figure 15

Since the vertical and horizontal tangent bundles of $\widetilde{V}_\Delta$ (which is the fiberwise covering of $V_\Delta = V \times V \to V$) are equal on the diagonal $\widetilde{\Delta}(V) \subset \widetilde{V}_\Delta$, the above cup-product after dualization satisfies

$$\begin{aligned}(\widetilde{\beta} \times \mathrm{id})_*\,\mathrm{PD}(\widetilde{\chi} \smile \mathrm{Tom}_{\widetilde{V}}\,\widetilde{B}_\Delta) &= ((\widetilde{\beta} \times \mathrm{id})_*\,\mathrm{PD}\,\widetilde{\chi}) \frown \widetilde{\Delta}[B] \\ &= p^{-1}(\beta_*(\mathrm{PD}\,\chi)) \frown \widetilde{\Delta}[B]\end{aligned}$$

for the pull-back (dual to Gys) homomorphism

$$p^{-1} : H_*(B) \to H_*(\widetilde{B}_\Delta)$$

for the fibration $p : \widetilde{B}_\Delta \to B$ and where $\widetilde{\beta} \times \mathrm{id} : \widetilde{B}_V \to \widetilde{B}_\Delta$ is the obvious map. Now the push-forward formula follows from the following general (and obvious) relation

$$p_*((p^{-1}(h)) \frown \widetilde{\Delta}[B]) = h$$

for all $h \in H_*(B)$. Q.E.D.

Thus the Novikov conjecture for $\beta : V \to B$, claiming the homotopy invariance of $\beta_*(\mathrm{PD}\, L_V) \in H_*(B)$ (which is Poincaré dual, if B is a manifold, to $\mathrm{Gys}_\beta\, L_V \in H^*_{\mathrm{comp}}(B)$) can be expressed in terms of the vertical tangent bundle $T_{\mathrm{vert}}\, \widetilde{V}_B$, namely, as an invariance of $\mathrm{Gys}_{\widetilde{q}}(\widetilde{L} \smile \mathrm{Tom}_{\widetilde{V}}\, \widetilde{B}_\Delta)$ for $\widetilde{L} = L(T_{\mathrm{vert}}\, \widetilde{V}_B)$.

This motivates the following *BN problem for submersions.* Let our submersion $\underline{\mathcal{C}} \to B$ be given a fiberwise metric and let $\mathcal{V} \to B$ be a fiberwise smooth and oriented submersion coming along with a proper morphism (i.e. a fiberwise proper map) to $\underline{\mathcal{C}}$ over B. We want to express as much as possible of the Pontryagin (or L) classes of the vertical tangent bundle of $\mathcal{V}$ in *B-homotopy* stable terms, where "B" now refers to "fiberwise bounded".

Short cohomology. A cohomology class $\theta \in H^*_{\mathrm{vc}}(\underline{\mathcal{C}})$ is called *ε-short* if there is a fiberwise proper and fiberwise ε-contracting map of $\underline{\mathcal{C}}$ onto the open unit ball in $\mathbb{R}^m$ such that the fundamental cohomology class $[\mathrm{Ball}]^{\mathrm{co}} \in H^m_{\mathrm{comp}}(\mathrm{Ball})$ pulls back to θ under this map.

Example. Let B be an m-dimensional parallelizable manifold with a complete metric of non-positive sectional curvature. Then the Thom class $\mathrm{Tom}\, \widetilde{B}_\Delta \in H^m_{\mathrm{vc}}(\widetilde{B})$ of the bundle $\widetilde{B}_\Delta \to B$ is ε-short for all $\varepsilon > 0$. In fact, the inverse exponential map gives us a contracting map of $\widetilde{B}_\Delta$ to $T(B) = B \times \mathbb{R}^m$ where the Thom class comes by the projection to (the unit ball in) $\mathbb{R}^m$.

Next, θ is called *multiply ε-short*, if it equals the push-forward of an ε-short class in some multi-domain $\widetilde{U}$ over $\underline{\mathcal{C}}$, i.e. a fiber-smooth submersion $\widetilde{U} \to B$ with a given locally homeomorphic fiberwise smooth morphism to $\underline{\mathcal{C}}$ over B where the implied (by the notion of shortness) fiberwise metric in $\widetilde{U}$ is the one induced from $\underline{\mathcal{C}}$.

Example. For the above parallelizable B with $K \leq 0$ the (trivial) fibration $B_\Delta = B \times B \to B$ has the class $\mathrm{PD}\, \Delta[B] \in H^m_{\mathrm{vc}}(B)$ multiply ε-short for all $\varepsilon > 0$ as is seen with domains $\widetilde{U} \subset \widetilde{B}_\Delta$ viewed as multi-domains over B.

Finally, call θ *stably multiply short* or *sms*, if there is a Euclidean vector bundle $Y' \to B$ and a class θ' in $H^*_{\mathrm{vc}}(\underline{\mathcal{C}}' = \underline{\mathcal{C}} \oplus Y')$ such that the push-forward of θ' to $\underline{\mathcal{C}}$ equals θ and such that θ' is multiply ε-short for all $\varepsilon > 0$. Here one could generalize by allowing Y' and/or θ' to depend on ε but this does not seem to bring in something new and interesting in specific cases.

Example. For every complete manifold B with $K(B) \leq 0$ the class $\mathrm{PD}\, \Delta[B]$ in $H^m_{\mathrm{vc}}(B_\Delta)$ is sms. Indeed one can make B parallelizable by taking the total space B' of some vector bundle over B complementary to $T(B)$. The curvature of this B' may be somewhere positive, but $\widetilde{B}'_\Delta \to B$ remains fiberwise hyper-Euclidean.

$9\frac{1}{3}$. Short cohomology, B-homotopy invariant Pontryagin classes and an elementary proof of NC for $K \leq 0$

Let $\underline{\mathcal{C}} \to B$ be a submersion with a fiberwise metric as earlier, and $\theta \in H^*_{\mathrm{vc}}(B)$ an sms *(stably multiply short and where "vc" stands for vertically (or fiberwise) compact support). Then for every fiberwise smooth submersion $q : \mathcal{V} \to B$ with a fiberwise proper morphism $\beta : \mathcal{V} \to \underline{\mathcal{C}}$ the push-forward class*

$$L_\theta = \mathrm{Gys}_q(L(T_{\mathrm{vert}}(\mathcal{V})) \smile \beta^*(\theta)) \in H^*(B)$$

is a fiberwise B-homotopy invariant.

Proof. Let us give a B-stable expression of the value of L_θ on a homology class $h \in H^{4i}(B)$. We assume (which is no big deal) that B is a polyhedron, realize h by a map of a stably parallelizable manifold, say $W \to B$, and denote by $\mathcal{V}_W \to B$ the submersion induced from $\mathcal{V} \to B$ by the map $W \to B$. This $\mathcal{V}_W$ is mapped to $\underline{\mathcal{C}}$ by the composition of maps $\mathcal{V}_W \to \mathcal{V} \to \underline{\mathcal{C}}$ and we may pull-back $\widetilde{U}$ from $\underline{\mathcal{C}}$ to $\mathcal{V}_W$ where it is called $\widetilde{\mathcal{V}}_W$ over $\mathcal{V}_W$. In fact, as we must work "stably" we first (Whitney) add a vector bundle $Y' \to B$ to $\mathcal{V} \to B$ as well as to $\underline{\mathcal{C}} \to B$ thus passing to the corresponding $\mathcal{V}'_W \to \mathcal{V}' \, \underline{\mathcal{C}}' \leftarrow U'$ and $\widetilde{\mathcal{V}}'_W$ over $\mathcal{V}'_W$. Since adding Y' changes the Pontryagin (and L) classes of all $\mathcal{V} \to B$ in the same way, it suffices to prove our theorem for $\theta' \in H^*_{\mathrm{vc}}(\underline{\mathcal{C}}')$ and we may as well keep our notations $\underline{\mathcal{C}}$, $\mathcal{V}$, $\mathcal{V}_W$ and $\widetilde{\mathcal{V}}_W$. Now this $\widetilde{\mathcal{V}}_W$ is mapped to the unit m-ball by the composition of maps $\widetilde{\mathcal{V}}_W \to \widetilde{U} \to m$-ball and, after slightly perturbing this composition, we may pull-back a regular value in the ball to a $4i$-dimensional submanifold, say $\widetilde{W}$ in $\widetilde{\mathcal{V}}_W$ with trivial normal bundle. Clearly $\sigma(\widetilde{W}) = L_\theta(h)$ and this signature $\sigma(\widetilde{W})$ admits a B-stable expression by means of our interesting neighbourhood $\mathbf{U}$ in the m-ball (i.e. the tubular neighbourhood of the product of surfaces) and a symplectic bundle over $\mathbf{U}$ exactly as in the above proof of BN for the fundamental class. Q.E.D.

Examples and corollaries. (a) Let $\mathcal{V}$ and $\mathcal{V}'$ be smooth fibrations (or just submersions) over B with fiberwise Riemannian metrics where $\mathcal{V}$ is fiberwise hyper-Euclidean i.e. it admits a fiberwise proper Lipschitz morphism onto some Euclidean vector bundle over B with fiberwise positive degree. Suppose that $\mathcal{V}$ and $\mathcal{V}'$ have contractible fibers and hence have well defined rational Pontryagin classes p_i and p'_i in $H^*(B)$.

If $\mathcal{V}$ and $\mathcal{V}'$ are fiberwise homotopy Lipschitz equivalent (i.e. homotopy equivalent in the category of fiberwise Lipschitz maps and homotopies; compare $7\frac{5}{6}$), then $p_i = p'_i$.

A particular example is where the fibers of $\mathcal{V}$ are complete manifolds with non-positive curvatures.

Proof. Lipschitz homotopies are bounded (where $\mathcal{V}$ serves here for $\underline{\mathcal{C}}$ as well).

(b) *Let B be a Riemannian manifold where the dual to the diagonal, $\theta =$ PD $\Delta[B]$ for the diagonal embedding $\Delta : B \to B \times B = B_\Delta \to B$ is* sms. *Then for every smooth V and a proper map $\beta : V \to B$ the push-forward L-class $\mathrm{Gys}_\beta(L_V) \in H^*(B)$ is B-homotopy invariant.*

This follows from the push-forward formula in $9\frac{2}{7}$ applied to the fibration $q : V_B \to B$ which shows that

$$\mathrm{Gys}_\beta(L_V) = \mathrm{Gys}_q(L(T_{\mathrm{vert}}(V_B)) \smile \theta^*)$$

where $V_B = V \times B \to B$ and θ^* is the pull-back of θ under $\beta \times \mathrm{id}$. (Notice that here V is non-compact and that we do not (have to) pass to $\widetilde{B}_\Delta$ and $\widetilde{V}_B$ as we did in the push-forward formula, but this causes no problems.)

(b') *The above B satisfies the ordinary Novikov conjecture, that if for every closed manifold V with a continuous map $\beta : V \to B$ the class $\beta_*(\mathrm{PD}\, L_V) \in H_*(B)$ is a homotopy invariant of (V, β).*

In fact, if V is compact, then all homotopies are bounded.

Notice that the class θ is sms for complete manifolds B with $K(B) \leq 0$ and so we obtain yet another (and the simplest of all) proof of NC for these manifolds.

On flatness and shortness. The present notion of (multiple!) shortness of cocycles is parallel to the ε-flatness of K-classes although the latter concerns area while the former belongs with length. Yet, the two notions do not seem to absorb one another; a manifold V with short fundamental class does not seem always to to have (at least not superficially) infinite K-area as K-classes are more choosy for maps suitable for push-forwards. (We indicate in the next section a notion of a "flat cocycle" generalizing both, shortness of cohomology classes and almost flatness of corresponding K-classes, that should imply the (bounded) Novikov conjecture as well as a bound on the scalar curvature.) Notice that ε-shortness of $[V]^{\mathrm{co}} \in H^n(V)$ prohibits $\mathrm{Sc}(V) \geq \varepsilon^2$ by a minimal surface argument applied to (non-complete!) $\widetilde{U}$ which, unfortunately, needs the unpublished result by Schoen-Yau to bypass the singularities for $\dim V \geq 8$.

However, if V (or at least $\widetilde{U}$) is spin and has uniformly bounded local geometry ($|K(V)| \leq \mathrm{const}$, $\mathrm{Inj\,Rad}\, V \geq \mathrm{const}^{-1}$). Then one can extend $\widetilde{U}$ (or rather $\widetilde{U} \times S^2$ for a large 2-sphere S^2) to a complete manifold with comparably large scalar curvature and follow the twisted Dirac operator approach. For example, if $\beta : V \to B$ *is a Lipschitz map where B as in* (b) *and* $\mathrm{Sc}\, V \geq \varepsilon > 0$, *then* $\mathrm{Gys}_\beta\, \widehat{A}_V = 0$.

$9\frac{1}{2}$. Almost flat bundles on open manifolds

We indicate here how to extend the results of $8\frac{3}{4}$ and $8\frac{4}{5}$ to open manifolds which would allow an alternative more elementary approach to the results in

$9\frac{1}{6}$ avoiding appearance of infinite dimensional bundles. However, as I did not check all this in detail, the statements in this section should be regarded as *conjectures.*

Combinatorial formula for L-classes. Just to start, let V be a closed combinatorial (or rational homology) manifold and let us define the L-class L_V by a formula for the values $(L_V \smile \operatorname{ch} X)[V]$ for all complex vector bundles $X \to V$. Such a bundle X over V will be given a piecewise smooth unitary connection so that we may speak of simplicial cochains with coefficients in X as in the π_1-free discussion in $8\frac{3}{4}$. We make a priori no assumptions on the flatness of X, but then we rescale V by a large constant which makes X ε-flat for small $\varepsilon > 0$. This amounts to subdivising V into small simplices of size about ε (and then regarding them as roughly of unit size). We allow only those subdivisions, where the maximal number c of neighbours a simplex may have remains bounded for ε-getting smaller and smaller. Thus the cochain "complex" of our ε-subdivision, say $C^*(V_\varepsilon, X)$, has $|\delta^2| \leq \varepsilon$ while operators involved (see diagram (D) in $8\frac{8}{9}$) are bounded by c. Thus for ε/c sufficiently small, one can extract the signature of the corresponding diagram $D(\varepsilon)$, satisfying

$$\sigma(D(\varepsilon)) = (L_V \smile \operatorname{ch} X)[V]\,, \tag{$*$}$$

which is invariant under subdivisions with controlled c. If V_ε is an ε-triangulation of a smooth manifolds, $\sigma(D(\varepsilon))$ appears as a combinatorial approximation to the index of the signature operator $\mathcal{L}$ twisted with X. This makes one ponder over a similar approximation of the Connes-Moscovici formula for $\mathcal{L}$ twisted with a straight cocycle.

Next, let us allow a non-compact V and let κ be a K^0-class on V with compact support. Then instead of a single diagram $D(\varepsilon)$ we have a Fredholm pair of these as in $9\frac{1}{6}$ and again the signature of this pair $D_F(\varepsilon)$ satisfies

$$\sigma(D_F(\varepsilon)) = (L_V \smile \operatorname{ch} \kappa)[V]\,. \tag{$**$}$$

Notice that ($**$) can be reduced to ($*$) once we know the excision property for $\sigma(D_F(\varepsilon))$, namely its independence of V outside the support of κ. For example let κ be given by $[X] - [\mathrm{Triv}]$ where X is trivialized outside the interior of a compact equidimensional submanifold $V_0 \subset V$ with boundary. Then we may take the double $V_0^* = V_0 + \overline{V}_0$ (where $\overline{V}_0$ denotes V_0 with the reversed orientation) with X extended to X^* on V_0^* trivially on $\overline{V}_0$. Then $\sigma(V_0'; \kappa')$ for $\kappa^* = [X^*] - [\mathrm{Triv}]$, equals $\sigma(V_0; \kappa) = \sigma(V; \kappa)$ by excision, while $\sigma(V_0^*; \kappa^*) = \sigma(V_0^*; X)$ as $\sigma(V_0^*; \mathrm{Triv}) = \operatorname{rank}(\mathrm{Triv})\, \sigma(V_0^*) = 0$ since V_0^* is a double.

On the B-homotopy invariance of $\sigma(V;\kappa)$. Here we assume V and V' are properly homotopy equivalent Riemannian manifolds where the implied maps $V \leftrightarrow V'$ as well as the homotopies $V \times [0,1] \to V$ and $V' \times [0,1]$ are

λ-Lipschitz for some $\lambda > 0$. Then we take some ε-flat K^0-class κ on V with compact support and we want to show that $\sigma(V;\kappa) = \sigma(V;\kappa')$ for κ' corresponding to κ whenever ε is small compared to λ. Notice that by scaling V and V' large, we can make κ and κ' as flat as we want, but this would correspondingly enlarge the Lipschitz constants of the homotopies as we do not scale the segment $[0,1]$. In fact, in order to prove the homotopy invariance of the signature of a diagram, the norms of the algebraic homotopy operators must be kept rather small compared to ε^{-1}. Taking all this into account we arrive at the following (not quite proven).

ε-flat B-invariance theorem. *Let* $\mathbf{B}$ *be a complete Riemannian manifold and* κ *be a* K^0*-class on* $\mathbf{B}$ *with compact support admitting an* ε*-flat representation with an arbitrarily small* $\varepsilon > 0$. *Then for every proper map* $\beta : V \to \mathbf{B}$ *the value*

$$(L_V \smile \beta^*(\operatorname{ch}\kappa))[V] = (\operatorname{ch}\kappa, \beta_*(\operatorname{PD} L_V)) \tag{+}$$

is a B*-homotopy invariant of* (V,β).

Application to **NC.** Take a closed manifold $\mathbf{B}$ whose universal covering $\mathbf{B} = \widetilde{\mathbf{B}}$ has infinite K-area. Then the above applied to $\widetilde{\mathbf{B}}$ yields the Novikov conjecture for the fundamental class of $\mathbf{B}$ (without resorting to infinite dimensional bundles as in $9\frac{1}{6}$). Similarly one can approach manifolds with Δ-area $= \infty$ by extending the above theorem to families of bundles. More generally, one may work with ε-flat C^*-algebra bundles and corresponding κ over B and V.

Multiply flat cocycles. Let us indicate a generalization involving (non-covering) multi-domains $\widetilde{U}$ over $\mathbf{B}$. Call a cohomology class $\rho \in H^*_{\text{comp}}(\mathbf{B})$ (ε, R)-flat if there exists such a $\widetilde{U}$ with a compactly supported K-class κ on $\widetilde{U}$ such that

(1) the push-forward of $\operatorname{ch}\kappa$ to $\mathbf{B}$ equals ρ;

(2) κ admits an ε-flat representative, i.e. $\kappa = [\widetilde{X}_1] - [\widetilde{X}_2]$ for ε-flat bundles over $\widetilde{U}$ with a connection preserving unitary isomorphism in the R-neighbourhood of the boundary (infinity) of $\widetilde{U}$ (i.e. every path of length $\leq R$ starting in $\operatorname{supp}\kappa$ stays in $\widetilde{U}$).

Take the subgroup $\widetilde{H}^*(R,\varepsilon) \in \widetilde{H}^*_{\text{comp}}(\mathbf{B})$ generated by these classes and define $\widetilde{H}^*_{\mathfrak{f}\ell}(\mathbf{B})$ as the intersection $\cap \widetilde{H}^*(R,\varepsilon)$ first over all $\varepsilon > 0$ and then over $R > 0$. Notice that this *multiply flat* cohomology generalizes the multiply short ones in $9\frac{2}{7}$ when the parameter space in B in $9\frac{2}{7}$ reduces to a single point (and the corresponding "flat notion" for families is suggested to the reader to work out by him/herself).

Now it seems that the ε-flat B invariance theorem remains true for the multiply flat classes $\rho \in H^*_{\text{comp}}(\mathbf{B})$ in the place of $\operatorname{ch} \kappa$, since all constructions in V can be limited to $\widetilde{V}$ over V which is the pull-back of $\widetilde{U}$. This, extended to families, appears the most general version of BN (and, for compact V, of NC) available with our macroscopic geometric techniques. One may also approach the problem of $\operatorname{Sc} > 0$ with such flat cocycles where it seems likely, for example, that the fundamental class of a spin manifold V with $\operatorname{Sc}(V) \geq \varepsilon^2 > 0$ cannot be multiply flat, but I feel less certain as the Dirac operator appears to me to have less inclination to excision than $\mathcal{L}$.

$9\frac{2}{3}$. Connes' index theorems for foliations and scalar curvature

Consider a space $\mathcal{V}$ foliated into leaves V which are smooth manifolds. Typically, $\mathcal{V}$ is a compact metric space, but the essential structure is a transversal measure (or a measure class) so that the topological structure in $\mathcal{V}$ (but not in V's) is not indispensable.

Example. Start with a compact manifold V_0 and let $\widetilde{V}_0$ be a Galois covering of $\widetilde{V}_0$ with Galois group Π. If Π acts on some space S one has the associated fiber space $\mathcal{V} = (\widetilde{V}_0 \times S)/\Gamma \to V_0$ which is naturally foliated into leaves isomorphic to coverings of V_0 below $\widetilde{V}_0$. In particular if the action of Π is free then all leaves are isomorphic to $\widetilde{V}_0$. Furthermore, if the action of Π on S preserves a measure, one has a natural transversal measure on $\mathcal{V}$.

Observe that every Π admits a non-trivial measure preserving action, for instance, the action on the space of functions $\Pi \to F$ where a F is a finite measure space. This space, called F^Π topologically, is the Cantor set and so the above $\mathcal{V}$ is locally $\mathbb{R}^n\times$ Cantor for $n = \dim V$.

In what follows, the leaves V are endowed with smooth complete Riemannian metrics which are continuous (or at least measurable) on $\mathcal{V}$, and we are interested in geometric differential operators along the leaves, namely Dirac, Hodge, and Dolbeault, which may be twisted with vector bundles $X \to \mathcal{V}$ with leafwise connections. Connes assigns to such an operator $\mathcal{D}$ its index $\kappa = \operatorname{ind} \mathcal{D}$ which is an element of K_0 of a suitable algebra of operators associated to the foliation $\mathcal{F}$ in question.

In the simplest case when the foliation has a transversal measure $d\mu$, this index gives rise to a real valued index (associated to the trace on the von Neumann algebra of $\mathcal{F}$) which admits a simple independent description as follows. Take the holonomy covering $\widetilde{V}$ of a leaf V and let $\widetilde{D}$ denote the differential operator over $\widetilde{V}$ corresponding to $\mathcal{D}$. Denote by $\widetilde{P}$ the orthogonal projection of the pertinent space of L_2-sections over $\widetilde{V}$ to $\operatorname{Ker} \widetilde{D}$ and observe that the trace function trace $\widetilde{P}(\widetilde{v}, \widetilde{v})$ is monodromy invariant and thus gives us a measurable function trace $\widetilde{P}(v,v)$ on $\mathcal{V}$ (compare $9\frac{1}{9}$). Then we define

$$\operatorname{ind}\mathcal{D} = \int_{\mathcal{V}} \operatorname{trace}\widetilde{P}(v,v)\,dv\,d\mu\,,$$

where dv denotes the leafwise Riemannian measure. This index can be expressed according to Connes as the integral of the differential n-form corresponding to $\mathcal{D}$. Namely, the Atiyah-Singer theorem expresses $D = \mathcal{D} \mid V$ as a certain characteristic number of $T(V)$ and X which can be represented by a differential n-form Ω_D on each V, expressed at each point $v \in V$ by some (Chern-Weil) polynomial of the curvatures of V and X at v. Thus we obtain a leafwise form $\Omega_{\mathcal{D}}$ on $\mathcal{V}$ which integrates with $d\mu$ to a number denoted $\int_{\mathcal{V}} \Omega_{\mathcal{D}}\,d\mu$.

The *first Connes index theorem* claims the equality

$$\operatorname{ind}\mathcal{D} = \int_{\mathcal{V}} \Omega_{\mathcal{D}}\,d\mu \tag{$*$}$$

under certain conditions on the foliation $\mathcal{F}$ on $\mathcal{V}$. Here is a suitable condition which makes both sides of $(*)$ well defined via absolutely convergent integrals:

The Riemannian curvatures of the leaves and the (leafwise) curvatures of X are bounded by a constant $C > 0$; furthermore the Riemannian metrics in the leaves are complete and the injectivity radii of the holonomy coverings of the leaves are bounded from below by C^{-1}; and the total mass of the measure $dv\,d\mu$ is finite.

Now $(*)$ extends the Atiyah L_2-index theorem in $9\frac{1}{9}$ (including the generalized version for manifolds of finite volume with the universal covering with bounded local geometry). In fact $(*)$, applied to the above example with the *atomic measure* at a fixed point of the action of Π on the space F^{Π} amounts to $(\star\star)$ in $9\frac{1}{9}$.

K-area and $\mathrm{Sc} > 0$ for foliations. We define the Chern numbers of an X with a leafwise connection by integrating the corresponding Chern-Weil forms, as in $(*)$. We can also speak of the leafwise norm of the curvature, $\|\mathcal{R}(X)\|$. With this we define K-area $\mathcal{V}$ or rather K-area $\mathcal{F}$ for the implied foliation $\mathcal{F}$. The Bochner-Lichnerowicz vanishing theorem extends without any problem to foliations and, in particular, we have the following twisted foliated version of the Lichnerowicz theorem concerning the leafwise scalar curvature of $\mathcal{V}$,

$(\star)$ *if* $\mathrm{Sc}_v \geq c_n\|\mathcal{R}_v(X)\|$ *for all* $v \in \mathcal{V}$ *then* $\int_{\mathcal{V}} \widehat{A}_{\mathcal{V}} \wedge \mathrm{ch}_X\,d\mu = 0$, *provided the holonomy coverings of the leaves are* spin, where $\widehat{A}_{\mathcal{V}}$ and ch_X denote the Chern-Weil forms corresponding to the $\widehat{A}$-genus of the leaves and the Chern character of X (along the leaves), respectively.

Remarks and corollaries. (a) One can replace everywhere the holonomy coverings by the universal coverings of the leaves which makes the spin requirement somewhat less demanding.

(b) If $\mathrm{Sc} \geq \varepsilon^2 > 0$ the above $(\star)$ shows that *the K-area of $\mathcal{V}$ (or $\mathcal{F}$) is finite.* This is already interesting for the above example where $\mathcal{V} = (V_0 \times S)/\Pi \to V_0$ as the condition K-area $\mathcal{V} = \infty$ is, a priori, less restrictive than K-area $V_0 = \infty$ while $\mathrm{Sc}\, V_0 \geq \varepsilon^2 \Rightarrow \mathrm{Sc}\, \mathcal{V} \geq \varepsilon^2$. For example, if the universal covering $\widetilde{V}_0$ has K-area $\widetilde{V}_0 = \infty$ then, (almost) obviously, K-area $\mathcal{V} = \infty$ for this $\mathcal{V} = (V_0 \times S)/\Pi$ and $S = F^{\Pi}$ which gives us an alternative approach to the K-area inequality in this case, where the foliated space $(V_0 \times S)/\Pi \to V_0$ plays the role of finite coverings $\widetilde{V}_i \to V_0$ needed in our first proof employing the residual finiteness of Π (see §5). Similarly, one may simplify (or at least, modify) the arguments concerning the homotopy invariance of the "almost flat" signature where Π is not residually finite (compare $8\frac{8}{9}$, $9\frac{1}{7}$).

c) $(\star)$ suggests a new definition of the K-area of a manifold V_0 appealing to the curvatures of bundles over $\mathcal{V} = (V_0 \times S)/\Pi$ for all S acted upon by Π, but probably this can be reduced to the K-area defined with almost flat bundles over C^*-algebras over V_0 itself. Yet, bringing in $\mathcal{V}$'s may be useful in specific examples for getting a lower bound on a (generalized) K-area of V_0.

d) It seems one can set up the Plateau problem for transversally measurable leaf-wise Riemannian foliations and construct stable minimal subfoliations $\mathcal{V}' \subset \mathcal{V}$ of leaf-wise codimension one under suitable conditions on $\mathcal{V}$. This would lead to Schoen-Yau style theorems without the spin requirement on the leaves.

Connes' vanishing theorem. Let $\mathcal{V}$ be a smooth closed manifold with a smooth foliation $\mathcal{F}$. Then if *$\mathcal{F}$ admits a Riemannian metric with (leaf-wise) positive scalar curvature then $\widehat{A}(\mathcal{V}) = 0$ provided $T(\mathcal{F})$ is spin* (where $\mathcal{V}$ does not have to be spin). Moreover, $(\widehat{A}_{\mathcal{F}} \smile \mathrm{ch}\,\nu)[\mathcal{V}] = 0$ *for every complex bundle associated to* the *normal bundle* $T(\mathcal{V})/T(\mathcal{F})$. Furthermore *if ρ is the Chern character of an almost flat bundle over $\mathcal{V}$* (in fact the a.f. condition is only needed along the leaves) *then* $(\rho \smile \widehat{A}_{\mathcal{F}} \smile \mathrm{ch}\,\nu)[\mathcal{V}] = 0$. In particular, *if K-area $\mathcal{V} = \infty$ then $\mathcal{V}$ admits no smooth spin foliation with* $\mathrm{Sc} > 0$ *(where the simplest example of such a $\mathcal{V}$ is a torus).* And much of this extends to open manifolds $\mathcal{V}$. For example, $\mathbb{R}^n$ *admits no (automatically spin) foliation with the induced metric in the leaves having* $\mathrm{Sc} \geq \varepsilon^2 > 0$, where moreover, instead of the original Euclidean metric on $\mathbb{R}^n$ one may use any hyper-Euclidean metric.

Let us indicate an approach to these theorems using the space $\mathcal{V}^*$ introduced in $1\frac{7}{8}$ (where our geometric picture of $\mathcal{V}^*$ in $1\frac{7}{8}$ mimics Connes' analysis). The simplest case is where $\mathcal{F}$ is coorientable and $\mathrm{codim}\,\mathcal{F} = 1$ and then $\mathcal{V}^*$ is obtained from $\mathcal{V} \times \mathbb{R}$ by rescaling the metric in the direction to $\mathcal{F}$ by the factor $\exp t$, $t \in \mathbb{R}$; so $\mathcal{V}^*$ is essentially as large as $\mathcal{V}$. For example, if $\mathcal{V}$ has infinite K-area then so does $\mathcal{V}^*$. But since $\mathcal{V}^*$ can be arranged with $\mathrm{Sc} > 0$, we conclude, for example, that $\operatorname{Sc}\mathcal{F} > 0 \Rightarrow K$-area $\mathcal{V} < \infty$, at least if $\mathcal{F}$ is spin and the rest of Connes' theorem (as we stated it) follows. Furthermore, one can use here the techniques of minimal varieties and show, for example, that $\operatorname{Sc}\mathcal{F} > 0$ prevents every (e.g. universal) covering $\widetilde{\mathcal{V}}$ of $\mathcal{V}$ from being hyper-Euclidean (where for $\dim \mathcal{V} \geq 7$ one should appeal to an unpublished result by Schoen and Yau while the case $\dim \mathcal{V} < 7$, and hence, $\dim \mathcal{V}^* \leq 7$ is covered by [G-L]$_{\mathrm{PSC}}$). In fact, one can prove here that $\mathcal{V}^*$ admits a hypersurface $\mathcal{V}_0$ homologous to $\mathcal{V} \subset \mathcal{V}^*$ and carrying a metric with $\operatorname{Sc} > 0$ (again with extra troubles for $\dim \mathcal{V} \geq 7$ due to possible singularities of minimal hypersurfaces). Then, if $\dim \mathcal{V} \geq 5$, one can apply surgery to $\mathcal{V}_0$ of codimension ≥ 3 and modify it back to $\mathcal{V}$ but now with a metric with positive scalar curvature on $\mathcal{V}$.

Foliations of codimension ≥ 2. The major difficulty with $\mathcal{V}^*$ is the (non-Abelian) holonomy which makes the Lipschitz geometry of $\mathcal{V}^*$ quite far from the product $\mathcal{V} \times M$ (while the problem of $\mathcal{U}^* \neq \mathcal{V}^*$ is a minor one). This difficulty disappears, for example, if the lift of the foliation $\mathcal{F}$ to the universal covering $\widetilde{\mathcal{V}}$ has neglegible holonomy, e.g. this lift is non-recurrent (which is very restrictive and so not truly interesting) or if the holonomy is proper on some transversal jet bundle which corresponds to the rigidity in the sense of [Gro]$_{\mathrm{RTG}}$. In any case, what one needs (to witness the largeness of the manifold $\mathcal{V}^*$ in the M-directions) is a *foliated* UAFl (virtual) bundle κ^* over $\mathcal{V}^*$, where "foliated" indicates that the implied flatness is required only along the leaves of $\mathcal{F}^*$, such that a pertinent Chern number of κ does not vanish. In fact, it is more logical to look for such a bundle κ over $\mathcal{V}$ starting from another bundle, say ν over $\mathcal{V}$, which has the required flatness along $\mathcal{F}$ but which is not unitary. For example, the normal bundle $\nu = T(\mathcal{V})/T(\mathcal{F})$ is flat along $\mathcal{F}$ and we want to unitarize it, i.e. find a unitary bundle κ flat along $\mathcal{F}$ with the same Chern number as ν.

The construction of Connes (already explained in $8\frac{1}{2}$) goes as follows. Take some action of the structure group G of ν on some symmetric space Z of non-compact type and let $\mathcal{Z} \to \mathcal{V}$ be the associated Z-fibered bundle. For example, if $\nu = T(\mathcal{V})/T(\mathcal{F})$ and $G = GL_k(k)$ for $k = \operatorname{rank} \nu$, then G acts on the space M so that $\mathcal{Z} = \mathcal{V}^*$ in this case. (Notice that M is *not* a symmetric space but it is $M^0 \times \mathbb{R}$ where $M^0 = SL_k\mathbb{R}/\mathcal{O}(k)$ is symmetric, and, in general, one must allow some non-symmetric spaces Z as well.) We take some Hilbert bundles $\mathcal{H}$ associated to $\mathcal{Z}$ where each fiber $H = H_v$ consists of L_2-sections on $Z = Z_v$ of a suitable bundle over Z satisfying some elliptic system, say $\Delta x = 0$, i.e. $H = \ker \Delta$. In fact, one needs a pair of such bundles H_+ and H_- but we are being rather sketchy here anyway.

Finally, we take some continuous section $v \mapsto z(v) \in Z_v$ and use the differentials of the fiberwise distance functions $d_z \operatorname{dist}_{Z_v}(z, z(v))$ to construct a family $\mathbf{F} = \{F_v\} : H_v \to H_v$ of Fredholm operators almost commuting with G so that $\operatorname{Ind}\mathbf{F}$ may serve for κ (compare $8\frac{1}{2}$ and see [Con]$_{\mathrm{CCTF}}$ and [Con]$_{\mathrm{NCG}}$

for the actual proof which also catches secondary characteristic classes). Once we have κ over $\mathcal{V}$, we may pass it over to $\mathcal{V}^*$ and apply a suitable index theorem there, or, which is more logical, we may stay (as Connes does) on $\mathcal{V}$, but then we need a *longitudinal* (i.e. leaf-wise) index theorem for $\mathcal{F}$ more powerful than the first Connes theorem and such is proven in [Co-Sk].

Remark and open question. (a) The above construction of κ can be performed for more general bundles $\mathcal{Z} \to \mathcal{V}$ where the fibers Z_v do not have to be symmetric or homogeneous, just complete Riemannian manifolds large in a suitable sense (e.g. being simply connected of $K \leq 0$ or hyper-Euclidean as in [C-G-M]$_{\text{GCLC}}$). Then one "unitarizes" $\mathcal{Z}$ by taking a suitable Hilbert bundle $\mathcal{H}$ of L_2-objects over the fibers Z_v with κ being the index of some Fredholm endomorphism of $\mathcal{H}$ (for which the largeness is needed). The important features of such construction are (a) the "bundle" κ is (at least) as flat (over all of $\mathcal{V}$ or along a given foliation) as the original $\mathcal{Z}$, and (b) by choosing Δ one can arrange κ to have $\operatorname{ch}\kappa$ as rich as that of $\mathcal{Z}$. This gives a different view on the similar construction of Fredholm representations in $8\frac{2}{3}$ and explain anew why strong Novikov forces $0 \in \operatorname{spec}\Delta_v$ for natural operators Δ_v on Z_v.

(a′) ***Example.*** Let $\mathcal{V}$ be a closed manifold and $\mathcal{Z} \to \mathcal{V}$ be a flat Riemannian bundle where the fibers Z_v are complete simply connected with non-positive curvatures. (An instance of that is the bundle $\widetilde{V}_\Delta \to V$, for $\widetilde{V}_\Delta = (\widetilde{V} \times \widetilde{V})/\Pi$ for a manifold V with $K(V) \leq 0$, compare $9\frac{2}{7}$.) Then *the Fredholm K-area of $\mathcal{V}$ (defined in $9\frac{1}{6}$) is infinite provided, $\mathcal{Z}$ has a non-zero Pontryagin number.* Furthermore, *both the Dirac and the signature operators on the universal covering $\widetilde{\mathcal{V}}$ have* $0 \in$ spec (where $\widetilde{\mathcal{V}}$ should be spin if we speak of Dirac).

Here one can separate two cases.

(1) The implied action of $\Pi = \pi_1(\mathcal{V})$ on Z is proper. Notice that we assume all fibers Z_v being mutually isometric. In fact, we may rather assume the image of Π in $\operatorname{Isom} Z$ is a discrete subgroup without torsion and then $W = Z/\operatorname{Im}\Pi$ is a complete manifold with $K(W) \leq 0$. The homomorphism $\Pi \to \operatorname{Im}\Pi = \pi_1(W)$ defines a (homotopy class of a) map $\mathcal{V} \to W$ which sends $[\mathcal{V}]$ to a non-zero class in $H_n(W;\mathbb{Q})$, $n = \dim\mathcal{V}$ (where $\mathcal{V}$ is assumed oriented); so the above statement can be derived from the corresponding properties of W.

(2) The action of Π is *non-proper*, which implies that the closure of $\operatorname{Im}\Pi \subset \operatorname{Isom} Z$ has positive dimension. Thus the essence of the problem becomes Lie theoretic (since Closure $\operatorname{Im}\Pi$ is a Lie group) and one, probably, can derive the general case from the two extremal ones, where either $\operatorname{Im}\Pi$ is discrete, or on the contrary has $\operatorname{Cl}\operatorname{Im}\Pi$ connected.

Now, look at a more general situation where the separation into two cases seems impossible. Namely suppose $\mathcal{Z}$ is *almost* flat rather than flat, which means that the fibers Z_v do not have to be mutually isometric anymore, but the monodromies should not distort the metric too much. (One may take, for instance, a small perturbation of the metrics in the fibers of the previous flat $\mathcal{Z} \to \mathcal{V}$ but more convincing examples are yet to be found.) Then the corresponding κ will be also almost flat (as flat as $\mathcal{Z}$) and we get a lower bound on the Fredholm K-area of $\mathcal{V}$ again.

(b) ***Fredholm K-area of foliated spaces and related invariants.*** Let $\mathcal{V}$ be a foliated manifold as earlier. Then one can define the (Fredholm) K-area of $\mathcal{V}$ (or rather of the implied foliation $\mathcal{F}$) with bundles X over $\mathcal{V}$ having non-trivial Chern numbers where the flatness of X is measured only along $\mathcal{F}$. Similarly, one may define various "norms" on homotopy classes of maps from $\mathcal{V}$ into standard spaces (spheres, Grassmannians etc.) by minimizing the dilation of these maps along the leaves (where "dilation" may refer to the norm of the differential on $\Lambda^p T(\mathcal{F})$, for example. And if $\mathcal{F}$ has a transversal measure $d\mu$ one may take integral norms such as $\left(\int_{\mathcal{V}} \|\Lambda^p\, df\|^q\, dv d\mu\right)^{\frac{1}{q}}$, but this is another story). For example, the above discussion shows (borowing from Connes) that *if the normal bundle* $\nu = T(\mathcal{V})/T(\mathcal{F})$ has a non-zero Pontryagin number, *then the Fredholm K-area of* $\mathcal{F}$ *is infinite* and *this K-area is also infinite if* $\mathcal{F}$ *has a metric with non-positive curvature.*

Question. What are relations between

(1) the (Fredholm) K-area of $\mathcal{V}$ disregarding $\mathcal{F}$,

(2) the (Fredholm) K-area of $\mathcal{F}$,

(3) the (Fredholm) K-area of the leaves V of $\mathcal{F}$?

And one may ask similar question for more general size characteristics of $\mathcal{F}$ using maps $\mathcal{V} \to$ standard spaces, such as $\operatorname{Rad} \mathcal{V}/S^n$, $\max\deg \ell\, \mathcal{V}/S^n$, maxchern, etc.

Intuitively, one expects the following implications.

The leaves V of $\mathcal{F}$ are "small" $\Rightarrow$ $\mathcal{F}$ is "small" $\Rightarrow$ $\mathcal{V}$ is "small", (where "small" may refer to the universal covering of the spaces in question) and some of these are obvious, such as

$$K\text{-area } \mathcal{F} < \infty \Rightarrow K\text{-area } \mathcal{V} < \infty.$$

But one may look deeper, for example, let the leaves V of $\mathcal{F}$ have (Fredholm) K-area $\leq$ const. Is then the (Fredholm) K-area of $\mathcal{V}$ finite? (Compare $2\frac{2}{3}$, where similar questions were raised for the macroscopic dimension of (some coverings of the leaves).

Finally, observe, that the opposite implication is also plausible:

the leaves are "large" $\Rightarrow$ the universal covering of $\mathcal{V}$ is "large".

For instance if the leaves have negative curvatures, then it seems the fundamental group $\pi_1(\mathcal{V})$ must be large (may be under extra assumptions such as the existence of a smooth ergodic transversal measure, smallness of codim $\mathcal{F}$ against largeness of dim $\mathcal{F}$, extra data on the geometry of the leaves etc.)

$9\frac{3}{4}$. Foliated max deg, Novikov-Shubin and related invariants

Consider a closed Riemannian manifold V, a Dirac-type operator D on V and an infinite Galois Π-covering $\widetilde{V} \to V$. We look for lower bounds on the von Neumann spectral density of the lift $\widetilde{D}$ to $\widetilde{V}$, i.e. for estimates

$$\dim_\Pi \operatorname{spec} \widetilde{D}[a,b] \geq \sigma_V(a,b) \qquad (*)$$

for some function σ expressible in terms of topology and macroscopic geometry of V (where, recall, spec $\widetilde{D}[a,b]$ denotes the subspace belonging to the spectrum of $\widetilde{D}$ in the interval $[a,b]$, so that $\dim_\Pi \operatorname{spec} \widetilde{D}[a,b] = \operatorname{Trace}_\Pi \psi_{[a,b]}(\widetilde{D})$, where $\psi_{[a,b]}$ is the characteristic function of the segment $[a,b]$).

If D is Hodge's $d+d^*$, then the spectral density of $\widetilde{D}$ near zero, i.e. in small intervals $[-a,a]$ with $a \to 0$, is a topological (even homotopical) invariant of V (see [No-Sh], [Gr-Sh]) and in standard examples $\dim_\Pi \operatorname{spec} \widetilde{D}[-a,a] \approx a^\alpha$ for some $\alpha > 0$ which is a homotopy (Novikov-Shubin) invariant $\alpha(V)$. In general, one may look for the maximal (open or closed) segment $I_\alpha = [0,\alpha]$ or $I_\alpha = [0,\alpha[$, depending on *topology* of V and (possibly) on a particular type of D, such that

$$\dim_\Pi \operatorname{spec} \widetilde{D}[-a,a] \geq \operatorname{const} a^\beta \ , \text{ for all } \beta \in I_\alpha \qquad (**)$$

where const may depend on the geometry of V. Thus every geometric operator D on V (not only $d+d^*$) gives us a topological invariant $I_\alpha = I(V,D)$ but, probably, this is independent of D for most geometric operators D. In fact, it seems logical to turn (α of) I_α into a (spin) bordism invariant of Π, say $I_\alpha = I(\varphi)$, $\varphi \in \operatorname{Brd} B\Pi$, by taking all V mapped to $B\Pi$ in the class of φ and maximizing the segements I_α satisfying $(**)$ for all these V.

If the group Π is residually finite, one can first estimate the spectra of finite coverings $\widetilde{V}_i \to V$ approximating $\widetilde{V} \to V$ in terms of geometric invariants of $\widetilde{V}_i$ (such as max deg $\lambda V/S^n$ and max ch λV; see $6\frac{5}{6}$) and then go to the limit $\widetilde{V}_i \to \widetilde{V}$ for $i \to \infty$ since the spectra are semi-continuous in the limit. Now, we indicate a similar geometric estimate using foliations over V rather than finite covering where we do not have the residual finiteness assumption. Namely, we make Π act on some probability space S preserving the probability measure (e.g. on the space F^Π of F-valued functions $\Pi \to F$ for a finite set F) and take the obvious foliation, say $\mathcal{F}$, on the space $\widetilde{V}_S = (\widetilde{V} \times S)/\Pi$ where Π acts diagonally on the product $\widetilde{V} \times S$. Now, for every measurable leaf-wise

Lipschitz map $f : \widetilde{V}_S \to S^n$, $n = \dim V$ (where S^n is the n-sphere unrelated to S) one may speak of the degree defined with the leaf-wise Jacobian by $\deg f = \int_{\widetilde{V}_S} \mathrm{Jac}\, f\, dv\, ds$ (this extends to more general space $\mathcal{V}$ measurably foliated into n-dimensional oriented (pseudo)manifolds V, where there is the fundamental foliated n-dimensional class $[\mathcal{V}]_{\mathrm{Fol}}$ functorial in a suitable category and behaving as an n-dimensional real homology class for certain maps $\mathcal{V} \to$ topological spaces, (compare [Sul], $[\mathrm{Gro}]_{\mathrm{FPP}}$) and then one defines $\max\deg(\ell\widetilde{V}_S/S^n)$ as the supremum of these over all ℓ-Lipschitz maps.

Finally, one can vary S and maximize max deg also over all possible probability spaces S with measure preserving Γ-actions, thus arriving at what is called $\max\deg(\ell V_{\mathrm{Fol}\Pi}/S^n)$. Similarly, one defines $\max\mathrm{ch}(\ell V_{\mathrm{Fol}\Pi}; N)$ and observes that the foliation of the Vafa-Witten argument leads to the following lower spectral bound on the spectrum of the lift $\widetilde{D}$ of a geometric Dirac type operator D on V to the Π-covering $\widetilde{V} \to V$ (compare §6 and $[\mathrm{Hur}]_{\mathrm{EITF}}$).

$$\dim_\Pi \operatorname{spec} \widetilde{D}[-a,a] \geq \delta_n N^{-1} \max\mathrm{ch}(\gamma_n\, a\, V_{\mathrm{Fol}\Pi}; N) \qquad (\star)$$

for every $N = 1, 2, \ldots$, and some positive constants δ_n and γ_n. Consequently

$$\dim_\Pi \operatorname{spec} \widetilde{D}[-a,a] \geq \delta'_n \max\deg(\gamma'_n\, a\, V_{\mathrm{Fol}\Pi}/S^n)\,. \qquad (\star\star)$$

Furthermore, if $n = \dim V$ is odd, one has similar bounds on the spectrum at all points ($\neq 0$), e.g.

$$\dim_\Pi \operatorname{spec} \widetilde{D}[a,b] \geq \delta'_n \max\deg(\gamma'_n(b-a)V_{\mathrm{Fol}\Pi}/S^n) \qquad (\star\star\star)$$

for all segments $[a,b] \subset \mathbb{R}$ and some universal $\gamma'_n > 0$.

Unfortunately, the known lower bounds on this foliated max deg are far from what is expected. For example, one does not know for manifolds V with non-positive sectional curvature whether $\max\deg(\ell V_{\mathrm{Fol}\Pi}/S^n) \gtrsim \ell^n$ for small $\ell \to 0$ (where $\Pi = \pi_1(V)$), and even the weaker bound $\max\deg \gtrsim \ell^\alpha$ for *some* $\alpha > 0$ is unavailable at the present moment.

On the positive side, let us indicate a lower bound on the foliated max deg by $\max\deg\,(\ell\widetilde{B}(R)/S^n)$, where the implied maps of the R-balls $\widetilde{B}(R) \subset \widetilde{V}$ to S^n are assumed constant on the boundary. We take a maximal foliated system of R-balls in $\widetilde{V}_S$ where the implied action of Π on S is a.e. free, so that the concentric $2R$-balls cover all $\widetilde{V}_S$. Then, clearly

$$\max\deg(\ell\widetilde{V}_S/S^n) \leq \frac{\infty \max\deg(\ell\widetilde{B}(R)/S^n)}{\sup \mathrm{Vol}\, \widetilde{B}(2R)}\,, \qquad (+)$$

where "inf" and "sup" are taken over all positions of the (centers of) the balls and where we may use arbitrary $R > 0$ and $\ell > 0$. Also observe that for large R, max deg and Vol are essentially independent of the positions of the balls and $\mathrm{Vol}\,2\widetilde{B}(R) \leq \exp cR$ for some $c = c(V) > 0$. So the key invariant here is $\max\deg \ell\widetilde{B}(R)/S^n$ as the function of R and ℓ, which has been already evaluated in some examples (see $6\frac{7}{8}$). Here we notice that if V has non-positive sectional curvature and $\Pi = \pi_1(V)$, i.e. $\widetilde{V} = \widetilde{V}_{\mathrm{univ}}$, then the R-balls in $\widetilde{V}$ are (at least) as large as the Euclidean balls and so

$$\max\deg(\ell\widetilde{B}(R)/S^n) \geq \mathrm{const}_n(R\,\ell)^n - 1\,. \tag{Eu}$$

This implies, together with the exponential bound on $\mathrm{Vol}\,B(2R)$, that

$$\max\deg(\ell\widetilde{V}_S/S^n) \gtrsim (R\,\ell)^n \exp -cR\,, \tag{ex}$$

and, consequently,

$$\dim_\Pi \operatorname{spec}\widetilde{D}[-a,a] \geq \mathrm{const}_n \exp(-ca^{-1}) \tag{ex$'$}$$

for $a \leq 1$, some $\mathrm{const}_n > 0$ and $c = c(V) > 0$.

All three estimates (Eu), (ex) and (ex$'$) appear highly non-efficient for non-flat manifolds V with $K(V) \leq 0$. Probably, (ex) and (ex$'$) can be freed of "exp" but (Eu) may admit only an insignificant improvement since for every *non-amenable* group Π

$$\max\deg(\ell\widetilde{B}(R)/S^n) \leq \mathrm{const}'_n\ \ell^n\ \mathrm{Vol}\,\widetilde{B}(R)/\exp\gamma\,\ell^{-1} \tag{$-$}$$

for some $\gamma = \gamma(V) > 0$. To see this, look at the pull-back $f^{-1}(S^n_+) \subset \widetilde{B}(R)$ of the hemisphere opposite to the f-image of the boundary $\partial\widetilde{B}$ and observe, using the non-amenability in the form of the linear isoperimetric inequality, that $\mathrm{Vol}\,f^{-1}(S^n_+) \leq \mathrm{Vol}\,\widetilde{B}(R)\exp\gamma\,\ell^{-1}$ since

$$\mathrm{dist}(f^{-1}(S^n_+), \partial\widetilde{B}(R)) \geq \pi/4\ell)\,.$$

Packing $\widetilde{V}$ by large balls and $\max\deg(V/S^n; \mathrm{Ar} \leq \ell^2)$**.** One could slightly improve (+) by using more efficient packing of $\widetilde{V}$ by R-balls so that $\mathrm{Vol}\,\widetilde{B}(2R)$ in the denominator of (+) could be replaced by $\mathrm{Vol}\,B(R + \mathrm{const})$.

Observation. Let $\widetilde{V}$ be a complete simply connected manifold with $K(V) \leq -\kappa^2 < 0$. Then, for every (arbitrarily large) $R > 0$, there exist disjoint R-balls $\widetilde{B}_i \subset \widetilde{V}$, $i = 1, 2, \ldots$, such that the concentric balls of radii $R + r$ cover $\widetilde{V}$, where $r = r(\kappa) > 0$ is a constant independent of R.

Proof. Start with some $\widetilde{B}_1 \subset \widetilde{V}$ and then add $\widetilde{B}_i$ with $i \geq 2$ layer after layer around $\widetilde{B}_1$. Namely, first take a maximal system of disjoint balls with the centers on the sphere $\widetilde{S}(2R)$ concentric to $\widetilde{B}_1$. Then add a maximal possible number of disjoint balls with centers on $\widetilde{S}(2R+1)$ so that the new balls do not intersect the old ones. Next use the balls with the centers on $\widetilde{S}(2R+2)$, etc. Then the δ-hyperbolicity of V (in the sense of $[\mathrm{Gro}]_{\mathrm{HG}}$) shows that all gaps will be of the size $\leq r = r(\delta) = r(\kappa)$. Q.E.D.

Corollary. If $\widetilde{V}$ has pinched curvature $-\infty < -\kappa_1^2 \leq K(\widetilde{V}) \leq -\kappa^2 < 0$, then it admits a packing by R-balls, for every $R > 0$, which cover a definite percentage of the total volume of $\widetilde{V}$ (as is also true for flat manifolds).

Questions. (a) Does the above corollary extends to manifolds with non-strictly negative curvature, e.g. to symmetric spaces? (Here one may allow not only balls, but other "ball-like" bodies such as product of balls in manifolds $V = V_1 \times V_2$.)

(b) If $\widetilde{V}$ is acted upon by Π, can one find a Π-*quasi-periodic* efficient packing? This means a Π-invariant measure in the space of such packings. If such exists, we obtain an efficient packing of some foliation $\widetilde{V}_S$ with a transversal measure by R-balls (or rather by R-plaques).

Let us modify the notion of $\max\deg(\ell V/S^n)$ by replacing ℓ-Lipschitz maps $V \to S^n$ by ℓ^2-area contracting ones (compare §4), denote this by $\max\deg(V/S^n; \mathrm{ar} \leq \ell^2)$ and recall (see §4) that the balls $\widetilde{B} = \widetilde{B}(R)$ in the complete simply connected manifolds $\widetilde{V}$ with $-\kappa_1^2 \leq K \leq -\kappa^2 < 0$ have

$$\max\deg(\widetilde{B}/S^n; \mathrm{ar} \leq \ell^2) \geq \mathrm{const}\, \ell^{\alpha n}\ \mathrm{Vol}\, \widetilde{B} - 1$$

for some const > 0 and $\alpha > 0$ depending on κ, κ_1 and $n = \dim \widetilde{V}$, where $\alpha = 1$ for $\kappa = \kappa_1 = 1$. Then the area version of (+) above implies that

$$\max\deg(V_{\mathrm{Fol}\Pi}/S^n; \mathrm{ar} \leq \ell^2) \geq \mathrm{const}\, \ell^{\beta}\,,$$

where $\beta = \beta(\kappa, \kappa_1, n) > 0$ and where we apply (the area version of) (+) to R-balls with $R \approx \log \ell^\gamma$ with a suitable γ (unpleasantly losing in precision because of the doubling of the radius in the denominator of (+) which could have been avoided with a quasi-periodic efficient packing discussed above). This gives us Π-quasi-periodic ε-flat bundles $\widetilde{X}$ over $\widetilde{V}$ with $\mathrm{ind}\, \widetilde{D}_{\widetilde{X}} \approx \varepsilon^\beta$ and since $K(\widetilde{V}) \leq -\kappa^2 < 0$, the ε-flatness of $\widetilde{X}$ implies ε'-straightness for $\varepsilon' \approx \kappa\, \varepsilon$. This suggests an approach to the lower bound of $\dim_\Pi \mathrm{spec}\, \widetilde{D}[-a, a]$ by a^β, but unfortunately the implied ε'-straight structure in $\widetilde{X}$, i.e. an ε'-parallel frame, is by no means Π-periodic or quasi-periodic. (One can recapture with such aperiodic frame our earlier exponetially non-efficient estimate but I failed to make it work for a^β facing the same difficulty as in the quasi-isometry invariance problem of the Novikov-Shubin invariants, see $8\mathrm{A}_6$ in $[\mathrm{Gro}]_{\mathrm{AI}}$.)

On mes-invariance of the foliated max deg. Two groups Π and Π' are called *mes-equivalent* if they admit mutually orbit equivalent ergodic actions on a probability space (see §4 in $[\mathrm{Gro}]_{\mathrm{RTG}}$ for an elementary introduction and further references), and one can show that the foliated max deg is invariant under such equivalence.

In fact, we shall only use a very special case of this, namely where we have discrete subgroups Π and Π' in the full isometry group $G = \operatorname{Isom} \widetilde{V}$ and we claim that $\max\deg(\ell V_{\mathrm{Fol}\Pi}/S^n) = \max\deg(\ell V'_{\mathrm{Fol}\Pi'}/S^n)$ and the same remains true with S^n replaced by another n-dimensional manifold, e.g. the n-torus. To see this, we use a "foliated correspondence" between $V = \widetilde{V}/\Pi$ and $V' = \widetilde{V}/\Pi'$ (where we assume Π' acts on $\widetilde{V}$ fixed point free to avoid a minor inconvenience), i.e. a foliated space $\mathcal{V}$ (with transversal measure) with projections $\mathcal{V} \to V$ and $\mathcal{V} \to V'$, such that the leaves of the implied foliation $\mathcal{F}$ on $\mathcal{V}$ cover V and V'. Such a correspondence can be made with $\mathcal{V}$ fibered over $V \times V'$ where the fiber at (v, v') equals the set of local isometries $V \to V'$ sending $v \mapsto v'$ (e.g. if $\widetilde{V}$ is G-homogeneous, then this fiber can be identified with the isotropy subgroup $G_v \subset G$ consisting of the isometries fixing v). This $\mathcal{V}$ naturally foliates into leaves which are graphs of the isometric immersions of $\widetilde{V}$ to V' and this foliation is exactly what we need.

Example. Suppose V' admits a map to the n-torus with positive degree. Then so does the foliation $\mathcal{F}$ on $\mathcal{V}$ over V and so $\max\deg(V_{\mathrm{Fol}\Pi}/T^n) > 0$ which implies that

$$\max\deg(\ell V_{\mathrm{Fol}\Pi}/T^n) \geq \mathrm{const}\, \ell^n$$

and hence,

$$\max\deg(\ell V_{\mathrm{Fol}\Pi}/S^n) \geq \mathrm{const}'\, \ell^n\,.$$

Consequently,

$$\dim_\Pi \operatorname{spec} \widetilde{D}[-a, a] \geq \mathrm{const}''\ a^n\,,$$

for $a \leq 1$.

To make it interesting, observe (following J. Millson) that the hyperbolic space H^n, for each $n \geq 2$, admits a cocompact lattice Π' for which $V' = H^n/\Pi'$ admits the above map $V' \to T^n$ with $\deg > 0$ and so every compact manifold V admitting a metric of constant negative curvature (or just a map of $\deg > 0$ to a manifold with $K = -1$ which may be quite different from V') has, for $a \leq 1$,

$$\dim_\Pi \operatorname{spec} \widetilde{D}[-a, a] \geq \mathrm{const}_V\ a^n \qquad (++)$$

for $\Pi = \pi_1(V)$.

The inequality (++) generalizes to manifolds V mapped to quotients of products of hyperbolic spaces. Furthermore, since we need at the initial stage only a "virtual map" (or mes-map in the language of $[\text{Gro}]_{\text{RTG}}$) $V' \to T^n$, one probably may extend the above to the complex hyperbolic spaces (and, possibly, to more general *a-T-menable* groups; see $[\text{Gro}]_{\text{AI}}$). On the other hand, this can not work for other non-compact symmetric spaces where Kazhdan's property T prohibits virtual homomorphisms into Abelian groups.

Exercises. (a) Generalize the above to non-compact complete manifolds V with $\operatorname{Vol} V < \infty$ and with $\widetilde{V}$ having bounded local geometry.

(b) Extend the (ex$'$)-bound to manifolds V admitting maps f to complete manifold W with $K(W) \leq 0$, such that $f_*[V] \neq 0$ in $H_n(W;\mathbb{Q})$.

Problems. (a) It is, probably, not hard to compute the spectral von Neumann densities of invariant geometric operators $\widetilde{D}$ on symmetric spaces $\widetilde{V}$ of non-compact type. Then one may ask if such density near zero (or at any point if n is odd) can be significantly diminished by a Π-periodic (or more general quasi-periodic) perturbation of the metric (with the expected answer "No").

(b) Find examples of manifolds V, where $\dim_\Pi \operatorname{spec} \widetilde{D}[-a,a] > 0$ for all $a > 0$ and all metrics on V, but yet, for some metric this $\dim_\Pi$ decays, for $a \to 0$, faster than a^β for all $\beta > 0$, or, even better, faster than $\exp -ca^{-1}$. In fact, nothing is known about the possible shape of the function $\dim_\Pi \operatorname{spec} \widetilde{D}[-a,a]$ near zero apart from a few simple examples. Probably, such examples are easier to construct if one drops the Π-periodicity (or quasi-periodicity) assumption and allows all complete manifolds $\widetilde{V}$ (possibly, required to be uniformly contractible and/or to have bounded local geometry). The spectral information concerning such (aperiodic) geometric operator $\widetilde{D}$ can be expressed with the Schwartzian kernel $K_\psi(\widetilde{v}_1, V_2)$ of the operators $\psi(\widetilde{D})$, e.g. for ψ being the characteristic function $\psi_{[-a,a]}$ of the interval $[-a,a]$, by the function $\operatorname{Tr}_\psi(\widetilde{v}) = \operatorname{Trace} K_\psi(\widetilde{v},\widetilde{v})$. For example, one may integrate $\operatorname{Tr}_a = \operatorname{Tr}_{\psi_{[-a,a]}}$ over the R-balls, look at

$$\sup(\operatorname{Vol} \widetilde{B}(R))^{-1} \int_{\widetilde{B}(R)} \operatorname{Tr}_a(\widetilde{v})\, d\widetilde{v}$$

with "sup" taken over all R-balls, and then go to the limit for $R \to \infty$ thus obtaining a function $\sigma(a)$ replacing the von Neumann spectral density for $a \to 0$. In particular, one may try this for the uniformly contractible example in [Fe-We] where some caution is needed as this has unbounded geometry. (Instead of Tr_a one may study maximal systems of sections $\varphi_i(\widetilde{v})$, $i = 1, 2, \ldots$, with mutually disjoint supports satisfying $\|\widetilde{D}\varphi_i\| \leq a\|\varphi_i\|$.)

Inflated manifolds. A horosphere H in a complete manifold V with $K(V) \leq 0$ can be indefinitely compressed by equidistant interior motion; see Figure 16 below

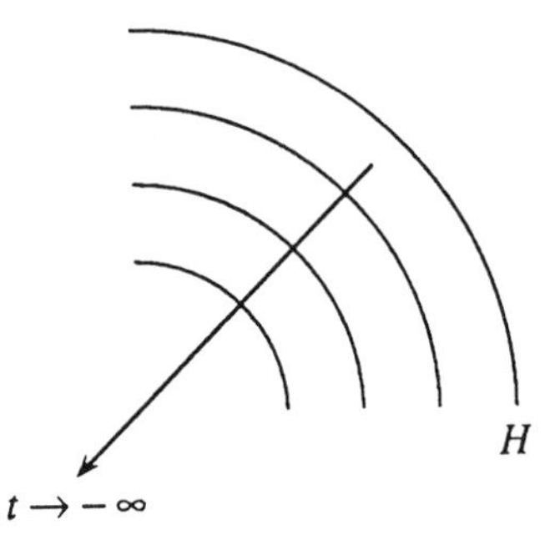

Figure 16

and so it can be thought of as the result of an infinite time inflating evolution. The strongest "inflated" condition (corresponding to the pinching $-\infty < -\kappa_1^2 \leq K(V) \leq -\kappa^2 < 0$) is as follows. A Riemannian manifold (H, g) is called *inflated* if there exists a sequence of Riemannian metrics $g_0 = g$, $g_{-1}, g_{-2}, \ldots$, on H, such that

$$2g_{-i-1} \leq g_{-i} \leq Cg_{_{-i-1}} \quad \text{for all } i = 0, 1, 2, \ldots$$

and some $C \geq 2$, where the local geometries of (H, g_{-i}) are uniformly bounded (i.e. $|K(g_{-i})| \leq \text{const}$ and $\operatorname{Inj Rad} g_i \geq (\text{const})^{-1}$).

Such inflated manifolds have "parabolic" geometry (compare $[\text{Gro}]_{\text{CCS}}$) mediating between $K < 0$ and $K > 0$. It is not hard to show that the R-balls B in such H have $\max \deg \ell B/S^n \geq (\ell R)^\alpha$ for $\alpha > 0$ and consequently, the geometric differential operators on H have the spectral density in $[-a, a]$ of order $\geq a^\alpha$ (where one can define the von Neumann dimension by averaging over the balls $B(R) \subset H$ as these have $\operatorname{Vol}_{n-1} \partial B(R)/\operatorname{Vol}_n B(R) \to 0$ for $R \to \infty$). One may expect, (but can not prove) that the above should hold true with $\alpha = n = \dim H$ (as for $\mathbb{R}^n$) but this is not quite known even for the (standard) examples of nilpotent groups with expanding maps (compare $6\frac{7}{8}$). Also one may think that all (or most) inflated H have $\operatorname{Vol} B(R) \geq R^n$ and $\infty \operatorname{Sc} H \leq 0$, but this (though known for the nilpotent case) remains unclear even for horospheres in compact manifolds with $K < 0$.

The above notion of "inflated" can be generalized in a variety of ways (e.g. the growth of g_{-i} may be less uniform: instead of g_{-i} on the same H one may have

$$(H_0, g_0) \to (H_{-1}\, g_{-1}) \to \cdots \to (H_{-i}, g_{-i}) \to \cdots ,$$

where the implied maps are contracting, etc.) and much of the above discussion generalizes as well thus leaving us with more conjectures on our hands.

$9\frac{4}{5}$. Perspectives, problems, omissions

Let us try to summarize what we were doing. We looked at a (typically) non-compact Riemannian manifold $\widetilde{V}$ which (in interesting cases) was rather symmetric. For example, it could be a covering of a compact manifold or a leaf of a compact foliation. This "large symmetry" was accompanied by some "homological largeness" of $\widetilde{V}$ which appears, for instance, if $\widetilde{V}$ Galois Π-covers a compact manifold V for which the classifying map $\beta : V \to B\Pi$ is "essential", e.g. $\beta_*[V] \neq 0$ in $H_n(B\Pi; \mathbb{Q})$, $n = \dim V$. Then we pursued the following implications where much remained conjectural.

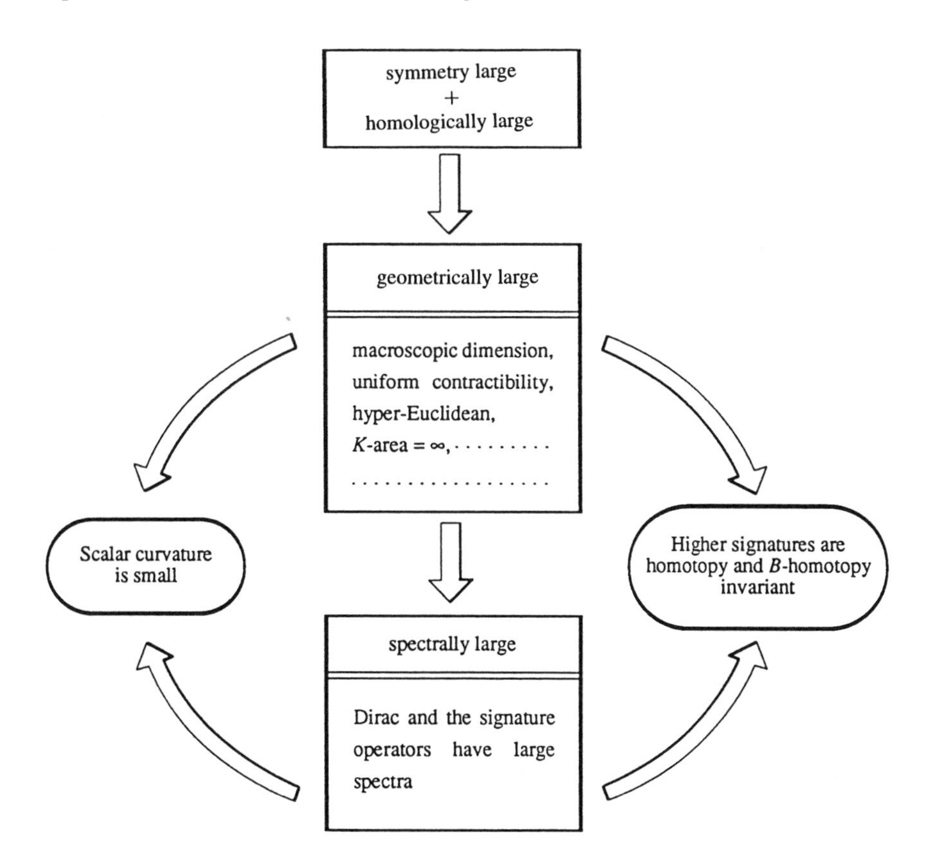

A-T-menability and related properties. The largeness of spaces and groups can be sometimes extracted somewhat paradoxally, from a possibility to "embed" such a space, say $\widetilde{V}$, into another (relatively standard) space W where "embedding" means a Lipschitz map $f : \widetilde{V} \to W$ such that $f(\widetilde{v}_1, \widetilde{v}_2) \geq c(d)$ for $d = \operatorname{dist}_{\widetilde{V}}(\widetilde{v}_1, \widetilde{v}_2)$ and where $c = c(d)$ is a function satisfying $c(d) \to \infty$ for $d \to \infty$. If $\widetilde{V}$ is acted upon by a group Π, e.g. if $\widetilde{V} = \Pi$, then one distinguishes equivariant "embeddings" for some isometric action of Π on W.

Examples. (a) Suppose $\widetilde{V} = \Pi$ and W is a Hilbert spaces. Then such an equivariant embedding $\Pi \to W$ amounts to an affine isometric action of Π on W which is *metrically proper*, i.e. for every bounded subset $B \subset W$ there are at most finitely many $\pi \in \Pi$ for which the intersection $\pi(B) \cap B$ is non-empty. The groups Π admitting such actions on a Hilbert space are called *a-T-menable* (as they strongly violate the T-property of Kazhdan claiming that every affine isometric action of Π on a Hilbert space has a fixed point) and one knows (see [B-C-V]) that amenable groups are *a-T*-menable. Yet we do not know if the *a-T*-menable groups satisfy NC or the related analytic property of approaching zero by the spectra of geometric operators $\widetilde{D}$ on $\widetilde{V}$ provided $\widetilde{V}$ is isometrically and cocompactly acted upon by such Π (where one may additionally assume that $\widetilde{V}/\Pi$ is "homologically Π-essential", e.g. $\widetilde{V}$ is contractible). All we can say in this regard is the inclusion $0 \in \operatorname{spec} \widetilde{\Delta}$ for Π *amenable* and $\widetilde{\Delta}$ acting on functions, which is one of the many equivalent definitions of the amenability.

(b) Let W be a complete simply connected manifold of non-positive curvature. Then one can show (see $[\text{G-L}]_{\text{PSC}}$ and compare $8\frac{4}{5}$) that every $\widetilde{V}$ "embeddable" in W is *stably hyperspherical* which means the existence of ε-contracting maps $\widetilde{V} \times \mathbb{R}^k \to S^N$ of non-zero degree for some N and $k = N - \dim \widetilde{V}$ and all $\varepsilon > 0$. In particular, $K\text{-area}_{\text{st}} \widetilde{V} = \infty$. This can be generalized (by allowing st to stand for ∞) to the situation where W is an *infinite* dimensional manifold of non-positive curvature (e.g. the Hilbert space $\mathbb{R}^\infty$) but it remains unclear at the moment how to carry over our analytic discussion to the infinite dimensional framework.

(c) The above suggests a classification of spaces (and/or groups) in some category where injections are "embedings" (with some equivariance assumption for groups) and where we may stabilize $X \rightsquigarrow \mathop{\times}\limits_{i=1}^{\infty} \lambda_i X$ with some $\lambda_i > 0$ and the Pythagorean (i.e. L_2) or more general L_p-metric on the products. (Thus the Hilbert space quasi-isometrically appears as $\mathop{\times}\limits_{i=1}^{\infty} \lambda_i \mathbb{Z}$ for some $\lambda_i \to 0$.) But we do not even know where the Lie group stand in this classification (compare 7.E in $[\text{Gro}]_{\text{AI}}$).

Our presentation of the ideas around NC by no means covered the whole research area. We said nothing about the cyclic cohomology and the Connes-Moscovici index theorem for differential operators twisted with straight

(Alexander Spanier) cocycles (rather than with vector bundles). This was extended to general open manifolds by Roe and applied to the problems of Sc > 0 by Yu who manage to solve it for $\widetilde{V}$ where the contractibility radius and the volume have polynomial growths (see [Con]$_{\mathrm{NCG}}$, [Co-Mo], [Roe], [Yu]). Also we had said very little about the ideal boundaries and coronas of large manifolds $\widetilde{V}$ introduced by Higson and studied further by Roe and Hurder (see [Hig], [Roe], [Hur]). And we barely touched the topological and algebraic approaches to NC and BC, i.e. the Borel conjecture claiming that the homotopy equivalence implies homeomorphism for closed aspherical manifolds V (see [Fa-Hi], [Fa-Jo], [Ran], [Wein], [NC+], and references therein).

Finally, just recently, a new (Seiberg-Witten) equation sprang to life providing an analytic key to the three basic "soft" structures in dimension four: the smooth structure, the symplectic one and Sc > 0 and suggesting a new jorney in a direction rather different from what we have taken in the present paper.

References

[Ang] N. Anghel, An abstract index theorem on non-compact Riemannian manifolds, *Houston J. Math.* 19:2 (1993), 223-237.

[A-P-S] M.F. Atiyah, V.K. Patodi and I.M. Singer, *Spectral asymmetry and Riemannian geometry*, Part I, Math. Proc. Camb. Phil. Soc. 77 (1975), 43-69, Part II, Math. Proc. Camb. Phil. Soc. 78 (1975), 405-432, Part III, Math. Proc. Camb. Phil. Soc. 79 (1976), 71-99.

[Alm] S.C. de Almeida, *The geometry of manifolds of nonnegative scalar curvature*, dissertation, Stony Brook, 1982.

[At]$_{\mathrm{EDO}}$ M.F. Atiyah, *Eigenvalues of the Dirac operator*, In Proceedings of the 25th Math. Arbeitstagung, Bonn 1984, 251-260; *Lecture Notes in Math.*, (1985) Vol. 1111.

[At]$_{\mathrm{GAE}}$ M.F. Atiyah, *Global aspects of the theory of elliptic differential operators*, Proc. ICM-1966 (1968), 57-66, Moscow.

[At]$_{\mathrm{EPDG}}$ M.F. Atiyah, *Elliptic operators, discrete groups and von Neumann algebras*, Asterisque (1976) no. 32-33, 43-72.

[At]$_{\mathrm{SFB}}$ M.F. Atiyah, *The signature of fibre bundles*, Global Analysis, papers in honor of K. Kodaira, Princeton Univ. Press, 1969, 73-84.

[Ba-Do] P. Baum and R. Douglas K-theory and index theory, Operator algebras and applications, , *Proc. Symp. Pure Math.*, 38 (I) (1982), 117-173.

[Bav] C. Bavard, Courbure presque négative en dimension 3, *Comp. Mathematica* 63 (1987), 223-236.

[B-C-G]$_{\mathrm{ER}}$ G. Besson, G. Courtois, S. Gallot, *Entropies et rigidités des espaces localement symétriques de courbure strictement négative*, preprint Institut Fourier, (1994) no. 28.

[B-C-G]$_{\mathrm{VE}}$ G. Besson, G. Courtois, S. Gallot, Le volume et l'entropie minimal des espaces localement symétriques, *Invent. Math.* 103 (1991), 417-445.

[B-C-V] M.E.B. Bekka, P.-A. Cherix and A. Valette, *Proper affine isometric actions of amenable groups*, preprint.

[Bera] P.H. Berard, The Bochner technique revisited, *Bull. AMS*, Vol. 19 (1988), no. 2, 371-406.

[BerBe] L. Bérard-Bergery, *La courbure scalaire des variétés riemanniennes*, Séminaire Bourbaki, 32e année, no. 556, 1979-80.

[B-G-P] Y. Burago, M. Gromov and G. Perelman, A. D. Alexandrov's spaces with curvature bounded below, *Uspekhi Mat. Nauk* 47, 3-51 (1992); English transl. in Russian Math. Surveys 47 (1992).

[Bla] D.E. Blair, *The "total scalar curvature" as a symplectic invariant and related results*, preprint.

[Bu-Gr] P . Buser and D. Gromoll, *Gromov's examples of almost negatively curved metrics on* $\mathbb{S}^3$, notes.

[C-G-M]$_{\mathrm{ppl}}$ A. Connes, M. Gromov and H. Moscovici, Conjecture de Novikov et fibrés prsque plats, *C.R. Acad. Sci. Paris*, t. 310, Série I, 273-277.

[C-G-M]$_{\mathrm{GCLC}}$ A. Connes, M. Gromov and H. Moscovici, *Group cohomology with Lipschitz control and higher signatures*, Geometric and Functional analysis, 3 (1990), 1-78.

[Ca-Pe] G. Carlson and E. Pedersen, *Controlled algebra and the Novikov conjectures for* K*- and* L*-theory*, preprint, 1993.

[Cap] S. Cappell, On homotopy invariance of higher signatures, *Inventiones Math.* 33 (1976), 171-179.

[Ch-Gr]$_{\mathrm{CN}}$ J. Cheeger, M. Gromov, On the characteristic numbers of complete manifolds of bounded curvature and finite volume, in *Diff. Geom. and Complex Analysis*, Rauch Memorial Volume, I. Chavel and H.M. Farkas, eds., Springer, Berlin (1985).

[Ch-Gr]$_{\mathrm{BVND}}$ J. Cheeger, M. Gromov, Bounds on the Von Neumann dimension of L^2-cohomology and the Gauss-Bonnet theorem for open manifolds, *J. Diff. Geom.* 21 (1985), 1-34.

[Ch-Gr]$_{L^2}$ J. Cheeger, M. Gromov, L^2-cohomology and group cohomology, *Topology*, vol. 25, no 2 (1986), 189-215.

[Cho]$_1$ A.W. Chou, *Remarks on the Dirac operator of pseudomanifolds*, preprint.

[Cho]$_2$ A.W. Chou, *The Dirac operator on spaces with conical singularities and positive scalar curvatures*, preprint.

[CdV] Y. Colin de Verdière, *Spectres de variétés riemanniennes et spectres de graphes*, Proc. ICM-1986 in Berkeley, A.M.S. Vol. 1 (1987), 522-530.

[Co-Hi]$_1$ A. Connes and N. Higson, *Déformations, morphismes asymptotiques et K-théorie bivariante*, preprint.

[Co-Hi]$_2$ A. Connes and N. Higson, *Almost homomorphisms and KK-theory*, preprint.

[Co-Mo] A. Connes and H. Moscovici,*Cyclic cohomology, the Novikov conjecture and hyperbolic groups*, Topology, 29, 345 (1990).

[Co-Sk] A. Connes and G. Skandalis, *The longitudinal index theorem for foliations*, Publ. Res. Inst. Math. Sci. Kyoto Univ., 20 (1984), 1139-1183.

[Con]$_{\mathrm{NCG}}$ A. Connes, *Non Commutative Geometry*, Academic Press, 1994.

[Con]$_{\mathrm{CCTF}}$ A. Connes, Cyclic cohomology and the transverse fundamental class of a foliation, in *Geometric methods in Operator Algebras*, H. Araki and E.G. Effros, eds., Pitman, Research Notes in Math. Series 123, 1986, 52-144.

[Dem] J.-P. Demailly, Champs magnétiques et inégalités de Morse pour la d''-cohomology, *Ann. Inst. Fourier* 35, 4 (1985), 189-229.

[Fa-Hs] F.T. Farrell and W.C. Hsiang, On Novikov's conjecture for non-positively curved manifolds, *Ann. Math.* 113 (1981), 197-209.

[Fa-Jo] F.T. Farrell and L.E. Jones, *Rigidity in Geometry and Topology*, Proc. 1990 I.C.M., Kyoto, 1991, 653-663 .

[FC-Sch] D . Fischer-Colbrie and R. Schoen, The structure of complete stable minimal surfaces in 3-manifolds of nonnegative scalar curvature, *Comm. Pure and Appl. Math* 33 (1980), 199-211.

[Fe-We]$_{\mathrm{CAN}}$ S.C. Ferry and S. Weinberger, *A coarse approach to the Novikov conjecture*, preprint, 1993.

[Fe-We]$_{\mathrm{FNAM}}$ S.C. Ferry and S. Weinberger, *A flexible uniformly aspherical manifold*, preprint.

[Gal] S. Gallot, Inégalités isopérimétriques, courbure de Ricci et invariants géométriques, I and II, *C.R. Acad. Sci. Paris* 296 (1983), 333-336 and 365-368.

[Ge-Mi] I.M. Gelfand and A.S. Miščenko, Quadratic forms over commutative group rings and K-theory, *Funkcional. Anal. i Priložen* 3 (1969), 28-33 (Russian).

[G-L-P] M. Gromov, J. Lafontaine and P. Pansu, *Structures métriques pour les variétés riemanniennes*, Cedic-Fernand Nathan, Paris, 1981.

[G-L]$_{\mathrm{PSC}}$ M. Gromov and H.B. Lawson, *Positive scalar curvature and the Dirac operator on complete Riemannian manifolds*, Publ. Math. I.H.E.S., 58 (1983), 83-196.

[G-L]$_{\mathrm{SSC}}$ M. Gromov and H.B. Lawson, Spin and scalar curvature in the presence of a fundamental group, *Ann. of Math.* 111 (1980), 209-230.

[Gol] W. Goldman, *Representations of fundamental groups of surfaces*, Geometry and Topology, Proceedings, Univ. of Maryland 1983-84, J. Alexander and J. Harer eds., Lecture Notes in Math. 1167 (1985) , 95-117.

[Gr-Sh] M. Gromov and M. Shubin, *Von Neumann spectra near zero*, GAFA 1:4(1991), 375-404.

[Gro]$_{\mathrm{AFM}}$ M. Gromov, Almost flat manifolds, *J. Diff. Geom.* 13 (1978), 231-241.

[Gro]$_{\mathrm{AI}}$ M. Gromov, *Asymptotic invariants of infinite groups*, Lond. Math. Soc. Lecture notes 182, ed. Niblo and Koller, Cambridge Univ. Press 1993.

[Gro]$_{\mathrm{CCS}}$ M. Gromov, *Carnot-caratheodory spaces seen from within*, preprint IHES (1994).

[Gro]$_{\mathrm{FPP}}$ M. Gromov, *Foliated plateau problem, parts I, II*, Geometric and Functional Analysis 1:1, 14-79 (1991); 253-320 (1991).

[Gro]$_{\mathrm{FRM}}$ M. Gromov, *Filling Riemannian manifolds*, Journal of diff. Geometry 18, 1-147 (1983).

[Gro]$_{\mathrm{HG}}$ M. Gromov, Hyperbolic groups, in *Essays in group theory*, S.M. Gersten, ed. Math. Sciences Research Institute Publication, 8, Springer-Verlag, New York, Inc., 1987, 75-264.

[Gro]$_{\mathrm{KH}}$ M. Gromov, Kähler hyperbolicity and L_2-Hodge theory, *J. Diff. Geom.* 33 (1991), 263-292.

[Gro]$_{\mathrm{LRM}}$ M. Gromov, *Large Riemannian manifolds*, Lecture Notes in Math., no 1201 (1985), 108-122.

[Gro]$_{\mathrm{MIK}}$ M. Gromov, *Metric invariants of Kähler manifolds*, preprint IHES.

[Gro]$_{\mathrm{PDR}}$ M. Gromov, *Partial differential relations*, Springer-Verlag (1986).

[Gro]$_{\mathrm{PHC}}$ M. Gromov, Pseudoholomorphic curves in symplectic manifolds, *Invent. Math.* 82 (1985), 307-347.

[Gro]$_{\mathrm{PLI}}$ M. Gromov, *Paul Levy's isoperimetric inequality*, Preprint IHES (1980).

[Gro]$_{\mathrm{RTG}}$ M. Gromov, Rigid transformation groups, in *Géométrie Différentielle*, Paris (1986), 65-139; (Bernard, Choquet-Bruhat), Travaux en cours, 33 (1988), Hermann, Paris.

[Gro-Sh] M. Gromov and M. Shubin, Von Neumann spectra near zero, *GAFA* 1:4 (1991), 375-404.

[Gro]$_{\mathrm{Sig}}$ M. Gromov, *Sign and geometric meaning of curvature*, Sem. Mat. e Fis. di Milano, LXI (1991), 10-123, Pavia 9 (1994).

[Gro]$_{\mathrm{VBC}}$ M. Gromov, *Volume and bounded cohomology*, Publ. Math. IHES 56, 5-100 (1982).

[Gro]$_{\mathrm{Wid}}$ M. Gromov,Width and related invariants of Riemannian manifolds, *Astérisque* 163-164, 93-109 (1988).

[Her] J. Hersch, *Quatre propriétés isopérimétriques des membranes sphériques homogènes,* C.R. Acad. Sci. Paris 270 (1970), 1645-1648 .

[Hit] N. Hitchin, *Harmonic spinors*, Adv. in Math. 14 (1974), 1-55.

[Hi-Ro] N. Higson and J. Roe, *A homotopy invariance theorem in coarse cohomology and K-theory*, preprint (1993).

[Hi-Sk] M. Hilsum and G. Skandalis, Invariance par homotopie de la signature à coeffcients dans un fibré presque plat, *Jour. Reine Angew.* Math., 423 (1992), 73-99.

[Hig]$_{\mathrm{KH}}$ N. Higson, *K-homology and operators on non-compact manifolds*, preprint.

[Hij] O. Hijazi, A conformal lower bound for the smallest eigenvalue of the Dirac operator and killing spinors, *Comm. Math. Phys.* 104 (1986), 151-162.

[H-S-U] H. Hess, R. Schrader, D. Uhlenbrock, Kato's inequality and the spectral distribution of Laplacians on compact Riemannian manifold, *J. Diff. Geom.*, 15 (1980), 27-39.

[Hur]$_{\mathrm{CGF}}$ S. Hurder, *Coarse geometry of foliations*, preprint IHES, 1994.

[Hur]$_{\mathrm{EIOI}}$ S. Hurder, Eta invariants and the odd index theorem for coverings, *Cont. Math.*, (1989).

[Hur]$_{\mathrm{EIT}}$ S. Hurder, *Exotic index theory and the Novikov conjecture*, preprint.

[Hur]$_{\mathrm{EITF}}$ S. Hurder, *Exotic index theory for foliations*, preprint, Univ. of Illinois (1993).

[Hur]$_{\mathrm{TCST}}$ S. Hurder, Topology of covers and spectral theory of geometric operators, In *Proc. Conference on K-homology and index theory*, J. Fox, ed., Providence, Amer. Math. Soc. *Contemp. Math.* Vol. 148 (1993).

[Ka-Mi] J. Kaminker and J.G. Miller, Homotopy invariance of the analytic index of signature operators over C^*-algebras, *J. Operator Theory*, 14 (1985), 113-127.

[Kas] G. Kasparov, Novikov's conjecture on higher signatures : The operator K-theory approach, *Contemporary Math.*, Vol. 145 (1993).

[Ka-Sca] G. Kasparov and G. Skandalis, *Groupes "boliques" et conjecture de Novikov*, preprint.

[Katz] M. Katz, The first diameter of 3-manifolds of positive scalar curvature, *Proc. of the Amer. Math. Soc.*, vol. 104 (1988), no 2.

[Ka-Wa] J. Kazdan and F. Warner, Prescribing curvatures, *Proc. Symp. in Pure Math.* 27 (1975), 309-319.

[Kazd] J.L. Kazdan, *Positive energy in general relativity*, Séminaire Bourbaki, 34e année, no 593, Astérisque (1982).

[Ki-Si] R. Kirby and L. Siebenmann, Foundational essays on topological manifolds smoothings and triangulations, *Ann. Math. Stud.* 88 (1977), Princeton.

[La-McD] F. Lalonde and D. McDuff, The geometry of symplectic energy, to appear in *Annals of Math.*

[La-Mi] H.B. Lawson, Jr. and M.-L.Michelsohn, *Spin Geometry*, Vol.38, Princeton Math. Series, Princeton Univ. Press, 1989.

[Lla]$_{\mathrm{SCE}}$ M. Llarull, *Scalar curvature estimates for $(n+4k)$-dimensional manifolds*, preprint.

[Lla]$_{\mathrm{SEDO}}$ M. Llarull, *Sharp estimates and the Dirac operator*, preprint.

[Loh]$_{\mathrm{ChP}}$ J. Lohkamp, Curvature h-principles, to appear in *Ann. Math.*

[Loh]$_{\mathrm{GLC}}$ J. Lohkamp, *Global and local curvatures*, preprint.

[Loh]$_{\mathrm{GNR}}$ J. Lohkamp, *On the geometry of negative Ricci and scalar curvature*, Max-Planck-Institut für Math., Bonn.

[Lot] J. Lott, Heat kernels on covering spaces and topological invariants, *J. Diff. Geom.*, 35 (1992), 471-510.

[Lus] G. Lusztig, Novikov's higher signature and families of elliptic operators, *J. Diff. Geom.* 7 (1971), 229-256.

[Math] V. Mathai, *Non negative scalar curvatures*, preprint, Univ. of Adelaide.

[Mey] W. Meyer, Die Signatur von lokalen Koeffizientensystemen und Faserbündlen, *Bonn. Math. Schr.* 53 (1972).

[M-M] M. Micallef, J. Moore, Minimal two-spheres and the topology of manifolds with positive curvature on totally isotropic two planes, *Ann. Math.* 127:1 (1988), 199-227.

[Mi-Fo] A.S. Mishchenko and A.T. Fomenko, The index of elliptic operators over C^*-algebras, *Math. USSR-Izv.* 15 (1980), 87-112.

[Mi-Hu] J. Milnor and D. Husemoller, *Symmetric bilinear forms*, Springer Verlag, Heidelberg, New York, 1973.

[Mi-St] J. Milnor and J. Stasheff, Characteristic classes, *Ann. Math. Stud.* 76 (1974), Princeton .

[Min]$_{\mathrm{PETL}}$ M. Min-Oo, *A partial extension of a theorem of Llarull to other compact symmetric spaces*, preprint.

[Min]$_{\mathrm{SCR}}$ M. Min-Oo, *Scalar curvature rigidity of asymptotically hyperbolic spin manifolds*, Math. Ann. 285, 527-539 (1989).

[Mis] A.S. Mishchenko, *Infinite dimensional representations of discrete groups and higher signatures*, English transl., Math. USSR-Izv 8, 85-112 (1974).

[NC+] *Novikov conjectures, index theorems and rigidity*, Proc. of an Oberwollfach meeting held on Sept. 1993. To be published by Cambridge Univ. Press, in the London Math. Soc. Lecture Notes series.

[No-Sh] S.P. Novikov, M.A. Shubin, Morse theory and von Neumann invariants on non-simply connected manifolds, *Uspekhi Matem. Nauk*, 41 (1986), 222-223 (in Russian.

[Ono] K. Ono, The scalar curvature and the spectrum of the Laplacian on spin manifolds, *Math. Ann.*, (1988), 163-168.

[Per] G. Perelman, *Spaces with curvature bounded below*, Proc. ICM-1994, Zurich, to appear.

[Pe-Ro-We] E.K Pedersen, J. Roe and S. Weinberger, *On the homotopy invariance of the boundedly controlled analytic signature of a manifold over an open cone*, preprint, 1994.

[Ran]$_{\mathrm{ALT}}$ A. Ranicki, *Algebraic L-theory and topological manifolds*, 102, Cambridge Tracts in Mathematics, Cambridge University Press, 1992.

[Ran]$_{\mathrm{Haup}}$ A. Ranicki, *The Hauptvermutung book*, a collection of papers by Casson, Sullivan, Armstrong, Rourke, Cooke and Ranicki, K-theory Journal book series, to appear.

[Ran]$_{\mathrm{LKLT}}$ A. Ranicki, *Lower K- and L-theory*, London Math. Soc., Lecture Note Series 178, Cambridge University Press.

[Ran]$_{\mathrm{NC}}$ A. Ranicki, *On the Novikov conjecture*, preprint.

[Roe]$_{\mathrm{CCIT}}$ J. Roe, Coarse cohomomology and index theory on complete Riemannian manifolds, *Memoirs of the Amer. Math. Soc.*, 497 (1993), Vol. 104.

[Roe]$_{\mathrm{PNM}}$ J. Roe, *Partitioning non-compact manifolds and the dual Toeplitz problem*, preprint.

[Ros]$_{\mathrm{ANFT}}$ J. Rosenberg, *Analytic Novikov for topologists*, preprint.

[Ros]$_{C^*\mathrm{APS}}$ J. Rosenberg, *C^*-algebras, positive scalar curvature, and the Novikov conjecture*, Part I, IHES Publ. Math. 58, 197-212 (1983), Part III, Top. Vol.25 (1986), no.3, 319-336.

[Ros]$_{\mathrm{KKK}}$ J. Rosenberg, *K and KK : Topology and operator algebras*, preprint.

[Ro-St] J. Rosenberg and S. Stolz, *Manifolds of positive scalar curvature*, Algebraic Topology and its applications (G. Carlsson, R. Cohen, W.-C. Hsiang and J.D.S. Jones eds.), M.S.R.I. Publ., Vol. 27 (1994), Springer, New York, 241-267.

[Sch] R. Schoen, *Minimal manifolds and positive scalar curvature*, Proc. Internat. Congress of mathematicians, Warsaw, 1983, Polish Scientific Publishers and North Holland, Warsaw, 1984, 575-578 .

[Sc-Ya]$_{\mathrm{EIMS}}$ R. Schoen and S.T. Yau, Existence of incompressible minimal surfaces and the topology of manifolds with nonnegative scalar curvature, *Ann. of Math.* 110 (1979), 127-142.

[Sc-Ya]$_{\mathrm{PA}}$ R. Schoen and S.T. Yau, Proof of the positive action conjecture in quantum relativity, *Phys. Rev. Let.* 42 (1979) (9), 547-548.

[Sc-Ya]$_{\mathrm{PM}}$ R. Schoen and S.T. Yau, On the proof of the positive mass conjecture in general relativity, *Comm. Math. Phys.* 65 (1979), 45-76.

[Sc-Ya]$_{\mathrm{SMPS}}$ R. Schoen and S.T. Yau,On the structure of manifolds with positive scalar curvature, *Man. Math.* 28 (1979), 159-183.

[Sto] S. Stolz, to appear in Proc. ICM-1994, Zurich.

[Sul] D. Sullivan, Cycles for the dynamical study of foliated manifolds and complex manifolds, *Inv. Math.* 36 (1976), 225-255.

[Va-Vi] C. Vafa and E. Witten, *Eigenalue inequalities for fermions in gauge theories*, Comm. Math. Phys. 95, 257-276 (1984).

[Var-Sa-Co] N. Varopoulos, L. Saloff-Coste and Th. Coulhon, Analysis and geometry on groups, Cambridge Univ. Press, 1993.

[VDD-Wi] L. van den Dries and A.J. Wilkie, Gromov's Theorem on groups of polynomial growth and elementary logic, *Journal of Algebra* 89 (1984), 349-374.

[Wei]$_{\text{ANC}}$ S. Weinberger, *Aspect of the Novikov conjecture*, preprint.

[Wit] E. Witten, *A New proof of the positive energy theorem*, Comm. Math. Phys., 80 (1981), 381-402.

[Yau] S.-T. Yau, *Minimal surfaces and their role in differential geometry*, Global Riemannian Geometry, T.J. Willmore and N.J. Hitchin, eds., Ellis Horwood and Halsted Press, Chichester, England, and New York, 1984, 99-103 .

[Yu] G. Yu, *Cyclic cohomology and higher indices for noncompact complete manifolds*, preprint (1994).

Institut des Hautes Etudes Scientifiques
35 Route de Chartres
91440 Bures-sur-Yvette, France
and
University of Maryland
College Park, MD 20742, USA

Received May, 1995

Geometric Construction of Polylogarithms, II

Masaki Hanamura[1] *and Robert MacPherson*[1]

Contents

Introduction

The purpose of this paper is to construct *Grassmannian p-cocycles* (and in particular *Grassmannian p-logarithms*) for every positive integer p. A Grassmannian p-cocycle is a collection of holomorphic differential forms, each one on a Grassmannian variety, satisfying a cocyle condition. (See §9 for the precise definition. We refer to [HM-M] and the references there for background on Grassmannian p-cocycles, and for a description of their utility.)

The existence of Grassmannian p-cocycles was conjectured in [BMS]. There have been several partial results toward this conjecture. The classical logarithm function may be viewed as a Grassmannian 1-cocycle, the cocycle condition being the functional equation $\log xy = \log x + \log y$. The classical dilogarithm function, when properly reformulated, gives a Grassmannian 2-cocycle. Grassmannian 3-cocycles were first constructed in [HR-M] using Hodge theoretic methods. "Grassmannian p-cocycles with singularities", i.e. differential forms with singularities satisfying the same cocycle condition, were constructed in [HR] for all p. Polylogarithm functions on real Grassmannians were constructed in [GM] for all p, but it is unknown whether they fit into Grassmannian p-cocycles.

We construct the differential forms constituting the Grassmannian p-cocycles by integration over certain geometric figures called *$\mathcal{P}$-figures.* We

[1]Both authors supported in part by the National Science Foundation

that the technology of $\mathcal{P}$-figures developed here may be useful for constructing differential forms in other contexts. The method of constructing holomorphic forms using $\mathcal{P}$-figures was introduced in [HM-M], where it was used to construct Grassmannian p-cocycles for $p = 1, 2$, and 3. The paper [HM-M] contains an intuitive introduction to $\mathcal{P}$-figures. We recommend that paper to the reader as an introduction to this one.

We will give explicit formulas for the differential forms constituting the Grassmannian p-cocycles as integrals over certain $\mathcal{P}$-figures. The arguments in the case $p \geq 4$ are more involved and require subtler analysis than those in [HM-M]. Because of this, some foundational materials need to be discussed more thoroughly. The self-contained account of the foundations takes §§1-8 of this paper. It contains most of §§1 – 4 of [HM-M].

Let $\mathcal{P}$ be a compact convex polyhedron in real Euclidean space. A $\mathcal{P}$-figure M in a complex projective space $\mathbb{P}^n$ is an assignment of a linear subspace $M(F)$ of $\mathbb{P}^n$ to each face F of $\mathcal{P}$ (including $\mathcal{P}$ itself) such that

(i) the complex dimension of the subspace $M(F)$ is the (real) dimension of F, and

(ii) if $F \subset F'$, then $M(F) \subset M(F')$.

Generalities on $\mathcal{P}$-figures are collected in §1.

Part of this paper (§§1-5) is concerned with defining the integration along the fibers for *admissible* $\mathcal{P}$-figures in general. Let M be a holomorphic family of admissible $\mathcal{P}$-figures in $\mathbb{P}^n$ parametrized by a complex manifold T (see (1.6) for the precise definition). The admissibility is a certain genericity condition on $M(F)$'s with respect to the simplex formed by the coordinate hyperplanes in $\mathbb{P}^n$. Suppose one has a (topological) "cycle" adapted to M, namely a smooth map $f : T \times \mathcal{P} \to \mathbb{P}^n_T$ such that $f(T \times F) \subset M(F)$ for any face F of $\mathcal{P}$. Let the *volume form* on $\mathbb{P}^n$ be the meromorphic form given by

$$von_n = \frac{dy_1}{y_1} \wedge \frac{dy_2}{y_2} \wedge \cdots \wedge \frac{dy_n}{y_n} ,$$

where $y_i = x_i/x_0$ are the affine coordinates associated to the homogeneous coordinates $(x_0, \cdots, x_n)$. Then one may integrate the pull-back of the volume form f^*vol_n along the fibers of the projection $T \times \mathcal{P} \to T$ to obtain a differential form on T. To make this rigorous, one has to deal with the following problems:

(1) The figure may partly lie in the coordinate simplex where vol_n has singularities. (The integral still converges when the figure is admissible.)

(2) One has to show that there is a natural way to produce a "cycle" adapted to M, and that the associated integral is well-defined independent of a (natural) choice.

Taking these into consideration, in §4, we make a more elaborate definition of a "cycle" adapted to a $\mathcal{P}$-figure.

Sections 2 and 3 are technical preliminaries to §4. In §2, we consider smooth manifolds $\mathfrak{X}$ and T, a proper submersive map $\mathfrak{X} \to T$, and a collection of submanifolds $\{\mathfrak{X}_\alpha\}$ of $\mathfrak{X}$. We define the notion of *good crossings* for $\{\mathfrak{X}_\alpha\}$ and show that such a collection gives rise to a stratified C^∞-bundle. §3 is concerned with *a locally linear configuration*, a collection of submanifolds of a complex manifold which, in appropriate local coordinates, looks like a collection of linear subspaces. We study the blow-up of a locally linear configuration, and in particular show an appropriate succession of blow-ups resolves the singularities of the locally linear configuration, namely, makes it of good crossings.

In §4, we construct a space $\mathfrak{X}$ with a map $\mathfrak{X} \to T$; this involves the blow-up of $\mathbb{P}^n$ along a locally linear configuration related to M. The $\mathfrak{X}$ can be considered to be a refinement of $\mathbb{P}^n_T$ in which the collection of submanifolds consisting of $\{M(F)\}$ and the coordinate simplex become good crossings. Using $\mathfrak{X}$, we make the definition of a "cycle" adapted to M. The problem (1) has been solved since the pull-back of vol_n to $\mathfrak{X}$ has no poles on a "cycle". As consequences of §§2 and 3, we show the following: suppose given a "cycle" adapted to M at $p \in T$.

a) For a smooth path $\gamma : [0,1] \to T$, $\gamma(0) = p, \gamma(1) = q$, one can "transport" the given "cycle" to another "cycle" adapted to M at q. This process is unique up to isotopy.

b) One can extend (uniquely up to isotopy) the "cycle" to a neighborhood of p in T.

These reduce the problem (2) to the question of naturally producing a "cycle" at a base point of T.

Section 5 develops the theory of integration along the fibers for admissible $\mathcal{P}$-figures, assuming given a "cycle" adapted to M at a base point of T. The results of §4 allow us to define the integral as a holomorphic form on the simply connected covering on T. It will be shown that the integral is determined only by the isotopy class of the "cycle" at the base point. Also shown is the analyticity of the form generated in this manner. This was used without a proof in [HM-M].

In §6, we show that there is a canonical choice of a cycle adapted to any *real* admissible $\mathcal{P}$-figure which takes values in a connected component of the complement of the coordinate simplex. (A real $\mathcal{P}$-figure is defined similarly to a complex $\mathcal{P}$-figure.) This implies that, given an admissible $\mathcal{P}$-figure M over T, which, at a base point $b \in T$, is real and contained in a connected component, the integral

$$\int [M \mid vol_n]$$

can be canonically defined as a holomorphic differential form on the simply connected covering of T.

Unfortunately, we have to consider not only admissible figures, but also *admissible collections* of figures. An admissible collection of figures consists of several non-admissible figures which, only considered together, can generate a differential form.

An admissible pair of line segments in $\mathbb{P}^2$, for example, consists of a pair of line segments $\vec{PQ}, \vec{PR}$ parametrized by T, with Q, R generic and the common end point P lying in L_0. Here L_i denotes the i-th coordinate line with respect to the homogeneous coordinates of $\mathbb{P}^2$.

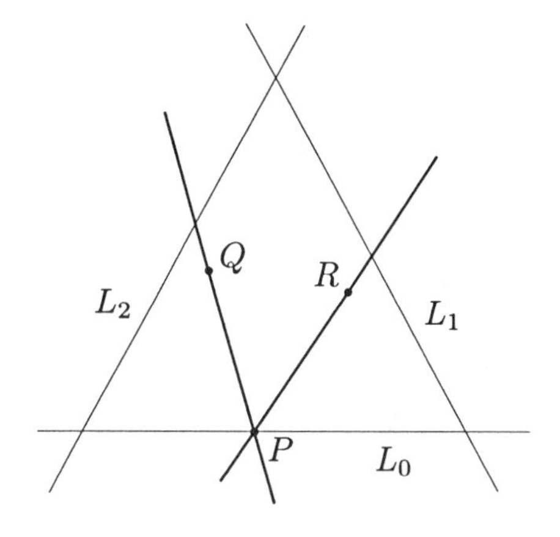

Figure 1

One cannot define integration along the fibers $\int[\vec{PQ}]$ or $\int[\vec{PR}]$ separately, since $\vec{PQ}, \vec{PR}$ are not admissible. One can, however, define a 1-form, which we may denote by

$$\int [\vec{PQ} - \vec{PR}] \ .$$

Consider a family of lines E_ϵ containing $L_0 \cap L_1$ and approaching L_0 as $\epsilon \to 0$. Let $R_\epsilon = \overline{RP} \cap E_\epsilon$; $\quad Q_\epsilon = \overline{QP} \cap E_\epsilon$. The pair of generic line segments $\vec{Q_\epsilon Q}$ and $\vec{R_\epsilon R}$ is the "ϵ-excision" of $\vec{PQ} - \vec{PR}$. We have the convergence of a 1-form on T

$$\int [\vec{Q_\epsilon Q}] - \int [\vec{R_\epsilon R}] \to \int [\vec{PQ} - \vec{PR}] \qquad (\epsilon \downarrow 0) \ .$$

We formulate the notion of admissible collections of figures in general, define the associated integrals and discuss their properties in §7. For an admissible collection of figures M, one can take its ϵ-excision M_ϵ, which is a *generic* figure; one then has the *limit formula*

$$\int [M_\epsilon \mid vol_n] \to \int [M \mid vol_n] \quad (\epsilon \downarrow 0)$$

as a form. Also proven is the *Stokes formula* of the form

$$d \int [M \mid vol_n] = \pm \int [\partial M \mid vol_n] \ ;$$

here ∂M is an admissible collection of figures suitably defined out of the "boundaries" of M. (This is to be made precise in §7.) Along with these formulas, the cancellation lemmas in §8 are the keys to the proof of the cocycle condition in §9. Special cases of the cancellation lemmas appeared in [HM-M].

Finally in §9, we present the construction of admissible collections of figures on Grassmannians. The associated forms are proved to satisfy the required cocycle condition by first taking some excisions and then applying a cancellation lemma.

This work was completed in 1991.

Notation and conventions.

$\mathbb{K} := \mathbb{R}$ or $\mathbb{C}$.

$*$: join of linear subspaces.

$< \ >$: linear span of a set.

$\overset{\circ}{F}$: interior of a manifold or a polyhedron F.

$A \backslash B$: the complement of B in A.

$\tilde{T}$: simply connected covering of T.

1. $\mathcal{P}$-figures

We will define the basic notion of $\mathcal{P}$-figures and parametrized families of $\mathcal{P}$-figures. In (1.7) the genericity and admissibility of a $\mathcal{P}$-figure is defined. These conditions will be important when we consider integration over figures in Sections 4 and 5.

1.1 By definition, a *polyhedron* is a compact convex set in a Euclidean space defined as the intersection of a finite number of closed half spaces; we always equip a polyhedron with an orientation. We will denote a polyhedron by $\mathcal{P}$. The dimension of $\mathcal{P}$ is defined to be the dimension of the linear span of $\mathcal{P}$.

For a polyhedron $\mathcal{P}$, let

$$\mathcal{F}(\mathcal{P}) = \text{ the set of faces other than } \mathcal{P}; \quad \mathcal{F}^+(\mathcal{P}) = \mathcal{F}(\mathcal{P}) \cup \{\mathcal{P}\}.$$

The set $\mathcal{F}^+(\mathcal{P})$ is partially ordered by the inclusion. A face $F \in \mathcal{F}^+(\mathcal{P})$ is itself a polyhedron.

A polyhedron may be given as a subset of a real projective space. In fact, the complement of a hyperplane is a Euclidean space. A polyhedron given as a subset of the complement of a hyperplane in a projective space is referred to as an *embedded* polyhedron.

1.2 Let $\mathbb{P}^n_{\mathbb{K}}$ be projective n-space over $\mathbb{K}$ with homogeneous coordinates $(x_0 : \cdots : x_n)$. The i-th coordinate hyperplane is defined by

$$L_i = \{x_i = 0\},\ i = 0, \ldots, n.$$

More generally, we will also consider a finite collection of hyperplanes $\{N_j\}$. We let, for a finite non-empty set $J = \{j_1, \ldots, j_k\}$,

$$N_J = N_{j_1} \cap \cdots \cap N_{j_k}\ ,$$

and $\mathcal{N} = \{N_J\}_J$. The *support* of $\mathcal{N}$ is defined to be the set

$$|\mathcal{N}| = \cup_j N_j\ .$$

For the collection $\{L_i\}$ of coordinate hyperplanes, we denote $\mathcal{L} = \{L_I\}_I$ where I varies over the subsets of $\{0, \cdots, n\}$; this is the *coordinate simplex.*

1.3 Definition. Let $\mathcal{P}$ be a polyhedron of dimension d. A *$\mathcal{P}$-figure* with values in $\mathbb{P}^n_{\mathbb{K}}$ is an assignment

$$\mathcal{F}^+(\mathcal{P}) \ni F \mapsto M(F),$$

where $M(F)$ is a linear subspace of $\mathbb{P}^n_{\mathbb{K}}$, such that
(1) $\dim_{\mathbb{R}} F = \dim_{\mathbb{K}} M(F)$;
(2) If $F \subset F'$, then $M(F) \subset M(F')$.

1.4 Definition. Let $\{N_j\}$ be a family of hyperplanes in $\mathbb{P}^n_{\mathbb{K}}$ and let M be a $\mathcal{P}$-figure in $\mathbb{P}^n_{\mathbb{K}}$. The *type* of a $\mathcal{P}$-figure M (with respect to $\mathcal{N}$) is the set of numbers $\dim M(F) \cap N_J$, indexed by the pairs (F, J) where F varies over all the faces $\in \mathcal{F}^+(\mathcal{P})$, J over the subsets of the indexing set $\{j\}$.

1.5 Let T be a connected $\mathbb{K}$-analytic manifold, and $\mathbb{P}^n_T := \mathbb{P}^n_{\mathbb{K}} \times T$ with $p : \mathbb{P}^n_T \to T$ the projection. For brevity we denote $\{N_j \times T\}$ still by $\{N_j\}$ and the collection of their intersections by $\mathcal{N}$.

As an example, we may consider the collection of the coordinate hyperplanes $L_i = L_i \times T$.

1.6 Definition. Let $\mathbb{P}^n_T$ and $\mathcal{N}$ be as above. A *relative $\mathcal{P}$-figure M over T with values in $\mathbb{P}^n_T$* (of constant type with respect to $\mathcal{N}$) is a $\mathcal{P}$-figure $M_t : F \mapsto M_t(F)$ in each fiber $\mathbb{P}^n \times \{t\}$ of p such that
(i) The subspace $M_t(F)$ varies $\mathbb{K}$-analytically with respect to t;
(ii) The type of M_t with respect to $\mathcal{N}$ stays constant as t varies in T;
(iii) For faces F, F' such that $M(F), M(F') \subset |\ \mathcal{N}\ |$, $M(F) \cap M(F')$ is equi-dimensional over T.

We set $M(F) = \cup_t M_t(F)$. We let $\mathbb{H} = M(\mathcal{P})$ and call it the *supporting subspace* of M.

Remarks. (a) The notion of a $\mathcal{P}$-figure depends only on the combinatorics of the faces of $\mathcal{P}$, more precisely on the poset $\mathcal{F}^+(\mathcal{P})$ (with each element labeled by its dimension).

(b) The d-simplex Δ^d may be regarded a polyhedron; then we may call a Δ^d-figure a "d-simplex" (or "line segment", "triangle", if $d = 1, 2$), for simplicity.

(c) In the case $\mathbb{K} = \mathbb{R}$, $\mathcal{P}$-figures associated to families of embedded polyhedra play an important role (cf. §6).

1.7 Definition. Let M be a relative $\mathcal{P}$-figure with values in $\mathbb{P}^n_T$ and the supporting subspace $\mathbb{H}$. M is called *generic* (with respect to $\mathcal{N}$) if for each $F \in \mathcal{F}^+(\mathcal{P})$, one has $M(F) \not\subset |\mathcal{N}|$.

We say that M is *admissible* (with respect to $\mathcal{N}$) if for any non-empty subset J and for any proper face F of $\mathcal{P}$, one has

$$N_J \cap \mathbb{H} \neq M(F) .$$

Remark. The admissibility condition appears in [BGSV] in the case $d = n$.

1.8 Let M be a $\mathcal{P}$-figure and $F \in \mathcal{F}^+(\mathcal{P})$. Since $\mathcal{F}^+(F) \subset \mathcal{F}^+(\mathcal{P})$, we obtain a F-figure by restriction; we denote it by $M\,|_F$.

1.9 Although we will not use it, the formulations and the results in §§1–5 are valid as well in the following situation (with obvious modifications in the statements):

T is a connected $\mathbb{K}$-analytic manifold, and $p : \mathbb{P}_T \to T$ is a $\mathbb{K}$-analytic $\mathbb{P}^n_{\mathbb{K}}$-bundle; $\{N_j\}$ is a finite collection of relative hyperplanes of $\mathbb{P}_T$ *of constant type* (over T).

Note, for example, Definition (1.6) makes sense as well under these assumptions.

2. π-distributions and stratified local triviality

2.1 In this section we consider only C^∞- manifolds (possibly with corners) and C^∞- (=smooth) maps between them. Let $\mathfrak{X}$ be a manifold with corners, T another manifold *without corners* and $\pi : \mathfrak{X} \to T$ a surjective C^∞-map. Recall that π is said to be a submersive if for any point $P \in \mathfrak{X}$, the tangential map $(d\pi)_p : T_P\mathfrak{X} \to T_{\pi(P)}T$ is surjective. Let $\dim \mathfrak{X} = n + r, \dim T = r$.

Throughout this section, we are given a collection of closed submanifolds $\{\mathfrak{X}_\alpha\}$ of $\mathfrak{X}$ such that for each $\alpha, \pi\,|_{\mathfrak{X}_\alpha}: \mathfrak{X}_\alpha \to T$ is submersive. In addition, we assume $\mathfrak{X}_\alpha$ is of π-good crossings, a notion to be defined below.

After (2.7), we will also assume π is proper, and show that π can be "trivialized" (including $\mathfrak{X}_\alpha$) along a path in T and in a neighborhood of a point.

Definition. We say that $\{\mathfrak{X}_\alpha\}$ is of *π-good crossings* (or *relative good crossings*) if for any $P \in \mathfrak{X}$, there exist coordinate neighborhoods

$$U \cong \mathbb{R}_+^s \times \mathbb{R}^{n-s} \times \mathbb{R}^r \text{ with coordinates } (x_1, \ldots, x_n, t_1, \ldots, t_r)$$

of P (where $\mathbb{R}_+ = \{x \in \mathbb{R} \mid x \geq 0\}$) and

$$V \cong \mathbb{R}^r \text{ with coordinates } (t_1, \ldots, t_r)$$

of $\pi(P)$ such that

(i) $V = \pi(U)$ and $\pi\ |_U\colon U = \mathbb{R}_+^s \times \mathbb{R}^{n-s} \times \mathbb{R}^r \to V = \mathbb{R}^r$ is the projection to the third factor.

(ii) For any α, there exists a subset $I_\alpha \subset \{1, \ldots, n\}$ such that

$$\mathfrak{X}_\alpha \cap U = \{(x_1, \cdots, x_n, t_1, \cdots, t_r) \mid x_i = 0 \quad \text{for} \quad i \in I_\alpha\}\ .$$

Remark. When $T = pt$ and $\mathfrak{X}_\alpha$ are of codimension one, this is analogous to the notion of "normal crossings" for divisors in a complex manifold.

2.2 Definition. Let $\pi : \mathfrak{X} \to T, \{\mathfrak{X}_\alpha\}$ be as before. A *π-distribution* (or *relative distribution*) on $\mathfrak{X}$ adapted to $\{\mathfrak{X}_\alpha\}$ is a C^∞-subbundle $\mathcal{D}$ of the tangent bundle $T\mathfrak{X}$ of $\mathfrak{X}$ such that $(d\pi)_* : \mathcal{D} \xrightarrow{\sim} T(T)$(= tangent bundle of T) and for each $\alpha, \mathcal{D}\ |_{\mathfrak{X}_\alpha} \subset T\mathfrak{X}_\alpha$. In other words, it is an assignment

$$\mathfrak{X} \ni P \mapsto \mathcal{D}_P \subset T_P\mathfrak{X}$$

where $\mathcal{D}_P$ is a linear subspace (the "horizontal subspace" at P) such that

(i) $(d\pi)_P : \mathcal{D}_P \xrightarrow{\sim} T_{\pi(P)}(T)$.

(ii) $\mathcal{D}_P$ varies smoothly with respect to P.

(iii) $\mathcal{D}_P \subset T_P(\mathfrak{X}_\alpha)$ if $P \in \mathfrak{X}_\alpha$.

Remark. $\mathcal{D}$ induces, for $P \in \mathfrak{X}$, the *horizontal lifting*

$$L_P : T_{\pi(P)}(T) \to T_P(\mathfrak{X})\ .$$

2.3 If $\mathfrak{X} = T \times \mathfrak{X}_0$ ($\mathfrak{X}_0$ is a manifold with corners) and $\pi : \mathfrak{X} \to T$ is the first projection, there is a π-distribution $\mathcal{D}$ given by

$$\mathcal{D}_{(x,t)} = 0 \oplus T_t(T) \subset T_{(x,t)}(\mathfrak{X}_0 \times T)$$

for $(x, t) \in \mathfrak{X}_0 \times T$. We refer to this as the *horizontal* π-distribution.

One may construct π-distributions in the following way. Let $\mathfrak{X} = \cup_\lambda U_\lambda$ be a locally finite open covering, $\{\rho_\lambda\}$ a partition of unity subordinate to it, and $\mathcal{D}_\lambda$ a π-distribution on U_λ. Denoting by $L_{\lambda,P}$ the horizontal lifting corresponding to $\mathcal{D}_\lambda$, the maps

$$\sum_\lambda \rho_\lambda L_{\lambda,P} : T_{\pi(P)}(T) \to T_P(\mathfrak{X})$$

give the horizontal lifting of a π-distribution on $\mathfrak{X}$.

2.4 Let $f : T' \to T$ be a C^∞-map between manifolds without corners. Given a submersive map $\pi : \mathfrak{X} \to T$, one can form the fiber product $\mathfrak{X} \times_T T'$, which we denote by $f^*\mathfrak{X}$. Let $\pi' : f^*\mathfrak{X} \to T'$ be the induced projection. If $\{\mathfrak{X}_\alpha\}$ is a π-good crossing family on $\mathfrak{X}$, so is $\{f^*\mathfrak{X}_\alpha\}$ on $f^*\mathfrak{X}$.

If $\mathcal{D}$ is a π-distribution on $\mathfrak{X}$ adapted to $\{\mathfrak{X}_\alpha\}$, then the pull-back $f^*\mathcal{D}$ is a π'-distribution on $f^*\mathfrak{X}$ adapted to $\{f^*\mathfrak{X}_\alpha\}$.

2.5 Definition. Let π and $\{\mathfrak{X}_\alpha\}$ be as above. Two π-distributions $\mathcal{D}$ and $\mathcal{D}'$ adapted to $\{\mathfrak{X}_\alpha\}$ are said to be *isotopic* if the following holds. There exists a relative distribution D adapted to $\{\mathfrak{X}_\alpha \times I(\sigma)\}$ on $\pi \times id_{I(\sigma)} : \mathfrak{X} \times I(\sigma) \to T \times I(\sigma)$ (here $I(\sigma) = [0,1]$ with σ as the parameter) which restricts to $\mathcal{D}$, $\mathcal{D}'$ on $\mathfrak{X} \times \{0\}$, $\mathfrak{X} \times \{1\}$, respectively. (More precisely, the distribution is to exist on $\mathfrak{X}\times$ (neighborhood of $I(\sigma)$).)
(One may show that isotopy is an equivalence relation.)

2.6 Proposition. *(1) Given a family $\pi : \mathfrak{X} \to T$ and $\{\mathfrak{X}_\alpha\}$ as above, there exists a π-distribution $\mathcal{D}$ adapted to $\{\mathfrak{X}_\alpha\}$.*

(2) Any two π-distributions are isotopic.

Proof. (1) Let $(U; (x_1, \dots, x_n; t_1, \dots, t_r))$ be a coordinate neighborhood as in Definition (2.1). Define a horizontal lifting at a point $P \in U$ by

$$T_{\pi(P)}(T) \to T_P(\mathfrak{X}),$$

$$\Sigma c_j \frac{\partial}{\partial t_j} \mapsto \Sigma c_j \frac{\partial}{\partial t_j}.$$

Here $\{\frac{\partial}{\partial t_j}\}$ is the local frame of $T(T)$ (resp. part of the local frame of $T(\mathfrak{X})$) associated with the local coordinates $(t_1, \cdots, t_r)$ (resp. $(t_1, \cdots, t_r, x_1, \cdots, x_n)$).

Take an open covering of $\mathfrak{X}$ by coordinate neighborhoods, and define the horizontal lifting as above on each of the neighborhoods. As in (2.3), one patches those horizontal liftings together according to a partition of unity to get a global π-distribution adapted to $\{\mathfrak{X}_\alpha\}$.

(2) Let $L_P, L'_P : T_{\pi(P)}(T) \to T_P\mathfrak{X}$ be the horizontal liftings corresponding to $\mathcal{D}, \mathcal{D}'$, respectively. Define a smooth family L_σ ($\sigma \in [0, 1]$) of horizontal

liftings by

$$(L_\sigma)_P := (1-\sigma)L_P + \sigma L'_P.$$

The corresponding π-distributions $\{D_\sigma\}$ form a required D.

2.7 For the rest of this section, we make the following assumption: $\pi : \mathfrak{X} \to T$ and $\{\mathfrak{X}_\alpha\}$ as above. $\mathcal{D}$ is a π-distribution on $\mathfrak{X}$ adapted to $\{\mathfrak{X}_\alpha\}$. In addition, π is a *proper* map.

Given a smooth path $\gamma : [0,1] \to T$ and a point $P \in \mathfrak{X}$ over $\gamma(0)$, there exists a unique smooth path in $\mathfrak{X}$ along $\mathcal{D}$ (the integral curve) which lifts γ.

2.8 Let $\gamma : [0,1] \to T$ be a smooth path with $\gamma(0) = p, \gamma(1) = q$. A *trivialization of* $\pi : \mathfrak{X} \to T$ *over* γ is a diffeomorphism

$$F_\gamma : \gamma^*\mathfrak{X} \xrightarrow{\sim} [0,1] \times \mathfrak{X}_p$$

over $[0,1]$, which sends the family $\{\gamma^*\mathfrak{X}_\alpha\}$ to the family $\{[0,1] \times \mathfrak{X}_{\alpha,p}\}$ ($\mathfrak{X}_{\alpha,p} =$ fiber of $\mathfrak{X}_\alpha$ over p).

Given a π-distribution adapted to $\{\gamma^*\mathfrak{X}_\alpha\}$ on $\gamma^*(\mathfrak{X})$, there is the induced trivialization of $\mathfrak{X}$ over γ by means of the integral curves along the π-distribution. Conversely, a trivialization over γ is induced from a π-distribution on $\pi : \gamma^*\mathfrak{X} \to [0,1]$.

Two trivializations (over γ) F_γ, $F'_\gamma : \gamma^*\mathfrak{X} \xrightarrow{\sim} [0,1] \times \mathfrak{X}_p$ are said to be *isotopic* if there exists a diffeomorphism

$$\gamma^*\mathfrak{X} \times I(\sigma) \xrightarrow{\sim} [0,1] \times \mathfrak{X}_p \times I(\sigma)$$

over $[0,1] \times I(\sigma)$ (where $I(\sigma)$ is the unit interval $[0,1]$ with σ as the parameter) which preserves $\{\mathfrak{X}_\alpha\}$ and restricts to F_γ, F'_γ on the fibers of $\sigma = 0, 1$, respectively.

2.9 Proposition. *(1) There exists a trivialization of π over γ.*

(2) Any two trivializations over γ are isotopic.

Proof. (1) There exists a π-distribution adapted to $\{\mathfrak{X}_\alpha\}$, and any such induces a trivialization of π over γ.

(2) Let F_γ, F'_γ be induced from the π-distributions $\mathcal{D}, \mathcal{D}'$ on $\gamma^*\mathfrak{X}$, respectively. By Proposition (2.6), (2), there exists an isotopy D between $\mathcal{D}$ and $\mathcal{D}'$. The integral curves along D provide an isotopy $F_\gamma \simeq F'_\gamma$.

2.10 Let $\gamma, \gamma' : [0,1] \to T$ be two smooth paths with the initial point p and the end point q. Assume that the two paths are smoothly homotopic, namely there exists a smooth map

$$G : [0,1] \times I(\sigma) \to T$$

satisfying

$$G(\{0\} \times I(\sigma)\ ) = p, \quad G(\{1\} \times I(\sigma)\ ) = q \ ;$$
$$G|_{[0,1]\times\{0\}} = \gamma \ , \quad G|_{[0,1]\times\{1\}} = \gamma' \ .$$

An *isotopy* over G between two trivializations

$$F_\gamma : \gamma^*\mathfrak{X} \xrightarrow{\sim} [0,1] \times \mathfrak{X}_p \ ; F'_{\gamma'} : {\gamma'}^*\mathfrak{X} \xrightarrow{\sim} [0,1] \times \mathfrak{X}_p \ ,$$

is a diffeomorphism over $[0,1] \times I(\sigma)$

$$H : G^*\mathfrak{X} \xrightarrow{\sim} [0,1] \times \mathfrak{X}_p \times I(\sigma)$$

which sends $\{G^*\mathfrak{X}_\alpha\}$ to $\{[0,1] \times \mathfrak{X}_{\alpha,p} \times I(\sigma)\}$, and restricts to F_γ and $F'_{\gamma'}$ over the fibers $\sigma = 0, 1$.

2.11 Proposition. *Given two homotopic smooth paths γ and γ', and trivializations (of $\mathfrak{X}$) $F_\gamma, F'_{\gamma'}$ over them respectively, there exist a homotopy $G : \gamma \simeq \gamma'$ and an isotopy (over G) $H : F_\gamma \simeq F'_{\gamma'}$.*

Proof. Let $G : [0,1] \times I(\sigma) \to T$ be a homotopy $\gamma \simeq \gamma'$ such that:

$$G|_{[0,1]\times\{\sigma\}} = \gamma \quad \text{if} \quad \sigma \in \quad \text{a neighborhood of} \quad [0,1/3];$$
$$G|_{[0,1]\times\{\sigma\}} = \gamma' \quad \text{if} \quad \sigma \in \quad \text{a neighborhood of} \quad [2/3,1] \ .$$

Let $\mathcal{D}$ (resp. $\mathcal{D}'$) be a π-distribution on $\gamma^*\mathfrak{X}$ (resp. ${\gamma'}^*\mathfrak{X}$) corresponding to F_γ (resp. $F'_{\gamma'}$); also take any π-distribution $\bar{\mathcal{D}}$ on $\mathfrak{X}$.

We have

the distribution on $G^*\mathfrak{X}|_{[0,1]\times\{0\leq\sigma\leq1/3\}}$ which restricts to $\mathcal{D}$ over each $[0,1] \times \{\sigma\}$;
the distribution on $G^*\mathfrak{X}|_{[0,1]\times\{2/3\leq\sigma\leq1\}}$ which restricts to $\mathcal{D}'$ over each $[0,1] \times \{\sigma\}$;
the distribution on $G^*\mathfrak{X}|_{[0,1]\times\{1/3\leq\sigma\leq2/3\}}$ which is the pull-back $G^*\bar{\mathcal{D}}$.

By the patching construction (2.3), we can produce a π-distribution on $G^*\mathfrak{X}$ such that

$$D|_{[0,1]\times\{\sigma=0\}} = \mathcal{D}, \quad D|_{[0,1]\times\{\sigma=1\}} = \mathcal{D}' \ .$$

This gives an isotopy $F_\gamma \simeq F'_{\gamma'}$.

2.12 Let $p \in T$. A *trivialization* of $\mathfrak{X}$ *over an open neighborhood* U of p is a diffeomorphism

$$F_U : \mathfrak{X}_U \xrightarrow{\sim} U \times \mathfrak{X}_p$$

over U preserving $\{\mathfrak{X}_\alpha\}$ and restricting to the identity over p.

Two trivializations $F_U, F'_U : \mathfrak{X}_U \xrightarrow{\sim} U \times \mathfrak{X}_p$ are *isotopic* if there exists a diffeomorphism over $U \times I(\sigma)$

$$\mathfrak{X}_U \times I(\sigma) \xrightarrow{\sim} U \times \mathfrak{X}_p \times I(\sigma)$$

which preserves $\{\mathfrak{X}_\alpha\}$ and restricts to F_U, F'_U over the fibers of $\sigma = 0, 1$, respectively.

2.13 Proposition. *(1) There exist a coordinate neighborhood U of p and a trivialization of $\mathfrak{X}$ over U.*

(2) Let U be a polydisk coordinate neighborhood of p. Then any two trivializations over U are isotopic.

Proof. (1) Take a π-distribution $\mathcal{D}$ on $\mathfrak{X}$ adapted to $\{\mathfrak{X}_\alpha\}$. A trivialization F_U is obtained via integral curves along $\mathcal{D}$ which lift radial curves centered at p.

(2) There exist local coordinates $(t_1, \ldots, t_r)$ of T around p such that $U = \{(t_1, \ldots, t_r) \mid \; |t_i| < \epsilon \text{ for all } i\}$ (ϵ a positive number). By replacing (F_U, F'_U) with $(F_U \circ F'^{-1}_U, id)$, we may also assume $\mathfrak{X}_U = U \times \mathfrak{X}_p$, $F'_U : \mathfrak{X}_U \xrightarrow{\sim} U \times \mathfrak{X}_p$ is the identity map, and $F_U : U \times \mathfrak{X}_p \xrightarrow{\sim} U \times \mathfrak{X}_p$ is a diffeomorphism over U whose restriction over p is the identity. Take an increasing smooth function $\rho(\sigma) : (-\infty, \infty) \to \mathbb{R}$ such that $\rho(\sigma) = 0$ for $\sigma \leq 0, \rho(\sigma) = 1$ for $\sigma \geq 1$. Let

$$s_\sigma : U \to U, \quad (t_1, \cdots, t_r) \mapsto (\rho(\sigma)t_1, \cdots, \rho(\sigma)t_r)$$

and $F_\sigma = F_U \circ (s_\sigma \times id) : U \times \mathfrak{X}_p \to U \times \mathfrak{X}_p$. This gives an isotropy $F_U \simeq id$.

3. Locally linear configurations and their blow-ups

3.1 Let $\mathbb{K} = \mathbb{R}$ or $\mathbb{C}, X$ be an analytic manifold of dimension n over $\mathbb{K}$.

Definition. A *locally linear configuration* on X is a finite collection of connected closed submanifolds of X

$$\mathcal{C} = \{F_\alpha\}_{\alpha \in A}$$

satisfying the following property: For any point $P \in X$, there exists a coordinate neighborhood $(U; \varphi)$ such that

(i) $\varphi : U \xrightarrow{\sim} \mathbb{K}^n$ sends P to 0.

(ii) $F_\alpha \cap U$ is mapped to a linear subspace of $\mathbb{K}^n$ for each $\alpha \in A$.

For example, a collection of affine subspaces of $X = \mathbb{A}^n$ is a locally linear configuration.

A member $F_\alpha \in \mathcal{C}$ is called a *face* of $\mathcal{C}$, or an r-face of $\mathcal{C}$ if $\dim_{\mathbb{K}} F_\alpha = r$. $\mathcal{C}$ is *saturated* if for $F_\alpha, F_{\alpha'} \in \mathcal{C}$, any connected component of $F_\alpha \cap F_{\alpha'}$ is also a face of $\mathcal{C}$. Given any locally linear configuration $\mathcal{C}$, its *saturation* $\mathcal{C}^{sat}$ is defined to be the locally linear configuration consisting of all the connected components of finite intersections $F_1 \cap \cdots \cap F_r, F_i \in \mathcal{C}$. Note that $\mathcal{C}$ is a locally linear sub-configuration of $\mathcal{C}^{sat}$.

3.2 Definition. Let $\mathcal{C}$ be a locally linear configuration on X and $P \in X$. $\mathcal{C}$ is said to be of *good crossings* at P if there exists a coordinate neighborhood U of P with coordinates $(x_1, \ldots, x_n)$ such that, for each α,

$$F_\alpha \cap U = \{(x_1, \cdots x_n) \mid x_i = 0 \text{ for } i \in I_\alpha\}$$

for a subset $I_\alpha \subset \{1, \ldots, n\}$. By definition, $\mathcal{C}$ is of *good crossings* (on X) if it is so at any point $P \in X$.

Note that $\mathcal{C}$ is of good crossings if and only if $\mathcal{C}^{sat}$ is.

3.3 For a locally linear configuration $\mathcal{C}$, define its *singular locus* to be the closed set

$$\Sigma = \Sigma(\mathcal{C}) = \{P \in X \mid \mathcal{C} \text{ is } \textit{not} \text{ of good crossings at } P\}.$$

We say that a face $F \in \mathcal{C}$ is *singular* (resp. *regular*) if $F \subset \Sigma$ (resp. $F \not\subset \Sigma$). Equivalently, F is singular (resp. regular) if $\mathcal{C}$ is not of good crossings (resp. of good crossings) at a general point of F.

3.4 Let C be a closed smooth submanifold of X. Then we denote by $B\ell_C$, the blow-up of X along C:

$$\begin{array}{rcl} B\ell_C : \tilde{X} & \to & X \\ \cup & & \cup \\ E & \to & C \end{array}$$

The exceptional divisor E, defined to be $B\ell_C^{-1}(C)$, is isomorphic to the projective bundle $\mathbb{P}(N_{C/X})$ over C, where $N_{C/X}$ is the normal bundle of C in X. For an irreducible subvariety $Z \subset X, Z \not\subset C$, its strict transform is the closure of $B\ell_C^{-1}(Z - C)$.

In the following we will study the behavior of locally linear configurations under blow-ups.

3.5 Let $\mathcal{C} = \{F_\alpha\}_{\alpha \in A}$ be a locally linear configuration on X. Let $F \in \mathcal{C}, \dim_{\mathbb{K}} F = m$, and assume: for any $F_\alpha \in \mathcal{C}$, either $F \subset F_\alpha$ or $F_\alpha \cap F = \emptyset$. Let $\mu : \tilde{X} \to X$ be the blow-up along F. Consider the family (called the pull-back of $\mathcal{C}$ by μ)

$$\mu^{-1}(\mathcal{C}) := \{\tilde{F}_\alpha\}_{\alpha \in A}$$

where $\tilde{F}_\alpha$ is the strict transform of F_α (resp. the exceptional divisor of μ) if $F_\alpha \neq F$ (resp. $F_\alpha = F$). Then

Proposition. *(1) $\mu^{-1}(\mathcal{C})$ is a locally linear configuration on $\tilde{X}$.*

(2) The strict transforms of the $(m+1)$-faces of $\mathcal{C}$ are disjoint in the inverse image of a neighborhood of F.

Remark. Clearly one has a parallel statement in the case where the center of the blow-up is a disjoint union of some F's with $\dim F = m$. We will use the notation $\mu^{-1}(\mathcal{C})$ in this case also.

3.6 Let $\mathcal{C}$ be a locally linear configuration on X and $\mathcal{C}^{sat}$ its saturation. Assume that we are given a saturated sub-configuration Γ of $\mathcal{C}^{sat}$. Then we can form a sequence of blow-ups

$$\tilde{X} = X_{n-1} \xrightarrow{\mu_{n-2}} \cdots \xrightarrow{\mu_1} X_1 \xrightarrow{\mu_0} X_0 = X$$

and a pair of locally linear configurations Γ_k, $\mathcal{C}_k$ on X_k satisfying the following properties:

(i) $\mathcal{C}_0 = \mathcal{C}$; $\Gamma_0 = \Gamma$. Γ_k is a locally linear sub-configuration of $\mathcal{C}_k$.

(ii) The faces of Γ_k are of dimension $\geq k$. The k-faces of Γ_k are disjoint.

(iii) $\mu_k : X_{k+1} \to X_k$ is the blow-up of all the k-faces of Γ_k; this makes sense because of (ii). One has $\mathcal{C}_{k+1} = \mu_k^{-1}(\mathcal{C}_k)$ and $\Gamma_{k+1} = \mu_k^{-1}(\Gamma_k)$ with the notation (3.5).

We denote the composition $\mu_0 \circ \cdots \circ \mu_{n-2} : \tilde{X} \to X$ by $B\ell_\Gamma$ and call it the *blow-up* of X *along* Γ. The locally linear configurations Γ_{n-1}, $\mathcal{C}_{n-1}$ on $\tilde{X}$ are also denoted $B\ell_\Gamma^{-1}(\Gamma)$, $B\ell_\Gamma^{-1}(\mathcal{C})$; the former consists of divisors.

3.7 Proposition. *Let $\mathcal{C}$ be a locally linear configuration on X, and let Γ be a saturated sub-configuration of $\mathcal{C}^{sat}$ containing all the singular faces of $\mathcal{C}^{sat}$. Let $B\ell_\Gamma : \tilde{X} \to X$ be the blow-up along Γ. Then the locally linear configuration $B\ell_\Gamma^{-1}(\mathcal{C})$ on $\tilde{X}$ is of good crossings.*

3.8 For the proof of the above, we need some preliminaries. Let $C \subset \Pi \subset \mathbb{K}^n$ be linear subspaces with $\dim C = m, \dim \Pi = n-1$. We retake the linear coordinates $(x_1, \cdots, x_n)$ so that

$$C = \{x_{m+1} = \cdots = x_n = 0\} = \mathbb{K}^m \times \{0\} \subset \Pi = \{x_n = 0\} \subset \mathbb{K}^n .$$

Let $\mu = B\ell_C : \widetilde{\mathbb{K}}^n \to \mathbb{K}^n$, E be the exceptional divisor, and Π' be the strict transform of Π by μ. Then one can has an isomorphism

$$\widetilde{\mathbb{K}}^n \backslash \Pi' \cong \mathbb{K}^n$$

in which the standard coordinates $(y_1, \cdots, y_n)$ of $\mathbb{K}^n$ and (the pull-backs of)

x_i are related by:

$$\begin{cases} x_i &= y_i \,, \quad 1 \le i \le m \\ x_i &= y_i y_n \,, \quad m+1 \le i \le n-1 \\ x_n &= y_n, \end{cases}$$

and $E \backslash \Pi' = \{y_n = 0\} \subset \mathbb{K}^n$. One shows, with this notation:

Proposition. *For a linear subspace $H \subset \mathbb{K}^n$ with $C \subset H \not\subset \Pi$, let H' be its strict transform. Then $H' \backslash \Pi' \subset \mathbb{K}^n$ is an affine subspace of the form*

$$\mathbb{K}^m \times \bar{H}' \times \mathbb{K} \quad \textit{for an affine subspace} \quad \bar{H}' \subset \mathbb{K}^{n-m-1} \,.$$

(Note $\dim \bar{H}' = \dim H - \dim C - 1$.*)*

The correspondence $H \mapsto \bar{H}'$ respects the inclusions and intersections. (In particular, if $H_1 \cap H_2 = C$, then the corresponding $\bar{H}_1', \bar{H}_2'$ are disjoint.)

The proofs of Propositions (3.5), (3.6) proceed by induction on m, reductions to the local questions, and the applications of the above proposition.

3.9 All of (3.1)–(3.7) can be naturally generalized to a relative situation as follows.

We let $\pi : \mathfrak{X} \to T$ be a submersive surjective map of connected $\mathbb{K}$-analytic manifolds (without boundary) of relative dimension n; let $r = \dim T$. By a *relative locally linear configuration* on $\pi : \mathfrak{X} \to T$ we mean a finite collection of connected closed submanifolds $\mathcal{C} = \{F_\alpha\}_{\alpha \in A}$ of $\mathfrak{X}$ such that

(i) Any non-empty intersection $F_1 \cap \cdots \cap F_r$, $F_i \in \mathcal{C}$ maps submersively onto T (so any connected component of $F_1 \cap \cdots \cap F_r$ has equi-dimensional fibers over T);

(ii) The family $\mathcal{C}$ is a locally linear configuration as defined in (3.1) on each fiber of π.

A relative locally linear configuration $\mathcal{C}$ on $\pi : \mathfrak{X} \to T$ is of π-*good crossings* (or *relative good crossings*) at a point $P \in \mathfrak{X}$ if a relative distinguished coordinate neighborhood $(U; (x_1, \ldots, x_n; t_1, \ldots, t_r))$ of P can be chosen so that, for each $\alpha \in A$,

$$F_\alpha \cap U = \{(x_1, \ldots, x_n; t_1, \ldots, t_r) \mid x_i = 0 \text{ for } i \in I_\alpha\}$$

for a subset $I_\alpha \subset \{1, \ldots, n\}$. $\mathcal{C}$ is of π-*good crossings* on $\mathfrak{X}$ if it is so at any point $P \in \mathfrak{X}$.

One can define saturatedness, the singular locus and the singular faces of a relative locally linear configuration.

By considering relative blow-ups, Proposition (3.5) continues to hold (with the assumption $\dim F = m$ replaced by $\dim(F/T) = m$). In parallel to (3.6), one can consider the relative blow-up of a relative linear configuration, and the obvious analogue of Proposition (3.7) also holds. We will extend the notations such as $B\ell_\Gamma$ to the relative case also.

3.10 *Blow-up of a polyhedron.*

We define the canonical blow-up of a polyhedron. Let $\mathcal{P} \subset \mathbb{R}^n$ be a polyhedron of dimension d. For $F \in \mathcal{F}(\mathcal{P})$, let $< F >$ be the linear span of F. Take an open neighborhood $V \supset \mathcal{P}$ small enough so that $< F > \cap < F' > \cap V = \emptyset$ if $F \cap F' = \emptyset$. Consider then the locally linear configuration $\{V \cap < F >\}_{F \in \mathcal{F}(\mathcal{P})}$ on V. We let $\mu : \hat{V} \to V$ be the blow-up along $\{V \cap < F >\}$, and $\hat{\mathcal{P}}$ = the closure of $\nu^{-1}(\overset{\circ}{\mathcal{P}})$. The restriction of ν to $\hat{\mathcal{P}}$, still denoted $\nu : \hat{\mathcal{P}} \to \mathcal{P}$, is the canonical blow-up of $\hat{\mathcal{P}}$.

Note that $\hat{\mathcal{P}}$ is a manifold with corners (this follows from (3.7)).

4. $\mathcal{P}$-cycles and their transports

4.1 We keep the notation in §1, in particular: $\mathcal{P}$ is an oriented polyhedron of dimension d. T is a connected $\mathbb{K}$-analytic manifold of dimension r, and $\mathbb{P}^n_T := \mathbb{P}^n_{\mathbb{K}} \times T$. $\{N_j\}$ is a family of hyperplanes in $\mathbb{P}^n_{\mathbb{K}}$, which gives rise to a family of relative hyperplanes in $\mathbb{P}^n_T$, still denoted by $\{N_j\}$. M is a $\mathcal{P}$-figure with values in $\mathbb{P}^n_T$ (of constant type with respect to $\{N_j\}$); $\mathbb{H} = \mathbb{H}_T$ is the supporting subspace.

In this section, we will first define the notion of topological cycles adapted to an admissible $\mathcal{P}$-figure M in $\mathbb{P}^n_T$. One might make a naive definition as follows (in the case $T = pt$): A $\mathcal{P}$-cycle adapted to M is a smooth map $f : \mathcal{P} \to \mathbb{P}^n_{\mathbb{K}}$ such that $f(F) \subset M(F)$ for $F \in \mathcal{F}^+(\mathcal{P})$. One would then face the following problems:

(a) The image of f may meet $|\mathcal{N}|$; however, over such a cycle one wants to integrate a differential form having logarithmic singularities along $\mathcal{N}$.

(b) One cannot parallel transport such "cycles" along a path in T; nor can one extend a "cycle" in the fiber of a point $p \in T$ to a neighborhood of p.

(c) One has trouble in showing additivity properties, an example of which is this:

Let, for instance, $M_1 = (M_0, M_1, M_3)$, $M_2 = (M_3, M_1, M_2)$ be two admissible triangles as in Fig. (4.1), and $M = (M_0, M_1, M_2)$ be the "union" of M_1 and M_2. Then one has

$$\int [M_1 \mid \omega] + \int [M_2 \mid \omega] = \int [M \mid \omega] \, .$$

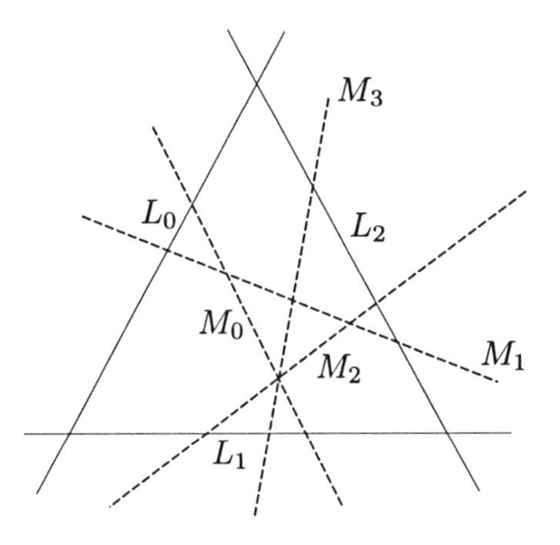

Figure 4.1

To overcome these difficulties, we proceed as follows:

(1) We replace $\mathbb{P}^n_T$ by a certain blow-up $\tilde{\mathbb{H}}_T$ of $\mathbb{H}_T$; then we form $\mathfrak{X} = \tilde{\mathbb{H}}_T \times \hat{\mathcal{P}}$. In $\mathfrak{X}$, the subspaces relevant to M and N_j are of good crossings.

(2) Using $\mathfrak{X}$, we make a definition of $\mathcal{P}$-cycle adapted to M, and isotopy between $\mathcal{P}$-cycles.

(3) For a $\mathcal{P}$-cycle over p, we show that its parallel transport along a curve and its extension to a neighborhood exist (uniquely).

(4) The additivity properties will be formulated and proved in §6.

4.2 A *linear division* $S = \{\mathcal{P}_\lambda\}$ of $\mathcal{P}$ is a finite set of sub-polyhedra $\{\mathcal{P}_\lambda\}$ of dimension d such that $\mathcal{P} = \cup \mathcal{P}_\lambda$ and $\overset{\circ}{\mathcal{P}}_\lambda \cap \overset{\circ}{\mathcal{P}}_{\lambda'} = \emptyset$ if $\lambda \neq \lambda'$ ($\circ$ denotes the interior).

Let $\nu : \hat{\mathcal{P}} \to \mathcal{P}$ be the blow-up of all the faces of codimension ≥ 2 in $\mathcal{P}$ as in (3.10). $\hat{\mathcal{P}}$ is a manifold with corners, and the codimension one faces of $\hat{\mathcal{P}}$ are in one-to-one correspondence with the faces of $\mathcal{P}$ (other than $\mathcal{P}$ itself). For a face F of $\mathcal{P}$, we denote by $\hat{F}$ the corresponding codimension one face of $\hat{\mathcal{P}}$. When a linear division $S = \{\mathcal{P}_\lambda\}$ is given, we let $\hat{\mathcal{P}}_\lambda$ be the closure of $\nu^{-1}(\overset{\circ}{\mathcal{P}}_\lambda)$.

4.3 Let Γ be the relative locally linear configuration on $\mathbb{H}$ consisting of the following relative subspaces:

(1) Those $M(F)$ which are contained in $|\mathcal{N}|$;

(2) Those $N_J \cap \mathbb{H}$ which are contained in $M(F)$ for some proper face F of $\mathcal{P}$;

(3) Those $N_J \cap \mathbb{H}$ along which $\{N_j \cap \mathbb{H}\}_j$ are not of normal crossings.

(Observe that from Definition (1.6) it follows that Γ is a relative locally linear configuration.) We let $\mu : \tilde{\mathbb{H}} \to \mathbb{H}$ be the relative blow-up of Γ. Let $\tilde{M}(F)$ be either the exceptional divisor over $M(F)$ or the strict transform of $M(F)$

(depending on whether $M(F)$ was blown up under μ); similarly let $\tilde{N}_J$ be either the exceptional divisor over, or the strict transform of $N_J \cap \mathbb{H}$. We put $|\tilde{\mathcal{N}}| = \cup_J \tilde{N}_J$.

We set

$$\mathfrak{X} = \tilde{\mathbb{H}} \times \hat{\mathcal{P}}$$

and $\pi : \mathfrak{X} \to T$ be the projection, which factors into the two projections

$$\mathfrak{X} \xrightarrow{q} T \times \hat{\mathcal{P}} \xrightarrow{pr_1} T .$$

For an open subset $U \subset T$, let $\mathfrak{X}_U = \tilde{\mathbb{H}}_U \times \hat{\mathcal{P}} = \pi^{-1}(U)$.

We consider the following collection of closed submanifolds of $\mathfrak{X}$:

$$\mathcal{S} = \{\tilde{M}(F) \times \hat{F}\} \cup \{\tilde{N}_J \times \hat{\mathcal{P}}\}$$

where F varies over the faces of $\mathcal{P}$ and J over the subsets of $\{j\}$. The restriction of $\mathcal{S}$ to a fiber $\mathfrak{X}_p = \pi^{-1}(p)$ $(p \in T)$ is denoted $\mathcal{S}_p$.

4.1 Proposition. *The collection $\mathcal{S}$ on $\mathfrak{X}$ is of π-good crossings.*

Proof. Let $(P, Q) \in \tilde{\mathbb{H}}_T \times \hat{P}$ be any point at which we examine the good crossing property of $\mathcal{S}$. Let $\{\hat{F}_k\}$ be all the codimension faces of $\hat{\mathcal{P}}$ containing Q. By renumbering if necessary, we may assume the corresponding faces $F_j \subset \mathcal{P}$ form a partial flag $F_1 \subset F_2 \subset \cdots \subset F_l$ of $\mathcal{P}$. The members of $\mathcal{S}$ which possibly contain (P, Q) are:

$$\tilde{M}(F_k) \times \hat{F}_k \quad \text{and} \quad \{\tilde{N}_J \times \hat{\mathcal{P}}\} .$$

We apply (the relative version of) Proposition (3.7) to the relative linear configuration

$$\mathcal{C} := \{M(F_k)\}_k \cup \{N_J \cap \mathbb{H}\}_J$$

and its sub-configuration $\mathcal{C} \cap \Gamma$ (check that $\mathcal{C} \cap \Gamma$ is a saturated sub-configuration of $\mathcal{C}^{sat}$ containing all of its singular faces). We find that $\{\tilde{M}(F_k)\}_k \cup \{\tilde{N}_J\}$ is of relative good crossings on $\tilde{\mathbb{H}}$. Since $\{\hat{F}_k\}$ is clearly of good crossings on $\hat{\mathcal{P}}$, we conclude that the family $\{\tilde{M}(F_k) \times \hat{F}_k\} \cup \{\tilde{N}_J \times \hat{\mathcal{P}}\}$ is of relative good crossings at (P, Q).

4.5 By the above proposition, there exists a π-distribution adapted to the family $\mathcal{S}$. From the construction one sees that $\mathcal{D}$ can be taken to be a lift of the horizontal relative distribution on $T \times \hat{\mathcal{P}} \to T$ (2.3); we call such a $\mathcal{D}$ a *special π-distribution.* For a special π-distribution, the induced trivializations in (2.8), (2.12) are both ones over the projection q. Specifically, using the notations introduced there, the following diagrams commute:

$$\begin{array}{ccc} \gamma^*\mathfrak{X} & \xrightarrow{F_\gamma} & [0,1]\times\mathfrak{X}_p \\ {\scriptstyle q}\downarrow & & \downarrow{\scriptstyle q} \\ [0,1]\times\hat{\mathcal{P}} & = & [0,1]\times\hat{\mathcal{P}} \end{array}$$

$$\begin{array}{ccc} \mathfrak{X}_U & \xrightarrow{F_U} & U\times\mathfrak{X}_p \\ {\scriptstyle q}\downarrow & & \downarrow{\scriptstyle q} \\ U\times\hat{\mathcal{P}} & = & U\times\hat{\mathcal{P}} \end{array}$$

4.6 A *locally polyhedral set* F of a manifold with corners M is defined to be a closed subset such that, locally on F, there is a a coordinate neighborhood U of M such that

$$F\cap U=\{\ell_i\geq 0\}$$

for a finite set of affine functions $\{\ell_i\}$ on U. Note that the boundary ∂F of F is a union of locally polyhedral sets. Let $F\subset M$, $F'\subset M'$ be locally polyhedral sets of manifolds with corners. A map $F\to F'$ is said to be *smooth* if it extends to a smooth map from a neighborhood of F into M'. A diffeomorphism $F\to F'$ is a smooth map with a smooth inverse map.

A *locally polyhedral division* of F is a finite collection of locally polyhedral subsets $\{F_\lambda\}$ of F such that $\overset{\circ}{F}_\lambda\cap\overset{\circ}{F}_{\lambda'}=\emptyset$ if $\lambda\neq\lambda'$ and $F=\cup F_\lambda$.

Note, in (4.2), the collection $\{\hat{\mathcal{P}}_\lambda\}$ is a locally polyhedral division of $\hat{\mathcal{P}}$.

4.7 Definition. Let $U\subset T$ be an open subset. A *$\mathcal{P}$-cycle over U adapted to M* is a continuous map over U

$$\alpha: U\times\hat{\mathcal{P}}\to\tilde{\mathbb{H}}_U\backslash|\tilde{\mathcal{N}}|$$

such that

(i) $\alpha(U\times\hat{F})\subset\tilde{M}(F)$ for any face F of $\mathcal{P}$ (other than $\mathcal{P}$ itself);

(ii) There exists a linear division $S:\mathcal{P}=\cup\mathcal{P}_\lambda$ such that for each λ the restriction $\alpha|_{U\times\hat{\mathcal{P}}_\lambda}: U\times\hat{\mathcal{P}}_\lambda\to\tilde{\mathbb{H}}_U\backslash|\tilde{\mathcal{N}}|$ is smooth.

By considering the graph $Z = Z_\alpha \subset \tilde{\mathbb{H}}_U \times \hat{\mathcal{P}}$, one sees that giving a $\mathcal{P}$-cycle α is equivalent to giving a closed subset $Z \subset \tilde{\mathbb{H}}_U \times \hat{\mathcal{P}}$ such that

(i) $Z \cap (|\tilde{\mathcal{N}}| \times \hat{\mathcal{P}}) = \emptyset$;

(ii) $q\,|_Z\colon Z \to U \times \hat{\mathcal{P}}$ is a homeomorphism; moreover, there exists a linear division $S : \mathcal{P} = \cup \mathcal{P}_\lambda$ such that for each λ, $Z_\lambda := (q|_Z)^{-1}(U \times \hat{\mathcal{P}}_\lambda)$ is a locally polyhedral set of $\tilde{\mathbb{H}}_U \times \hat{\mathcal{P}}$ and $q\,|_{Z_\lambda}\colon Z_\lambda \to U \times \hat{\mathcal{P}}_\lambda$ is a diffeomorphism.

(iii) For the boundary ∂Z of Z we have

$$\partial Z \subset \bigcup_F \tilde{M}(F) \times \hat{F}$$

(on the right hand side F varies over all the proper faces of $\mathcal{P}$).

Remark. In [HM-M], the definition of $\mathcal{P}$-cycle was more restricted; we only allowed those maps which satisfy the above conditions with the trivial linear division. The allowance of "piecewise smooth" maps is of use in proving the additivity of the integral $\int[M|\omega]$ with respect to M, and also in the proof of the existence of $\mathcal{P}$-cycles. For the examples of $\mathcal{P}$-figures in [HM-M], "smooth" $\mathcal{P}$-cycles were seen to exist (at the base point), so there was no necessity to consider the more refined definition above.

4.8 Definition. Let $\alpha, \alpha' : U \times \hat{\mathcal{P}} \to \tilde{\mathbb{H}}_U \backslash |\tilde{\mathcal{N}}|$ be two $\mathcal{P}$-cycles/U adapted to M. An *isotopy* between α and α' is a continuous map over $U \times [0,1]$

$$A : U \times \hat{\mathcal{P}} \times [0,1] \longrightarrow (\tilde{\mathbb{H}}_U \backslash |\tilde{\mathcal{N}}|) \times [0,1]$$

satisfying:

(i) There exists a linear division $S : \mathcal{P} = \cup \mathcal{P}_\lambda$ such that the restriction of A to each $U \times \hat{\mathcal{P}}_\lambda \times [0,1]$ is smooth;

(ii) For each $\sigma \in [0,1]$ the restriction of A to the fibers over σ $A_\sigma : U \times \mathcal{P} \to \tilde{\mathbb{H}}_U \backslash |\tilde{\mathcal{N}}|$ is a $\mathcal{P}$-cycle adapted to M;

(iii) $A_0 = \alpha$ and $A_1 = \alpha'$.

4.9 Let $\gamma : [0,1] \to T$ be a smooth path. A trivialization $F_\gamma : \gamma^* \mathfrak{X} \xrightarrow{\sim} [0,1] \times \mathfrak{X}$ over $[0,1] \times \hat{\mathcal{P}}$ induces, by restriction, a diffeomorphism

$$F_{\gamma;p,q} : \mathfrak{X}_p \xrightarrow{\sim} \mathfrak{X}_q$$

over $\hat{\mathcal{P}}$. Given a $\mathcal{P}$-cycle α adapted to M_p with graph $Z_p \subset \mathfrak{X}_p$, let $Z_q \subset \mathfrak{X}_q$ be the image of Z_p under $F_{\gamma;p,q}$; it is a $\mathcal{P}$-cycle adapted to M_q (in view of the second definition). We say that Z_q is *the transport of Z_p along γ*.

In view of Propositions (2.9) and (2.11), we have the following

4.10 Proposition. *The isotopy class of the transport Z_q is uniquely determined by the isotopy class of Z_p and the homotopy class of the path γ.*

(If $[Z_p], [Z_q]$ denote the isotopy classes of Z_p, Z_q, and $[\gamma]$ the homotopy class of γ, then one has "$[Z_q] = [\gamma]_*[Z_p]$".)

4.11 Let $F_U : \mathfrak{X}_U \xrightarrow{\sim} U \times \mathfrak{X}_p$ be a trivialization of $\mathfrak{X}$ over $U \times \hat{\mathcal{P}}$. By means of F_U, a $\mathcal{P}$-cycle $\alpha_p : \hat{\mathcal{P}} \to \tilde{\mathbb{H}}_p$ can be extended to a $\mathcal{P}$-cycle over U $\alpha : U \times \hat{\mathcal{P}} \to \tilde{\mathbb{H}}_U$. In terms of the graphs Z_p, Z_U, one has $Z_U = F_U^{-1}(U \times Z_p)$. Proposition (2.13) implies the following

4.12 Proposition. *Assume that U is a polydisk coordinate neighborhood. Then the isotopy class of an extension α of α_p is uniquely determined by the isotopy class of α_p.*

5. Integration along the fiber for admissible $\mathcal{P}$-figures

5.1 In this section, in addition to (4.1), we make the following assumption: A $\mathcal{P}$-cycle $\alpha_p : \hat{\mathcal{P}} \to \tilde{\mathbb{H}}_p$ is given over a point $p \in T$; Also a $\mathbb{K}$-meromorphic m-form ω on $\mathbb{P}^n$ is given with only logarithmic singularities along $\{N_j\}$.

In the case the divisors are the coordinate hyperplanes $\{L_i\}$, a particular choice for ω is the *volume form*:

$$von_n = \frac{dy_1}{y_1} \wedge \frac{dy_2}{y_2} \wedge \cdots \wedge \frac{dy_n}{y_n},$$

where $y_i = x_i/x_0$ are the affine coordinates associated to the homogeneous coordinates $(x_0, \cdots, x_n)$.

Let U be a neighborhood of p and $\alpha : U \times \hat{\mathcal{P}} \to \tilde{\mathbb{H}}_U$ be an extension of α_p.

We will define the integral over the fibers, $\int [M_U \mid \omega]$, which is an $(m-d)$-form on U. We then show: this form is analytic and well-defined, namely depends only on the isotopy class of α_p, Propositions (5.9) and (5.10).

Combing this with the results in §4, we can define a holomorphic differential form $\int [M \mid \omega]$ on the simply connected covering of T, (5.13).

5.2 Let $\pi : E \to T$ be a C^∞-fiber bundle with fiber F where T is a manifold without boundary, F is a compact manifold with corners, and E is a manifold with corners. We denote by $\mathcal{A}_E^p$ (resp. $\mathcal{A}_T^p$) the sheaf of smooth p-forms on E (resp. T), and $\mathcal{A}_{E/T}^p$ the sheaf of smooth relative p-forms on E. If $m \geq d(= \dim F)$, we have

$$\mathcal{A}_E^m \;/\; \mathcal{A}_E^{d-1} \wedge \pi^* \mathcal{A}_T^{m-d+1} = \mathcal{A}_{E/T}^d \otimes \pi^* \mathcal{A}_T^{m-d} .$$

So any section of $\mathcal{A}_E^m$ induces a section of $\mathcal{A}_{E/T}^d \otimes \pi^* \mathcal{A}_T^{m-d}$.

One has the base change property: given a map $g : S \to T$, letting

$$\begin{array}{ccc} E_S & \xrightarrow{h} & E \\ \downarrow & & \downarrow \\ S & \xrightarrow{g} & T \end{array}$$

be the induced Cartesian diagram, one has $h^* \mathcal{A}^m_{E/T} = \mathcal{A}^m_{E_S/S}$.

5.3 If $\pi : X \to T$ is a submersive map of relative dimension d between $\mathbb{K}$-analytic manifolds, one can define the sheaves of analytic relative forms $\Omega^m_{X/T}$.

A reduced divisor $D = \sum D_i$ (with irreducible components D_i) on X is of *relative normal crossings* if locally on X there are local coordinates $(x_1, \cdots, x_d, t_1, \cdots, t_r)$ (as in (2.1)) such that $D_i = \{x_{\alpha(i)} = 0\}$. In the following we consider more generally D satisfying

5.3.1 Any subdivisor D consisting of at most d irreducible components is of relative normal crossings. Then one defines the sheaf of relative differential forms by

$$\Omega^m_{X/T} < D > = \sum_{D' \subset D} \Omega^m_{X/T} < D' >$$

where D' varies over subdivisors of relative normal crossings.

5.4 Proposition. *Let $i : Y \hookrightarrow X$ be an analytic submanifold, submersive over T such that*

(i) Each $D_i \cap Y$ is a divisor;

(ii) For any subdivisor $\sum D_i$ of D consisting of at most $\dim Y/T$ *components, $\sum (D_i \cap Y)_{red}$ satisfies the condition (5.3.1).*

Then there is a natural restriction map

$$i^* \Omega^m_{X/T} < D > \to \Omega^m_{Y/T} < D \cap Y > \ .$$

5.5 Proposition. *Let C be a submanifold of X (of codimension ≥ 2) such that: if $D_{i_1}, \ldots, D_{i_k}$ are all the irreducible components of D meeting C, then $C \subset D_{i_1} \cap \cdots \cap D_{i_k}$. Let $\mu : \tilde{X} \to X$ be the blow-up along C, D'_i the strict transform of D_i, and E the exceptional divisor. Set*

$$\tilde{D} = \begin{cases} \Sigma D'_i \text{ if } C \subsetneqq D_{i_1} \cap \cdots \cap D_{i_k}; \\ \Sigma D'_i + E \text{ if } C = D_{i_1} \cap \cdots \cap D_{i_k}. \end{cases}$$

Then for $m \geq d$ there is a natural map

$$\mu^* \Omega^m_{X/T} < D > \to \Omega^m_{\tilde{X}/T} < \tilde{D} > \ .$$

5.6 We first recall the formalism of integration along the fibers in the (usual) context of smooth forms; then variants will be discussed; we also fix some notation.

(1) Let $\pi : E \to T$ be a C^∞-fiber bundle with fiber F where T is a manifold without boundary, F an oriented manifold with corners (not necessarily compact), and E a manifold with corners. The boundary ∂E of E is a union of manifolds with corners, and the restriction of π to any of the closed face of ∂E satisfies the same assumption as π. (Note that the fiber of $\partial E \to T$ is ∂F, which is equipped with the induced orientation from F.)

Let φ be a global section of $\mathcal{A}^m_{E/T}$. The integral of φ along π is, by definition, a smooth $(m - \dim F)$-form on T denoted by $\pi_*\varphi$ such that for any $\psi \in \mathcal{A}_T^{\dim E - m}$ with compact support, one has

$$\int_T \pi_*\varphi \wedge \psi = \int_E \varphi \wedge \pi^*\psi$$

(the absolute convergence of the right hand side is part of the condition); it is unique if it exists. For later purposes, the notation

$$\int [E \mid \varphi]$$

in place of $\pi_*\varphi$ will be more convenient.

Note that

— The existence of $\pi_*\varphi$ is a local question on T.

— If $\varphi = \sum \varphi_\lambda$ (locally finite), $\pi_*\varphi_\lambda$ exists and $\sum \pi_*\varphi_\lambda$ converges, then $\pi_*\varphi$ exists and $\pi_*\varphi = \sum \pi_*\varphi_\lambda$.

— $\pi_*\varphi$ exists if *Supp*$(\varphi) \to T$ is a proper map (for example, if F is compact).

(2) We will need a variant of this. We restrict ourselves to the case of the trivial fiber bundle $E = F \times T$; this is not an essential restriction.

We further assume:

(i) F is a locally polyhedral subset (cf. §4) of a manifold with corners N; also given a finite partition $F = \cup F_\lambda$ into locally polyhedral subsets;

(ii) On each F_λ given a smooth m-form φ_λ on F_λ such that $\varphi_\lambda \mid_{F_\lambda \cap F_{\lambda'}} = \varphi_{\lambda'} \mid_{F_\lambda \cap F_{\lambda'}}$ for $\lambda \neq \lambda'$. (Denote this collection of forms by $\varphi = \{\varphi_\lambda\}$)

Then one can ask if $\pi_*\varphi$ exists so that the condition

$$\int_T \pi_*\varphi \wedge \psi = \sum \int_{F_\lambda \times T} \varphi_\lambda \wedge \pi^*\psi$$

holds. It does exist if F is compact. Then $\overset{\circ}{\pi}_*\overset{\circ}{\varphi}$ also exists where $\pi : \overset{\circ}{E} =$

$\overset{\circ}{F} \times T \to T$, $\overset{\circ}{\varphi} = \varphi | \overset{\circ}{F} \times T$, and equals $\pi_* \varphi$:

$$\int [E \mid \varphi] = \int [\overset{\circ}{E} \mid \overset{\circ}{\varphi}] \,. \tag{5.6.1}$$

5.7 Proposition. *In the situation (5.6) (1) or (2), assume further that F is compact. One has the following Stokes formula*

$$d \int [E \mid \varphi] = (-1)^{\dim F + 1} \int [\partial E \mid \varphi] + (-1)^{\dim F} \int [E \mid d\varphi] \,;$$

here by definition

$$\int [\partial E \mid \varphi] = \sum_G \int [G \mid \varphi] \,,$$

the sum being taken over codimension one faces $G \subset F$.

5.8 We now return to the assumptions (5.1). The m-form ω on $\mathbb{P}^n_T$ defines a section of $\Omega^m_{\mathbb{P}^n_T/T} < \mathcal{N} >$, which we denote by $\bar{\omega}$. Applying (5.3), (5.4) to the composition

$$\mu : \widetilde{\mathbb{H}}_T \to \mathbb{H}_T \hookrightarrow \mathbb{P}^n_T$$

we obtain $\mu^* \bar{\omega} \in \Omega^m_{\widetilde{\mathbb{H}}_T/T} < \tilde{\mathcal{N}} >$, which will still be denoted $\bar{\omega}$ for simplicity.

We apply (5.6) (2) to the trivial bundle $U \times \hat{\mathcal{P}} \to U$ and the pull-back $\alpha^* \bar{\omega}$. The assumptions are seen to be met; note that $\{\hat{\mathcal{P}}_\lambda\}$ is a division of $\hat{\mathcal{P}}$ into a locally polyhedral subsets. We can define a smooth $(m-d)$-form on U

$$\int [U \times \hat{\mathcal{P}} \mid \alpha^* \bar{\omega}].$$

We denote it also by $\int [M_U \mid \omega]$, $\int [U \times \hat{\mathcal{P}} \mid \bar{\omega}]$ $\int [\alpha]$, etc.

These integrals have the properties the following propositions state. Proposition (5.9) will be proven later in a more general framework.

5.9 Proposition. *The $(m-d)$-form $\int [U \times \hat{\mathcal{P}} \mid \alpha^* \bar{\omega}]$ is $\mathbb{K}$-analytic.*

5.10 Proposition. *The $(m-d)$-form $\int [U \times \hat{\mathcal{P}} \mid \alpha^* \bar{\omega}]$ is determined only by the isotopy class $[\alpha_p]$ of α over p and by the m-form ω; it is independent of the choice of an extension α of α_p.*

Proof. Let $\alpha, \alpha' : U \times \hat{\mathcal{P}} \to \widetilde{\mathbb{H}}_U \backslash |\tilde{\mathcal{N}}|$ be two $\mathcal{P}$-cycles such that $\alpha_p \simeq \alpha'_p$. By Proposition (4.12) there exists an isotopy $A : U \times \hat{\mathcal{P}} \times I(\sigma) \to \widetilde{\mathbb{H}}_U \backslash |\tilde{\mathcal{N}}|$ between

α and α'. To claim $\int[\alpha] = \int[\alpha']$, we may assume $\dim U = m - d, \dim \tilde{\mathbb{H}}_U = m$. Note

$$\partial\left(U \times \hat{\mathcal{P}} \times I(\sigma)\right) = -(-1)^d U \times \hat{\mathcal{P}} \times \{0\} + (-1)^d U \times \hat{\mathcal{P}} \times \{1\} + \partial\hat{\mathcal{P}} \times I(\sigma)$$

(signs indicating the orientations). As $\dim \tilde{\mathbb{H}}_U = m$, we have $d\bar{\omega} = 0$. We apply Stokes formula (5.7) to get:

$$d\int[U \times \hat{\mathcal{P}} \times I(\sigma)|A^*\bar{\omega}] = -(-1)^{d+1}\int[U \times \hat{\mathcal{P}}|\alpha^*\bar{\omega}]$$
$$+(-1)^{d+1}\int[U \times \hat{\mathcal{P}}|\alpha'^*\bar{\omega}] + (-1)^{d+1}\int[U \times \partial\hat{\mathcal{P}} \times I(\sigma)|A^*\bar{\omega}].$$

We claim that the third term on the right-hand side is zero. This is because $\alpha(U \times \partial\hat{\mathcal{P}} \times I(\sigma)) \subset \bigcup_{F\in\mathcal{F}(\mathcal{P})} \tilde{M}(F)$ and $\dim \tilde{M}(F) < m$, on which the m-form $A^*\bar{\omega}$ restricts to zero.

The left-hand side is also zero. In fact we show $\int[U \times \hat{\mathcal{P}} \times I(\sigma) \mid A^*\bar{\omega}] = 0$. This is a $\mathbb{K}$-analytic $(m-d-1)$-form. Thus we have to prove that its restriction to $U' \subset U$, an arbitrary submanifold of dimension $(m-d-1)$, is zero. For any point $p \in U'$,

$$\alpha(\{p\} \times \hat{\mathcal{P}} \times I(\sigma)) \subset \tilde{\mathbb{H}}_p,$$

hence $\alpha(U' \times \hat{\mathcal{P}} \times I(\sigma))$ is contained in a $\mathbb{K}$-analytic submanifold of dimension $(m-d-1)+d = m-1$. The m-form $A^*\bar{\omega}$ restricts to zero on this submanifold. We are done.

5.11 To prove Proposition (5.8), we consider a slightly more general situation where an inductive argument works.

Let P be a compact C^∞-manifold with corners, $d = \dim P$, and $\mathcal{F}^+(P)$ the set of connected faces of P. Let $U = \mathbb{K}^r$ and $\pi : \mathcal{Y} \to U$ be a submersive map of $\mathbb{K}$-analytic manifolds. Suppose φ is a $\mathbb{K}$-meromorphic section of $\Omega^m_{\mathcal{Y}/U}$ on $\mathcal{Y}$. Let A be an assignment

$$\mathcal{F}^+(P) \ni G \mapsto A(G)$$

where

(i) $A(G)$ is a closed $\mathbb{K}$-analytic submanifold $\subset \mathcal{Y}$, submersive over U, with $\dim_{\mathbb{K}} A(G)/G \leq \dim_{\mathbb{R}} G$;

(ii) If $G \subset G'$, then $A(G) \subset A(G')$.

Note that the hypothesis is inductive; for any $G \in \mathcal{F}^+(P)$, the assignment $\mathcal{F}^+(G) \ni G' \mapsto A(G')$ also satisfies the above conditions, with P replaced by G.

Suppose also given a "P-cycle/U adapted to A", which is by definition a continuous map

$$\alpha : U \times P \to \mathcal{Y}$$

satisfying

(i) $\alpha(U \times G) \subset A(G)$ for $G \in \mathcal{F}^+(P)$;

(ii) For a locally polyhedral division $P = \cup P_\lambda$ of P, the restriction $\alpha|_{U\times P_\lambda} : U \times P_\lambda \to \mathcal{Y}$ is smooth;

(iii) the $\mathbb{K}$-meromorphic form φ is regular in a neighborhood of $\alpha(U \times P)$.

5.12 Proposition. *Under the assumptions (5.11), the $(m-d)$-form*

$$\int [U \times P | \alpha^* \varphi]$$

is $\mathbb{K}$-analytic.

Proof. It is enough to consider the case $\mathbb{K} = \mathbb{C}$. We proceed by induction on d, the case $d = 0$ being obvious.

Let $(t_1, \dots, t_r)$ be coordinates of U and write

$$\varphi = \sum_{|J|=m-d} \psi_J \wedge dt_J \ ,$$

with ψ_J a meromorphic section of $\Omega^d_{\mathcal{Y}/U}$. We have

$$\int [U \times P \mid \alpha^* \varphi] = \sum_J \Big(\int [U \times P \mid \alpha^* \psi_J] \Big) dt_J \ .$$

To show that the coefficients $\int [U \times P \mid \alpha^* \psi_J]$ are holomorphic, we may assume $\varphi = \psi$ is a d-form. In that case, by applying Stokes formula (5.7), we have

$$d \int [U \times P \mid \alpha^* \psi] = (-1)^{d+1} \int [U \times \partial P \mid \alpha^* \psi] + (-1)^d \int [U \times P \mid \alpha^* d\psi].$$

The first term on the right-hand side is a holomorphic 1-form by induction hypothesis. We claim that the second term is zero. For this, we may assume that T is a curve. Then since the holomorphic $(d+1)$-form $d\psi$ restricts to zero on $A(P)$, $\alpha^* d\psi = 0$. This completes the proof.

5.13 Given a base point $b \in T$ and an isotopy class $[Z_b]$ of a $\mathcal{P}$-cycle adapted to M over b, we can produce a $\mathbb{K}$-analytic differential form on the space $\tilde{T}$ of homotopy classes of smooth paths in T with the initial point b ($\tilde{T}$ is a simply connected covering of T).

For a point $\tilde{p} \in \tilde{T}$ which is the homotopy class of a smooth path $\gamma : [0,1] \to T, \gamma(0) = b, \gamma(1) = p$, let Z_p be the parallel transport of Z_b along γ. Take an open neighborhood U of p, and an extension Z_U of Z_p. Let $\int [Z_U \mid \bar{\omega}]$ be the analytic form to be produced. This is well-defined in view of (5.9), since the isotopy class $[Z_p]$ is well-defined. The resulting holomorphic $m-d$-form on $\tilde{T}$ will be denoted by

$$\int [M \mid \omega] .$$

6. Real $\mathcal{P}$-figures and integrations

6.1 In this section, we consider only real figures. We keep the assumptions and notation from (4.1), (5.1) (with $\mathbb{K} = \mathbb{R}$). We denote a small coordinate neighborhood of p by V. We will shrink V around p at our convenience.

6.2 We take a connected component Ω of $\mathbb{P}_T \backslash |\mathcal{N}|$; its closure is denoted $\bar{\Omega}$. If one takes a hyperplane $\Pi \subset \mathbb{P}^n_{\mathbb{R}}$ disjoint from $\bar{\Omega}$, one has $\bar{\Omega} \subset \mathbb{P}^n_{\mathbb{R}} \backslash \Pi \cong \mathbb{R}^n$. (The latter isomorphism is canonical up to an affine transformation of $\mathbb{R}^n$.) Therefore the set $\bar{\Omega}$ has the affine structure induced from that of $\mathbb{R}^n$. This means that for any points $P, Q \in \bar{\Omega}$, there exists a smooth map

$$[0,1] \to \bar{\Omega}, \tau \mapsto \text{“}(1-\tau)P + \tau Q\text{”};$$

the image of which we denote by $[P, Q]$ and call the *line segment* joining P, Q (in $\bar{\Omega}$).

More generally, given $d+1$ points $P_0, \ldots, P_d \in \bar{\Omega}$, one has a unique affine map

$$\Delta^d[v_0, \ldots, v_d] \to \bar{\Omega}$$

such that $v_i \mapsto P_i$. The image of this map is denoted $Hull(P_0, \ldots, P_d)$ and called the *convex hull* of $P_0, \ldots, P_d$. A set of this form is called an *embedded (convex) polyhedron* in $\bar{\Omega}$.

The affine structure depends on the choice of Π; we denote it by $Aff(\Pi)$. But the convex hull is independent of the choice.

We let $\bar{\Omega}_V = \bar{\Omega} \times V \subset \mathbb{P}^n_V$.

6.3 We will show, for an admissible $\mathcal{P}$-figure M which takes values in $\bar{\Omega}$, there is a canonical choice of a cycle adapted to it. We then establish (1) the additivity property of the integral $\int [M]$ with respect to refinements of M and (2) the relationship between the coefficient functions (with respect to local coordinates of T) of $\int [M \mid \omega]$ and the volumes (with respect to certain measures) of some convex domains related to M. Using this, we derive limit formulas and continuity properties of integrals.

We say that a linear division $S = \{\mathcal{P}_\lambda\}$ is *simplicial* if each $\mathcal{P}_\lambda$ is a d-dimensional simplex. The following fact is obvious; we will make use of the proof in (6.6).

6.4 Proposition. *Given a linear division of $\mathcal{P}$, there exists a simplicial linear division refining it. Given two linear divisions of $\mathcal{P}$, there exists a common simplicial refinement.*

Proof. We may only prove the first statement in the case of the trivial division; namely we claim that any polyhedron $\mathcal{P}$ has a simplicial linear division.

For each face F, take a point

$$c_F \in F - \bigcup_{F' \subsetneqq F} F' .$$

To each sequence of faces $F_0 \subset \cdots F_d = \mathcal{P}$ with $\dim F_i = i$, associate the d-simplex $[c_{F_0}, \cdots, c_{F_d}]$. They form a simplicial division of $\mathcal{P}$.

6.5 Definition. Given a linear division $S : \mathcal{P} = \cup \mathcal{P}_\lambda$, set

$$\mathcal{F}^+(\mathcal{P}; S) = \mathcal{F}^+(\mathcal{P}) \cup \bigcup_\lambda \mathcal{F}^+(\mathcal{P}_\lambda) .$$

This is a partially ordered set by inclusion relation.

A *$(\mathcal{P}; S)$-figure M (over V) with values in* $\bar{\Omega}$ is an assignment

$$\mathcal{F}^+(\mathcal{P}; S) \ni F \mapsto M(F)$$

where $M(F)$ is a relative subspace of $\mathbb{P}^n_V$ (of constant type with respect to $\mathcal{N}$) such that

(i) $\dim_{\mathbb{R}} F = \dim_{\mathbb{R}} M(F)$;

(ii) If $F \subset F'$, then $M(F) \subset M(F')$ (We do *not* require that $< F > \subset < F' >$ implies $M(F) \subset M(F')$ where $< F >$ is the span of F);

(iii) For any 0-face F, $M(F) \subset \bar{\Omega}_V$.

Note that M has the underlying $\mathcal{P}$-figure and the $\mathcal{P}_\lambda$-figures, the latter denoted by M_λ. If S is the trivial linear division, a $(\mathcal{P}; S)$-figure in $\bar{\Omega}$ is the same as a $\mathcal{P}$-figure with the additional condition (iii).

We say that M is an *admissible* $(\mathcal{P}; S)$-figure if moreover

(iv) For $F \in \mathcal{F}^+(\mathcal{P}; S)$, and any J, $M(F) \neq N_J \cap M(\mathcal{P})$.

In this case each M_λ is an admissible $\mathcal{P}_\lambda$-figure and M is admissible as a $\mathcal{P}$-figure.

6.6 Proposition. *Let M be a $\mathcal{P}$-figure (over V) with values in $\bar{\Omega}$. After shrinking V around p if necessary, there exists a simplicial linear division S : $\mathcal{P} = \cup \mathcal{P}_\lambda$ and a $(\mathcal{P}; S)$-figure (over V) which refines M.*

Proof. We may assume that S is as in the proof of Proposition (6.4). In this case, note that a face of a $\mathcal{P}_\lambda$ is of the form $[c_{F_{i_0}}, \cdots, c_{F_{i_k}}]$ which is associated to a sequence of faces (of $\mathcal{P}$) $F_{i_0} \subset \cdots \subset F_{i_k}$ with $\dim F_{i_r} = i_r$. Choose, for each $F \in \mathcal{F}^+(\mathcal{P})$, a general family of points $M(c_F)$ so that

$$M(c_F) \subset \big(M(F) - \bigcup_{F' \subsetneqq F} M(F')\big) \cap \bar{\Omega} .$$

Then define

$$M([c_{F_{i_0}}, \cdots, c_{F_{i_k}}]) = M(c_{F_{i_0}}) * \cdots * M(c_{F_{i_k}}) .$$

One sees that its type with respect to $\mathcal{N}$ is constant over V (if V is small enough). This gives a $(\mathcal{P}; S)$-figure which refines M.

We make a choice of Π (and $Aff(\Pi)$); it will be fixed throughout (6.7)–(6.9).

6.7 Definition. Let M be an admissible $\mathcal{P}$-figure over V with values in $\bar{\Omega}$. A $\mathcal{P}$-*cycle* over V *in the strong sense* adapted to M with values in $\bar{\Omega}$ is a continuous map

$$\beta : V \times \mathcal{P} \to \bar{\Omega}$$

satisfying the following conditions for *some* linear division $S = \{\mathcal{P}_\lambda\}$ of $\mathcal{P}$.

(i) For each face $F \in \mathcal{F}^+(\mathcal{P})$, $\beta(V \times F) \subset M(F)$;

(ii) For each $\mathcal{P}_\lambda$, the restriction $\beta|_{V \times \mathcal{P}_\lambda} : V \times \mathcal{P}_\lambda \to \bar{\Omega}$ is smooth;

(iii) If $\mathcal{P}_\lambda$ meets some face $F \in \mathcal{F}(\mathcal{P})$ such that $M(F) \subset |\mathcal{N}|$, then $\mathcal{P}_\lambda$ is a d-simplex Δ^d and the map $\beta|_{V \times \mathcal{P}_\lambda} : V \times \mathcal{P}_\lambda \to \bar{\Omega}$ is an affine map (with respect to $Aff(\Pi)$ on $\bar{\Omega}$) on each each fiber over V.

Definition. Let $\beta, \beta' : V \times \mathcal{P} \to \bar{\Omega}$ be two $\mathcal{P}$-cycles $/V$ in the strong sense adapted to M. An *isotopy* between them is defined to be a continuous map

$$B : V \times \mathcal{P} \times I(\sigma) \to \bar{\Omega}$$

(where $I(\sigma) = [0, 1]$ with parameter σ) satisfying the following conditions for some linear division $S = \{\mathcal{P}_\lambda\}$ of $\mathcal{P}$.

(i) For each $\mathcal{P}_\lambda$, the restriction $B|_{V \times \mathcal{P}_\lambda \times I(\sigma)} : V \times \mathcal{P}_\lambda \times I(\sigma) \to \bar{\Omega}$ is smooth.

(ii) For each $\sigma \in I(\sigma)$, the restriction B_σ of B to the fiber over σ, together with S, satisfying the definition of a $\mathcal{P}$-figure/V in the strong sense adapted to M.

(iii) One has $B_0 = \beta, B_1 = \beta'$.

6.8 Proposition. *Let S be a simplicial linear division of $\mathcal{P}$, and M be a $\mathcal{P}$-figure which can be refined to a $(\mathcal{P}; S)$-figure. Then there exits a $\mathcal{P}$-cycle in the strong sense adapted to M, the conditions in the definition (6.7) being satisfied with the same S.*

Any two $\mathcal{P}$ -cycles in the strong sense are isotopic.

Proof. For $\beta : V \times \mathcal{P} \to \bar{\Omega}$ to be a $\mathcal{P}$-cycle in the strong sense, one must have: for each 0-face $F \in \mathcal{F}^+(\mathcal{P}; S)$, $\beta(V \times F) = M(F)$. For each $\mathcal{P}_\lambda$, one extends this map to an affine map $\beta : V \times \mathcal{P}_\lambda \to \bar{\Omega}$ uniquely using $Aff(\Pi)$.

Let $\beta, \beta' : V \times \mathcal{P} \to \bar{\Omega}$ be two $\mathcal{P}$-cycles in the strong sense. By taking a common refinement, we may assume that the two cycles satisfy the conditions of (6.7) for the same S. Let $B : V \times \mathcal{P} \times I(\sigma) \to \bar{\Omega}$ be the continuous map whose restriction B_σ to the fiber over σ is given by

$$B_\sigma = (1-\sigma)\beta + \sigma\beta'$$

using $Aff(\Pi)$. Then B gives an isotopy $\beta \simeq \beta'$.

6.8.1 Remark. A $\mathcal{P}$-figure M is said to be associated to a family of embedded polyhedra if there is a family of polyhedra $\{\mathcal{P}_t\}$ in $\bar{\Omega}$ such that $M(F) \cap \mathcal{P}_t$ is a face of $\mathcal{P}_t$ and any face of $\mathcal{P}_t$ is of this form. (Notice this allows $\dim \mathcal{P}_t < d$ for some t.)

In this case, one may take S to be the simplicial linear division in (6.4); in the proof (6.6), one may take $M(c_F) \subset \cup_t (M(F) \cap \mathcal{P}_t)^\circ$ so that M_λ are themselves associated to families of embedded polyhedra. The above proof produces a $\mathcal{P}$-cycle in the strong sense $\beta : V \times \mathcal{P} \to \bar{\Omega}$ which induces a homeomorphism $V \times \mathcal{P} \xrightarrow{\sim} \cup \mathcal{P}_t$.

6.9 Proposition. *Let M be an admissible $\mathcal{P}$-figure.*

(1) For a $\mathcal{P}$-cycle /V in the strong sense $\beta : V \times \mathcal{P} \to \bar{\Omega}$, there exists a unique $\mathcal{P}$-cycle /V $\alpha : V \times \hat{\mathcal{P}} \to \tilde{\mathbb{H}}_V$ such that the following diagram commutes:

$$\begin{array}{ccc} V \times \hat{\mathcal{P}} & \xrightarrow{\alpha} & \tilde{\mathbb{H}}_V \\ \nu \downarrow & & \mu \downarrow \\ V \times \mathcal{P} & \xrightarrow{\beta} & \mathbb{H}_V \end{array}$$

(We call α the lift of β.)

(2) Let $\beta, \beta' : V \times \mathcal{P} \to \bar{\Omega}$ be two $\mathcal{P}$-cycles $/V$ in the strong sense and $B : V \times \mathcal{P} \times I(\sigma) \to \bar{\Omega}$ be an isotopy $\beta \simeq \beta'$. Let $\alpha, \alpha' : V \times \hat{\mathcal{P}} \to \tilde{\mathbb{H}}_V$ be the lifts of β, β'. Then there exists an isotopy $A : V \times \hat{\mathcal{P}} \times I(\sigma) \to \tilde{\mathbb{H}}_V$ between α and α' which lifts B.

Proof. (1) For $F \in \mathcal{F}(\mathcal{P})$ such that $M(F) \subset |\mathcal{N}|$, either $F \cap \mathcal{P}_\lambda = \emptyset$ or $F \cap \mathcal{P}_\lambda$ is to be blown up under ν. So $\beta|_{\mathcal{P}_\lambda} : V \times \mathcal{P}_\lambda \to \mathbb{H}_V$ extends to $V \times \hat{\mathcal{P}}_\lambda \to \tilde{\mathbb{H}}_V$. (2) is similar.

6.10 Let M be an admissible $\mathcal{P}$-figure. Take a $\mathcal{P}$-cycle (over V) in the strong sense $\beta : V \times \mathcal{P} \to \bar{\Omega}$, and let $\alpha : V \times \hat{\mathcal{P}} \to \tilde{\mathbb{H}}_V$ be the $\mathcal{P}$-cycle (over V) lifting β. One obtains a real analytic $(m-d)$-form

$$\int [V \times \hat{\mathcal{P}} | \alpha^* \bar{\omega}]$$

on V; by a remark in (5.6), this also equals $\int [V \times \mathring{\mathcal{P}} | \beta^* \bar{\omega}] = \int [V \times \mathring{\mathcal{P}} | \alpha^* \bar{\omega}]$. Once $Aff(\Pi)$ is chosen, this is well-defined independent of the choice of β. It will be shown to be also independent of the choice of $Aff(\Pi)$.

When there is no possible confusion, we denote it by

$$\int [M|\omega] \quad \text{or} \quad \int [M] \,.$$

6.11 Let $(t_1, \ldots, t_r)$ be local coordinates for V which is to be fixed. One can uniquely write

$$\bar{\omega}|_{\mathbb{H}_V} = \sum_{|J|=m-d} \bar{\omega}_J \wedge dt_J \tag{6.11.1}$$

with $\bar{\omega}_J$ a section of $\Omega^d_{\mathbb{H}_V/V} < \mathcal{N} >$. For a point $t \in V$, $\bar{\omega}_{J,t} := \bar{\omega}_J|_{\mathbb{H}_t}$ is a section of $\Omega^d_{\mathbb{H}_t} < \mathcal{N} >$.

Let $\mu_{J,t}$ be the measure which $\bar{\omega}_{J,t}$ defines on $\mathbb{H}_t$. In the case ω is a d-form, denote the measure it defines by $\mu_t (= \mu_{\emptyset,t})$. A measurable subset $A \subset \mathbb{H}_V$ induces a function on V given by

$$V \ni t \mapsto \mu_{J,t}(A \cap \mathbb{H}_t) \,,$$

which we denote $\mu_J(A)$ (or $\mu(A)$ if ω is a d-form).

If $A \subset \mathbb{H}_V$ is a locally closed set, letting $\mathring{A}$ denote the interior of A, one has

$$\mu_J(A) = \int [\mathring{A} - |\mathcal{N}| \mid \omega_J] \,. \tag{6.11.2}$$

6.12 Proposition. *Let M be an admissible Δ^d-figure. Then we have*

$$\int [M|\omega] = \sum_J \pm\mu_J(Hull(M)\ )dt_J\ .$$

where $Hull(M) := Hull\{M(F)|F \quad \text{is a 0-face of} \quad \Delta^d\}$, and the sign before μ_J is $+$ (resp. $-$) if β is orientation preserving (resp. reversing).

Proof. Since β induces a homoemorphism $V \times \mathring{\Delta}^d \xrightarrow{\sim} Hull(M)^\circ$,

$$\begin{aligned}\int [M \mid \omega] &= \int [V \times \mathring{\Delta}^d \mid \beta^*\bar{\omega}] && \text{by (6.10)}\\ &= \pm \int [Hull(M)^\circ \mid \bar{\omega}]\\ &= \sum_J \pm \int [Hull(M)^\circ \mid \bar{\omega}_J]dt_J && \text{by (6.11.1)}\\ &= \sum_J \pm\mu_J(Hull(M)^\circ\ )dt_J && \text{by (6.11.2)}\ .\end{aligned}$$

6.12.1 Remark. More generally the same proof shows:

If M is the admissible $\mathcal{P}$-figure associcated to a family of embedded polyhedra $\{\mathcal{P}_t\}$, then one has

$$\int [M|\omega] = \sum \pm\mu_J(\{\mathcal{P}_t\})dt_J\ .$$

6.13 Proposition. (1) (*additivity*) Let S be a linear division of $\mathcal{P}$, and M be an admissible $(\mathcal{P};S)$-figure (6.5). Then we have

$$\int [M|\omega] = \sum_\lambda \int [M_\lambda|\omega]\ .$$

(2) If, furthermore, S is simplicial, the above is also equal to

$$\sum_{J,\lambda} \pm\mu_J(Hull(M_\lambda)\ )dt_J\ .$$

In particular, the integral is independent of the choice of $Aff(\Pi)$.

Proof. (1) Take $\beta : V \times \mathcal{P} \to \bar{\Omega}$ to be a $\mathcal{P}$-cycle $/V$ in the strong sense which is affine on each $V \times \mathcal{P}_\lambda$ with respect to an affine structure $Aff(\Pi)$. Then,

$$\begin{aligned}\int [M|\omega] &= \int [V \times \mathring{\mathcal{P}} | \beta^* \bar{\omega}] \\ &= \sum_\lambda \int [V \times \mathring{\mathcal{P}}_\lambda | \beta^* \bar{\omega}] = \sum_\lambda \int [M_\lambda \mid \omega] \, .\end{aligned}$$

(2) Applying (6.12) to each $\int [M_\lambda \mid \omega]$, we obtain the assertion.

6.14 Canonical excision.

Let $\mathcal{P} \subset \mathbb{R}^d$ be a polyhedron of dimension d with a distinguished face F_0 of codimension ≥ 2. The set of the faces $\mathcal{F}^+(\mathcal{P})$ can be divided into three disjoint sets according to their *types* defined as follows:

Type I: F such that $F \cap F_0 = \emptyset$.

Type II: F such that $F \cap F_0 \neq \emptyset$ and $F \not\subset F_0$.

Type III: F such that $F \subset F_0$.

Note that the face $\mathcal{P}$ is of type II. We denote these three sets of faces by $\mathcal{F}_I(\mathcal{P})$, $\mathcal{F}_{II}^+(\mathcal{P}) = \mathcal{F}_{II}(\mathcal{P}) \cup \{\mathcal{P}\}$, $\mathcal{F}_{III}(\mathcal{P})$.

Let ℓ be an affine function on $\mathbb{R}^d$ and $D = \{\ell = 0\}$ be the hyperplane it defines. We set

$$D^{>0} = \{\ell > 0\}; \quad D^{\geq 0} = \{\ell \geq 0\}$$

(and similarly $D^{<0}$ and $D^{\leq 0}$) so that $\mathbb{R}^d = D^{>0} \cup D^{<0} \cup D$.

We say that $(D; \ell)$ *excises* $\mathcal{P}$ along F_0 if

(i) $D^{<0}$ contains a neighborhood of F_0 in $\mathcal{P}$;

(ii) D meets each 1-face of type II in a point.

For a face F of type II, let $F^+ = F \cap D^{\geq 0}$ and $F^- = F \cap D^{\leq 0}$. There arises a linear division $\mathcal{P} = \mathcal{P}^+ \cup \mathcal{P}^-$, which is called the *canonical linear division* of $\mathcal{P}$ about F_0. We have:

$$\begin{aligned}\mathcal{F}^+(\mathcal{P}^+) &= \mathcal{F}_I(\mathcal{P}) \cup \{F^+ \mid F \in \mathcal{F}_{II}^+(\mathcal{P})\} \cup \{F \cap D \mid F \in \mathcal{F}_{II}^+(\mathcal{P})\}; \\ \mathcal{F}^+(\mathcal{P}^-) &= \{F^- \mid F \in \mathcal{F}_{II}^+(\mathcal{P})\} \cup \{F \cap D \mid F \in \mathcal{F}_{II}^+(\mathcal{P})\} \cup \mathcal{F}_{III}(\mathcal{P}) \, .\end{aligned}$$

(The unions are disjoint ones.)

Fig.(6.14) illustrates the case where $\mathcal{P}$ is a prism with vertices as labeled, and $F_0 = [v_0, v_0']$; if $F = [v_0, v_1, v_1', v_0']$, F^- is the shaded quadrangle.

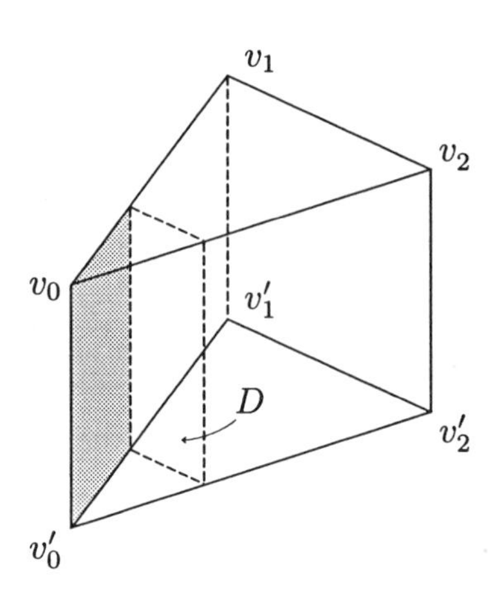

Figure 6.14

6.15 Exhausting family. In the sequel we assume that M is an admissible $\mathcal{P}$-figure satisfying:

(6.15.1) *$M(F_0) \subset N_1$ for some $N_1 \in \{N_j\}$; For a face F, one has $M(F) \subset |\mathcal{N}|$ if and only if $F \subset F_0$.*

Let $\{E_\epsilon\}$ be a real analytic family of hyperplanes in $\mathbb{P}^n_{\mathbb{R}}$ parametrized by $\epsilon \in [0, \epsilon_0)$. Let, for $\epsilon \in (0, \epsilon_0)$,

$$\bar{\Omega} = (\bar{\Omega} \cap E_\epsilon) \cup \bar{\Omega}_\epsilon^{>0} \cup \bar{\Omega}_\epsilon^{<0}$$

be the disjoint partition of $\bar{\Omega}$ which E_ϵ gives rise to so that $\cup \bar{\Omega}_\epsilon^{>0}$ is an open subset of $\bar{\Omega} \times (0, \epsilon_0)$. We say the $\{E_\epsilon\}$ *exhausts M towards N_1 inside $\bar{\Omega}$* if the following holds.

(i) $E_0 = N_1$; $\bar{\Omega}_0^- = \emptyset$, $\bar{\Omega}_0^+ = \bar{\Omega} - N_1$;

(ii) The type of E_ϵ (with respect to $\mathcal{N}$ and M) stays constant as $\epsilon \in (0, \epsilon_0)$ varies, namely, the dimensions of the fibers of $M(F) \cap N_J \cap E_\epsilon$ are constant;

(iii) For any 0-face $F \subset F_0$, and any $\epsilon \in (0, \epsilon_0)$, $M(F) \subset \bar{\Omega}_\epsilon^{<0}$;

(iv) For each 1-face $F \in \mathcal{F}_{II}(\mathcal{P})$, $M(F)$ intersects transversally with E_ϵ and $M(F) \cap E_\epsilon \subset \Omega$.

The following is an admissible Δ^2-figure M, and a family of lines $\{E_\epsilon\}$ which exhausts it.

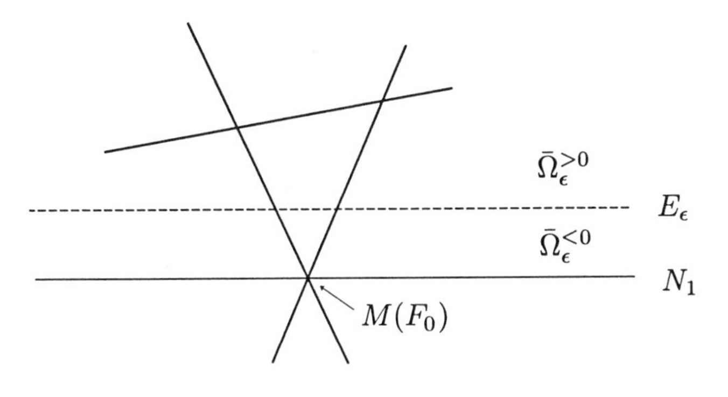

Figure 6.15

6.16 Let $(\mathcal{P}; \mathcal{P}^+ \cup \mathcal{P}^-)$ be the canonical linear division of $\mathcal{P}$ about F_0. Given a $\mathcal{P}$-figure M, an exhausting family $\{E_\epsilon\}$, and $\epsilon \in (0, \epsilon_0)$, we can define a $(\mathcal{P}; \mathcal{P}^+ \cup \mathcal{P}^-)$-figure M_ϵ which refines M as follows. For $F \in \mathcal{F}_{II}(\mathcal{P})$, we let

$$M_\epsilon(F^+) = M_\epsilon(F^-) = M(F);$$
$$M_\epsilon(F \cap D) = M(F) \cap E_\epsilon \ .$$

In particular, we obtain a $\mathcal{P}^-$-figure denoted by M_ϵ^- and a $\mathcal{P}^+$-figure denoted by M_ϵ^+. Note that M_ϵ^+ is a generic $\mathcal{P}^+$-figure and that M_ϵ^- is admissible if M is. In this case the integrals $\int[M]$, etc. are defined, and the following holds:

$$\int [M|\omega] = \int [M_\epsilon^+|\omega] + \int [M_\epsilon^-|\omega] \ .$$

6.17 Proposition (Limit formula). *We have*

$$\int [M_\epsilon^+ \mid \omega] \to \int [M \mid \omega]$$

uniformly on V as an $(m-d)$-form.

6.17.1 Lemma. *Let M be an admissible $\mathcal{P}$-figure over $V \times (0, \epsilon_0]$ such that*

(i) the condition (6.15.1) is satisfied;

(ii) it is associated to a family of embedded polyhedra $\{\mathcal{P}_{\epsilon,t}\}$ $(t \in V)$ such that if $\epsilon > \epsilon'$, then $\mathcal{P}_{\epsilon,t} \supset \mathcal{P}_{\epsilon',t}$;

(iii) for a 0-face F of $\mathcal{P}$ and $t \in V$, the corresponding $M_{\epsilon,t}(F)$ approaches to a point in N_1.

(In words, M_ϵ is a $\mathcal{P}$-figure monotonely shrinking to N_1 as $\epsilon \downarrow 0$.)

Then one has (letting M_ϵ be the restriction of M to $\{\epsilon\} \times V$): $\int[M_\epsilon \mid \omega] \to 0$.

[This lemma follows from Remark (6.12.1).]

Proof of Proposition. Let $S = \{\mathcal{P}_\lambda\}$ be a simplicial linear division of $\mathcal{P}$, and M still denote an admissible $(\mathcal{P}; S)$-figure refining M. Let $(D; \ell)$ be a pair as in (6.13); it gives rise to a linear division which refines S. For $\mathcal{P}_\lambda$ such that $\mathcal{P}_\lambda \cap F_0 \neq \emptyset$, the $(D; \ell)$ gives a canonical excision of $\mathcal{P}_\lambda$ about $\mathcal{P}_\lambda \cap F_0$. In particular, we have the $\mathcal{P}_\lambda^\pm$-figures $M_{\lambda,\epsilon}^\pm$.

By (6.13), one has

$$\int [M_\epsilon^-] = \sum_\lambda \int [M_{\lambda,\epsilon}^-] \,.$$

We have only to show each $\int[M_{\lambda,\epsilon}^-] \to 0$. Thus we may assume $\mathcal{P} = \Delta^d$.

If ϵ is small enough, the figure M_ϵ^- is associated to the embedded polyhedron $E_\epsilon^- \cap Hull(M)$. By Lemma (6.17.1), one has $\int[M_\epsilon^- | \omega] \to 0$.

6.18 Proposition (Stokes formula for generic figures). *Let M be a generic $\mathcal{P}$-figure with values in Ω. Define*

$$\int [\partial M \mid \omega] = \sum_F \int [M_F \mid \omega]$$

where F varies over the codimension one faces of $\mathcal{P}$ (with the induced orientations) and M_F is the F-figure obtained by restriction. Then we have

$$d \int [M \mid \omega] = (-1)^{\dim \mathcal{P}+1} \int [\partial M \mid \omega] + (-1)^{\dim \mathcal{P}} \int [M \mid d\omega] \,.$$

Proof. By Stokes formula (5.7), the exterior differential of $\int[M \mid \omega] = \int[V \times \hat{\mathcal{P}} \mid \alpha^* \bar{\omega}]$ is equal to:

$$(-1)^{\dim \mathcal{P}+1} \int [V \times \partial\hat{\mathcal{P}} \mid \alpha^* \bar{\omega}] + (-1)^{\dim \mathcal{P}} \int [V \times \hat{\mathcal{P}} \mid d\, \alpha^* \bar{\omega}]$$

Let F be a face of $\mathcal{P}$ and $\hat{F}$ be the corresponding codimension one face of $\hat{\mathcal{P}}$. If F is of codimension one, then ν restricts to the canonical blow-up of F: $\hat{F} \to F$; hence

$$\int [V \times \hat{F} \mid \alpha^* \bar{\omega}] = \int [M_F \mid \omega].$$

Assume now that codim $F \geq 2$. We claim $\int[V \times \hat{F} \mid \alpha^*\bar{\omega}]$ is zero. For this, we may assume that $\dim V = m - d + 1$. Since $\dim M(F) = \dim T + \dim F < m$, $\bar{\omega}$ restricts to zero on $M(F)$ and so also on $\tilde{M}(F)$, hence the claim.

6.19 Proposition. *(1) If the restriction of ω to the supporting subspace $M(\mathcal{P})$ is zero, then $\int[M \mid \omega] = 0$.*

(2) If $M(\mathcal{P})$ is contained in $Y \times V$ for a hyperplane $Y \subset \mathbb{P}^n_{\mathbb{R}}$, and ω is the pull-back of a form on $\mathbb{P}^n_{\mathbb{R}}$, then $\int[M \mid \omega] = 0$.

Proof. (1) is trivial. (2) is a special case of (1).

6.20 Let $\{M_\epsilon\}$ be a real analytic family of admissible $\mathcal{P}$-figures over V of constant type (in $\bar{\Omega}$), parametrized by $\epsilon \in [0, \epsilon_0)$. In other words, we assume that an admissible $\mathcal{P}$ -figure M on $V \times [0, \epsilon_0)$ of constant type in $\bar{\Omega}$ is given; let $M_\epsilon = M\,|_{V \times \{\epsilon\}}$. Then $\int[M \mid \omega]$ is a real analytic $(m-d)$-form on $V \times [0, \epsilon_0)$, which restricts to $\int[M_\epsilon \mid \omega]$ over $V \times \{\epsilon\}$. Hence follows:

Proposition (continuity). *Under the above assumptions,*

$$\int [M_\epsilon \mid \omega] \xrightarrow{\epsilon \downarrow 0} \int [M_0 \mid \omega]$$

uniformly on V as an $(m-d)$-form.

6.21 Let $\{D_\epsilon\}_{\epsilon \in [0,\epsilon_0)}$ be a real analytic family of hyperplanes in $\mathbb{P}^n_{\mathbb{R}}$. Let $\mathcal{P}_1$ be a polyhedron of dimension $d-1$ and $\mathcal{P} = \mathcal{P}_1 \times [0,1]$. Consider a generic $\mathcal{P}$-figure with values in Ω such that $M(\mathcal{P}_1 \times \{0\}) = D_0$, and such that for a face $F \subset \mathcal{P}_1$ and $\epsilon \in [0, \epsilon_0)$, $M(F \times [0,1])$ meets D_ϵ transversally.

Let M_ϵ be the $\mathcal{P}$-figure such that $M_\epsilon(F) = M(F)$ if $F \not\subset \mathcal{P}_1 \times \{0\}$ and $M(F \times [0,1]) \cap D_\epsilon$ if $F \subset \mathcal{P}_1$.

Proposition (continuity). *Under the above assumptions, M_ϵ is a generic $\mathcal{P}$-figure in Ω and*

$$\int [M_\epsilon \mid \omega] \to \int [M \mid \omega] \quad (\epsilon \downarrow 0)\,.$$

Proof. Let D_0 be the convex hull of $M(F)$'s for 0-faces $F \subset \mathcal{P}_0$. Let $N_\rho(D_0)$ be the ρ-neighborhood (in a metric on Ω) of D_0.

For each ϵ, the "difference" of M and M_ϵ is a $\mathcal{P}_1 \times [0,1]$-figure M'_ϵ which takes values in $N_{\rho(\epsilon)}(D_0)$. Here $\rho(\epsilon)$ is a non-negative increasing function such that $\rho(0) = 0$.

Take simplicial linear divisions of $\mathcal{P}_1$ and $\mathcal{P}_1 \times [0,1]$, and consider the corresponding refinement of M'_ϵ into a sum of d-"simplices" $M'_{i,\epsilon}$, $i = 1, \cdots, \ell$. One has

$$\int [M'_\epsilon \mid \omega] = \sum \pm \mu_J(M'_{i,\epsilon}) dt_J$$

and $\mu_J(M'_{i,\epsilon}) \to 0$ $(\epsilon \downarrow 0)$ since $\mu_J(N_{\rho(\epsilon)}(D_0)) \to 0$.

7. Admissible collections of figures and their integrations

7.1 In this section, we make the following assumptions on the parameter space. $T_{\mathbb{R}}$ is a real analytic manifold with a base point b; T is a complexification of $T_{\mathbb{R}}$. Let V be a small coordinate neighborhood of b in $T_{\mathbb{R}}$, which may be shrunk around b at our convenience. W denotes a connected and relatively compact open neighborhood of b in T containing V. For any smooth path $\gamma : [0,1] \to T$ with $\gamma(0) = b$, one can take W so that $W \supset \gamma$.

$$\begin{array}{ccccc} & T_{\mathbb{R}} & \subset & T \\ & \cup & & \cup \\ b \in & V & \subset & W \end{array}$$

We assume we are given a collection of real hyperplanes $\{N_j\}$ of $\mathbb{P}^n_{\mathbb{R}}$, and a *closed* real meromorphic m-form ω on $\mathbb{P}^n_{\mathbb{R}}$ with logarithmic singularities along $\{N_j\}$. One of them, say N_1, will play a distinguished role in the following. By the same $\{N_j\}$ we denote the induced real or complex (relative) hyperplane of $\mathbb{P}^n_{\mathbb{C}}$, $\mathbb{P}^n_V = \mathbb{P}^n_{\mathbb{R}} \times V$, or $\mathbb{P}^n_W = \mathbb{P}^n_{\mathbb{C}} \times W$; the pull-backs of ω to these spaces are all denoted by the same ω. The closedness of ω implies, in Stokes formula (5.7), the second term on the right hand side is zero. Integrations along the fibers will be always with respect to ω.

By Ω we denote a connected component of $\mathbb{P}^n_{\mathbb{R}} \backslash |\mathcal{N}|$. We let $\Omega_V = \Omega \times V$, etc.

In this section, we define admissible collections of figures, or admissible $\mathcal{C}$-figures and show that there are associated a well-defined integrals. We show the limit formula (7.9), which approximates the integral of a $\mathcal{C}$-figure by the integral of a generic figure. The Stokes formulas for admissible figures and for admissible $\mathcal{C}$-figures, (7.10), (7.10), both involve (related) admissible $\mathcal{C}$-figures. The integral of an admissible $\mathcal{C}$-figure has the additivity property (7.12) similar to the one for an admissible figure.

7.2 We keep the notation from the previous section, (6.14), (6.15). $\mathcal{P}$ is a polyhedron of dimension d, $F_0 \subset \mathcal{P}$, etc. Let

$$\mathcal{C} = \mathcal{C}(\mathcal{P}; F_0) = \Big(\bigcup_{F \in \mathcal{F}_{II}(\mathcal{P}), F \supset F_0, F \neq \mathcal{P}} \mathcal{F}^+(F) \Big) \cup \{ F^- \mid F \in \mathcal{F}_{II}(\mathcal{P}), F \not\supset F_0 \} \ ;$$

$$\mathcal{C}_D = \mathcal{C}_D(\mathcal{P}; F_0) = \mathcal{C}(\mathcal{P}; F_0) \cup \{F \cap D \mid F \in \mathcal{F}_{II}(\mathcal{P})\}.$$

Note that a face $\in \mathcal{C}$ (or $\mathcal{C}_D$) is of dimension $\leq d-1$.

7.3 Definition. An *admissible* $\mathcal{C}$*-figure* $M = M_{\mathcal{C}}$ *over* W *with values in* $\bar{\Omega}$ consists of

(a) a relative subspace $\mathbb{H} \subset \mathbb{P}^n_W$ (called the *supporting subspace*) of constant type, and

(b) an assignment $\mathcal{C} \ni F \mapsto M(F) \subset \mathbb{P}^n_W$ of relative subspaces subject to the following conditions:

(i) $\dim_{\mathbb{C}} \mathbb{H}/W = d-1$ or d. The restriction $\mathbb{H} \times_W V$ is defined over $\mathbb{R}$, namely the complexification of the real subspace $\mathbb{H} \cap \mathbb{P}^n_V$ is $\mathbb{H} \times_W V$. The type of $\mathbb{H}$ with respect to $\mathcal{N}$ is constant over W.

(ii) For $F \in \mathcal{C}$ we have: $M(F) \subset \mathbb{H}$; $\dim_{\mathbb{R}} F = \dim_{\mathbb{C}} M(F)/W$; if $F \subset F'$, then $M(F) \subset M(F')$; $M(F) \times_W V$ is defined over $\mathbb{R}$; If F is a 0-face, then $M(F) \subset \bar{\Omega}$.

(iii) We have $M(F_0) \subset N_1$. For a face $F \in \mathcal{C}$, $F \subset |\mathcal{N}|$ if and only if $F \subset F_0$.

(iv) The type of M is constant over W, namely an intersection $M(F) \cap N_J$ has equidimensional fibers over W, where F varies over $\mathcal{F}^+(\mathcal{P})$, $J \subset \{j\}$.

(v) There exists a component N_2 of $\{N_j\}$ such that for any 0-face $F \in \mathcal{C}$, $M(F) \not\subset N_{12} (:= N_1 \cap N_2)$.

(vi) (admissibility) Let $F \in \mathcal{C}$ be any face in the case $\dim \mathbb{H}/V = d$, and any face $\not\supset F_0$ in the case $\dim \mathbb{H}/V = d-1$; let J be any subset. Then one has: $M(F) \neq N_J \cap \mathbb{H}$.

For $F \in \mathcal{F}_{II}(\mathcal{P}), F \supset F_0$, the restriction $M_{\mathcal{C}} \mid_F$ is a F-figure which is not necessarily admissible. In §§8 and 9, the following notation will also be useful:

$$M_{\mathcal{C}} = \sum_{F \in \mathcal{F}_{II}(\mathcal{P}), F \supset F_0} M_{\mathcal{C}} \mid_F$$

For a further justification of this notation, see Proposition (7.9).

An admissible $\mathcal{C}_D$-figure (over $\cdots$) can be defined similarly using $\mathcal{C}_D$ in place of $\mathcal{C}$.

7.4 Examples. Let $\mathcal{P}, F_0$ be as in Figure (6.14). A $\mathcal{C}$-figure with values in $\mathbb{P}^3$ consists of:

a) a pair of $[0,1]^2$-figures (or "quadrangles") $M \mid_{[v_0, v_1, v_1', v_0']}$, $M \mid_{[v_0, v_2, v_2', v_0']}$, and

b) a pair of planes $M([v_0, v_1, v_2]), M([v_0', v_1', v_2'])$

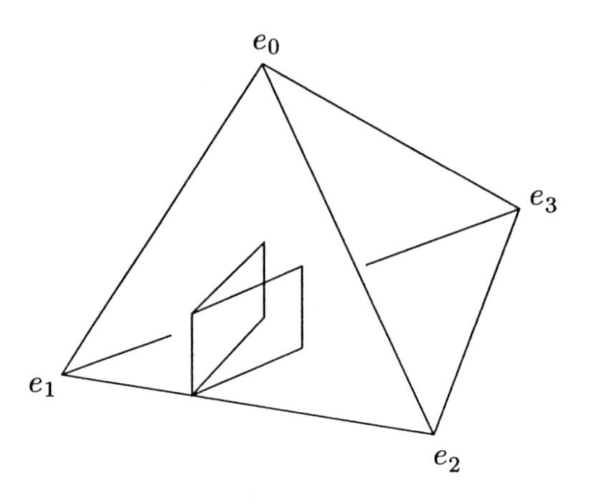

Figure 7.4.1

such that $M([v_0, v_1, v_2]) \supset M([v_0, v_1]), M([v_0, v_2])$ and $M([v_0', v_1', v_2'] \supset M([v_0', v_1']), M([v_0', v_2'])$.

One may call an admissible $\mathcal{C}$-figure a pair of "quadrangles". Fig. (7.4.1) shows what M looks like if $\mathbb{H}_W = \mathbb{P}^3_W$ (case $\dim \mathbb{H}/W = d$). If M is supported on a plane $\mathbb{H}_T$ (case $\dim \mathbb{H}/W = d-1$), then M looks as in Figure 7.4.2.

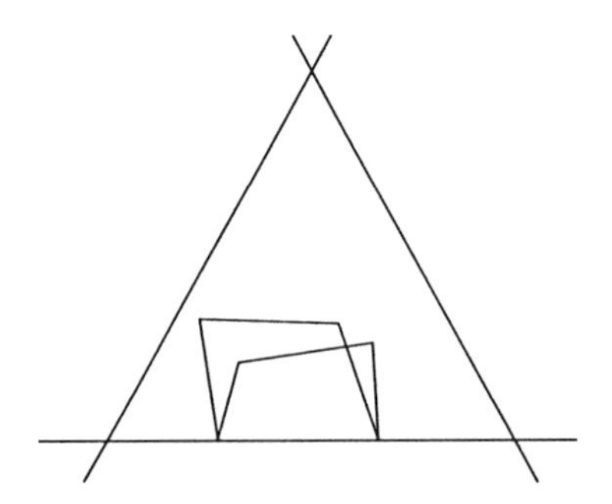

Figure 7.4.2

Note there are other types of pairs of "quadrangles"; for example, the polyhedron $\mathcal{P}$ and the edge F_0 in Fig. (7.4.3) give another such pair.

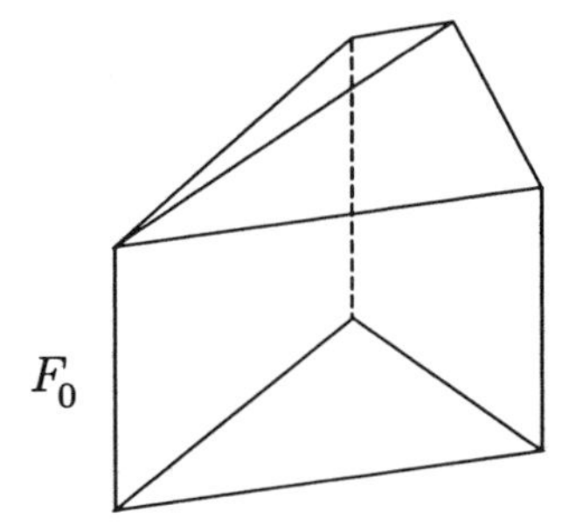

Figure 7.4.3

7.5 Definition. Let M_C be as above. A real analytic family of hyperplanes $\{E_\epsilon\}$ is said to *exhaust* M_C *towards* N_1 *inside* $\bar{\Omega}$ if it satisfies the following (with the same notations $\bar{\Omega}_\epsilon^{<0}$ etc, as in (6.15)):

(0) If $F \not\subset F_0$, $M(F)$ meets $E_\epsilon^{\mathbb{C}}$ transversally over any point of W, where $E_\epsilon^{\mathbb{C}}$ is the complexification of E_ϵ.

(i) $E_0 = N_1$; $\bar{\Omega}_0^- = \emptyset$, $\bar{\Omega}_0^+ = \bar{\Omega} - N_1$;

(ii) The type of E_ϵ (with respect to $\mathcal{N}$ and M_C) stays constant as $\epsilon \in (0, \epsilon_0)$ varies.

(iii) For any 0-face $F \subset F_0$, and any $\epsilon \in (0, \epsilon_0)$, $M_C(F) \cap \mathbb{P}_V^n \subset \bar{\Omega}_\epsilon^{<0}$;

(iv) For each 1-face $F \in \mathcal{F}_{II}(\mathcal{P})$, we have $M_C(F) \cap E_\epsilon \cap \mathbb{P}_V^n \subset \Omega$.

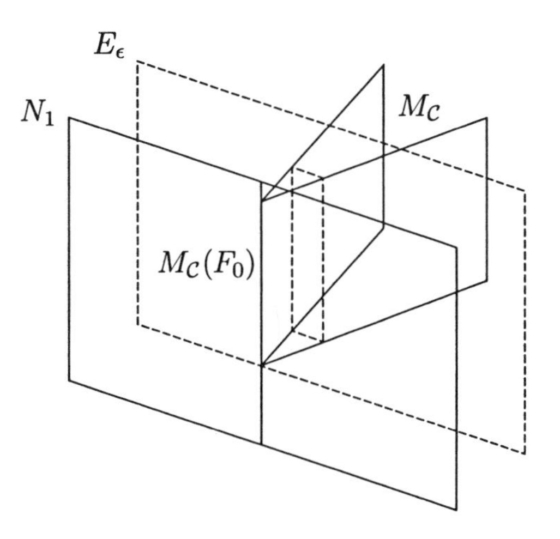

Figure 7.5

Notice that an exhausting family $\{E_\epsilon\}$ for M_C exists; one may take E_ϵ a perturbation of N_1 containing N_{12}. (For small ϵ the condition (0) is satisfied since W is relatively compact.)

7.6 Proposition. *Let $M_{\mathcal{C}}$ and $\{E_\epsilon\}$ be as above. Then for small enough ϵ, $M_{\mathcal{C}}$ can be refined to a $\mathcal{C}_D$-figure (over W) $M_{\mathcal{C},\epsilon}$ in which D corresponds to E_ϵ.*

Proof. For a face $F \in \mathcal{F}_{II}(\mathcal{P})$, let

$$M_{\mathcal{C},\epsilon}(F \cap D) = M_{\mathcal{C}}(F) \cap E_\epsilon^{\mathbb{C}} \; .$$

Since the intersection is transversal, it defines a $\mathcal{C}_D$-figure. (7.5) (iii)(iv) implies it takes values in $\bar{\Omega}$.

7.7 With the above notation, the $\mathcal{C}_D$-figure/W $M_{\mathcal{C},\epsilon}$ induces, in particular,

(a) (In the case $\dim \mathbb{H}/W = d$) an admissible $\mathcal{P}^-$-figure $/W$, which we denote by $M_{\mathcal{P}^-}$ (whose supporting subspace is $\mathbb{H}$);

(b) For each $F \in \mathcal{F}_{II}(\mathcal{P})$ such that $F \not\supset F_0$, an admissible F^--figure $/W$, which we denote by $M_{F^-,\epsilon}$;

(c) For each $F \in \mathcal{F}_{II}(\mathcal{P})$ such that $F \supset F_0$, a generic F^+-figure $/W$, which we denote by M_{F^+}.

To each of these figures, applying the construction in §5, one can associate the integral. We define the integration along the fiber of the $\mathcal{C}$-figure $M_{\mathcal{C}}$ to be the holomorphic $(m-d+1)$-form on $\tilde{W}$ (defined as in (5.13)) given by:

$$\text{(7.7.1)} \quad \int [M_{\mathcal{C}}] = (-1)^{d+1}\, d \int [M_{\mathcal{P}^-,\epsilon}] - \sum_{F \in \mathcal{F}_{II}(\mathcal{P}), F \not\supset F_0} \int [M_{F^-,\epsilon}] + \sum_{F \in \mathcal{F}_{II}(\mathcal{P}), F \supset F_0} \int [M_{F^+,\epsilon}] \; .$$

(If $\dim \mathbb{H}/W = d-1$, interpret the first term to be zero. In the second and the third terms, the sum is taken over codimension one faces $F \in \mathcal{F}_{II}(\mathcal{P})$.)

7.8 Proposition. *The integral $\int [M_{\mathcal{C}}]$ is well-defined independent of the choice of an exhausting family $\{E_\epsilon\}$.*

Proof. We have only to prove the independence of its restriction to V. Take an excision of $\mathcal{P}$ along F_0 by another hyperplane $\bar{D}$ which is "closer" to F_0 than D. Let $\mathcal{P} = (\bar{\mathcal{P}})^+ \cup (\bar{\mathcal{P}})^-$ be the resulting division. For any $F \in \mathcal{F}_{II}(\mathcal{P})$ also, we have $F = (\bar{F})^+ \cup (\bar{F})^-$. Let $F_m = F^- \cap (\bar{F})^+$; then one has a linear division $F = (\bar{F})^- \cup F_m \cup F^+$.

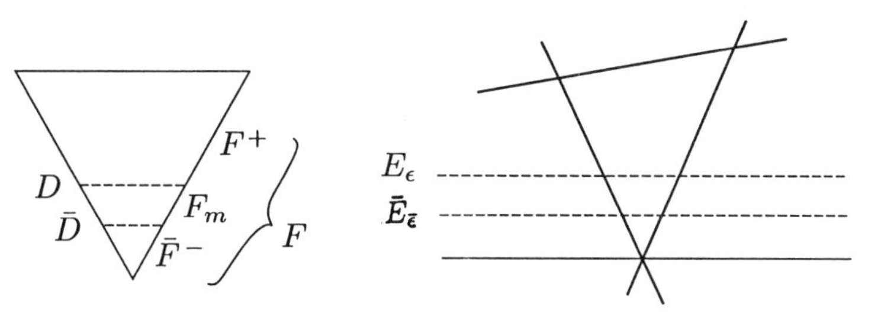

Figure 7.8

Let $\mathcal{C}_{D,\bar{D}} = \mathcal{C}_D \cup \mathcal{C}_{\bar{D}}$. The notion of a $\mathcal{C}_{D,\bar{D}}$-figure is defined as in Definition (7.4) with $\mathcal{C}$ replaced by $\mathcal{C}_{D,\bar{D}}$.

Given two exhausting families $\{E_\epsilon\}$ and $\{\bar{E}_\epsilon\}$, we take $\bar{\epsilon} \ll \epsilon$ so that $M_{\mathcal{C}}$ can be refined to a $\mathcal{C}_{D,\bar{D}}$-figure by the correspondence (for $F \in \mathcal{F}_{II}(\mathcal{P})$)

$$F \cap D \mapsto M_{\mathcal{C}}(F) \cap (E_\epsilon \times W) ;$$
$$F \cap \bar{D} \mapsto M_{\mathcal{C}}(F) \cap (\bar{E}_{\bar{\epsilon}} \times W) .$$

By the additivity (6.13), for a codimension one face $F \in \mathcal{F}_{II}(\mathcal{P})$, the following equality of $(m-d+1)$-forms on V holds:

$$\int [M_{F^-}] - \int [M_{(\bar{F})^-}] = \int [M_{F_m}] \qquad \text{if} \quad F \not\supset F_0 ;$$
$$-\int [M_{F^+}] + \int [M_{(\bar{F})^+}] = \int [M_{F_m}] \qquad \text{if} \quad F \supset F_0 .$$

One has $\partial \mathcal{P}_m = (\mathcal{P} \cap D) \cup (\mathcal{P} \cap \bar{D}) \cup (\cup_F F_m)$ where F varies over the codimension one faces $F \in \mathcal{F}_{II}(\mathcal{P})$. Since ω is closed, and $E_\epsilon \times W$, $\bar{E}_{\bar{\epsilon}} \times W$ are constant over W, by (5.7), (6.19), we have

$$\sum_F \int [M_{F_m}] = (-1)^{d+1} d \int [M_{\mathcal{P}^-}] - (-1)^{d+1} d \int [M_{(\bar{\mathcal{P}})^-}]$$

in the case $\dim \mathbb{H} = d$, while $\sum \int [M_{F_m}] = 0$ in the case $\dim \mathbb{H} = d-1$. The claim hence follows.

7.9 Proposition (Limit formula for admissible $M_{\mathcal{C}}$-figures). *We have the convergence of a real analytic $(m-d+1)$-form on V*

$$\sum_{F \in \mathcal{F}_{II}(\mathcal{P}), F \supset F_0} \int [M_{F^+,\epsilon}] \to \int [M_{\mathcal{C}}] \qquad (\epsilon \downarrow 0) .$$

(The restrictions of the integrals to V are being considered.) In particular, the integral $\int[M_{\mathcal{C}}]$ depends only on the F-figures $M_{\mathcal{C}}|_F$ for codimension one faces $F \supset F_0, F \in \mathcal{F}_{II}(\mathcal{P})$.

Proof. By the limit formula for admissible $\mathcal{P}$-figures (6.17), we have

$$\int [M_{\mathcal{P}^-,\epsilon}] \to 0 \quad \text{(in the case} \quad \dim \mathbb{H}/W = d), \quad \text{and}$$

$$\int [M_{F^-,\epsilon}] \to 0\ .$$

The claim follows from these and (7.7.1).

7.10 Proposition (Stokes formula for an admissible $\mathcal{P}$-figure). *Let $M_{\mathcal{P}}$ be an admissible $\mathcal{P}$-figure with values in $\bar{\Omega}$; assume that it satisfies the condition (6.15.1). Then we have*

$$(-1)^{d+1} d \int [M_{\mathcal{P}}] \\ = \int [M_{\mathcal{C}(\mathcal{P};F_0)}] + \sum_{F \in \mathcal{F}_{II}(\mathcal{P}), F \not\supset F_0} \int [M_F] + \sum_{F \in \mathcal{F}_I(\mathcal{P})} \int [M_F]\ .$$

(In the second and the third terms on the right hand side, we take the sums over codimension one faces F.) This justifies defining $(-1)^{d+1}\int[\partial M_{\mathcal{P}}]$ to be the right hand side of the above formula, a useful notation for quotations in the sequel.

Proof. We take and fix some ϵ, which will be dropped from the notation. The formula follows from the equalities $d\int[M_{\mathcal{P}}] = d\int[M_{\mathcal{P}^+}] + d\int[M_{\mathcal{P}^-}]$,

$$(-1)^{d+1} d \int [M_{\mathcal{P}^+}] = \sum_{F \in \mathcal{F}_{II}(\mathcal{P}), F \not\supset F_0} \int [M_{F^+}] \\ + \sum_{F \in \mathcal{F}_{II}(\mathcal{P}), F \supset F_0} \int [M_{F^+}] + \sum_{F \in \mathcal{F}_I(\mathcal{P})} \int [M_F]$$

(Stokes for a generic figure (6.18)) and (7.7.1).

7.11 Proposition (Stokes formula for an admissible $\mathcal{C}$-figure). *With the notation (7.3), one has*

$$(-1)^d d \int [M_{\mathcal{C}(\mathcal{P};F_0)}] = \sum_{F \in \mathcal{F}_{II}(\mathcal{P}), F \not\supset F_0} \int [M_{\mathcal{C}(F; F \cap F_0)}] + \sum_G \int [M_G]\ ;$$

in the last sum, G varies over the set

$$\{G \in \mathcal{F}_I(\mathcal{P}) \mid \dim G = d-2, \text{ and there exists } F \in \mathcal{F}_{II}(\mathcal{P}), F \supset F_0 \text{ and } G \text{ is a face of } F\}.$$

Also G is given the induced orientation from F.

Proof. Take the exterior differential of (7.7.1), and apply (7.10) to the admissible F^--figures M_{F^-} for $F \not\supset F_0$.

7.12 Keeping the notation (7.2), assume also given a linear division $S : \mathcal{P} = \cup_\lambda \mathcal{P}_\lambda$. Let

$$\mathcal{C}_\lambda := \mathcal{C}(\mathcal{P}_\lambda; \mathcal{P}_\lambda \cap F_0); \quad \mathcal{C}_{\mathcal{P};S;F_0} := \cup_\lambda \mathcal{C}_\lambda$$

and define the notion of an admissible $\mathcal{C}_{\mathcal{P};S;F_0}$-figure by replacing $\mathcal{C}$ with $\mathcal{C}_{\mathcal{P};S;F_0}$ in Definition (7.3).

An admissible $\mathcal{C}_{\mathcal{P};S;F_0}$-figure induces an admissible $\mathcal{C}$-figure $M_\mathcal{C}$ and admissible $\mathcal{C}_\lambda$-figures $M_{\mathcal{C}_\lambda}$. For the associated integrals, we have:

Proposition (Additivity for admissible $\mathcal{C}$-figures). *The following equality holds:*

$$\int [M_\mathcal{C}] = \sum_\lambda \int [M_{\mathcal{C}_\lambda}] \, .$$

Proof. Take an excision of $\mathcal{P}$ along F_0 which induces excisions of $\mathcal{P}_\lambda$ along $\mathcal{P}_\lambda \cap F_0$. The formula follows from (7.7.1) for $\int [M_\mathcal{C}]$, $\int [M_{\mathcal{C}_\lambda}]$, and the additivity (6.13).

8. The cancellation lemmas of differential forms

8.1 We will consider real figures with values in $\mathbb{P}^n_V$, where V is a real anaytic manifold. Throughout we take $\omega = vol_n$, and the integrals are with respect to it. As in §6, we make a choice of a connected component Ω of $\mathbb{P}^n_\mathbb{R} \backslash |\mathcal{L}|$.

8.2 Let ξ be a family of lines, generic with respect to $\{L_i\}$, in $\mathbb{P}^{n+1}_V$. Put

$$K = < e_{01}, e_2, \cdots, e_{n+1} >$$

where $e_{01} = (1 : -1 : 0 : \cdots : 0)$. Let $Q_i = \xi \cap (L_i \times V)$, $Q_i' =$ the projection of from $e_1$onto$K$ and $Q_0' = Q_1' = R$. By means of the canonical identifications $L_i \times V \cong \mathbb{P}^n_V$, we may view $Q_i \subset \mathbb{P}^n_V$. Then the following makes sense: *we also assume $Q_i \subset \Omega \subset \mathbb{P}^n_V$.*

Let $\vec{Q'_iQ_i}$ $(i = 0, 1, \cdots, n-2)$ denote the $\Delta^1 = [v_0, v_1]$-figure ("line segment") for which

$$v_0 \mapsto Q'_i, \quad v_1 \mapsto Q_i, \quad [v_0, v_1] \mapsto (e_1 * \xi) \cap L_i \ .$$

This takes values in $\bar{\Omega}$. The pair $\{\vec{RQ_0}, \vec{RQ_1}\}$ constitutes an admissible pair of line segments denoted by $\vec{RQ_0} - \vec{RQ_1}$. Associated to these are real analytic $(n-1)$-forms on V

$$\int [\vec{Q'_iQ_i}] \quad (i = 2, \cdots, n); \quad \int [\vec{RQ_0} - \vec{RQ_1}] \ .$$

Proposition. *With the above notation, one has*

$$\int [\vec{RQ_0} - \vec{RQ_1}] + \sum_{i=2}^{n} (-1)^i \int [\overrightarrow{Q'_iQ_i}] = 0. \tag{8.2.1}$$

The proof is the same as in [HM-M, §5]. If we denote the first term by $\int [\vec{Q'_0Q_0}] - \int [\vec{Q'_1Q_1}]$, the above reads $\sum_{i=0}^{n} (-1)^i \int [\overrightarrow{Q'_iQ_i}] = 0$. This is in line with (but not a consequence of) the following proposition.

Proposition. *Let $\mathcal{P}$ be a d-dimensional polyhedron,*

$$\mathcal{F}^+(F) \ni F \mapsto \mathcal{M}(F) \subset \mathbb{P}_V^{n+1}$$

be an assignment of relative subspaces such that

(i) $\dim_{\mathbb{R}} F = \dim_{\mathbb{R}} \mathcal{M}(F)/V \ + 1$;

(ii) If $F \subset F'$, then $\mathcal{M}(F) \subset \mathcal{M}(F')$;

(iii) For each F, $\mathcal{M}(F)$ meets L_i transversally. (Therefore, for $i = 0, \cdots, n$, one obtains a $\mathcal{P}$-figure $\mathcal{M}_i : F \mapsto \mathcal{M}(F) \cap L_i$ in $\mathbb{P}_V^n$);

(iv) The $\mathcal{P}$-figures $\mathcal{M}_i$ are all generic with values in Ω.

Then we have the following equality of $(n-d)$-forms:

$$\sum_{i=0}^{n} (-1)^i \int [\mathcal{M}_i \mid vol_n] = 0 \ .$$

[The figure illustrates the case $\mathcal{P} = \Delta^2$ and $n = 2$.]

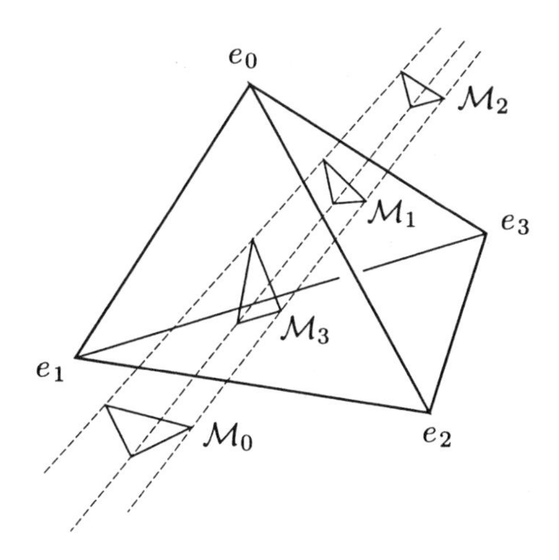

Figure 8.3.1

Proof. Take a simplicial subdivision $S : \mathcal{P} = \cup_\lambda \mathcal{P}_\lambda$ and a $(\mathcal{P}; S)$-figure which refines $\mathcal{M}$ (after shrinking V). We may assume that each of the induced $\mathcal{P}_\lambda$-figures $\mathcal{M}_\lambda$ satisfies the assumptions (i)–(iv) of the proposition. Thus we may assume $\mathcal{P} = \Delta^d$, and moreover

$$v := \bigcap_{F \in \mathcal{F}^+(\mathcal{P})} M(F) \notin |\mathcal{L}| \ .$$

For $i \neq n$, let $pr_i : L_i \times V \to L_n \times V$ be the projection from v, which is an isomorphism. Under pr_i the $\mathcal{P}$-figures M_i and M_n correspond to each other (see Fig.(8.3.2) for the case $d = 2$, $n = 2$).

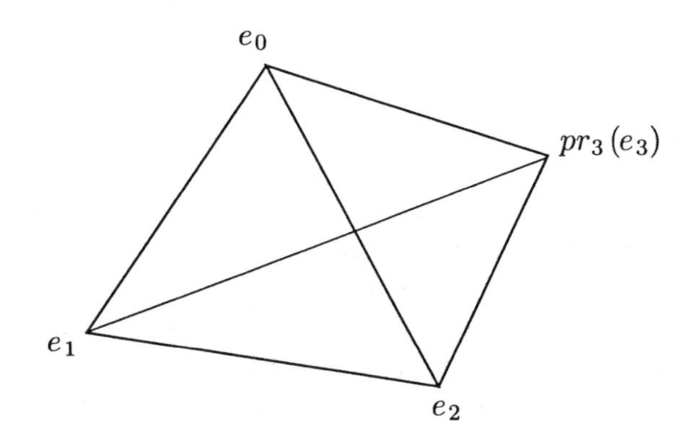

Figure 8.3.2

We thus have

$$\int [M_i \mid vol_n] = \int [M_n \mid (pr_i)_* vol_n] \ .$$

We are thus reduced to the following lemma.

8.4 Proposition. *Let $S = \mathbb{P}^n_V \backslash |\mathcal{L}|$, $pr_i : L_i \times S \to L_n \times S$ be the projection with center $\Delta_S \subset S \times S$ (the diagonal). Then we have*

$$\sum_{i=0}^{n} (-1)^i (pr_i)_* vol_n = 0$$

as meromorphic sections of $\Omega^n_{L_n \times S/S}$.

For the proof one reduces to the case $S = pt$, and observes that the poles of the forms $(pr_i)_* vol_n$ cancel one another. Alternatively, see [HM-M, §5] where the stronger claim (in which $\Omega^n_{L_n \times S/S}$ is replaced with $\Omega^n_{L_n \times S}$) is proved.

9. The construction of figures over the Grassmannian cosimplicial scheme

9.1 We define the *generic part of the Grassmannian* G^p_q of dimension q and codimension p to be the variety of q-dimensional projective subspaces of $\mathbb{P}^{p+q}_{\mathbb{C}}$ which meet the configuration of coordinate hyperplanes $\{L_i\}$ transversely. (This is a smooth quasi-projective variety.)

We define the i-th restriction map $A_i : G^p_q \to G^p_{q-1}$, $\quad i = 0, \cdots, p+q$, to be the map which sends a q-plane $\xi \subset \mathbb{P}^{p+q}$ to its intersection with L_i. (The hyperplane L_i may be canonically identified with $\mathbb{P}^{p+q-1}$ by omitting the i-th coordinate.)

We will produce a Grassmannian p–cocycle as follows:

Over the Grassmannian G^p_q with $(0 \leq q \leq p-1)$, we construct M^p_q, which is a Δ^1-figure in $\mathbb{P}^p$ if $q = 0$, and an admissible pair of $(\Delta^q \times [0,1])$-figures in $\mathbb{P}^p$ if $q \geq 1$ (see (9.6), (9.7)). By integration along the fiber with respect to the volume form vol_p, we obtain a holomorphic $(p-q-1)$-form $\int [M^p_q]$ on $\tilde{G}^p_q$, the universal covering space of G^p_q. These differential forms are proved to satisfy the cocycle condition, which is the defining property of a Grassmannian p-cocycle, in the rest of this section. For this, we will take excisions of the figures and reduce the problem to the cancellation for generic figures, which was discussed in §8.

9.2 We fix here notation for some linear constructions in $\mathbb{P}^p$. Denote by $*$ the join of two linear subspaces. Recall that e_i denotes the "i-th vertex". We let

$$< i_1, \ldots, i_k > = < e_{i_1}, \ldots, e_{i_k} > := e_{i_1} * \cdots * e_{i_k} \quad (i_1 < \cdots < i_k);$$

$$e_{ij}(a) := (0 : \cdots : 0 : \overset{i}{1} : 0 : \cdots : 0 : \overset{j}{a} : 0 : \cdots : 0) \ ; \quad e_{ij} := e_{ij}(-1);$$

$$K = K < i_0, \cdots, i_k > = e_{i_0 i_1} * e_{i_2} * \cdots * e_{i_k} \ .$$

For any k-face $< i_0, \ldots, i_k >$ $(i_0 < i_1 < \cdots < i_k)$, there exists a canonical isomorphism $< i_0, \ldots, i_k > \cong \mathbb{P}^k$ via which the vertex e_{i_j} corresponds to e_j. Thus any two k-faces of $\mathbb{P}^p$ are canonically identified.

9.3 We refer the reader to [HR-M, §5] for the following:

(1) There exist base points of G^p_q in a compatible way with the A's. More precisely, there exist real points

$$b_q \in G^p_q(\mathbb{R})$$

and paths from $A_i(b_q)$ to b_{q-1} in $G^p_{q-1}(\mathbb{R})$ which satisfy the obvious compatibilities. Let $V_q \subset G^p_q(\mathbb{R})$ be a small coordinate neighborhood of b_q.

(2) The base point $b_0 \in G^p_0$ is contained in the *distinguished real simplex* of $\mathbb{P}^p_{\mathbb{R}}$; by definition, the distinguished real simplex is

$$\Omega := \{(x_0 : \cdots : x_p) \in \mathbb{P}^p_{\mathbb{R}} \mid \text{ all } x_i \neq 0, \frac{x_1}{x_0}, \frac{x_2}{x_1}, \cdots, \frac{x_p}{x_{p-1}} \text{ are all negative}\}.$$

This is a connected component of $\mathbb{P}^p_{\mathbb{R}} \backslash |\mathcal{L}|$. We denote by $\bar{\Omega}$ the closure of Ω.

(3) Note that, for $\xi \in V_q$, $\xi \cap < i_0, \cdots, i_p > \in \Omega < i_0, \cdots, i_p >$ for any p-face $< i_0, \cdots, i_p >$. Here $\Omega < i_0, \cdots, i_p >$ denotes the distinguished real simplex of $< i_0, \cdots, i_p >$.

9.4 (cf. [HR-M], §2) Given a complex algebraic variety with a base point (X, b), denote by $\tilde{\Omega}^p(X)$ the $\mathbb{C}$-vector space of holomorphic p-forms on $\tilde{X}$:= the space of homotopy classes of smooth paths in X with the initial point b. One obtains the de Rham complex $\tilde{\Omega}^\bullet(X)$.

Given another such (X', b'), a map $f : (X, b) \to (X', b')$ is defined to be a pair of a holomorphic map $f : X \to X'$ and the homotopy class of a smooth path from $f(b)$ to b' in X'. Given such a map f, one can define the pull-back $f^* : \tilde{\Omega}^p(X') \to \tilde{\Omega}^p(X)$ in the obvious manner.

9.5 By (9.3), we have the maps $A_i : (G^p_q, b_q) \to (G^p_{q-1}, b_{q-1})$; thus by (9.4) the pull-backs

$$A_i^* : \tilde{\Omega}^k(G^p_{q-1}) \to \tilde{\Omega}^k(G^p_q)$$

are induced. For simplicity let $A^* = \sum_i (-1)^i A_i^*$.

Definition. A *Grassmannian p-cocycle* is a collection of multi-valued differential forms $\psi_q \in \tilde{\Omega}^{p-q-1}(G_q^p)$ $(0 \le q < p)$ such that

$$d\psi_q = A^* \psi_{q-1} \ (0 < q < p)$$

and

$$d\psi_0 = vol_p \ .$$

Here vol_p denotes the volume form on $G_0^p = (\mathbb{C}^*)^p \subset \mathbb{P}^p$.

9.6 *Construction of the Δ^1-figure M_0^p over G_0^p with values in $\mathbb{P}_{\mathbb{C}}^p$.*
Let $\Delta^1 = [v_0, v_1]$ be the 1-simplex with vertices v_0, v_1. For a point

$$Q \in G_0^p = \mathbb{P}^p \backslash |\mathcal{L}| \ ,$$

we let

$$M_0^p(v_1) = Q \ ; M_0^p(v_0) = Q' := (e_1 * Q) \cap K_{01} \ ;$$
$$M_0^p(\Delta^1) = e_1 * Q \ .$$

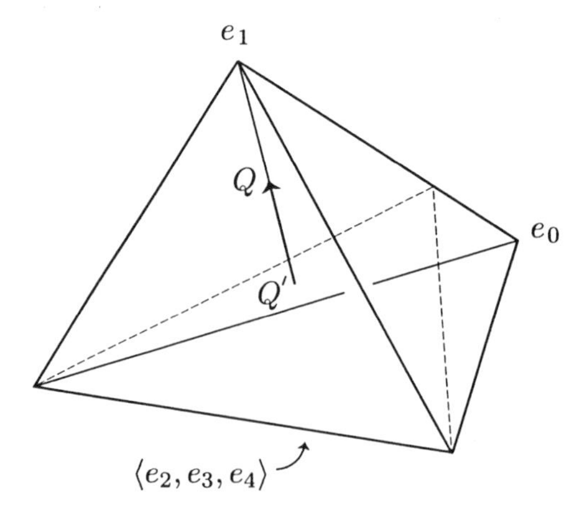

Figure 9.6

This defines a Δ^1-figure M_0^p over G_0^p with values in $\mathbb{P}_{\mathbb{C}}^p$ (of constant type with respect to $\mathcal{L}$). Note that there exists a neighborhood V_0 of b_0 in $G_0^p(\mathbb{R})$ such that if $Q \in V_0$, then $M_0^p(v_0), M_0^p(v_1) \in \Omega$. Thus M_0^p induces a real Δ^1-figure over V_0 with values in Ω.

9.7 *Construction of the admissible pair of $\mathcal{P}_{q+1}$-figures M_q^p over G_q^p with values in $\mathbb{P}_{\mathbb{C}}^p$.*

Let $q \geq 1$ and $\Delta^q[v_0, \cdots, v_q]$ be a q-simplex with the canonical orientation. Let

$$\mathcal{P}_{q+1} = \Delta^q[v_0, \cdots, v_q] \times [0,1] ,$$

which is a "prism" of dimension $q+1$; we equip it with the product orientation. The 0-faces of $\mathcal{P}_{q+1}$ consist of

$$v_k \times \{1\}, \quad v_k \times \{0\} ,$$

which we denote by v_k, v_k', respectively. (See the following figure for $\mathcal{P}_3$.)

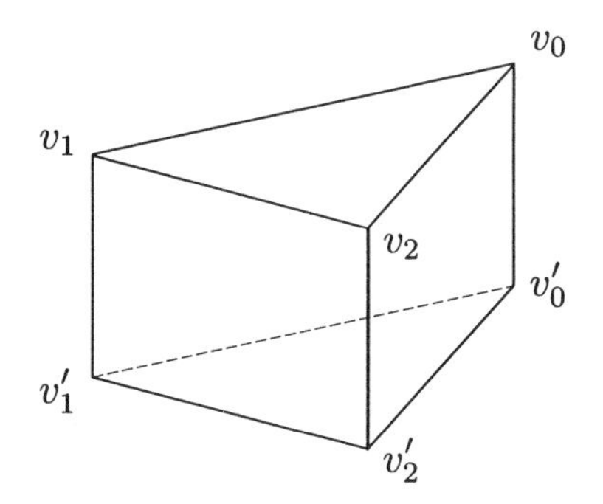

Figure 9.7.1

One has (with orientations taken into account)

$$(9.7.1) \qquad \partial\mathcal{P}_{q+1} = \sum_{k=0}^{q}(-1)^k\big(\Delta^{q-1}[v_0, \cdots, \widehat{v_k}, \cdots v_q] \times [0,1]\big) \\ + (-1)^q\Delta^q[v_0, \cdots, v_q] - (-1)^q\Delta^q[v_0', \cdots, v_q'] .$$

Let

$$M_q^p = M_q^p\{\xi\} \qquad (\xi \in G_q^p)$$

be an admissible pair of $\mathcal{P}_{q+1}$-figures over G_q^p defined as follows. Throughout this section we fix a p and often drop the superscript p from the notation. The faces of $\mathcal{P}_{q+1}$ consist of (where $1 \leq k_1 < \cdots k_j \leq q$):

(a) $[v_0, v_{k_1}, \cdots, v_{k_j}]$;
(b) $[v_0, v_{k_1}, \cdots, v_{k_j}] \times [0,1]$;
(c) $[v_{k_1}, \cdots, v_{k_j}]$;

(d) $[v_{k_1}, \cdots, v_{k_j}] \times [0,1]$;
(e) $[v'_0, v'_{k_1}, \cdots, v'_{k_j}]$;
(f) $[v'_{k_1}, \cdots, v'_{k_j}]$.

For $\xi \in G^p_q$ and $r = 0, 1$, we let ${}_rM_q\{\xi\}$ be the $\mathcal{P}_{q+1}$-figure with values in the p-face $< r, q+1, \cdots, p+q >$ which assigns subspaces as follows to each type of the faces (a)–(f) (where $B_r :< r, q+1, \cdots, p+q > \hookrightarrow \mathbb{P}^{p+q}$ is the inclusion and $K = K < r, q+1, \cdots, p+q >$):

(a) $B^*_r(\xi * < k_1, \cdots, k_j >)$;
(b) $B^*_r(\xi * < k_1, \cdots, k_j, q+1 >)$;
(c) $B^*_r(\xi * < k_1, \cdots, k_j >) \quad \cap < q+1, \cdots, p+q >$;
(d) $B^*_r(\xi * < k_1, \cdots, k_j, q+1 >) \quad \cap < q+1, \cdots, p+q >$;
(e) $B^*_r(\xi * < k_1, \cdots, k_j, q+1 >) \quad \cap \quad K$;
(f) $B^*_r(\xi * < k_1, \cdots, k_j, q+1 >) \quad \cap \quad < q+1, \cdots, p+q > \cap K$.

For convenience we also use the notation

$$ {}_rM_q\{\xi\} = \mathcal{P}_{q+1}(\xi; < r, q+1, \cdots, p+q >; \overset{\underline{v_1, \cdots, v_q; v'_0}}{e_1, \cdots, e_q; e_{q+1}}; \overset{\underline{[v_1, \cdots, v_q] \times [0,1];}}{< q+1, \cdots, p+q >}; \overset{\underline{[v'_0, v'_1, \cdots, v'_q]}}{K}) $$

which manifests the vertices and the subspaces (and the corresponding vertices and faces in $\mathcal{P}_{q+1}$) involved in the construction. This notation is justified by observing that $M =_r M_q$ is determined only by $M(v_0)$, $M([v_0, v_k])$ $(k = 1, \cdots, q)$, $M([v_0, v'_0])$, $M([v_1, \cdots, v_q] \times [0,1])$, and $M([v'_0, \cdots, v'_q])$. Notice ${}_rM_q \mid_{[v_1, \cdots, v_q] \times [0,1]}$ for $r = 0, 1$ are the same figures in the common $(p-1)$-face $< q+1, \cdots, p+q >$. In the case $p = 4$, the figures ${}_rM_q$ $(q = 1, 2)$ are pictured in Figure 9.7.2.

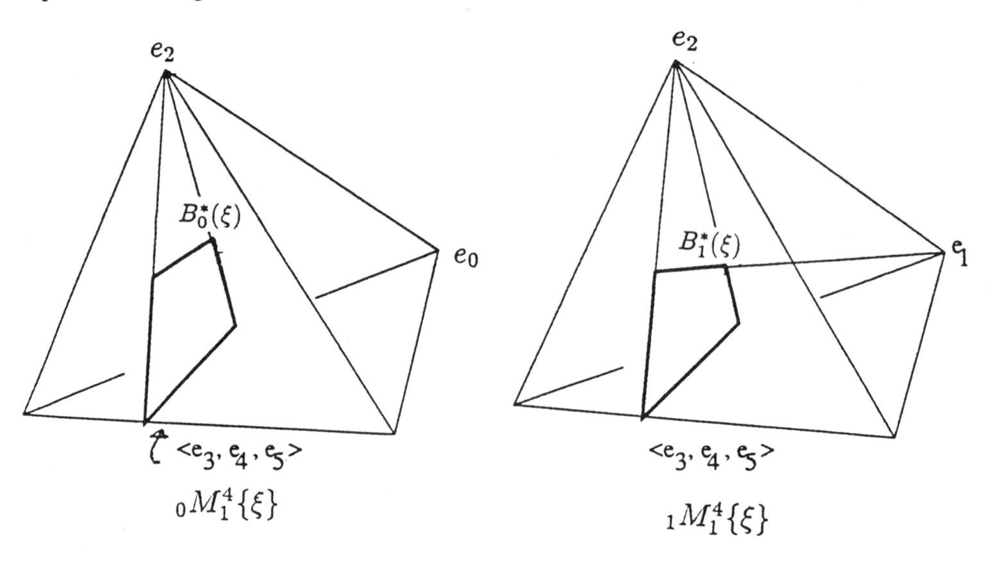

Figure 9.7.2a **Figure 9.7.2b**

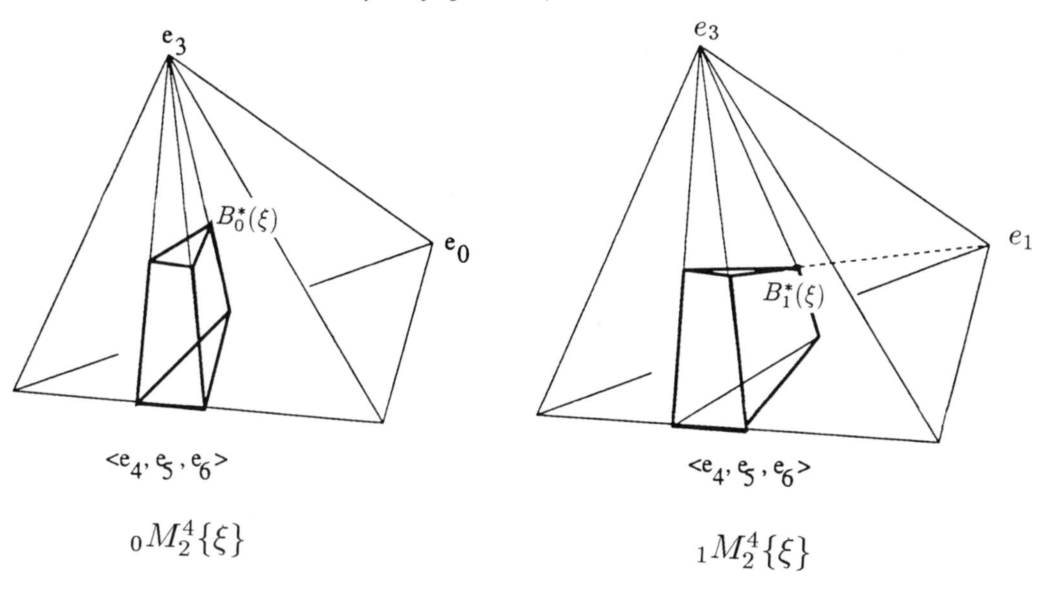

Figure 9.7.2c **Figure 9.7.2d**

The construction may be phrased as follows: Consider the point $B_r^*(\xi)$ in the p-simplex $< r, q+1, \cdots, p+q >$. "Project" that point from the centers $e_1, \cdots, e_q$ onto the $(p-1)$-simplex $< q+1, \cdots, p+q >$, and also from the center e_{q+1} onto K. (Since e_1 (in the case $r=0$) and $e_2, \cdots, e_q \notin < r, q+1, \cdots, p+q >$, the meaning of "projection" should be understood to be the join of ξ with the center, followed by B_r^*.)

Proposition. *Via the canonical identifications* $\mathbb{P}^p \cong < 0, q+1, \cdots, p+q > \cong < 1, q+1, \cdots, p+q >$, ${}_rM_q\{\xi\}$ $(r = 0, 1)$ *can be regarded as* $\mathcal{P}_{q+1}$*-figures with values in* $\mathbb{P}^p$, *which jointly form an admissible pair of* $\mathcal{P}_{q+1}$*-figures (over* G_q^p*) with values in* $\mathbb{P}^p$*:*

$$M_q\{\xi\} = \sum_{r=0,1} (-1)^r \; {}_rM_q\{\xi\} \; .$$

For the proof, we form the two polyhedra $\mathcal{R}_{q+1}$ and $\mathcal{Q}_{q+2}$ (with the subscripts indicating the dimensions). The polyhedron $\mathcal{Q}_{q+2}$ will have a q-face F_0 such that the two faces containing F_0 are $\mathcal{P}_{q+1}$'s. We will show that the above $M_q\{\xi\}$ can be extended to an admissible $\mathcal{C}(\mathcal{Q}_{q+2}; F_0)$-figure.

The following construction of a polyhedron is to be used. Given a polyhedron $\mathcal{P}$, a codimension one face F, a face $f \subset F$. Assume that there exists a unique face F_1 such that $\dim F_1 = \dim f + 1$ and $F_1 \cap F = f$. Take a point $v_1 \in \overset{\circ}{f}$, and another point $v \in < F_1 > - \mathcal{P}$ close to v_1; let $\mathcal{P}'$ be the convex hull of $\mathcal{P}$ and v.

Note that the process is the gluing of $\mathcal{P}$ and $v * F$ along F:

$$\mathcal{P}' = \mathcal{P} \cup (v * F) .$$

The faces of $\mathcal{P}'$ are described as follows:

(case 1) If G is a face of $\mathcal{P}$ with $f \subset G \subset F$, then G is *not* a face of $\mathcal{P}'$.

(case 2) If G is a face of $\mathcal{P}$ with $f \not\subset G \subset F$, then G and $v * G$ are both faces of $\mathcal{P}'$.

(case 3) If G is a face of $\mathcal{P}$ with $G \not\subset F$, $G \supset F_1$ then G *extends* to the face $G \cup \ v * (G \cap F)$. (Note $G \cap F$ is a face in case 1.)

(case 4) If G is a face of $\mathcal{P}$ with $G \not\subset F$, $G \not\supset F_1$ then G is also a face of $\mathcal{P}'$. (Note $G \cap F$ is either empty or a face in case 2.)

Moreover, any face of $\mathcal{P}'$ arises this way from a face G of $\mathcal{P}$ in one of the cases 2–4.

We construct a polyhedron $\mathcal{R}_{q+1}$ of dimension $q+1$ by repeatedly applying the gluing process as follows. We start with a $q+1$-simplex $\mathcal{R}^0_{q+1} = [u_0, u_1] * [v_1, \cdots, v_q]$, and, for $k = 1, \cdots, q-1$, inductively construct $\mathcal{R}^k_{q+1}$ with vertices $u_0, \cdots, u_{k+1}, v_1, \cdots, v_q$. $\mathcal{R}^k_{q+1}$ is obtained from $\mathcal{R}^{k-1}_{q+1}$ by the above gluing process (call this the *k-th gluing*) using the following data of faces:

$$\begin{aligned} F_k &= [u_0, u_1, \cdots, u_k] * [v_{k+1}, \cdots v_q], \\ f_k &= [u_0, u_1] * [v_{k+1}], \\ F_{1,k} &= [u_0, u_1] * [v_1, v_{k+1}] . \end{aligned}$$

(Here we keep the notation F, f, F_1 with the added subscripts k.) The new vertex in the k-th gluing is to be u_{k+1}, so

$$\mathcal{R}^k_{q+1} = \mathcal{R}^{k-1}_{q+1} \cup \ [u_0, \cdots, u_{k+1}] * [v_{k+1}, \cdots, v_q] .$$

We put

$$\mathcal{R}_{q+1} = \mathcal{R}^{q-1}_{q+1} = \cup_{k=1,\cdots,q}[u_0, u_1, \cdots, u_k] * [v_k, v_{k+1}, \cdots v_q] ,$$

$\mathcal{Q}_{q+2} = \mathcal{R}_{q+1} \times [0,1]$ and $F_0 := [v_1, \cdots, v_q] \times [0,1]$.
The orientation of $\mathcal{Q}_{q+2}$ is given so that it induces the orientations $(-1)^r \Delta^q(u_r, v_1, \cdots, v_q) \times [0,1]$, $r = 0, 1$, on the two faces. Let

$$F(0,1,i_2,\cdots,i_s) := \bigcup_{r=1,\cdots,s} [u_0, u_1, u_{i_2}, \cdots, u_{i_r}] * [v_{i_r}, \cdots, v_{i_s}]$$
$$(1 = i_1 < i_2 < \cdots < i_s) .$$

Lemma. *The faces of $\mathcal{R}_{q+1}$ consist of the following (mutually disjoint):*

(a) $F(0,1,i_2,\cdots,i_s)$ where $1 < i_2 < \cdots < i_s$;

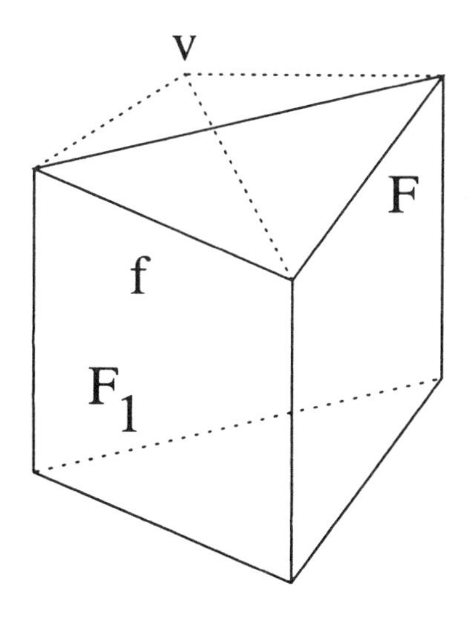

Figure 9.7.3

(b) $[u_{i_0}, \cdots, u_{i_r}] * [v_{i_{r+1}}, \cdots, v_{i_s}]$ *where* $(i_0, i_1) \neq (0,1)$ *and* $0 \leq i_0 < i_1 < \cdots < i_s \leq q$;

(c) $[u_{i_0}, \cdots, u_{i_r}] * [v_{i_r}, \cdots, v_{i_s}]$ *where* $(i_0, i_1) \neq (0,1)$ *and* $0 \leq i_0 < i_1 < \cdots < i_s \leq q$.

Proof. Consider the face $[u_0, u_1] * [v_1, v_{i_2}, \cdots, v_{i_s}]$ of $\mathcal{R}^0_{q+1}$. It continues to be a face of $\mathcal{R}^k_{q+1}$ for $k = 0, \cdots, i_2 - 2$. (This is seen by induction: suppose it is a face of $\mathcal{R}^{k-1}_{q+1}$. Since it is not contained in F_k and does not contain $F_{1,k}$ — the case 4 in the above description — it is a face of $\mathcal{R}^k_{q+1}$ also. In the $(i_2 - 1)$-th gluing, since this face is not contained in F_{i_2-1} but contains F_{1,i_2-1} — the case 3 — it extends to the face

$$[u_0, u_1] * [v_1, v_{i_2}, \cdots, v_{i_s}] \cup [u_0, u_1, u_{i_2}] * [v_{i_2}, v_{i_3}, \cdots, v_{i_s}]$$

of $\mathcal{R}^{i_2-1}_{q+1}$. This, in turn, continues to be a face of $\mathcal{R}^k_{q+1}$ for $k = i_2 - 1, \cdots, i_3 - 2$. This pattern persists and shows that $[u_0, u_1] * [v_1, v_{i_2}, \cdots, v_{i_s}]$ extends to the face $F(0, 1, i_2, \cdots, i_s)$ of $\mathcal{R}_{q+1}$.

Take the face $[u_0, u_1] * [v_{i_2}, \cdots, v_{i_s}]$ of $\mathcal{R}^0_{q+1}$. It continues to be a face until $\mathcal{R}^{i_2-2}_{q+1}$. In $\mathcal{R}^{i_2-2}_{q+1}$ this face contains f_{i_2-1} and is contained in F_{i_2-1} —the case 1. Thus in the $(i_2 - 1)$-th gluing the face disappears.

One also sees that the faces in (b), (c) continues to be faces through $\mathcal{R}_{q+1}$.

Lemma. *(1) A face of* $\mathcal{Q}_{q+2}$ *is of the form* $\mathfrak{f} \times \{0\}$, $\mathfrak{f} \times \{1\}$, *or* $\mathfrak{f} \times [0,1]$ *where* $\mathfrak{f}$ *is a face of* $\mathcal{R}_{q+1}$.

(2) The faces of $\mathcal{Q}_{q+2}$ *containing* F_0 *are* $u_0 * [v_1, \cdots, v_q] \times [0,1]$ *and* $u_1 * [v_1, \cdots, v_q] \times [0,1]$.

(3) The faces of $\mathcal{C}(\mathcal{Q}_{q+2}; F_0)$ consist of:
— those involving at least one v_k, and
— the faces of $u_0 * [v_1, \cdots, v_q] \times [0,1]$ and $u_1 * [v_1, \cdots, v_q] \times [0,1]$.

Proof of Proposition. We will define an assignment of a relative subspace to each face in $\mathcal{C}(\mathcal{Q}_{q+2}; F_0)$.

Let $B_i :< i, q+1, \cdots, p+q > \hookrightarrow \mathbb{P}^{p+q}$ be the inclusion of the p-face. For $(i_0, i_1) = (0,1)$ or not, let

$$[u_{i_0}, \cdots, u_{i_r}] * [v_{i_r}, \cdots, v_{i_s}] \times \{1\} \mapsto B_{i_0}^*(\xi * < i_0, \cdots, i_s >);$$

For $(i_0, i_1) \neq (0,1)$,

$$[u_{i_0}, \cdots, u_{i_r}] * [v_{i_{r+1}}, \cdots, v_{i_s}] \times \{1\} \mapsto B_{i_0}^*(\xi * < i_1, \cdots, i_s >).$$

For the other faces i.e. those of the form $\mathfrak{f} \times \{0\}$, $\mathfrak{f} \times [0,1]$, the subspaces to be assigned are related to the above, but invove either e_{q+1}, or e_{q+1} and K; for example,

$$[u_{i_0}, \cdots, u_{i_r}] * [v_{i_r}, \cdots, v_{i_s}] \times [0,1] \mapsto B_{i_0}^*(\xi * < i_0, \cdots, i_s, q+1 >);$$
$$[u_{i_0}, \cdots, u_{i_r}] * [v_{i_r}, \cdots, v_{i_s}] \times \{0\} \mapsto B_{i_0}^*(\xi * < i_0, \cdots, i_s, q+1 >) \cap K.$$

One checks that

— An s-face corresponds to a subspace of relative dimension s. The subspaces are all of constant type with respect to $\mathcal{L}$.

— A well-defined assignment to each face of $\mathcal{C}(\mathcal{Q}_{q+2}; F_0)$ is induced. More specifically, for a face of the form $\mathfrak{f} \times \{1\}$, $\mathfrak{f} \times \{0\}$, or $\mathfrak{f} \times [0,1]$ where $\mathfrak{f} = F(0, 1, i_2, \cdots, i_s)$, all subpolyhedra of the form $\mathfrak{f}' \times \{1\}$ (resp. $\mathfrak{f}' \times \{0\}$, etc.) with $\mathfrak{f}' = [u_0, u_1, \cdots, u_{i_r}] * [v_{i_r}, \cdots, v_{i_s}]$ (resp. $\mathfrak{f}' \times \{0\}$, etc.) corresponds to the same relative subspace.

— The required inclusion relations are satisfied. We show an example. For $(i_0, i_1) \neq (0,1)$ and $k < r+1$, there is an inclusion of faces

$$[u_{i_0}, \cdots, \widehat{u_{i_k}}, \cdots, u_{i_{r+1}}] * [v_{i_{r+1}}, \cdots, v_{i_s}] \times \{1\} \subset [u_{i_0}, \cdots, u_{i_{r+1}}] * [v_{i_{r+1}}, \cdots, v_{i_s}] \times \{1\}$$

which is required to imply $B_{i_0}^*(\xi * < i_0, \cdots, \widehat{i_k}, \cdots, i_s >) \subset B_{i_0}^*(\xi * < i_0, \cdots, i_s >)$ if $k \neq 0$, and $B_{i_1}^*(\xi * < i_1, \cdots, i_s >) \subset B_{i_0}^*(\xi * < i_0, \cdots, i_s >)$ if $k = 0$. This is obvious if $k \neq 0$. If $k = 0$, one notes $B_{i_0}^*(\xi * < i_0, i_1, \cdots, i_s >) = B_{i_1}^*(\xi * < i_0, i_1, \cdots, i_s >)$.

—Over V_q, M_q takes values in $\bar{\Omega}$. This will be part of Lemma (9.12) (2).

We are in a position to apply the method in the previous sections to the admissible collection of figures $M_q\{\xi\}$, the volume form vol_p, and the connected component Ω in (9.3). Let $\int[M_q\{\xi\}]$ be the holomorphic $(p-q-1)$-form on

$\tilde{G}^p_q$ produced by integration. We now state the theorem the proof of which will take the rest of this section.

9.8 Theorem. *We have the equality of holomorphic $(p-q)$-forms on $\tilde{G}^p_q$:*

$$d\int[M_0\{\xi\}] = vol_p \qquad (q=0);$$

$$(-1)^{q+2}d\int[M_q\{\xi\}] = \sum_{s=0,\cdots,p+q}(-1)^s\int[M_{q-1}\{A_s\xi\}] \qquad (1\le q\le p-1)\,.$$

The case $q=0$ is immediate from Stokes formula. The case $q=1$ is parallel to [HM-M 1, (6.7.2)] and follows from Proposition (8.2). In the following we assume $q\ge 2$. We have only to show the equality on V_q, on which both sides are real analytic forms.

9.9 Proposition. *We have the following equality of formal sums of $\mathcal{P}_q$-figures*

$$(9.9.1)\quad \sum_{k=1,\cdots q}(-1)^k M_q\{\xi\}\mid_{[v_0,v_1,\cdots,\hat{v}_k,\cdots,v_q]\times[0,1]} = \sum_{s=0,\cdots q}(-1)^s M_{q-1}\{A_s\xi\}\,.$$

Here each term of each side, which is an admissible pair of $\mathcal{P}_q$-figures, is regarded as a formal sum of two $\mathcal{P}_q$-figures. (Note that, there is an identification $[v_0,v_1,\cdots,\hat{v}_k,\cdots,v_q]\times[0,1]=\mathcal{P}_q(:=[v_0,v_1,\cdots,v_{q-1}]\times[0,1])$ which sends $v_0,v_1,\cdots,\hat{v}_k,\cdots,v_q$ to $v_0,v_1,\cdots,v_{q-1}$ in this order. Thus each term $M_q\{\xi\}\mid_{[v_0,v_1,\cdots,\hat{v}_k,\cdots,v_q]\times[0,1]}$ on the left hand side can be regarded as an admissible pair of $\mathcal{P}_q$-figures.)

Proof. The left hand side of (9.9.1) equals

$$\sum_{r,k}(-1)^r(-1)^k\ \mathcal{P}_q(\xi;<r,q+1,\cdots,p+q>;$$
$$e_1,\cdots,\hat{e}_k,\cdots,e_q;e_{q+1};<q+1,\cdots,p+q>;K)$$

where the sum is taken over $r=0,1$ and $k=1,\cdots,q$. Note that these $\mathcal{P}_q$-figures appear in the two p-faces $<r,q+1,\cdots,p+q>$, $r=0,1$.

On the other hand, one sees that $(-1)^s M_{q-1}\{A_s\xi\}$ on the right hand side is equal to

$$(-1)^s\mathcal{P}_q(\xi;<0,q+1,\cdots,p+q>;e_1,\cdots,\hat{e}_s,\cdots,e_q;e_{q+1};<q+1,\cdots,p+q>;K)$$
$$-(-1)^s\mathcal{P}_q(\xi;<1,q+1,\cdots,p+q>;e_1,\cdots,\hat{e}_s,\cdots,e_q;e_{q+1};<q+1,\cdots,p+q>;K)$$

if $s=2,\cdots,q$;

$$(-1)^1\mathcal{P}_q(\xi;<0,q+1,\cdots,p+q>;\hat{e}_1,\cdots,\cdots,e_q;e_{q+1};<q+1,\cdots,p+q>;K)$$
$$-(-1)^1\mathcal{P}_q(\xi;<2,q+1,\cdots,p+q>;\hat{e}_1,\cdots,\cdots,e_q;e_{q+1};<q+1,\cdots,p+q>;K)$$

if $s=1$;

$$(-1)^0\mathcal{P}_q(\xi;<1,q+1,\cdots,p+q>;\hat{e}_1,\cdots,\cdots,e_q;e_{q+1};<q+1,\cdots,p+q>;K)$$
$$-(-1)^0\mathcal{P}_q(\xi;<2,q+1,\cdots,p+q>;\hat{e}_1,\cdots,\cdots,e_q;e_{q+1};<q+1,\cdots,p+q>;K)$$

if $s=0$.

Among these are the two $\mathcal{P}_q$-figures in the p-face $<2,q+1,\cdots,p+q>$, which cancel each other. The other $\mathcal{P}_q$-figures correspond to the $\mathcal{P}_q$-figures on the left hand side.

9.10 Proposition. *We have* $\int[M_q\{\xi\}\mid_{[v_0',\cdots,v_q']}]=0$.

Proof. Obvious since $M_q\{\xi\}([v_0',\cdots,v_q'])$ is contained in the hyperplane K on which the volume form restricts to zero.

9.11 Proposition. *We have*

$$(-1)^{q+2}\int[M_q\{\xi\}\mid_{[v_0,\cdots,v_q]}]=\sum_{s=q+1,\cdots,p+q}(-1)^{s-q}\int[M_{q-1}\{A_s\xi\}]\,.$$

Note that Theorem 9.8 follows from Propositions (9.9), (9.10), (9.11) and Stokes formula:

$$\begin{aligned}(-1)^{q+2}d\int[M_q\{\xi\}]&=\int[\partial M_q\{\xi\}]\\ &\qquad\text{by Stokes formula for an admissible } \mathcal{C}\text{-figure (7.11)}\\ &=(-1)^q\int[M_q\{\xi\}\mid_{[v_0,\cdots,v_q]}]\\ &\quad+\sum_{k=1}^{q}(-1)^k\int[M_q\{\xi\}\mid_{[v_0,\cdots,\hat{v}_k,\cdots,v_q]\times[0,1]}]\\ &\qquad\text{by (9.10), (9.7.1)}\\ &=(-1)^q\sum_{s=q+1,\cdots,p+q}(-1)^{s-q}\int[M_{q-1}\{A_s\xi\}]\\ &\quad+\sum_{s=0,\cdots,q}(-1)^s\int[M_{q-1}\{A_s\xi\}]\,.\end{aligned}$$

For the proof of Proposition (9.11), let $\bar{M}_q\{\xi\} = M_q\{\xi\}\ |_{[v_0,\cdots,v_q]}$; this is an admissible pair of Δ^q-figures which we may also express

$$\sum_{r=0,1} (-1)^r \Delta^q(\xi; < r, q+1, \cdots, p+q >; e_1, \cdots, e_q; < q+1, \cdots, p+q >)$$

clarifying the vertices and subspaces involved. More precisely, $\bar{M}_q$ is an admissible $\mathcal{C}(\mathcal{R}_{q+1}; [v_1, \cdots, v_q])$-figure where $\mathcal{R}_{q+1}$ is as in (9.7). Recall that for $s = q+1, \cdots, p+q$, $M_{q-1}\{A_s\xi\}$ is an admissible pair of $\mathcal{P}_q$-figures with values in $\mathbb{P}^p$ given by

$$\underline{v_1}, \cdots, \underline{v_q};\quad \underline{v_0'};\quad \underline{[v_1, \cdots, v_q] \times [0,1]};\quad \underline{[v_0', \cdots, v_q']}$$

$$\sum_{r=0,1} (-1)^r \quad \mathcal{P}_q(\xi; < r, q, \cdots, \hat{s}, \cdots, p{+}q >; e_1, \cdots, e_{q-1}; e_q; < q, \cdots, \hat{s}, \cdots, p{+}q >; \quad K)\,.$$

Let $q \geq 2$, δ be a fixed small positive number. For ϵ which is small, positive and to be varied, let

$$\begin{aligned}\Pi(\epsilon) &=< e_q, e_{q+1}, \cdots, e_{p+q-2}, e_{p+q-1,0}((-1)^p\delta^{-1}\epsilon), e_{p+q,0}((-1)^p\epsilon) > \\ &= \{x_0 - (-1)^p\delta\epsilon^{-1}x_{p+q-1} - (-1)^p\epsilon^{-1}x_{p+q} = 0\}\end{aligned}$$

be a hyperplane in the $p+1$-face

$$< 0, q, q+1, \cdots, p+q >=< 1, q, q+1, \cdots, p+q > .$$

See (9.2) for the notation $e_{ij}(a)$. Note $e_{ij}(a)$ approaches e_i as $a \to 0$. Also recall the notation $\Omega < 0, 1, \cdots, p >$. The following remark will be useful: $\Omega < 0, 1, \cdots, p >$ and $\Omega < 1, \cdots, p+1 >$ are both contained in $\bar{\Omega} < 0, 1, \cdots, p, p+1 >$.

9.12 Lemma. *(1) The hyperplane $\Pi(\epsilon)$ approaches the p-face*

$$< q, q+1, \cdots, p+q > \quad \text{as} \quad \epsilon \to 0.$$

(2) We have

$$\Pi(\epsilon) \cap < q, q+1, \cdots, p+q >=< e_q, \cdots, e_{p+q-2}, e_{p+q-1,p+q}(-\delta) >$$

(note it is independent of ϵ) and $e_{p+q-1,p+q}(-\delta) \in \bar{\Omega} < q, q+1, \cdots, p+q >$.

(3) For $r = 0, 1$, $i = 1, \cdots, q$, and $\xi \in V_q$, we have

$$\begin{aligned}{}_rM_q\{\xi\}(v_i) &\in \Omega < q+1, \cdots, p+q > ; \\ {}_rM_q\{\xi\}(v_i') &\in \Omega < q+2, \cdots, p+q > \ .\end{aligned}$$

The intersection $\Pi(\epsilon)\cap < 0, q+1, \cdots, p+q >$ *exhausts both the admissible pair of* $\mathcal{P}_{q+1}$*-figures* $M_q\{\xi\}$ *and the admissible pair of* Δ^q*-figures* $\bar{M}_q\{\xi\}$ *inside* $\bar{\Omega} < 0, q+1, \cdots, p+q >$ *towards the* $(p-1)$*-face* $< q+1, \cdots, p+q >$*. (See §§6 and 7 for exhausting hyperplanes.)*

(4) For $r = 0, 1$, $i = 1, \cdots, q-1$, $s = q+1, \cdots, p+q$, *and* $\xi \in V_q$, *we have*

$$
\begin{aligned}
{}_rM_{q-1}\{A_s\xi\}(v_i) &\in \Omega < q, \cdots, \hat{s}, \cdots, p+q > ;\\
{}_rM_{q-1}\{A_s\xi\}(v_i') &\in \Omega < q+1, \cdots, \hat{s}, \cdots, p+q > .
\end{aligned}
$$

The intersection $\Pi(\epsilon)\cap < 0, q, \cdots, \hat{s}, \cdots, p+q >$ *exhausts the admissible pair of* $\mathcal{P}_q$*-figures* $M_{q-1}\{A_s\xi\}$ *inside* $\bar{\Omega} < 0, q, \cdots, \hat{s}, \cdots, p+q >$ *towards the* $(p-1)$*-face* $< q, \cdots, \hat{s}, \cdots, p+q >$*.*

Proof. (1) Obvious.

(2) This follows since $< e_{p+q-1,0}\big((-1)^p\delta^{-1}\epsilon)\big), e_{p+q,0}\big((-1)^p\epsilon\big) >$ and $< e_{p+q-1}, e_{p+q} >$ meets in the point $e_{p+q-1,p+q}(-\delta)$.

(3) We have

$$
\begin{aligned}
{}_rM_q\{\xi\}(v_i) &= (\xi * e_i)\cap < q+1, \cdots, p+q >\\
&= \big(\xi\cap < i, q+1, \cdots, p+q >\big) * e_i \quad \cap < q+1, \cdots, p+q > \in \Omega\\
&\quad < q+1, \cdots, p+q >
\end{aligned}
$$

since $\xi\cap < i, q+1, \cdots, p+q > \in \Omega < i, q+1, \cdots, p+q >$ by (9.3) (3) and $\Omega < q+1, \cdots, p+q > \subset \bar{\Omega} < i, q+1, \cdots, p+q >$. Also,

$$
{}_rM_q\{\xi\}(v_i') = \big({}_rM_q\{\xi\}(v_i) * e_{q+1}\big) \quad \cap < q+2, \cdots, p+q > \in \Omega < q+2, \cdots, p+q > .
$$

The component $\bar{\Omega} < 0, q+1, \cdots, p+q >$ contains the point $e_{p+q,0}\big((-1)^p\epsilon\big)$ but not the point $e_{p+q-1,0}\big((-1)^p\delta^{-1}\epsilon)\big)$. The intersection $\Pi(\epsilon)\cap < q+1, \cdots, p+q-2 > = < q+1, \cdots, p+q > * e_{p+q-1,p+q}(-\delta)$. Hence the $(p-1)$-plane $\Pi(\epsilon)\cap < 0, q+1, \cdots, p+q >$ approaches the $(p-1)$-face $< q+1, \cdots, p+q >$ and divides $\bar{\Omega} < 0, q+1, \cdots, p+q >$ into $\bar{\Omega}^{<0}_{\epsilon}$ and $\bar{\Omega}^{>0}_{\epsilon}$ in such a way that $\bar{\Omega}^{<0}_{\epsilon}$ contains the vertex e_{p+q} as well as ${}_rM_q\{\xi\}(F)$ where $F = v_i, v_i'$. (See Figure 9.12.1).

(4) The first statement repeats (3) applied to $A_s\xi$. If $s = p+q-1$,

$$
\Pi(\epsilon)\cap < 0, q, q+1, \cdots \widehat{p+q-1}, p+q > = < e_q, e_{q+1}, \cdots, \widehat{e_{p+q-1}}, e_{p+q,0}\big((-1)^p\epsilon\big) > .
$$

Since the point $e_{p+q,0}\big((-1)^p\epsilon\big) \in \bar{\Omega} < 0, q, q+1, \cdots \widehat{p+q-1}, p+q >$, the claim follows (see Fig. (9.12.2)). The case $s = p+q$ is similar; note that $e_{p+q,0}\big((-1)^p\delta^{-1}\epsilon\big) \in \bar{\Omega} < 0, q, \cdots, p+q-1 >$. If $s = q+1, \cdots, p+q-2$, we argue as in (3).

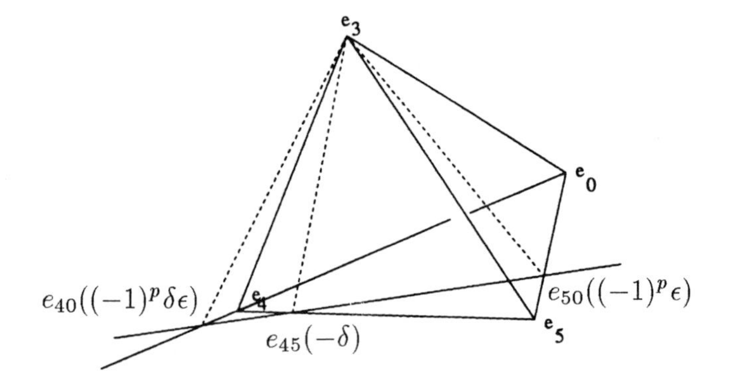

$< 0, q+1, \cdots, p+q >$ and $\Pi(\epsilon)$ (where $p = 3$, $q = 2$)

Figure 9.12.1

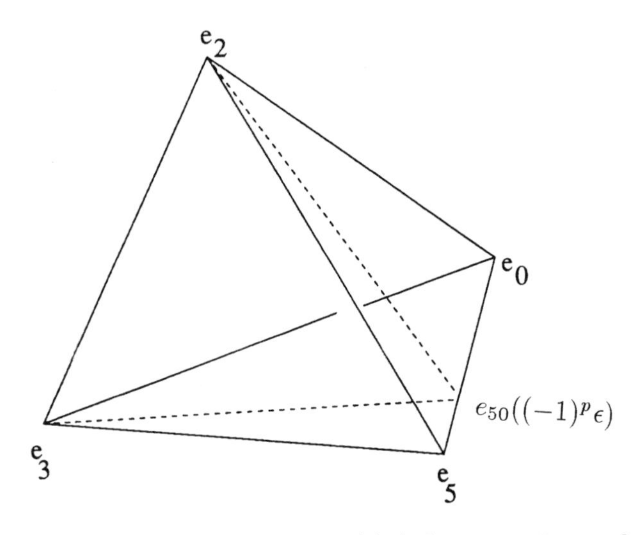

$< 0, q, \cdots, \hat{s}, \cdots, p+q >$ and $\Pi(\epsilon)$ (where $p = 3$, $q = 2$ and $s = 4$).

Figure 9.12.2

9.13 Lemma. *For $r = 0, 1$ and $k = 1, \cdots, q-1$, we have*

$$_rM_q\{\xi\}([v_k, v_q]) \quad \cap \Pi(\epsilon) \in \Omega < q+1, \cdots, p+q > \ .$$

(This point is independent of ϵ since $\Pi(\epsilon) \cap < q+1, \cdots, p+q >$ is.)

Proof. We have $_rM_q\{\xi\}([v_k, v_q]) = \Big(\xi * < e_k, e_q >\Big) \cap < q+1, \cdots, p+q >$.

We only need to show

(9.13.1) $(\xi * < e_k, e_q >) \cap < q+1, \cdots, p+q-1 > \subset \Omega < q+1, \cdots, p+q-1 >$.

In fact, by $\bar{\Omega} < q+1, \cdots, p+q > \supset \Omega < q+1, \cdots, p+q-1 >$ and (9.12)(2), it will then follow that the line $(\xi * < e_k, e_q >) \cap < q+1, \cdots, p+q >$ meets $\Pi(\epsilon)$ in a point $\in \Omega < q+1, \cdots, p+q >$. For (9.13.1), we look at

$$\begin{aligned}(\xi * < e_k, e_q >) \quad &\cap < k, q, q+1, \cdots, p+q-1 > \\ &= (\xi \cap < k, q, q+1, \cdots, p+q-1 >) * < e_k, e_q >\end{aligned}$$

which meets $< q+1, \cdots, p+q-1 >$ in $\Omega < q+1, \cdots, p+q-1 >$. (We have used $\xi \cap < k, q, q+1, \cdots, p+q-1 > \in \Omega < k, q, q+1, \cdots, p+q-1 >$ and $\bar{\Omega} < k, q, q+1, \cdots, p+q-1 > \supset \Omega < q+1, \cdots, p+q-1 >$.)

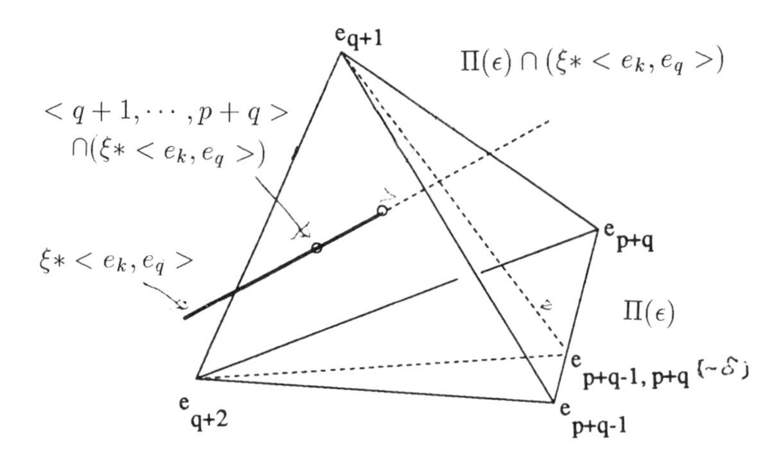

Figure 9.13 $< q+1, \cdots, p+q >$ and $\xi * < e_k, e_q >$

9.14 We will consider an admissible pair of $\mathcal{P}_q$-figures which is a perturbation of $\bar{M}_q\{\xi\}$. To define this, take the canonical subdivision of Δ^q along the 1-face $[v_0, v_q]$ by a hyperplane $\mathcal{D}$. Let (for $k = 1, \cdots, q-1$)

$$w_k = [v_0, v_k] \cap \mathcal{D}; \; w'_k = [v_q, v_k] \cap \mathcal{D}.$$

Let the division be $\Delta^q = \mathcal{P}'_q \cup \mathcal{S}_q$ where $\mathcal{P}'_q \ni v_0, v_q$ and $\mathcal{S}_q \ni v_1, \cdots, v_{q-1}$. There exists an identification $\mathcal{P}'_q = \mathcal{P}_q$ which sends $v_0, w_1, \cdots, w_{q-1}, v_q, w'_1, \cdots, w'_{q-1}$ to $v_0, v_1, \cdots, v_{q-1}, v'_0, v'_1, \cdots, v'_{q-1}$ (in the notation (9.7)), respectively. This identification is orientation-*reversing*. (Note that $\Delta^q[v_0, \cdots, v_q]$ induces the orientation $(-1)^q \Delta^{q-1}[v_0, \cdots, v_{q-1}]$ while $\Delta[v_0, w_1, \cdots, w_{q-1}] \times [0,1]$ induces the orientation $(-1)^{q-1}\Delta^{q-1}[v_0, w_1, \cdots, w_{q-1}] \times \{1\}$.) On the other hand, $\mathcal{S}_q \cong \Delta^{q-2} \times \Delta^2$ combinatorially via which the $(q-2)$-face $[v_1, \cdots v_{q-1}] \cong \Delta^{q-2} \times \{\text{a vertex of } \Delta^2\}$.

We take the refinement of the Δ^q-figure ${}_r\bar{M}_q\{\xi\}$ according to the subdivision, in which the face $\mathcal{D}$ corresponds to the hyperplane $\Pi(\epsilon)$; if F is a face of Δ^q meeting $\mathcal{D}$, we let

$${}_r\bar{M}_q\{\xi\}\ (F\cap\mathcal{D}) ={}_r \bar{M}_q\{\xi\}\ (F)\ \cap\Pi(\epsilon)\ .$$

This gives us a $\mathcal{P}'_q$-figure ($=\mathcal{P}_q$-figure) denoted ${}_r\bar{M}^\epsilon_q\{\xi\}$ (we employ the orientation of $\mathcal{P}_q$), and a $\mathcal{S}_q$-figure denoted ${}_r S^\epsilon_q\{\xi\}$. That these take values in $\bar{\Omega} < r, q+1,\cdots,p+q >$ follows from Lemma (9.12),(3) and Lemma (9.13). They respectively yield an admissible pair of $\mathcal{P}_q$-figures

$$\bar{M}^\epsilon_q\{\xi\} = \sum_{r=0,1}(-1)^r\ \mathcal{P}_q\big(\xi;< r,q+1,\cdots,p+q >; \overset{\underline{v_0},\cdots,\underline{v_{q-1}};\underline{v'_0}}{e_1,\cdots,e_{q-1};e_q}; \Pi(\epsilon);K\big)\ ,$$

(cf. (9.7) for the convention) and an admissible pair of $\mathcal{S}_q$-figures

$$S^\epsilon_q\{\xi\} = \sum_{r=0,1}(-1)^r\ {}_r S^\epsilon_q\{\xi\}\ .$$

To see rigorously that they form admissible pairs of figures, one proceeds as follows. Let $\mathcal{R}_{q+1} = \mathcal{R}'_{q+1}\cup\mathcal{R}''_{q+1}$ be the division obtained by excising $\mathcal{R}_{q+1}$ by a hyperplane $\mathcal{H}$ so that $\mathcal{H}$ excises $[u_0,u_1]*[v_1,\cdots v_q]$ about the 2-face $[u_0,u_1]*v_q$, and $\mathcal{R}'_{q+1}\ni u_0,u_1,v_q$ (see Fig. (9.14.3) in the case $q=2$). Note that on $u_0*[v_1,\cdots,v_q]$ and $u_1*[v_1,\cdots,v_q]$, the induced divisions coincide with the ones by $\mathcal{D}$ above, when u_0,u_1 are identified with v_0. By refining the $\mathcal{C}(\mathcal{R}_{q+1};[v_1,\cdots,v_q])$-figure $M_q\{\xi\}$ according to this division and $\Pi(\epsilon)$, we obtain the admissible $\mathcal{C}(\mathcal{R}'_{q+1};[v_1,\cdots,v_q]\cap\mathcal{R}'_{q+1})$-figure $\bar{M}^\epsilon_q$ and the admissible $\mathcal{C}(\mathcal{R}''_{q+1};[v_1,\cdots,v_q]\cap\mathcal{R}''_{q+1})$-figure S^ϵ_q.

By the additivity (7.12) (of admissible figures) we have

$$\int[\bar{M}_q] = -\int[\bar{M}^\epsilon_q] + \int[S^\epsilon_q]\ ;$$

(the negative sign occurs since we use the orientation opposite to that of $\mathcal{P}'_q$). The second term in the right hand side converges to zero:

9.15 Lemma *We have* $\int[S^\epsilon_q\{\xi\}]\to 0$ *as* $\epsilon\to 0$.

Proof. By the above construction, there is an admissible $\mathcal{R}''_{q+1}$-figure $T^\epsilon_q\{\xi\}$ such that

$$T^\epsilon_q\{\xi\}|_{\mathcal{R}''_{q+1}\cap u_0*[v_1,\cdots,v_q]} - T^\epsilon_q\{\xi\}|_{\mathcal{R}''_{q+1}\cap u_1*[v_1,\cdots,v_q]}$$

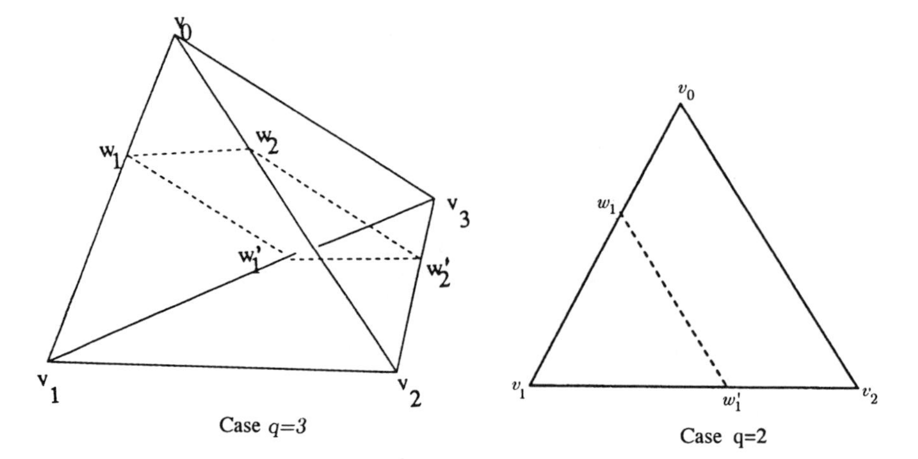

Figure 9.14.1

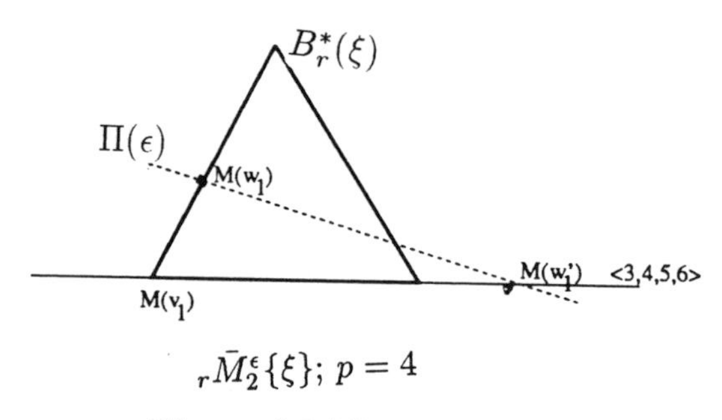

${}_r\bar{M}_2^{\epsilon}\{\xi\};\ p=4$

Figure 9.14.2

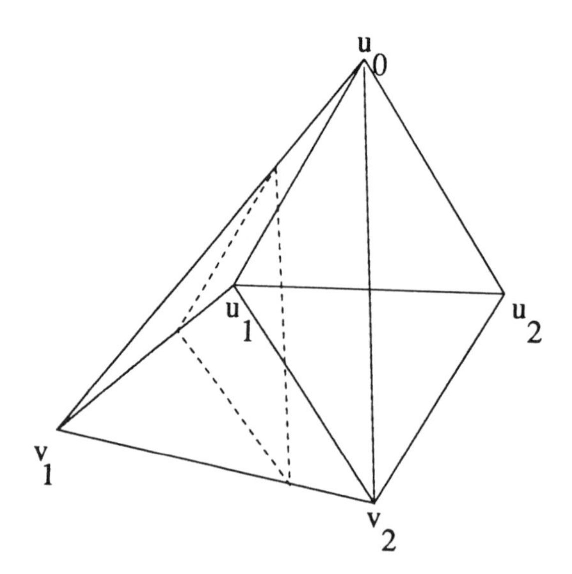

Figure 9.14.3

coincides with the admissible pair $S_q^\epsilon\{\xi\}$; $T_q^\epsilon\{\xi\}|_{\mathcal{R}''_{q+1}\cap\mathcal{H}}$ is supported on $\Pi(\epsilon)$, and $T_q^\epsilon\{\xi\}|_{\mathcal{R}''_{q+1}\cap[u_0,u_1]*[v_1,\cdots,v_{q-1}]}$ is an admissible figure. Since $\Pi(\epsilon)$ exhausts, for a fixed ϵ_1, the admissible figure $T_q^{\epsilon_1}\{\xi\}|_{\mathcal{R}''_{q+1}\cap[u_0,u_1]*[v_1,\cdots v_{q-1}]}$, by (6.17),

$$\int [T_q^\epsilon\{\xi\}|_{\mathcal{R}''_{q+1}\cap[u_0,u_1]*[v_1,\cdots v_{q-1}]}] \to 0 \ .$$

$T_q^\epsilon\{\xi\}$ is a union of the following two embedded polyhedra: the $\Delta^{q-1}\times\Delta^2$-figure with vertices

$$M(v_k) = (\xi * e_k)\cap < q+1,\cdots,p+q > \quad (k=1,\ldots,q);$$
$$B_r^*((\xi * e_k)\cap\Pi(\epsilon)) \quad (r=0,1; k=1,\ldots,q)$$

and the Δ^{q+1}-figure with vertices

$$M(v_q) = (\xi * e_q)\cap < q+1,\cdots,p+q >, \quad B_r^*((\xi * e_k)\cap\Pi(\epsilon)) \quad (r=0,1) \quad \text{and}$$
$$(\xi * < e_k, e_q >)\cap\Pi(\epsilon) \quad (q=1,\cdots,q-1).$$

Both of these embedded polyhedra that monotonely shrink to N_1. Applying Lemma (6.17.1), one has $\int[T_q^\epsilon\{\xi\}] \to 0$, and similarly for its differential. One applies Stokes formula to obtain the claim.

For $s = q+1,\cdots,p+q$, $M_{q-1}\{A_s\xi\}$ is an admissible pair of $\mathcal{P}_q$-figures with values in $< 0,q,\cdots,\hat{s},\cdots,p+q >$. We define its perturbation by

$$M_{q-1}^\epsilon\{A_s\xi\} = \sum_{r=0,1}(-1)^r\, \mathcal{P}_q(A_s\xi; < r,q,\cdots,\hat{s},\cdots,p+q >; e_1,\cdots,e_{q-1}; e_q; \Pi(\epsilon); K) \ ,$$

which is a pair of *generic* $\mathcal{P}_q$-figures in $< 0,q,\cdots,\hat{s},\cdots,p+q >$, see Figure 9.16.2.

9.16 Proposition. *We have*

(1) $\int[\bar{M}_q^\epsilon\{\xi\}] \to \int[\bar{M}_q\{\xi\}] \quad (\epsilon\downarrow 0)$.

(2) For $s = q+1,\cdots,p+q$,

$$\int[M_{q-1}^\epsilon\{A_s\xi\}] \to \int[M_{q-1}\{A_s\xi\}] \quad (\epsilon\downarrow 0) \ .$$

(3) For any small ϵ,

$$\int[\bar{M}_q^\epsilon\{\xi\}] = \sum_{s=q+1,\cdots,p+q}(-1)^{s-q-1}\int[M_{q-1}^\epsilon\{A_s\xi\}].$$

Note that Proposition (9.11) hence follows. (1) is a restatement of (9.15); (2) follows from (9.12), (4) and (7.9). We have been left only with the proof of (3). For the rest of this section, we fix an ϵ.

Define a p-plane in the $(p+1)$-face $< 0, q, q+1, \cdots, p+q >$ by

$$\mathcal{K} =< e_{0,q}(-1), e_{q+1}, \cdots, e_{p+q-1}, e_{p+q} > \ .$$

We then have

$$\mathcal{K} \cap < 0, q, \cdots \hat{s}, \cdots, p+q >= K < 0, q, \cdots \hat{s}, \cdots, p+q > \quad (s = q+1, \cdots p+q)$$

and

$$\mathcal{K} \cap < 0, q+1, \cdots, p+q >=< q+1, \cdots, p+q > .$$

Let $\mathcal{K}(\gamma)$, $0 < \gamma \ll 1$ be a real analytic family of p-planes in

$$< 0, q, q+1, \cdots, p+q >$$

such that

(i) $\mathcal{K}(\gamma) \to \mathcal{K}$ as $\gamma \to 0$;

(ii) The intersection $\mathcal{K}(\gamma) \cap < 0, q+1, \cdots, p+q >$ is a $(p-1)$-plane which exhausts $\bar{M}_q^\epsilon\{\xi\}$ towards $< q+1, \cdots, p+q >$ inside $\bar{\Omega} < 0, q+1, \cdots, p+q >$.

Consider the perturbations of $\bar{M}_q^\epsilon\{\xi\}$ and $M_{q-1}^\epsilon\{A_s\xi\}$, obtained by perturbing $\mathcal{K}$ to $\mathcal{K}(\gamma)$ as follows:

$$\bar{M}_q^{\epsilon,\gamma}\{\xi\} = \sum_{r=0,1} (-1)^r \ \mathcal{P}_q\Big(\xi; < r, q+1, \cdots, p+q >; e_1, \cdots, e_{q-1}; e_q; \Pi(\epsilon); \mathcal{K}(\gamma)\Big) \ ;$$

$$M_{q-1}^{\epsilon,\gamma}\{A_s\xi\} = \sum_{r=0,1} (-1)^r \ \mathcal{P}_q\Big(\xi; < r, q, q+1, \cdots, \hat{s}, \cdots, p+q >; e_1, \cdots, e_{q-1}; e_q; \Pi(\epsilon); \mathcal{K}(\gamma)\Big).$$

These are pairs of generic $\mathcal{P}_q$-figures. In the case $p = 4, q = 2$ they are pictured in Figures 9.16 (next page).

Proposition (9.16), (3) now follows from:

9.17 Proposition. *(1)* $\int[\bar{M}_q^{\epsilon,\gamma}\{\xi\}] \to \int[\bar{M}_q^\epsilon\{\xi\}] \quad (\gamma \downarrow 0)$.

(2) $\int[M_{q-1}^{\epsilon,\gamma}\{A_s\xi\}] \to \int[M_{q-1}^\epsilon\{A_s\xi\}] \quad (\gamma \downarrow 0)$.

(3) $\int[\bar{M}_q^{\epsilon,\gamma}\{\xi\}] + \sum_{s=q+1,\cdots,p+q}(-1)^{s-q} \int[M_{q-1}^{\epsilon,\gamma}\{A_s\xi\}] = 0$.

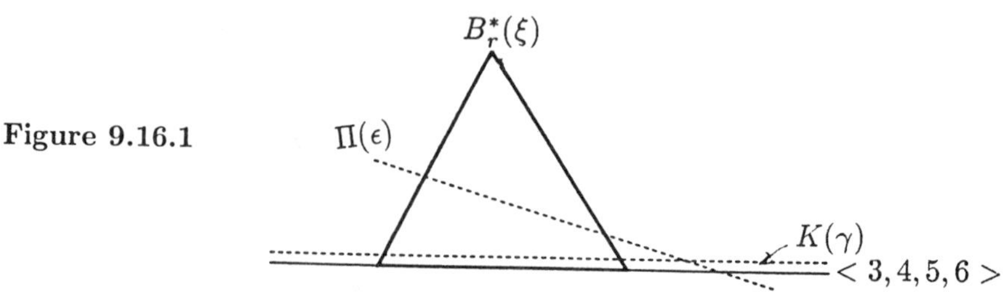

Figure 9.16.1

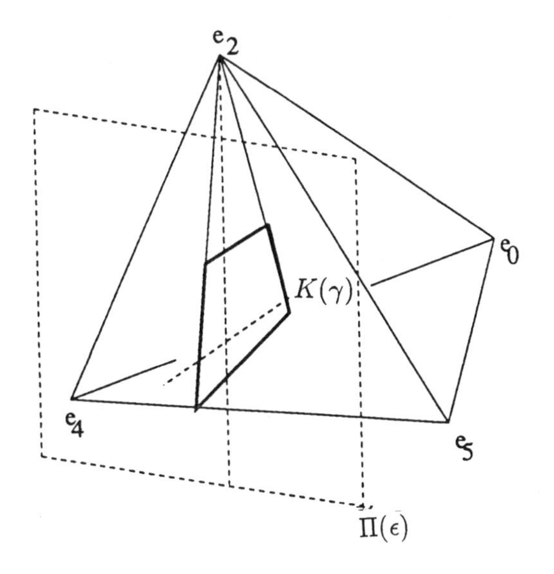

${}_0M_1^{\epsilon}\{A_3\xi\}$ in $< 0,2,4,5 >$

Figure 9.16.2

Proof. (1) follows from the limit formula (7.9).

(2) follows from the continuity (6.21).

(3) For $r = 0,1$, consider the assignment ${}_r\mathcal{M}\{\xi\}$ of relative subspaces of $\mathbb{P}^{p+1} \times V =< r,q,q+1,\cdots,p+q >$ to the faces of $\mathcal{P}_q$ given as follows:

$$\begin{aligned}
[v_0, v_{k_1}, \cdots, v_{k_j}] &\mapsto (\xi * < k_1, \cdots, k_j >) \cap < r,q,q+1,\cdots,p+q > ;\\
[v_0, v_{k_1}, \cdots, v_{k_j}] \times [0,1] &\mapsto (\xi * < k_1, \cdots, k_j, q >) \cap < r,q,q+1,\cdots,p+q > ;\\
[v_{k_1}, \cdots, v_{k_j}] &\mapsto (\xi * < k_1, \cdots, k_j >) \cap \Pi(\epsilon) ;\\
[v_{k_1}, \cdots, v_{k_j}] \times [0,1] &\mapsto (\xi * < k_1, \cdots, k_j, q >) \cap \Pi(\epsilon) ;\\
[v'_0, v'_{k_1}, \cdots, v'_{k_j}] &\mapsto (\xi * < k_1, \cdots, k_j >) \cap \mathcal{K}(\gamma) ;\\
[v'_{k_1}, \cdots, v'_{k_j}] &\mapsto (\xi * < k_1, \cdots, k_j >) \cap \Pi(\epsilon) \cap \mathcal{K}(\gamma) .
\end{aligned}$$

We see that $(s = q+1, \cdots, p+q)$

$$\begin{aligned}
{}_r\mathcal{M} \cap < r, q+1, \cdots, p+q > &=_r \bar{M}_q^{\epsilon,\gamma}\{\xi\} ;\\
{}_r\mathcal{M} \cap < r, q, \cdots, \hat{s}, \cdots, p+q > &=_r M_{q-1}^{\epsilon,\gamma}\{A_s\xi\} ;\\
{}_0\mathcal{M} \cap < q, q+1, \cdots, p+q > &=_1 \mathcal{M} \cap < q, q+1, \cdots, p+q > .
\end{aligned}$$

(The last holds since $< q, q+1, \cdots, p+q >$ is the common p-face of $< r, q, q+1, \cdots, p+q >$ $(r = 0,1)$.) Since ${}_r\bar{M}_q^{\epsilon,\gamma}\{\xi\}$ and ${}_rM_{q-1}^{\epsilon,\gamma}\{A_s\xi\}$ are generic figures, so are ${}_r\mathcal{M} \cap < q, q+1, \cdots, p+q >$. One applies to ${}_r\mathcal{M}$ the cancellation lemma of the differential forms associated with generic figures (8.3):

$$\int [{}_r\mathcal{M} \cap < q, q+1, \cdots, p+q >] - \int [{}_r\bar{M}_q^{\epsilon,\gamma}\{\xi\}] + \sum_{s=q+1,\cdots,p+q} (-1)^{s-q-1} \int [{}_r M_{q-1}^{\epsilon,\gamma}\{A_s\xi\}] = 0$$

Taking the difference for $r = 0, 1$, we obtain (3).

References

[BGSV] Beilinson, A., Goncharov, A., Schechtman, V., Varchenko, A., *Aomoto dilogarithms, mixed Hodge structures, and motivic cohomology of pairs of triangles on the plane*, Grothendieck Festschrift, Vol. I, Progress in Mathematics, 135–172.

[BMS] Beilinson, A., MacPherson, R., Schechtman, V., Notes on motivic cohomology, *Duke Math. J.* **54** (1987), 679—710.

[Bo] Bott, R., Lectures on characteristic classes and foliations, in *Lecture Notes in Mathematics* **279**, Springer.

[GM] Gelfand, I.M., MacPherson, R., Geometry of Grassmannians and a generalization of the dilogarithm, *Adv. in Math.* **44** (1982), 279–312.

[G] Goncharov, A., Geometry of configurations, polylogarithms and motivic cohomology, preprint.

[HR] Hain, R., The existence of higher logarithms, preprint.

[HR-M] Hain, R., MacPherson, R., Higher logarithms, Ill. *J. of Math.* **34** (1990), 392–475.

[HM] Hanamura, M., Dilogarithm, Grassmannian complex and scissors congruence groups of algebraic polyhedra, preprint.

[HM-M] Hanamura, M., MacPherson, R., Geometric construction of polylogarithms, *Duke Math.J.* **70** (1993), 481–516.

[L] Lewin, L. (ed.), Structural Properties of Polylogarithms.

[Y] Yang, J., The Hain-MacPherson trilogarithm, the Borel regulators and the value of the Dedekind zeta function at 3, preprint.

R. D. MacPherson
Institute for Advanced Study
Princeton, NJ 08540
rdm@ias.edu

M. Hanamura
Max Planck Institüt für Mathematik
53225 Bonn
Germany

Received May 1995

A Note on Localization and the Riemann-Roch Formula

Lisa C. Jeffrey and Frances C. Kirwan*

Introduction

Let M be a compact symplectic manifold of (real) dimension $2m$, equipped with the Hamiltonian action of a compact connected Lie group K with maximal torus T; we denote the moment map for this action by $\mu : M \to \mathbf{k}^*$. In this note, we shall treat some properties of the symplectic quotient $M_{\text{red}} = \mu^{-1}(0)/K$, whose symplectic structure ω_0 descends from the symplectic structure on M. (We assume that 0 is a regular value of μ, so that M_{red} has at worst finite quotient singularities.) We shall describe some applications of the main result of [16] (Theorem 8.1, the residue formula): this formula specifies the evaluation on the fundamental class of M_{red} of $\eta_0 e^{\omega_0}$, for any class $\eta_0 \in H^*(M_{\text{red}})$. The residue formula relates cohomology classes on M_{red} to the equivariant cohomology $H^*_K(M)$ of M, via the natural ring homomorphism $\kappa_0 : H^*_K(M) \to H^*(M_{\text{red}})$ whose surjectivity was proved in [20].

For any $\eta \in H^*_K(M)$, this formula expresses $\kappa_0(\eta)e^{\omega_0}[M_{\text{red}}]$ in terms of the restriction of η to the fixed point set of the maximal torus T in M: it is an application of the abelian localization formula of Berline and Vergne [5] (for which Atiyah and Bott gave a topological proof in [1]).

One application of the residue formula is the determination of the ring structure of the cohomology of the space M_{red}, in terms of data at the fixed point set of T. Because of the surjectivity of the map κ_0, the cohomology of M_{red} is determined if we know the kernel of κ_0. We may study this kernel via the observation that since $H^*(M_{\text{red}})$ satisfies Poincaré duality, an element η is in the kernel of κ_0 if and only if for all $\zeta \in H^*_K(M)$ we have

$$\kappa_0(\zeta)\kappa_0(\eta)[M_{\text{red}}] = 0. \tag{1.1}$$

Thus knowing the ring structure of $H^*(M_{\text{red}})$ reduces in principle to knowing the *intersection pairings* in the cohomology of M_{red}.

The residue formula is related to a result of Witten (the nonabelian localization theorem [25]): Witten's work indeed provided the starting point for our investigation. Like the residue formula, Witten's theorem expresses $\eta_0 e^{\omega_0}[M_{\text{red}}]$ in terms of appropriate data on M. Witten's objective was to characterize certain integrals associated to an equivariant cohomology class η on M: he did

* Partially supported by NSF grant DMS-9306029.

this in terms of the sum of a contribution from $\eta_0 e^{\omega_0}[M_{\rm red}]$ and certain additional contributions associated to higher critical points of the function $|\mu|^2$ on M. His result was phrased in terms of equivariant cohomology, and his proof was broadly analogous to the methods used by Berline and Vergne. Witten's motivation was to study the cohomology rings of moduli spaces of flat connections on a Riemann surface Σ; applying nonabelian localization *formally* to an appropriate infinite-dimensional manifold (the space of all connections on Σ), he was able to obtain formulas for a generating functional which in principle determines all the intersection pairings in the cohomology rings of these moduli spaces.

Here we turn our attention to a different problem, that of relating the Riemann-Roch number of a certain line bundle $\mathcal{L}_{\rm red}$ over $M_{\rm red}$ to appropriate information about M. Let us assume there exists a line bundle $\mathcal{L}$ on M for which $c_1(\mathcal{L}) = \omega$, and that the action of K on M lifts to an action on the total space of $\mathcal{L}$. Under the assumption that K acts freely on $\mu^{-1}(0)$, we get a line bundle $\mathcal{L}_{\rm red}$ on $M_{\rm red}$ whose first Chern class is ω_0. The characteristic class $\mathrm{ch}(\mathcal{L}_{\rm red})\mathrm{Td}(M_{\rm red})$ (which expresses the Riemann-Roch number of a line bundle $\mathcal{L}_{\rm red}$ on $M_{\rm red}$ whose first Chern class is ω_0) is naturally of the form $\eta_0 e^{\omega_0}[M_{\rm red}]$ which appears in our residue formula.

Suppose also that there exists a K-invariant Kähler structure on M, in other words a complex structure compatible with ω and preserved by the action of K. The bundle $\mathcal{L}$ then acquires a holomorphic structure in a standard manner, and we define the *quantizations* $\mathcal{H}$ and $\mathcal{H}_{\rm red}$ to be the virtual vector spaces

$$\mathcal{H} = \oplus_{j\geq 0}(-1)^j H^j(M,\mathcal{L}) \tag{1.2}$$

and

$$\mathcal{H}_{\rm red} = \oplus_{j\geq 0}(-1)^j H^j(M_{\rm red},\mathcal{L}_{\rm red}). \tag{1.3}$$

The space $\mathcal{H}$ is a virtual representation of K. The Riemann-Roch numbers $RR^K(\mathcal{L})$ and $RR(\mathcal{L}_{\rm red})$ are defined by

$$RR^K(\mathcal{L}) = \sum_{j\geq 0}(-1)^j \dim H^j(M,\mathcal{L})^K \tag{1.4}$$

$$RR(\mathcal{L}_{\rm red}) = \sum_{j\geq 0}(-1)^j \dim H^j(M_{\rm red},\mathcal{L}_{\rm red}). \tag{1.5}$$

Under the same conditions as above (and some additional positivity hypotheses on $\mathcal{L}$), Guillemin and Sternberg showed in [13] that

$$RR(\mathcal{L}_{\rm red}) = RR^K(\mathcal{L}). \tag{1.6}$$

A special case of our residue formula is equivalent to this result, as we shall show below in the case $K = U(1)$. Our original motivation for considering Riemann-Roch numbers was to provide a link between the residue we had defined and more standard definitions of residues in algebraic geometry (such as the Grothendieck residue [15]). Since we first began considering the application of the residue formula to Riemann-Roch numbers, several papers have appeared which extend the Guillemin-Sternberg result to a wider class of situations, and in which the main tool is localization in equivariant cohomology. There are two approaches, one due to Guillemin [12], the other due to Vergne [23]. What is important about all the proofs based on localization (including the one we shall present) is that they apply under considerably weaker hypotheses than Guillemin and Sternberg's original proof. Guillemin and Sternberg imposed a positivity hypothesis on the line bundle $\mathcal{L}$, and required that M have a K-invariant Kähler structure. However the application of the residue formula to yield a formula for $RR(\mathcal{L}_{\text{red}})$ requires only that there exist an *almost complex* structure on M compatible with the action of K (see Section 3 of [12]): such a structure enables one to define a spin-$\mathbb{C}$ Dirac operator which can be used to define the virtual vector space $\mathcal{H}$. Furthermore, one need not require any positivity hypothesis on the line bundle $\mathcal{L}$ in order for these proofs of Guillemin and Sternberg's result to be valid.

Guillemin's proof [12] uses the residue formula to reduce the verification of (1.6) to a combinatorial identity involvig lattice points in polyhedra. Guillemin then observes that this identity is known when K is a torus acting in a quasi-free manner. Meinrenken [22] has subsequently extended this proof to torus actions which need not be quasi-free.

Vergne [23] has given a different proof of the Guillemin-Sternberg conjecture when K is a torus, also using ideas based on localization in equivariant cohomology. Her proof likewise does not require a Kähler structure or positivity of the line bundle $\mathcal{L}$.

Although many features of the rank one case are quite special, and although the proofs of Vergne and Guillemin-Meinrenken described above apply in much greater generality, we felt nevertheless that it was instructive to give a written account of our approach to this case since it is simple and self-contained. Below we sketch our proof of the Guillemin-Sternberg conjecture when $K = U(1)$. The result is the following:

Theorem 4.5: *Suppose $K = U(1)$ acts in a Hamiltonian fashion on the Kähler manifold M, in such a way that 0 is a regular value of μ. Then $RR^K(\mathcal{L}) = RR(\mathcal{L}_{\text{red}})$.*

A more detailed account is given in [17], where we also treat the case $K = SO(3)$ (under a small hypothesis on the image of the moment map for the maximal torus T). As described above, the residue formula specifies the evaluation $\eta_0 e^{\omega_0}[M_{\rm red}]$, where η_0 is any class in $H^*(M_{\rm red})$; the special value of η_0 we consider is

$$\eta_0 = \mathrm{Td}(M_{\rm red}) \tag{1.7}$$

which comes from a particular equivariant cohomology class η on M. By the Riemann-Roch formula, we have[1]

$$RR(\mathcal{L}_{\rm red}) = \eta_0 e^{\omega_0}[M_{\rm red}]. \tag{1.8}$$

In Section 2, we shall apply the residue formula to give a formula for the right hand side of (1.8) as a sum over the components of the fixed point set of T. In Section 3, we correspondingly apply the holomorphic Lefschetz formula to obtain a similar fixed point sum for $RR^K(\mathcal{L})$. Finally, in Section 4 we identify the two expressions.

We owe thanks to Miles Reid, who originally proposed that we should investigate the possibility of trying to reformulate the residue in [16] in more algebro-geometric terms, and suggested that there might be a relation between our residue formula and the Riemann-Roch theorem.

2. The residue formula

We now recall the main result (the residue formula, Theorem 8.1) of [16]. The residue formula is phrased in terms of equivariant cohomology: if M is a compact oriented manifold acted on by a compact connected Lie group K, the K-equivariant cohomology $H^*_K(M)$ of M may be identified with the cohomology of the chain complex (see Chapter 7 of [4])

$$\Omega^*_K(M) = \Big(S(\mathbf{k}^*) \otimes \Omega^*(M)\Big)^K \tag{2.1}$$

equipped with the differential[2]

$$D = d - i\iota_{X_M} \tag{2.2}$$

where X_M is the vector field on M generated by the action of $X \in \mathbf{k}$. Sometimes we shall use an appropriate formal completion $\hat{\Omega}^*_K(M)$ of $\Omega^*_K(M)$: it will turn

---

[1] This is true provided $M_{\rm red}$ is a manifold, which follows if K acts freely on $\mu^{-1}(0)$: in the more general case $RR(\mathcal{L}_{\rm red})$ is given by Kawasaki's Riemann-Roch theorem for orbifolds (Theorem 4.4).

[2] This (nonstandard) definition of the equivariant cohomology differential is different from that used in [16] but consistent with that used in [25]. We have found it convenient to introduce this definition to obtain consistency with the formulas in Section 4.

out to be convenient to make use of the $\Omega^*_K(M)$ module

$$\hat{\Omega}^*_K(M) = \{\eta e^{\omega_{\mathbf{k}}} : \quad \eta \in \Omega^*_K(M)\}, \tag{2.3}$$

where we have introduced the quantity $\omega_{\mathbf{k}}(X) = \omega + i\mu(X)$ which is the extension of the symplectic form ω to an equivariantly closed 2-form on M.

We shall make use of equivariant characteristic classes (see Section 7.1 of [4]). The most important for our purposes are the equivariant Chern character and the equivariant Euler class. Suppose F is a component of the fixed point set of the maximal torus T in M. We may (formally) decompose the normal bundle ν_F to F (using the splitting principle if necessary) as a sum of line bundles $\nu_F = \sum_{j=1}^N \nu_{F,j}$, in such a way that T acts on $\nu_{F,j}$ with weight $\beta_{F,j} \in \mathbf{t}^*$.[3] The T-equivariant Euler class e_F of the normal bundle ν_F is then defined for $X \in \mathbf{t}$ by

$$e_F(X) = \prod_{j=1}^N \big(c_1(\nu_{F,j}) + i\beta_{F,j}(X)\big). \tag{2.4}$$

In terms of the notation we have introduced, the residue formula is stated as follows:

Theorem 2.1 [[16]] *Let $\eta \in H^*_K(M)$ induce $\eta_0 \in H^*(M_{\mathrm{red}})$. Then we have*

$$\eta_0 e^{\omega_0}[M_{\mathrm{red}}] = n_0 C^K \mathrm{res}\left(\varpi^2(X) \sum_{F \in \mathcal{F}} r^\eta_F(X)[dX]\right), \tag{2.5}$$

where n_0 is the order of the subgroup of K that acts trivially on $\mu^{-1}(0)$, and the constant C^K is defined by

$$C^K = \frac{i^l}{(2\pi)^{s-l}|W| \operatorname{vol}(T)}. \tag{2.6}$$

Here, T is the maximal torus and $|W|$ is the order of the Weyl group. We have introduced $s = \dim K$ and $l = \dim T$. The meromorphic function r^η_F on $\mathbf{t} \otimes \mathbb{C}$ is defined by

$$r^\eta_F(X) = e^{i\mu_T(F)(X)} \int_F \frac{i^*_F(\eta(X)e^{\omega})}{e_F(X)}. \tag{2.7}$$

Here, $\mathcal{F}$ denotes the set of components of the fixed point set of T, $i_F : F \to M$ is the inclusion and e_F is the equivariant Euler class of the normal bundle

[3] Throughout this paper we shall use the convention that weights $\beta_{F,j} \in \mathbf{t}^*$ send the integer lattice $\Lambda^I = \mathrm{Ker}\,(\exp : \mathbf{t} \to T)$ to $\mathbb{Z}$.

to F in M, which was defined at (2.4). The polynomial $\varpi : \mathbf{t} \to \mathbb{R}$ is defined by $\varpi(X) = \prod_{\gamma>0} \gamma(X)$, where γ runs over the positive roots.

The general definition of the residue res was given in Section 8 of [16]. Here we shall treat the case where $K = U(1)$, for which the results are as follows. See Footnote 3 for our conventions on weights.

Corollary 2.2 ([16]; [18], [26]). *In the situation of Theorem 2.1, let $K = U(1)$. Then*

$$\eta_0 e^{\omega_0}[M_{\text{red}}] = i n_0 \text{res}_0 \Big(\sum_{F \in \mathcal{F}_+} r_F^\eta(X) d\lambda(X) \Big).$$

Here, the meromorphic function r_F^η on $\mathbb{C}$ was defined by (2.7), and res_0 *denotes the coefficient of the meromorphic 1-form $d\lambda(X)/\lambda(X)$ on $\mathbf{k} \otimes \mathbb{C}$, where $X \in \mathbf{k}$ and λ is the generator of the weight lattice of $U(1)$. The set $\mathcal{F}_+$ is defined by $\mathcal{F}_+ = \{F \in \mathcal{F} : \mu_T(F) > 0\}$. The integer n_0 is as in Theorem 2.1.*

If h is the restriction to $\mathbf{t}$ of a meromorphic function on $\mathbf{t} \otimes \mathbb{C}$, the residue $\text{res}(h dX)$ is defined ([16], Definition 8.5 and Proposition 8.7) as the value at zero of the Fourier transform of $h dX$ (under appropriate hypotheses on h, and with an appropriate regularization of the Fourier transform). The residue formula is proved (when the action of K on $\mu^{-1}(0)$ is effective, from which the general result follows easily) by first identifying

$$\eta_0 e^{\omega_0}[M_{\text{red}}] = C_1^K F_K(\Pi_* \eta e^{\omega_{\mathbf{k}}})(0) \tag{2.8}$$

([16], Proposition 8.10(a)). The constant C_1^K is defined as $C_1^K = (2\pi)^{-s/2} \operatorname{vol}(K)^{-1}$; we denote the Fourier transform on $\mathbf{k}$ by F_K, and the Fourier transform on $\mathbf{t}$ by F_T (in the conventions of (3.3) and (3.4) of [16]). The equation

$$\Pi_*(\eta e^{\omega_{\mathbf{k}}})(X) = \int_M \eta(X) e^{\omega_{\mathbf{k}}(X)}$$

defines a C^∞ function of $X \in \mathbf{k}$. Equation (2.8) follows from the *normal form theorem* [14, 21], which gives a normal form for the symplectic form, the moment map and the K action in a neighbourhood of $\mu^{-1}(0)$.

The Weyl integration formula is then used to express $F_K(\Pi_* \eta e^{\omega_{\mathbf{k}}})(0)$ in terms of the restriction to the Lie algebra of the maximal torus T, so we have also

$$\eta_0 e^{\omega_0}[M_{\text{red}}] = C_2^K F_T\Big(\varpi^2 \Pi_*\big(\eta e^{\omega_{\mathbf{k}}}\big)\Big)(0) \tag{2.9}$$

where F_T is the Fourier transform over $\mathbf{t}$ and

$$C_2^K = \frac{(2\pi)^{l/2}}{(2\pi)^s |W| \operatorname{vol}(T)}$$

is another constant. Finally the abelian localization formula [5] is used to express $\Pi_*\big(\eta e^{\omega_{\mathbf{k}}}\big)(X)$ (for $X \in \mathbf{t}$) as a sum over the components F of the fixed point set of T in M_{red}: this results in Theorem 2.1.

Let us examine the case when T has rank 1: see the proof of Corollary 8.2 in [16]. We identify $\mathbf{t}$ with $\mathbb{R}$. Each term $r_F^\eta(X)$ (2.7) is a sum of terms

$$\tau_\alpha(X) = c_\alpha X^{-n_\alpha} e^{i\mu_T(F)X} \tag{2.10}$$

for some constants c_α and integers n_α. The residue is given (see [16], (8.28)) by

$$\operatorname{res}(\tau_\alpha) = \lim_{\epsilon\to 0} \frac{1}{2\pi i} \int_{X\in\mathbb{R}-i\zeta} \chi(\epsilon X)\tau_\alpha(X)dX \tag{2.11}$$

where one chooses $\zeta > 0$, and χ is the extension to a holomorphic function on $\mathbb{C}$ of a compactly supported function on $\mathbb{R}$. We see that $\operatorname{res}(\tau_\alpha) = 0$ when $n_\alpha \le 0$. When $n_\alpha > 0$, we complete the integral over $\mathbb{R} - i\zeta$ to a contour integral by adding a semicircular curve at infinity, which is in the upper half plane if $\mu_T(F) > 0$ and in the lower half plane if $\mu_T(F) < 0$. This choice of contour is made so that the function τ_α is bounded on the added contours, so the added semicircular curves contribute zero to the integral. Since only those contours corresponding to values of F for which $\mu_T(F) > 0$ enclose the pole at 0, application of Cauchy's residue formula gives the result.

We now specialize to the case $\eta_0 = \operatorname{Td}(M_{\text{red}})$ (see (1.7)). Recall that the Todd class of a vector bundle V is given in terms of the Chern roots x_l by

$$\operatorname{Td}(V) = \prod_l \frac{x_l}{1 - e^{-x_l}} = \sum_{j\ge 0} \operatorname{Td}_j(V),$$

where Td_j is a homogeneous polynomial of degree j in the x_l. If the Todd class is given in terms of the Chern roots by

$$\operatorname{Td} = \tau(x_1, \dots, x_N)$$

then the T-equivariant Todd class of the normal bundle ν_F is given for $X \in \mathbf{t}$ by

$$\operatorname{Td}_T(\nu_F)(X) = \tau\Big(c_1(\nu_{F,1}) + i\beta_{F,1}(X), \dots, c_1(\nu_{F,N}) + i\beta_{F,N}(X)\Big). \tag{2.12}$$

We may also define the K-equivariant Todd class $\mathrm{Td}_K(V)$ of any K-equivariant vector bundle V on M, and in particular the equivariant Todd class $\mathrm{Td}_K(M) = \mathrm{Td}_K(TM)$ of M. Because we wish to work with $\hat{\Omega}^*_K(M)$ (as defined by (2.3)), it will be convenient to introduce the *truncated* equivariant Todd class of V, defined by

$$\mathrm{Td}_K^{<n>}(V) = \sum_{j=0}^{n} (\mathrm{Td}_K)_j(V) \in H^*_K(M), \tag{2.13}$$

for any $n > 0$.

Assume that T acts at the fixed point F with weights $\beta_{F,j} \in \mathbf{t}^*$. From now on we assume that the action of K on $\mu^{-1}(0)$ is effective, so that $n_0 = 1$ in Theorem 2.1.

Proposition 2.3 *Suppose that K is abelian. We then have*

$$\int_{M_{\mathrm{red}}} \mathrm{ch}(\mathcal{L}_{\mathrm{red}})\mathrm{Td}(M_{\mathrm{red}})$$
$$= C^K \mathrm{res}\left(\sum_{F\in\mathcal{F}} e^{i\mu_T(F)(X)} \times \int_F \frac{e^{\omega}\mathrm{Td}_K^{<n>}(\nu_F)(X)\mathrm{Td}(F)}{e_F(X)} \right). \tag{2.14}$$

This is equal to $RR(\mathcal{L}_{\mathrm{red}})$ provided K acts freely on $\mu^{-1}(0)$.

Here, the constant C^K was define in (2.6). We have decomposed the restriction to F of the K-equivariant Todd class of M as

$$\mathrm{Td}_K(M)(X) = \mathrm{Td}_K(\nu_F)(X)\mathrm{Td}(TF). \tag{2.15}$$

We have used the multiplicativity of the Todd class and the fact that the action of T on TF is trivial. Then the Proposition follows immediately from Theorem 2.1.

The special case of Proposition 2.3 when $K = U(1)$ is:

Proposition 2.4 *If $K = U(1)$, we have*

$$\int_{M_{\mathrm{red}}} \mathrm{ch}(\mathcal{L}_{\mathrm{red}})\mathrm{Td}(M_{\mathrm{red}})$$
$$= i\mathrm{res}_0\Big(\sum_{F\in\mathcal{F}_+} e^{i\mu_T(F)(X)} \int_F \frac{e^{\omega}\mathrm{Td}_K^{<n>}(\nu_F)(X)\mathrm{Td}(F)}{e_F(X)} d\lambda(X)\Big) \tag{2.16}$$

This is equal to $RR(\mathcal{L}_{\mathrm{red}})$ provided K acts freely on $\mu^{-1}(0)$.

Here, $X \in \mathbf{k}$ and res_0 denotes the coefficient of the meromorphic 1-form $d\lambda(X)/\lambda(X)$ on $\mathbf{k}\otimes\mathbb{C}$, where the element $\lambda \in \mathbf{k}^*$ is the generator of the weight lattice of $\mathbf{k}$.

3. The holomorphic Lefschetz formula

The next important ingredient in the argument is the holomorphic Lefschetz formula:

Theorem 3.1 (Holomorphic Lefschetz formula) *Let $t \in T$ be such that the fixed point set of t in M is the same as the fixed point set $\cup_{F\in\mathcal{F}} F$ of T in M; then the character $\chi(t)$ of the action of t on $\mathcal{H}$ is given by*

$$\chi(t) = \sum_{F\in\mathcal{F}} \chi_F(t),$$

where

$$\begin{aligned}\chi_F(t) &= \int_F \frac{i_F^*\mathrm{ch}_T(\mathcal{L})(t)\mathrm{Td}(F)}{\prod_j (1 - t^{-\beta_{F,j}} e^{-c_1(\nu_{F,j})})} \\ &= \sum_{F\in\mathcal{F}} t^{\mu_T(F)} \int_F \frac{e^{\omega}\mathrm{Td}(F)}{\prod_j (1 - t^{-\beta_{F,j}} e^{-c_1(\nu_{F,j})})}\end{aligned} \tag{3.1}$$

Here, the $\beta_{F,j} \in$ Hom $(T, U(1)) \subset \mathbf{t}^*$ are the weights of the action of T on the normal bundle ν_F of F in M.

The theorem is proved by Atiyah and Singer ([3], Theorem 4.6), and is based on results of Atiyah and Segal [2]; an exposition of the general result from which the theorem follows is given in Theorem 6.16 of [4]. A more general equivariant index theorem involving equivariant cohomology is proved by Berline and Vergne in [6]. The statement given above is in a form that will be convenient for us. For any weight β, we define t^β as $\exp(2\pi i\beta(X)) \in U(1) \subset \mathbb{C}^\times$, where $t = \exp(X)$ and the weights β have been chosen to send the integer lattice Λ^I in $\mathbf{t}$ to $\mathbb{Z} \subset \mathbb{R}$.

When the T action has isolated fixed points, (3.1) reduces to

$$\chi_F(t) = \frac{t^{\mu_T(F)}}{\prod_j (1 - t^{-\beta_{F,j}})} \tag{3.2}$$

When the action of T is quasi-free (in other words, free off the fixed point set of T), the formulae (3.1) and (3.2) are valid for $t \neq 1$. Otherwise they hold for all t which act freely off $M - M^T$.

In the general case, the structure of the right hand side of (3.1) is given as follows:

Lemma 3.2 *The expression*

$$\frac{1}{\prod_j (1 - t^{-\beta_{F,j}} e^{-c_1(\nu_{F,j})})}$$

appearing in (3.1) is given by

$$\prod_j \sum_{r_j \geq 0} \frac{t^{-r_j \beta_{F,j}} (e^{-c_1(\nu_{F,j})} - 1)^{r_j}}{(1 - t^{-\beta_{F,j}})^{r_j+1}}. \tag{3.3}$$

In particular the only poles occur at $t^{\beta_{F,j}} = 1$.

Proof. This follows by examining for each j

$$\frac{1}{1 - t^{-\beta_{F,j}} e^{-c_1(\nu_{F,j})}} = \frac{1}{1 - y(1+u)} = \frac{1}{1-y} \sum_{r \geq 0} \frac{y^r u^r}{(1-y)^r}$$

where $y = t^{-\beta_{F,j}}$ and $u = e^{-c_1(\nu_{F,j})} - 1$. □

We restrict from now on to the case $T = U(1)$, which is regarded as embedded in $\hat{\mathbb{C}}$ in the standard way. We identify the weights with integers by writing them as multiples of the generator λ of the weight lattice of $U(1)$.

Proposition 3.3 *The character* $\chi(t)$ *extends to a holomorphic function on* $\mathbb{C}^\times = \hat{\mathbb{C}} - \{0, \infty\}$.

Proof. This follows since χ is the character of a finite dimensional (virtual) representation of $U(1)$, so it is of the form $\chi(t) = \sum_{m \in \mathbb{Z}} c_m t^m$ for some integer coefficients c_m, finitely many of which are nonzero. □

The following is immediate:

Proposition 3.4 *The expression* χ_F *given in (3.1) defines a meromorphic function on* $\hat{\mathbb{C}}$ *such that* $\sum_{F \in \mathcal{F}} \chi_F(t)$ *agrees with* $\chi(t)$ *on the open subset of* $U(1)$ *consisting of those* t *whose action does not fix any point of* $M - M^T$. *Hence, by analyticity,* $\chi(t) = \sum_{F \in \mathcal{F}} \chi_F(t)$ *on an open set in* $\hat{\mathbb{C}}$ *containing* $\mathbb{C}^\times - U(1)$.

Proposition 3.5 *The dimension of the* T*-invariant subspace of* $\mathcal{H}$ *is given by*

$$\dim \mathcal{H}^T = \frac{1}{2\pi i} \int_{|t| \in \Gamma} \frac{dt}{t} \sum_{F \in \mathcal{F}} \chi_F(t), \tag{3.4}$$

where χ_F *was defined after (3.1). Here, for any* $\epsilon > 0$, $\Gamma = \{t \in \hat{\mathbb{C}} : |t| = 1 + \epsilon\} \subset \Omega$ *is a cycle in* $\hat{\mathbb{C}}$ *on which the* χ_F *have no poles.*

Proof. This follows since

$$\begin{aligned}\dim \mathcal{H}^T &= \frac{1}{2\pi i}\int_{|t|=1}\frac{dt}{t}\chi(t)\\ &= \frac{1}{2\pi i}\int_{|t|\in\Gamma}\frac{dt}{t}\chi(t),\end{aligned}$$

and by applying Proposition 3.4 to identify χ with $\sum_{F\in\mathcal{F}}\chi_F$ on Γ. □

Remark. One obtains an equivalent formula by defining $\Gamma = \{t \in \mathbb{C} : |t| = 1-\epsilon\}$ for $0<\epsilon<1$.

Let us now regard

$$h_F = \chi_F(t)\frac{dt}{t} = \frac{dt}{t}t^{\mu_T(F)}\int_F \frac{e^{\omega}\mathrm{Td}(F)}{\prod_j(1-t^{-\beta_{F,j}}e^{-c_1(\nu_{F,j})})} \tag{3.5}$$

as a meromorphic 1-form on $\hat{\mathbb{C}}$, whose poles may occur only at 0, ∞ and $s\in\mathcal{W}_F$, where we define

$$\mathcal{W}_F = \{s\in U(1) : s^{\beta_{F,j}} = 1 \text{ for some } \beta_{F,j}\}. \tag{3.6}$$

(This is true by inspection of (3.2) when the fixed point set of the action of T consists of isolated points. In the general case it follows from Lemma 3.2.) The integral (3.4) then yields

$$\dim \mathcal{H}^T = -\sum_{F\in\mathcal{F}}\mathrm{res}_\infty h_F. \tag{3.7}$$

Let us examine the poles of h_F on $\hat{\mathbb{C}}$. We have

Lemma 3.6 *For a given F, let $n_{F,\pm}$ be $\sum_{j:\pm\beta_{F,j}>0}|\beta_{F,j}|$. If $\mu_T(F) > -n_{F,+}$ then* $\mathrm{res}_0 h_F = 0$*, while if $\mu_T(F) < n_{F,-}$ then* $\mathrm{res}_\infty h_F = 0$.

Proof. To study the residue at 0, we assume $|t|<1$, so that $(1-t)^{-1} = \sum_{n\geq 0}t^n$ and $(1-t^{-1})^{-1} = -t\sum_{n\geq 0}t^n$. For $r\geq 1$ we examine

$$\frac{t^{\mu_T(F)}}{\prod_j(1-t^{-\beta_{F,j}})^r}\frac{dt}{t}$$

$$= t^{\mu_T(F)}(-1)^{l_+}t^{rn_{F,+}}\prod_j\Big(\sum_{n_j\geq 0}t^{|\beta_{F,j}|n_j}\Big)^r\frac{dt}{t}, \tag{3.8}$$

where l_+ is the number of $\beta_{F,j}$ that are positive. It follows that if $n_{F,+} + \mu_T(F) > 0$ then the residue at 0 is zero. A similar calculation yields the result for the residue at ∞. □

Recall that the action of T on M is said to be *quasi-free* if it is free on the complement of the fixed point set of T in M. The following is shown in [9]:

Lemma 3.7 *The action of $T = U(1)$ on M is quasi-free if and only if the weights are $\beta_{F,j} = \pm 1$.*

Proposition 3.8 *If the action of T is quasi-free, then we have*

$$\mathrm{res}_\infty \sum_{F\in\mathcal{F}} h_F = - \sum_{F\in\mathcal{F}_+} \mathrm{res}_1 h_F. \tag{3.9}$$

Here, $\mathcal{F}_+ = \{F \in \mathcal{F} : \mu_T(F) > 0\}$. More generally the result is true if $\mathrm{res}_1 h_F$ is replaced by $\sum_{s\in\mathcal{W}_F} \mathrm{res}_s h_F$, where the set $\mathcal{W}_F$ was defined in (3.6).

Proof. Assume for simplicity that the action of T is quasi-free: the proof of the general case is almost identical. Lemma 3.6 establishes that

$$\mathrm{res}_\infty \sum_{F\in\mathcal{F}} h_F = \sum_{F\in\mathcal{F}_+} \mathrm{res}_\infty h_F.$$

Also, if $F \in \mathcal{F}_+$ then $\mu_T(F) > -n_{F,+}$ so $\mathrm{res}_0 h_F = 0$; hence (3.9) follows because the meromorphic 1-form h_F has poles only at $0, 1$ and ∞ and their residues must sum to zero, so $\mathrm{res}_1 h_F = -\mathrm{res}_\infty h_F$ when $F \in \mathcal{F}_+$. □

Remark. Recall that $\mu_T(F)$ is never zero.

The following is an immediate consequence of combining Proposition 3.8 with Proposition 3.5:

Corollary 3.9 *If the action of $T = U(1)$ on M_{red} is quasi-free, we have $RR^T(\mathcal{L}) = \sum_{F\in\mathcal{F}_+} \mathrm{res}_1 h_F$. More generally we have $RR^T(\mathcal{L}) = \sum_{F\in\mathcal{F}_+} \sum_{s\in\mathcal{W}_F} \mathrm{res}_s h_F$, where $\mathcal{W}_F$ was defined by (3.6).*

4. Conclusion

In this final section we shall first indicate how to obtain the final result under the assumption that the action of $K = U(1)$ on M is quasi-free. We shall then treat the more general case using Kawasaki's Riemann-Roch theorem for orbifolds [19].

Let us examine the residue $\mathrm{res}_1 h_F$ in the case $K = U(1)$. We denote a generator of the weight lattice of $\mathbf{k}$ by λ, and replace the parameter t (in a small neighbourhood of $1 \in \hat{\mathbb{C}}$) by

$$t = e^{i\lambda(X)} \tag{4.1}$$

(where $X \in \mathbf{k} \otimes \mathbb{C}$ is in a small neighbourhood of 0 in $\mathbf{k} \otimes \mathbb{C}$), so that

$$\frac{dt}{t} = i d\lambda(X)$$

defines a meromorphic 1-form on $\mathbf{k} \otimes \mathbb{C}$. (The substitution (4.1) differs from the substitution used in Section 4, where we set $t = e^{2\pi i \lambda(X)}$; however, the value of the residue obviously is independent of which of these substitutions is used, and the substitution (4.1) yields the formulas in Section 3.)

We then find that

$$\mathrm{res}_1 h_F = i\mathrm{res}_0\Big(e^{i\mu_T(F)(X)} \int_F \frac{e^\omega \mathrm{Td}(F)}{\prod_j (1 - e^{-i\beta_{F,j}(X) - c_1(\nu_{F,j})})} d\lambda(X)\Big) \tag{4.2}$$

$$= i\mathrm{res}_0\Big(e^{i\mu_T(F)(X)} \int_F \frac{e^\omega \mathrm{Td}_K(\nu_F)(X)\mathrm{Td}(F)}{\prod_j (i\beta_{F,j}(X) + c_1(\nu_{F,j}))} d\lambda(X)\Big), \tag{4.3}$$

$$= i\mathrm{res}_0\Big(e^{i\mu_T(F)(X)} \int_F \frac{e^\omega \mathrm{Td}_K(\nu_F)(X)\mathrm{Td}(F)}{e_F(X)} d\lambda(X)\Big) \tag{4.4}$$

where res_0 denotes the coefficient of $d\lambda(X)/\lambda(X)$. Combining (4.4) with Proposition 2.4 one obtains

Proposition 4.1 *We have*

$$\int_{M_{\mathrm{red}}} \mathrm{ch}(\mathcal{L}_{\mathrm{red}})\mathrm{Td}(M_{\mathrm{red}}) = \sum_{F \in \mathcal{F}_+} \mathrm{res}_1 h_F. \tag{4.5}$$

This equals $RR(\mathcal{L}_{\mathrm{red}})$ provided K acts freely on $\mu^{-1}(0)$.

Comparing Proposition 4.1 with Corollary 3.9, we have

Proposition 4.2 *Let the action of $K = U(1)$ on M be quasi-free (which implies K acts freely on $\mu^{-1}(0)$). Then $RR^K(\mathcal{L}) = RR(\mathcal{L}_{\mathrm{red}})$.*

Thus we have obtained the Guillemin-Sternberg result when the action of $K = U(1)$ is quasi-free:

Theorem 4.3 *Suppose $K = U(1)$ acts in a Hamiltonian fashion on the Kähler manifold M, and that the action is quasi-free. We assume a moment* *map μ for the action of K has been chosen in such a way that 0 is a regular value of μ. Then $RR^K(\mathcal{L}) = RR(\mathcal{L}_{\rm red})$.*

We now sketch the proof of the result $RR(\mathcal{L}_{\rm red}) = RR^K(\mathcal{L})$ when $K = U(1)$, *without* the assumption that the action of K is quasi-free. In this more general case, $M_{\rm red}$ is an orbifold and $\mathcal{L}_{\rm red}$ an orbifold bundle. The Riemann-Roch number of $\mathcal{L}_{\rm red}$ is then given by applying Kawasaki's Riemann-Roch theorem for orbifolds. We state Kawasaki's result only as it applies in our particular situation:

Theorem 4.4 (Kawasaki [19]) *The Riemann-Roch number of the orbifold bundle $\mathcal{L}_{\rm red}$ is given by*

$$RR(\mathcal{L}_{\rm red}) = \int_{M_{\rm red}} {\rm ch}(\mathcal{L}_{\rm red}){\rm Td}(M_{\rm red}) + \sum_{s\in\mathcal{S}} \frac{1}{n_s} \sum_{a\in\mathcal{A}_s} \int_{M^a_{s,{\rm red}}} \mathcal{I}^{s,a}. \tag{4.6}$$

Here, $\mathcal{S}$ is the (finite) set of elements $s \in U(1)$ for which $s \neq 1$ and whose fixed point set M_s is strictly larger than the fixed point set of $U(1)$, and n_s is the order of $s \in \mathcal{S}$. The components of M_s are denoted M^a_s, where $a \in \mathcal{A}_s$; we introduce $M^{0,a}_s = M^a_s \cap \mu^{-1}(0)$, and $M^a_{s,{\rm red}} = M^{0,a}_s/S^1$. The class $\mathcal{I}^{s,a}$ $\in H^*(M^a_{s,{\rm red}})$ is defined by

$$\mathcal{I}^{s,a} = \frac{{\rm ch}(L)s^{\mu_a}{\rm Td}(M^a_{s,{\rm red}})}{\prod_{k\in\kappa_a}\left(1 - s^{-\beta_{s,a,k}}e^{-c_1(\nu_{s,a,k})}\right)}. \tag{4.7}$$

Here, μ_a is the weight of the action of s on the fibre of $\mathcal{L}$ over any point in M^a_s. We decompose the normal bundle $\nu(M^a_s)$ to M^a_s in M (which is the same as the normal bundle to $M^{0,a}_s$ in $\mu^{-1}(0)$ since K is abelian) as a formal sum of line bundles

$$\nu(M^a_s) = \oplus_{k\in\kappa_a}\nu_{s,a,k}, \tag{4.8}$$

and denote by $\beta_{s,a,k} \in \mathbb{Z}$ the weight of the action of s on $\nu_{s,a,k}$.

We shall use this theorem to prove Guillemin and Sternberg's result, by identifying the additional terms on the right hand side of (4.6) with the additional residues at the points $1 \neq s \in \mathcal{W}_F$ that appear in the statement of Proposition 3.8 when the action of K is not quasi-free. Meinrenken uses Kawasaki's theorem in a different way to eliminate the quasi-free action hypothesis from the proof given by Guillemin in [12]: see [22], Remark 1 following Theorem 2.1.

Our earlier results (Corollary 3.9) give

$$RR^K(\mathcal{L}) = \int_{M_{\rm red}} {\rm ch}(\mathcal{L}_{\rm red}){\rm Td}(M_{\rm red}) + \sum_{s\in\mathcal{S}}\sum_{a\in\mathcal{A}_s}\Big(\sum_{F\in\mathcal{F}_+:F\subset M^a_s} {\rm res}_s h_F\Big), \tag{4.9}$$

where the meromorphic 1-form h_F on $\hat{\mathbb{C}}$ was defined in (3.5). We know from Proposition 4.1 (a consequence of applying the residue formula to the class $\mathrm{ch}(\mathcal{L}_{\mathrm{red}})\mathrm{Td}(M_{\mathrm{red}})$ on M_{red}) that

$$\int_{M_{\mathrm{red}}} \mathrm{ch}(\mathcal{L}_{\mathrm{red}})\mathrm{Td}(M_{\mathrm{red}}) = \mathrm{res}_1 \sum_{F \in \mathcal{F}_+} h_F. \tag{4.10}$$

We may also apply the residue formula (Theorem 2.1) to the action of $U(1)$ on the symplectic manifold M_s^a: in the notation of that theorem, we introduce an appropriate equivariant cohomology class $\eta e^{\omega_{\mathbf{k}}} = \mathcal{I}_K^{s,a} \in H_K^*(M_s^a)$ which descends on the symplectic quotient $M_{s,\mathrm{red}}^a$ to $\mathcal{I}^{s,a} = \eta_0 e^{\omega_0}$. (Here, the quantity $\omega_{\mathbf{k}}$ was introduced after (2.8).) The class $\mathcal{I}_K^{s,a}$ is given by

$$\mathcal{I}_K^{s,a} = \mathrm{ch}_K(L) s^{\mu_a} \mathrm{Td}_K^{<n>}(M_s^a) \Big(\prod_{k \in \mathcal{K}_a} \big(1 - s^{-\beta_{s,a,k}} e^{-c_1(\nu_{s,a,k})_K}\big)^{-1} \Big)^{<n>} \tag{4.11}$$

where $c_1(\nu_{s,a,k})_K$ is the K-equivariant first Chern class of the virtual line bundle $\nu_{s,a,k}$. Here, the superscript $< n >$ denotes truncation at a sufficiently large value of n, as in Section 2. This yields for each $s \in \mathcal{S}$ and $a \in \mathcal{A}_s$ that

$$\sum_{F \in \mathcal{F}_+ : F \subset M_s^a} \mathrm{res}_s h_F = \frac{1}{n_s} \int_{M_{s,\mathrm{red}}^a} \mathcal{I}^{s,a}. \tag{4.12}$$

(Here, the factor n_s is the order of the subgroup of $U(1)$ that acts trivially on M_s^a: see the statement of Theorem 2.1.) Substituting (4.12) in (4.9) we recover the right hand side of (4.6). Thus we have finally obtained the Guillemin-Sternberg result in the special case $K = U(1)$:

Theorem 4.5 *Suppose $K = U(1)$ acts in a Hamiltonian fashion on the Kähler manifold M, in such a way that 0 is a regular value of μ. Then $RR^K(\mathcal{L}) = RR(\mathcal{L}_{\mathrm{red}})$.*

References

[1] M.F. Atiyah, R. Bott, The moment map and equivariant cohomology, *Topology* **23** (1984) 1-28.

[2] M.F. Atiyah, G.B. Segal, The index of elliptic operators II, *Ann. Math.* **87** (1968) 531-545.

[3] M.F. Atiyah, I.M. Singer, The index of elliptic operators III, *Ann. Math.* **87** (1968) 546-604.

[4] N. Berline, E. Getzler, M. Vergne, *Heat Kernels and Dirac Operators*, Springer-Verlag (Grundlehren vol. 298), 1992.

[5] N. Berline, M. Vergne, Classes caractéristiques équivariantes. Formules de localisation en cohomologie équivariante, *C. R. Acad. Sci. Paris* **295** (1982) 539-541; N. Berline, M. Vergne, Zéros d'un champ de vecteurs et classes caractéristiques équivariantes, *Duke Math. J.* **50** (1983) 539-549.

[6] N. Berline, M. Vergne, The equivariant index and Kirillov's character formula, *Amer. J. Math.* **107** (1985) 1159-1190.

[7] T. Bröcker, T. Tom Dieck, *Representations of Compact Lie Groups*, Springer-Verlag, 1985.

[8] H. Cartan, Notions d'algèbre différentielle; applications aux variétés où opère un groupe de Lie, in *Colloque de Topologie* (C.B.R.M., Bruxelles, 1950) 15-27; La transgression dans un groupe de Lie et dans un fibré principal, *op. cit.*, 57-71.

[9] R. De Souza, V. Guillemin, E. Prato, Consequences of quasi-free. *Ann. Global Anal. Geom.* **8** (1990) 77-85.

[10] M. Duflo, M. Vergne, Orbites coadjointes et cohomologie équivariante, in M. Duflo, N.V. Pedersen, M. Vergne (ed.), *The Orbit Method in Representation Theory* (Progress in Mathematics, vol. 82), Birkhäuser, (1990) 11-60.

[11] J.J Duistermaat, G. Heckman, On the variation in the cohomology of the symplectic form of the reduced phase space, *Invent. Math.* **69** (1982) 259-268; Addendum, **72** (1983) 153-158.

[12] V. Guillemin, Reduced phase spaces and Riemann-Roch, in *Lie Groups and Geometry, in honor of B. Kostant*, J.-L Brylinski, R. Brylinski, V. Guillemin, V. Kac (eds), Progress in Mathematics, Vol. 123, Birkhäuser, 1994, 305–334.

[13] V. Guillemin, S. Sternberg, Geometric quantization and multiplicities of group representations, *Invent. Math.* **67** (1982) 515-538.

[14] V. Guillemin, S. Sternberg, *Symplectic Techniques in Physics*, Cambridge University Press, 1984.

[15] R. Hartshorne, *Residues and Duality* (Lecture Notes in Mathematics v. 20), Springer, 1966.

[16] L.C. Jeffrey, F.C. Kirwan, Localization for nonabelian group actions, *Topology* **34** (1995), 291–327.

[17] L.C. Jeffrey, F.C. Kirwan, On localization and Riemann-Roch numbers for symplectic quotients, *Quart. J. Math.*, to appear.

[18] J. Kalkman, Cohomology rings of symplectic quotients, *J. Reine Angew. Math.* **485** (1995), 37–52.

[19] T. Kawasaki, The Riemann-Roch theorem for complex V-manifolds, *Osaka J. Math.* **16** (1979) 151-159.

[20] F.C. Kirwan, *Cohomology of Quotients in Symplectic and Algebraic Geometry*, Princeton University Press, 1984.

[21] C.-M. Marle, Modèle d'action hamiltonienne d'un groupe de Lie sur une variété symplectique, in: Rendiconti del Seminario Matematico, Università

e Politechnico, Torino **43** (1985) 227-251.

[22] E. Meinrenken, On Riemann-Roch formulas for multiplicities, *J. Amer. Math. Soc.*, to appear.

[23] M. Vergne, Quantification géométrique et multiplicités, *C.R. Acad. Sci. Paris Sér. 1. Math.* **319** (1994), 327–332; Geometric quantization and multiplicities I, *Duke Math. J.*, to appear.

[24] M. Vergne, A note on Jeffrey-Kirwan-Witten localization, *Topology*, to appear.

[25] E. Witten, Two dimensional gauge theories revisited, preprint hep-th/9204083; *J. Geom. Phys.* **9** (1992) 303-368.

[26] S. Wu, An integral formula for the squares of moment maps of circle actions, *Lett. Math. Phys.* **29** (1993), 311–328.

Lisa C. Jeffrey
Mathematics Department
Princeton University
Princeton, NJ 08544
email: jeffrey@math.princeton.edu

and

Frances C. Kirwan
Balliol College
Oxford OX2 3BJ, United Kingdom
email:fkirwan@vax.ox.ac.uk

Received July 1994; revised October 1995

A Note on ODEs from Mirror Symmetry ◇

A. Klemm, B.H. Lian, S.S Roan, and S. T. Yau

In Honor of Professor Israel M. Gelfand
on the occasion of his 80th birthday.

Abstract. We give close formulas for the counting functions of rational curves on complete intesection Calabi-Yau manifolds in terms of special solutions of generalized hypergeometric differential systems. For the one modulus cases we derive a differential equation for the Mirror map, which can be viewed as a generalization of the Schwarzian equation. We also derive a nonlinear seventh order differential equation which directly governs the Prepotential.

1. Introduction

In a seminal paper [1], physicists solved a problem in enumerative geometry, namely to count[1] the "number" n_d of rational curves[2] of arbitrary degree d on the quintic threefold X in $\mathbb{P}^4$. The answer was given in terms of the large volume expansion of the correlation function between three states $\mathcal{O}_J$ of the twisted $N = 2$ topological σ-model on X, which has the formal expansion [1], [2], [3]

$$K_{JJJ} = \langle \mathcal{O}_J \mathcal{O}_J \mathcal{O}_J \rangle = \int_X J \wedge J \wedge J + \sum_d \frac{d^3 n_d q^d}{(1 - q^d)}. \tag{1.1}$$

Here $q = e^{2\pi i t}$ and the modulus t parametrizes the complexified Kähler class of X, i.e. $(\mathrm{Im}(t))^3 \propto$ volume of X and $\mathrm{Re}(t)$ parametrizes an antisymmetric tensor field, which is the component of a harmonic $(1,1)$-form on X [4]. The correlation function (1.1), is sometimes referred to as an intersection number of the quantum (co)homology of X. In the large volume limit the contribution of the instantons is damped out and (1.1) approaches the classical self intersection number between the cycle dual to the Kähler form J.

◇ Research supported by grant DE-FG02-88-ER-25065.

[1] Only if the moduli space of the map from $\mathbb{P}^1$ to X, with three points fixed, is zero dimensional n_d counts the number of isolated rational curves. In general, n_d has to be understood as the integral of the top Chern class over the moduli space of that map.

[2] These are called instantons by physicists as they correspond to classical solutions of the σ-model equations of motion.

It is a remarkable fact that this counting function (1.1) is expressible in a closed form in terms of solutions of a generalized hypergeometric system. This has been used in [5], [6], [7] to predict the number of rational curves on various Calabi-Yau spaces.

In section 2 we review the physical reasoning, which explains that fact and give, as a generalization, closed formulas for the counting functions on nonsingular complete intersection Calabi-Yau spaces in products of weighted projective spaces. An important step in these calculations is the definition of the mirror map. We discuss therefore in sections 3 and 4 the differential equation which governs the mirror map. As we will see in section 2 the most important quantity is the prepotential from which the the correlation function (1.1) and the Weil-Peterson metric for the complex moduli space, can be derived. We will obtain in section 5 a differential equation for the prepotential.

2. Counting of rational curves and generalized hypergeometric functions

It was argued by Witten [3] that the states in the ($N = 2$) topological σ-model on an (arbitrary Calabi-Yau) space X are in one-to-one correspondence with the elements in the cohomology groups of X. For example, the state $\mathcal{O}_J$ above corresponds to the Kähler form in $H^{1,1}(X, \mathbf{Z})$. As there is a natural involution symmetry in the $N = 2$ topological σ-model on Calabi-Yau manifolds exchanging the states corresponding to the cohomology groups $H^{3-p,q}(X)$ and $H^{p,q}(X)$, physicists suspect that Calabi-Yau spaces occur quite generally in mirror pairs[3] X and X^*, in which the rôle of these cohomology groups are exchanged. In particular $h^{3-p,q}(X) = h^{p,q}(X^*)$ holds and the Euler number of X is therefore the negative of the Euler number of X^*. By the same token it is expected that the correlation functions among one type of states on X can be calculated by the same methods as its counterparts among the corresponding states on X^*. In particular the correlation function (1.1) is related by this argument [1], [3] to the correlation function

$$K_{J^*J^*J^*} = \langle \mathcal{O}_{J^*}\mathcal{O}_{J^*}\mathcal{O}_{J^*}\rangle = \int_{X^*} \Omega \wedge b^\alpha_{J^*} \wedge b^\beta_{J^*} \wedge b^\gamma_{J^*} \Omega_{\alpha\beta\gamma}, \qquad (2.1)$$

where Ω is the unique no-where vanishing holomorphic threeform on X^* and $b^\alpha_{J^*} \in H^1(X, TX) \cong H^{2,1}(X)$. In fact the expansion (1.1) and hence the successful prediction of rational curves on the quintic X was obtained in [1] by calculating the correlation function (2.1) on the mirror X^* using classical methods of the theory of complex structure deformation. The integral on the

---

[3] See [8][6] and references therein for geometrical constructions of mirror pairs, which support this expectation.

right-hand side of (2.1), introduced in [9] depends only, via Ω, on the choice of the complex structure modulus but not on the choice of complexified Kähler modulus. After calculating its dependence on the complex structure modulus using the Picard-Fuchs equation, the decisive steps are to find the correct point of expansion in this moduli space and to determine the map from the complex structure modulus in (2.1) to the Kähler structure modulus t in (1.1). This map (or its inverse) will be called the mirror map.

For the quintic hypersurface in $\mathbb{P}^4$, the mirror X^* can be constructed concretely as the canonical desingularized quotient $X^* = \widehat{X/(\mathbb{Z}_5^3)}$, where $\mathbb{Z}_5^3$ acts by phase multiplication on the homogenoeus coordinates $(x_1 : \ldots : x_5)$ of $\mathbb{P}^4$ and is generated by $g_i : (x_i, x_5) \mapsto (\exp(2\pi i/5)x_i, \exp(8\pi i/5)x_5)$ $i = 1, 2, 3$. Here (2.1) depends on the one-dimensional complex structure deformation of X^*, which can be studied by considering the deformations of the quintic X, but restricted to the unique $\mathbb{Z}_5^3$-invariant element $x_1x_2x_3x_4x_5$ in its local ring.

The Picard-Fuchs ODE can therefore be derived by the Dwork-Griffith-Katz reduction method from the standard residuum expression of the period [10] for $N = 5$

$$\tilde{\omega}_i(s_0, \ldots, s_N) = \int_\gamma \int_{\Gamma_i} \frac{\mu}{\sum_{i=1}^N s_i x_i^N - s_0 \prod_{i=1}^N x_i}, \tag{2.2}$$

where γ is a small cycle in $\mathbb{P}^{N-1}$, $\Gamma_i \in H_{N-2}(M)$ and the measure is

$$\mu = \sum_i (-1)^i x_i dx_1 \wedge \ldots \widehat{dx_i} \ldots \wedge dx_N.$$

Instead of using this generic alogarithm, let us consider of the symmetries of (2.2) directly. Obviously

$$\hat{\omega}_i(\lambda^N s_0, \ldots, \lambda^N s_N) = \lambda^{-N} \hat{\omega}_i(s_0, \ldots, s_N) \tag{2.3}$$

$$\hat{\omega}_i(s_0, \ldots, \lambda_j^N, \ldots, \lambda_i^{-N} s_N) = \hat{\omega}_j(s_0, \ldots, s_N), \quad \text{for } j = 1, \ldots, N-1, \tag{2.4}$$

with $\lambda, \lambda_i \in \mathbf{C}^*$. Writing (2.3), (2.4) in infinitesimal form we obtain the differential equations

$$\left\{ \sum_{i=0}^N s_i \frac{\partial}{\partial s_i} + 1 \right\} \hat{\omega}_i(s) = 0 \quad , \tag{2.5}$$

$$\left(s_i \frac{\partial}{\partial s_i} - s_N \frac{\partial}{\partial s_N} \right) \hat{\omega}_i(s) = 0 \text{ for } i = 1, \cdots, N-1\,. \tag{2.6}$$

The trivial relation $x_1^N \ldots x_N^N - (x_1 \ldots z_N)^N \equiv 0$ leads to a further differential equation

$$\left\{\prod_{i=1}^{N}\frac{\partial}{\partial s_i}-\left(\frac{\partial}{\partial s_0}\right)^{N}\right\}\hat{\omega}_i(s)=0\,. \tag{2.7}$$

This system of differential equations (2.5)-(2.7) is precisely the type of generalized hypergeometric system, which was investigated by Gelfand, Kapranov and Zelevinsky in [11] with the characters defined by

$$\begin{aligned}\chi_1=(1,\underbrace{0,\ldots,0}_{(N-1)-times}),\quad \chi_2=(1,1,0,\ldots,0),\ldots\\ \ldots,\chi_N=(1,0,\ldots,0,1),\quad \chi_{N+1}=(1,\underbrace{-1,\ldots,-1}_{(N-1)-times})\end{aligned} \tag{2.8}$$

and the exponents: $\vec{\beta}=(-1,0,\ldots,0)$. The generator of the lattice $\mathbf{L}$ of relations is

$$l=(-N;\underbrace{1,\ldots,1}_{N-times}). \tag{2.9}$$

The eqns. (2.5),(2.6) are satisfied identically by the ansatz

$$\hat{\omega}_i(s)=\frac{1}{s_0}\omega\left(\frac{\prod_{i=1}^{N}s_i}{s_0^N}\right). \tag{2.10}$$

By using the new coordinate for the complex structure modulus

$$z=(-1)^{l_0}\prod_{i=0}^{N}s_i^{l_i} \tag{2.11}$$

the eqn. (2.7) can be written in the following convenient form

$$\theta_z\left[\theta^{N-1}-Nz\prod_{i=1}^{N-1}(N\theta+i)\right]\omega_i=0, \tag{2.12}$$

where $\theta=z\frac{d}{dz}$. The generalized hypergeometric system defined by (2.8) and $\vec{\beta}$ is proven to be holonomic [11] and a formal power series expansion and (Euler) integral representations were likewise given. For the quintic ($N=5$) the system has 5 solutions, but it is semi-resonant, which implies that the monodromy on the full solutions space is reducible. On the other hand the monodromy for the 4 periods on X^* is known to be irreducible. The unique subsystem of the solutions of (2.12) on which the monodromy acts irreducible is given by the 4 solutions to

$$\left[\theta^{N-1} - Nz\prod_{i=1}^{N-1}(N\theta+i)\right]\omega_i = 0, \tag{2.13}$$

which identifies the later equation with the Picard-Fuchs equation of the mirror X^*.

The complex structure moduli space of a Calabi-Yau threefold exhibits special geometry, as it was explained in [12] using crucially the results of [13]. This structure is charaterized by the existence of a section $\tilde{F}$ of a holomorphic line bundle over the complex moduli space, which is a prepotential for structure constant(s) (2.1) and the Kähler potential K of the Weil-Peterson metric. There exists a special coordinate choice, given by a ratio of periods $\tilde{t} = \tilde{\omega}_1(z)/\tilde{\omega}_0(z)$ in which these relations read

$$\begin{aligned} K_{J^*J^*J^*} &= \partial_{\tilde{t}}^3 \tilde{F} \\ K &= -\log\left((\tilde{t}-\bar{\tilde{t}})(\partial_{\tilde{t}}\tilde{F} + \bar{\partial}_{\bar{\tilde{t}}}\bar{\tilde{F}}) + 2\,\bar{\tilde{F}} - 2\,\tilde{F}\right). \end{aligned} \tag{2.14}$$

These coordinates can equivalently be characterized by the property that the period vector is expressible in terms of the prepotential as

$$\Pi(z) = (\tilde{\omega}_0(z), \tilde{\omega}_1(z), \tilde{\omega}_2(z), \tilde{\omega}_3(z)) = \tilde{\omega}_0(1, \tilde{t}, \partial_{\tilde{t}}\tilde{F}, 2\tilde{F} - \tilde{t}\partial_{\tilde{t}}\tilde{F}) \tag{2.15}$$

and vice versa

$$\tilde{F}(\tilde{t}) = \frac{1}{2\tilde{\omega}_0^2}(\tilde{\omega}_3\tilde{\omega}_0 + \tilde{\omega}_1\tilde{\omega}_2). \tag{2.16}$$

It has been argued [1],[12] that the moduli space of the complexified Kähler structure of the the $N = 2$ topological σ-model exhibits also special geometry with (1.1) as structure constant(s) and that t is the special coordinate (especially for the last point see also [14]). Because of the analog of (2.14) for the Kähler structure modulus the prepotential $F(t)$ is determined by K_{JJJ} up to a quadratic polynomial in t:

$$F(t) = \frac{\int_X J\wedge J\wedge J}{3!}t^3 + \frac{a}{2}t^2 + bt + c + F_{inst}(q). \tag{2.17}$$

To identify t with $\tilde{t}$ and $F(t)$ with $\tilde{F}(\tilde{t})$, we must find the special point z_1 corresponding to the large volume limit $\mathrm{Im}(t) \to \infty$, in the complex structure moduli space. This can be done in the following heuristic way. First note the invariance of (1.1) under the shift symmetry $t \to t+1$. In fact more generally, shifting the parameter of the antisymmetric background $\mathrm{Re}(t)$ by an integer, is a symmetry of the σ-model in the large volume region [3], [4]. We require that the transformation of the "period" $(1, t, \partial_t F, 2F - t\partial_t F)$ under that symmetry should correspond to a monodromy operation on $\Pi(z)$ under counterclockwise analytic continuation around z_1. That is, we search a point $z = z_1$ in the complex modulus space with the specific monodromy action:

$$\vec{\Pi}(z) \mapsto \begin{pmatrix} 1 & 0 & 0 & 0 \\ 1 & 1 & 0 & 0 \\ \left(a+\frac{K}{2}\right) & K & 1 & 0 \\ \left(2b-\frac{K}{6}\right) & \left(a-\frac{K}{2}\right) & -1 & 1 \end{pmatrix} \vec{\Pi}(z), \quad \text{with } K = \int_X J \wedge J \wedge J \tag{2.18}$$

which is unipotent of order 4.

The importance of this monodromy requirement was pointed out in [15]. It is easy to see that among the three regular singular points $z = 0, 1/5^5, \infty$ of the ODE (2.13) with $N = 5$, the point that admits such a monodromy is $z = 0$, where the indicial equation is four-fold degenerate. Around this point, there is one power series solution given by

$$\omega_0(z) = \omega_0(z,\rho)|_{\rho=0} = \sum_{n\geq 0} c(n,\rho) z^{n+\rho}\bigg|_{\rho=0}, \tag{2.19}$$

where the coefficients $c(\rho, n)$ can be expressed in terms of gamma-functions from the l in (2.9) as

$$c(n,\rho) = \frac{\Gamma\left(l_0(n+\rho)+1\right)}{\prod_{i=1}^N \Gamma\left(l_i(n+\rho)+1\right)}. \tag{2.20}$$

The other solutions can be obtained by the well-known Frobenius method (see e.g.[16]):

$$\omega_p = \frac{1}{p!}\left(\frac{1}{2\pi i}\frac{\partial}{\partial_\rho}\right)^p \omega_0(z,\rho)\bigg|_{\rho=0}, \quad \text{for } p = 1, \ldots, N-2. \tag{2.21}$$

Their monodromy is dictated by the terms linear, quadratic und cubic in $\log(z)$. By comparing the monodromy of these solutions with (2.18) we conclude that the mirror map is given by

$$t = \frac{\omega_1(z)}{\omega_0(z)}. \tag{2.22}$$

Also from the monodromy requirement and using the special geometry relations (2.14) we get, indepentently of a, b, c, a unique expansion of (1.1) completely expressed in terms of special solutions to the GKZ system :

$$K_{JJJ} = \frac{\int_X J \wedge J \wedge J}{2} \partial_t^2 \left(\frac{\omega_2(z(t))}{\omega_0(z(t))} \right), \tag{2.23}$$

where we denote (the inverse of) the mirror map (2.22) by $z(t)$.

Let us finish this section with the generalization of the result (2.23) to nonsingular complete intersection Calabi-Yau spaces in products of k weighted projective spaces and give closed formulas for the large radius expansions of the triple intersection (1.1) $\langle \mathcal{O}_{J_i} \mathcal{O}_{J_j} \mathcal{O}_{J_k} \rangle$, where J_i is the Kähler class induced from the ith weighted projective space. From these expansions one can read off the numbers of rational curves of any multidegree spaces, with respect to the Kähler classes induced from the projective spaces. These results were obtained in [7].

We consider in the following complete intersections of l hypersurfaces in products of k projective spaces. Since most formulas allow for an incorporation of weights we will state them for the general case. Denote by $d_j^{(i)}$ the degree of the coordinates of $\mathbb{P}^{n_i}[\vec{w}^{(i)}]$ in the j-th polynomial p_j ($i = 1, \ldots, k$; $j = 1, \ldots, l$). The residuum expression for the periods [10], with k perturbations satisfies again a GKZ-system, where the lattice of relations $\mathbf{L}$ is generated by k generators $l^{(s)}$ ($s = 1, \ldots, k, j = 1, \ldots, l$)

$$l^{(s)} = (-d_1^{(s)}, \ldots, -d_l^{(s)}; \ldots, w_1^{(s)}, \ldots, w_{n_s+1}, 0, \ldots) \equiv \left(\{l_{0,j}^{(s)}\}; \{l_i^{(s)}\} \right), \tag{2.24}$$

from which one obtains k linear differential operators ($\theta_s = z_s \frac{d}{dz_s}$)

$$\begin{aligned} \mathcal{L}_s = & \prod_{j=1}^{n_s+1} \left(w_j^{(s)} \theta_s \right) \left(w_j^{(s)} \theta_s - 1 \right) \cdots \left(w_j^{(s)} \theta_s - w_j^{(s)} + 1 \right) \\ & - \prod_{j=1}^{l} \Big(\sum_{i=1}^{k} d_j^{(i)} \theta_i \Big) \cdots \Big(\sum_{i=1}^{k} d_j^{(i)} \theta_i - d_j^{(s)} + 1 \Big) z_s. \end{aligned} \tag{2.25}$$

The point[4] $z = 0$ is again a point of maximal unipotent monodromy, and the unique power series solution is given

$$\omega_0(z) = \sum_{n_s \geq 0} c(n, \rho) z^{n+\rho} \Bigg|_{\rho=0}, \text{with } c(n, \rho) = \frac{\prod_j \Gamma \left(-\sum_{s=1}^{k} l_{0j}^{(s)} (n_s + \rho_s) + 1 \right)!}{\prod_i \Gamma \left(\sum_{s=1}^{k} l_i^{(s)} (n_s + \rho_s) + 1 \right)!}. \tag{2.26}$$

[4] Here and in the following we denote by z, n and ρ the k-tuples $z_1 \ldots z_k, n_1, \ldots, n_k$ and $\rho_1, \ldots, \rho_k$. We use obvious abbreviations such as $z^n := \prod_{s=1}^{k} z_s^{n_s}$ etc.

Again the system is semi resonant and the monodromy of (2.25) is reducible. Therefore one has to specify the subset of solutions, which correspond to the $2(k+2)$ period integrals on X^*. This problem was solved in [7] by factorizing the differential operators and the following convenient basis for the period vector was found:

$$\Pi(z) = \begin{pmatrix} w_0(z) \\ D_i^{(1)} w_0(z,\rho)|_{\rho=0} \\ D_i^{(2)} w_0(z,\rho)|_{\rho=0} \\ D^{(3)} w_0(z,\rho)|_{\rho=0} \end{pmatrix}. \tag{2.27}$$

Here the $D_i^{(k)}$ are differentials with respect to the parameter ρ_i, which are defined in terms of the classical intersection numbers among the Kähler classes J_i induced from the i'th ambient space in the product space $\otimes_i \mathbb{P}_i^n$ as follows ($\partial_{\rho_i} := \left(\frac{1}{2\pi i}\right)\left(\frac{\partial}{\partial_{\rho_i}}\right)$):

$$\begin{aligned} D_i^{(1)} &:= \partial_{\rho_i}, \quad D_i^{(2)} := \frac{\int_X J_i \wedge J_j \wedge J_k}{2} \partial_{\rho_j} \partial_{\rho_k} \quad \text{and} \\ D^{(3)} &:= -\frac{\int_X J_i \wedge J_j \wedge J_k}{6} \partial_{\rho_i} \partial_{\rho_j} \partial_{\rho_k}. \end{aligned} \tag{2.28}$$

By a straightforward generalization of the monodromy requirement one finds the generalization of (2.22)

$$t_i = \frac{\omega_i(z)}{\omega_0(z)}. \tag{2.29}$$

The following explicit expansions for the correlation function (1.1), which generalize (2.23)

$$\begin{aligned} \langle \mathcal{O}_{J_i} \mathcal{O}_{J_j} \mathcal{O}_{J_k} \rangle &= \partial_{t_i} \partial_{t_j} \frac{D_k^{(2)} w_0|_{\rho=0}}{w_0}(t) \\ &= \int_X J_i \wedge J_j \wedge J_k + \sum_{d_1,\ldots,d_k} \frac{n^r_{d_1,\ldots,d_k} d_i\, d_j\, d_k}{1 - \prod_{i=1}^k q_i^{d_i}} \prod_{i=1}^k q_i^{d_i} \end{aligned} \tag{2.30}$$

can be read off from the period vector (2.27), after normalizing by $1/w_0(z)$ and transforming the period vector by the inverse of (2.29) to the t variables. The prepotential was also given in [7] as

$$F(t) = \frac{1}{2}\left(\frac{1}{w_0}\right)^2 \left\{ w_0 D^{(3)} w_0 + D_l^{(1)} w_0 D_l^{(2)} w_0 \right\}(t) \Big|_{\rho=0}. \tag{2.31}$$

These formulas apply immediatly to all nonsingular complete intersections in weighted projective spaces. Let us summarize the observations, made for these series in [7]:

a .) The mirror map (2.29) as well as its inverse have integral expansion.
b .) The numbers $n^r_{d_1,\ldots,d_n}$ in (2.30) are integers.
c .) The constants of the quadratic polynomial in t_i of multimoduli prepotential are $a_{ij} = 0$, $b_i = \left(\frac{1}{2\pi i}\right)^2 \int_X c_2 J_i \zeta(2)$ and $c = \left(\frac{1}{2\pi i}\right)^3 \int_X c_3 \zeta(3)$.
d .) In all cases the invariants $n^r_{d_1,\ldots d_r}$ coincide, as far as they can be checked, with the invariants of rational curves calculated with classical methods of algebraic geometry. For example, consider the Calabi-Yau manifolds defined by

$$p_1 = \sum a_{ijk} y_i y_j y_k = 0, \quad p_2 = \sum b_{ijk} x_i x_j x_k = 0, \quad p_3 = \sum c_{ij} y_i z_j = 0 \tag{2.32}$$

as complete intersections in $\mathbb{P}^3 \times \mathbb{P}^3$, where y_i are the homogeneous coordinates of the first $\mathbb{P}^3$ and z_i of the second. Then one obtains from (2.30) the following invariants $n^r_{d_1,d_2}$ for the rational curves of bidegree less than 6:

(0,1)	81
(0,2)	81
(0,3)	18
(0,4)	81
(0,5)	81
(0,6)	18

(1,1)	729
(2,2)	33534
(3,3)	5433399
(1,2)	2187
(2,4)	1708047

(1,3)	6885
(1,4)	18954
(1,5)	45927
(2,3)	300348

The invariants for bidegree less then three coincide with the ones calculated by classical methods in [17].

In the remaining sections we want to investigate both the mirror map $z(t)$ and the prepotential $F(t)$. An important question is: are there any natural differential equations which govern z and F? The answer to this questions is affirmative as we shall see.

3. Differential Equation for the Mirror Map by Examples

We will discuss in three examples in dimensions 1,2 and 3 respectively, the differential equation which governs the mirror map. We will state some general properties of the equation. Our original motivation for studying this equation was to understand the observations made experimentally on the mirror map and the Yukawa couplings.

3.1. Periods of Elliptic Curves

As a warm-up, we will first consider the most elementary example of Mirror Symmetry — for complex curves [18], [19]. This will be a brief exposition of some well-known classical construction — but in the context of Mirror Symmetry.

Consider the following one-parameter family of cubic curves in $\mathbb{P}^2$:

$$X_s : x_1^3 + x_2^3 + x_3^3 - s x_1 x_2 x_3 = 0. \tag{3.1}$$

We may transform X_s by a $PGL(3, \mathbf{C})$ transformation to an elliptic curve in the Weierstrass form:

$$y^2 = 4x^3 - g_2 x - g_3 \tag{3.2}$$

where

$$\begin{aligned} g_2 &= 3s(8+s^3) \\ g_3 &= 8 + 20s^3 - s^6. \end{aligned} \tag{3.3}$$

We would like to consider the variation of the period of the holomorphic 1-form $\frac{dx}{y}$ along a homology cycle Γ:

$$\omega_\Gamma = \int_\Gamma \frac{dx}{y}. \tag{3.4}$$

It can be shown that as a function of s, ω_Γ satisfies the second order ODE:

$$\frac{d^2\omega_\Gamma}{ds^2} + a_1(s)\frac{d\omega_\Gamma}{ds} + a_0(s)\omega_\Gamma = 0 \tag{3.5}$$

where

$$\begin{aligned} a_1 &= -\frac{d}{ds} log\left(\frac{3}{2\Delta}(2g_2\frac{dg_3}{ds} - 3\frac{dg_2}{ds}g_3)\right) \\ a_0 &= \frac{1}{12}a_1\frac{d}{ds}log\ \Delta + \frac{1}{12}\frac{d^2}{ds^2}log\ \Delta - \frac{1}{16\Delta}(g_2\frac{dg_2}{ds}^2 - 12\frac{dg_3}{ds}^2) \end{aligned}, \tag{3.6}$$

where $\Delta = g_2^3 - 27g_3^2$ is the discriminant of the above elliptic curve. By a change of coordinate $s \to z = s^{-3}$, equation (3.5) transforms into the hypergeometric equation (2.13) for $N = 3$ with regular singularities at $z = 0, 1/3^3, \infty$:

$$\left(\theta^2 - 3z(3\theta+2)(3\theta+1)\right)\omega_\Gamma = 0. \tag{3.7}$$

Thus the period ω_Γ is a linear combination of two standard hypergeometric functions.

We now do the following change of coordinates $s \to J = \frac{g_2^3}{\Delta}$, and write ω_Γ as $\sqrt{\frac{g_2}{g_3}}\Omega_\Gamma$. Then our equation (3.5) becomes

$$\frac{d^2\Omega_\Gamma}{dJ^2} + \frac{1}{J}\frac{d\Omega_\Gamma}{dJ} + \frac{31J-4}{144J^2(1-J)^2}\Omega_\Gamma = 0. \tag{3.8}$$

This equation has the following universal property: it is derived without the use of the explicit form of g_2, g_3 above, despite the fact that we began with a particular realization (as a cubic in $\mathbb{P}^2$) of an elliptic curve. This means that if we had started from any other model for an elliptic curve, we will have arrived at the same equation (3.8), ie. this is the universal form of the Picard-Fuchs equation for the periods of the elliptic curves. Note also that under the above transformation, the ratio t of two periods $\omega_\Gamma, \omega_{\Gamma'}$ (which are two hypergeometric functions) remains the same.

We can now ask for a differential equation which governs the function $t(J)$ (which is a Schwarzian triangular function). This is the well-known Schwarzian equation:

$$\{t, J\} = 2\left(\frac{3}{16(1-J)^2} + \frac{2}{9J^2} + \frac{23}{144J(1-J)}\right). \tag{3.9}$$

Here $\{z, x\}$ denotes the Schwarzian derivative $\frac{z'''}{z'} - \frac{3}{2}\left(\frac{z''}{z'}\right)^2$. Note that in this equation, by inverting $t(J)$ we may regard $J(t)$ as the dependent variable. Recall that the inverse function for the period ratio is precisely the mirror map. Thus $J(t)$ is our mirror map in this case and (3.9)is our differential equation which governs it. With a suitable choice of the period ratio t, $J(t)$ admits, up to overall constant, an integral q-series ($q = exp(2\pi i t)$) expansion

$$J(q) = \frac{1}{1728}(q^{-1} + 744 + 196884q + 21493760q^2 + \ldots). \tag{3.10}$$

We can also relate the J-function for different realizations of the elliptic curves in different ways to solutions of GKZ systems. For example there exist three realizations of the elliptic curves as hypersurfaces in weighted projective spaces $\mathbb{P}^2(1,1,1), \mathbb{P}^2(1,1,2)$ and $\mathbb{P}^2(1,2,3)$.

	constraint	diff. operator	$1728J(z)$
P_8	$x_1^3 + x_2^3 + x_3^3 - z^{-1/3}x_1x_2x_3 = 0$	$\theta^2 - 3z(3\theta+2)(3\theta+1)$	$\frac{(1+216z)^3}{z(1-27z)^3}$
X_9	$x_1^4 + x_2^4 + x_3^2 - z^{-1/4}x_1x_2x_3 = 0$	$\theta^2 - 4z(4\theta+3)(4\theta+1)$	$\frac{(1+192z)^3}{z(1-64z)^2}$
J_{10}	$x_1^6 + x_2^3 + x_3^2 - z^{-1/6}x_1x_2x_3 = 0$	$\theta^2 - 12z(6\theta+5)(6\theta+1)$	$\frac{1}{z(1-432z)}$

Here the differential operators are specified by factorizing the obvious differential operators from the general expression (2.25). By the expression for the $J(z)$-function, which were obtained by transfoming the contraints into the Weierstrass form, they can be brought in the form (3.8). The mirror map is related to the solutions of the GKZ system, by the formulas (2.19)-(2.22) using the generators of the lattice l given by (2.24). Concretely this yields, by inversion of (2.22), the following expansion for the functions $z(q)$

$$\begin{aligned} P_8 &: z(q) = q - 15q^2 + 171q^3 - 1679q^4 + 15054q^5 - 126981q^6 + \dots \\ X_9 &: z(q) = q - 40q^2 + 1324q^3 - 39872q^4 + 1136334q^5 - 31239904q^6 + \dots \\ J_{10} &: z(q) = q - 312q^2 + 87084q^3 - 23067968q^4 \\ &\qquad + 5930898126q^5 - 1495818530208q^6 + \dots \end{aligned} \tag{3.11}$$

The remarkable fact is that this expansions are already integer. Inserting them into the expressions for the $J(z)$ functions yields of course the expansion (3.10).

The above construction (ie. the periods, the Picard-Fuchs equation and the Schwarzian equation for the elliptic curves) is of course classical. We will now give a similar construction for K3 surfaces (using quartics in $\mathbb{P}^3$) and for the quintics in $\mathbb{P}^4$. At the end, we will have a Schwarzian equation which governs the period ratio (hence the Mirror map) in each of the cases. To our knowledge, this equation is new. Actually, we also have a similar construction for any Calabi-Yau complete intersection in a toric variety. But for the purpose of exposition, we must restrict ourselves to the above simple examples. Details for the general cases will be given in our forthcoming papers [20].

3.2. Periods of K3 surfaces

We consider the following one-parameter family of quartic hypersurfaces in $\mathbb{P}^3$:

$$X_s : W_s(x_1, x_2, x_3, x_4) = x_1^4 + x_2^4 + x_3^4 + x_4^4 - s x_1 x_2 x_3 x_4 = 0. \tag{3.12}$$

The period of a holomorphic 2-form along a homology 2-cycle Γ_i in X_s is given by (2.2) with $N = 4$. The Picard-Fuchs equation (2.13) for the K_3 case is a third order ODE of Fuchsian type and has singularities at $z = 0, 1/4^4, \infty$. Thus the period ω_{Γ_i} is a linear combination of three generalized hypergeometric functions. There is one solution which is regular at $z = 0$. The other two given by (2.21) have singular behavior $log\ z$ and $(log\ z)^2$ respectively.

What is the analogue of the universal equation (3.8) in the case of K3 surfaces, ie. the Picard-Fuchs equation which is independent of the model for the K3 surfaces? To answer this, we should first interpret (3.8)as follows. Given a topological type of complex n-folds X, there is a universal moduli space M of complex structures on X. In the case of Calabi-Yau (or elliptic curves), there is a flat Gauss-Manin connection ∇_M on M. The period vector $\boldsymbol{\Omega}$ of the holomorphic n-form of X is then a section which satisfies

$$\nabla_M \boldsymbol{\Omega} = 0 \tag{3.13}$$

on a vector bundle $H^n(X, \mathbf{C}) \to E \to M$.

In the case of the elliptic curves, we may view J as the coordinate on M. The universal Picard-Fuchs equation (3.8) should be thought of as the equation (3.13) in the local coordinate. It is an interesting problem to derive the analogue of such an equation in the case of K3 surfaces.

However in the absence of such an equation, we can still ask for the analogue of the Schwarzian equation, ie. a differential equation for the mirror map which in this case is the local inverse of the function $t(z) = \omega_1(z)/\omega_0(z)$. To write this equation, it is convenient to first transform the Picard-Fuchs equation (2.13) to the form $(\frac{d^3}{dz^3} + q_1(z)\frac{d}{dz} + q_0(z))f$. This is obtained from (2.13) by a suitable change of dependent variable $\omega_\Gamma \to f$. Then for the quartic model of K3 surfaces above, the Schwarzian equation is the following fifth order ODE:

$$\begin{aligned}\{z,t\}_3 =&(-24T_2^2 + 6T_4)q_1 z'^2 - 18T_2 q_1^2 z'^4 - 4q_1^3 z'^6 + 12T_3(\partial_z q_1) z'^3 \\ &+ 3(\partial_z q_1)^2 z'^6 - 12T_2(\partial_z^2 q_1) z'^4 - 6q_1(\partial_z^2 q_1)^2 z'^6 - 54T_3 q_0 z'^3 \\ &- 27q_0^2 z'^6 + 36T_2(\partial_z q_0) z'^4 + 18(\partial_z q_0) q_1 z'^6\end{aligned} \tag{3.14}$$

where

$$\begin{aligned}\{z,t\}_3 :=& -8\,T_2^3 - 15\,T_3^2 + 12\,T_2\,T_4 \\ T_i :=&\nabla^{i-2}\{z,t\}\end{aligned} \tag{3.15}$$

and $\nabla := \left(\frac{d}{dt} - k\frac{z''}{z'}\right)$. Note that prime here means $\frac{d}{dt}$. For each k the object $T_k dt^k$ is a rank k tensor under linear fractional transformations $t \to \frac{at+b}{ct+d}$, with $a, b, c, d \in \mathbb{C}$. Then ∇ above is a covariant derivative on this tensor. The eqn (3.14)has a solution given by the mirror map:

$$z(q) = q - 104q^2 + 6444q^3 - 311744q^4 + 13018830q^5 - 493025760q^6 + \ldots \tag{3.16}$$

As for the classical Schwarzian equation, the new equation (3.14) is of course $SL(2,\mathbf{C})$ invariant. This implies that if $z(t)$ solves the equation, so does $z((at+b)/(ct+d))$ where a, b, c, d are entries of a usual $SL(2,\mathbb{C})$ matrix. Beside the invariance under this linear fractional transformation the differential equation (3.14) exhibits also invariances under nonlinear transformations, which were used in [20] to fix the numerical coefficients in (3.14)(3.16) uniquely. Once again we have observed experimentally that the q-series expansion of the mirror map z which satisfies (3.14) is in fact integral. In the case of elliptic curves (using the Weierstrass model), we have seen that the mirror map is given by the J function which is well-known to have an integral expansion. It would be interesting to establish a similar statement for $z(q)$ in the case of K3.

3.3. Periods of Quintic Threefolds

The periods of the quintic hypersurface in $\mathbb{P}^4$ were studied in the last section. Special geometry introduces the prepotential F as the new object of interest. The Weil-Peterson metric on the complex structure moduli space of mirror of the quintic X^* is described by F. Moreover the mirror hypothesis asserts that there is a special coordinate transformation given by a ratio of periods $t = \omega_1(z)/\omega_0(z)$, in which $\partial_t^3 F(t)$ gives the generating function for the number of rational curves in a generic quintic. It is therefore important to understand both the mirror map z and the prepotential F. Thus a relevant question is: are there natural differential equations which govern z and F?

For the mirror map z, there is a natural generalization of the Schwarzian equations (3.9)(3.14). Specifically, we claim that the mirror map $z(t)$ defined above satisfies the following seventh order ODE (see next section):

$$\{z,t\}_4 = -256 q_0^3 z'^{12} + 128 q_0^2 q_2^2 z'^{12} + \ldots \tag{3.17}$$

where

$$\begin{aligned}\{z,t\}_4 :=& -64T_2^6 - 560T_2^3T_3^2 - 1275T_3^4 + 448T_2^4T_4 + 2040T_2T_3^2T_4 - 192T_2^2T_4^4 - \\ & 504T_4^3 + 1120T_2^2T_3T_5 + 840T_3T_4T_5 - 280T_2T_5^2 + 20\,T_6\,\{z,t\}_3.\end{aligned} \tag{3.18}$$

As in the cases of K3 surfaces and elliptic curves, this Schwarzian equation is also manifestly $SL(2,\mathbb{C})$ invariant. In the case of the quintic hypersurface, the mirror map which satisfies this equation has the q-expansion:

$$z(q) = q - 770q^2 + 171525q^3 - 81623000q^4 - 35423171250q^5 - 54572818340154q^6 + \ldots \tag{3.19}$$

For the prepotential F, we have also derived a similar (seventh order) but a considerably more complicated polynomial differential equation. We will discuss this in the last section.

4. Construction of the Schwarzian equations

We now give an exposition for the construction of the differential equation which governs our mirror map $z(t)$ in each case.

Note that in each case we begin with an nth order ODE of Fuchsian type:

$$Lf := \left(\frac{d^n}{dz^n} + \sum_{i=0}^{n-1} q_i(z) \frac{d^i}{dz^i} \right) f = 0 \tag{4.1}$$

(n being 2,3 and 4 respectively for the elliptic curves, K3 surfaces and Calabi-Yau 3-folds.) In particular, the $q_i(z)$ are rational functions of z. Let f_1, f_2 be two linearly independent solutions of this equation and consider the ratio $t := f_2(z)/f_1(z)$. Inverting this relation (at least locally), we obtain z as a function of t. Our goal is to derive an ODE, in a canonical way, for $z(t)$.

We first perform a change of coordinates $z \to t$ on (4.1) and obtain:

$$\sum_{i=0}^{n} b_i(t) \frac{d^i}{dt^i} f(z(t)) = 0 \tag{4.2}$$

where the $b_i(t)$ are rational expressions of the derivatives $z^{(k)}$ (including $z(t)$). For example we have $b_n(t) = a_n(z(t))z'(t)^{-n}$. It is convenient to simplify the equation by writing (gauge transformation) $f = Ag$, where $A = exp(-\int \frac{b_{n-1}(t)}{nb_n(t)})$, and multiplying (4.2) by $\frac{1}{Ab_n}$ so that it becomes

$$\tilde{L}g := \left(\frac{d^n}{dt^n} + \sum_{i=0}^{n-2} c_i(t) \frac{d^i}{dt^i} \right) g(z(t)) = 0 \tag{4.3}$$

where c_i is now a rational expression of $z(t), z'(t), .., z^{(n-i+1)}$ for $i = 0, .., n-2$. Now $g_1 := f_1/A$ and $g_2 := f_2/A = tg_1$ are both solution to the equation (4.3). In particular we have

$$\begin{aligned} P :=& \tilde{L}g_1 = \left(\frac{d^n}{dt^n} + \sum_{i=0}^{n-2} c_i(t) \frac{d^i}{dt^i} \right) g_1 = 0 \\ Q :=& \tilde{L}(tg_1) - t\tilde{L}g_1 = \left(n\frac{d^{n-1}}{dt^{n-1}} + \sum_{i=0}^{n-3} (i+1)c_{i+1}(t) \frac{d^i}{dt^i} \right) g_1 = 0. \end{aligned} \tag{4.4}$$

Note that since c_i is a rational expression of $z(t), z'(t), .., z^{(n-i+1)}$, it follows that P involves $z(t), .., z^{(n+1)}$ while Q involves only $z(t), .., z^{(n)}$. Eqns (4.4) may be viewed as a coupled system of differential equations for $g_1(t), z(t)$. Our goal is to eliminate $g_1(t)$ so that we obtain an equation for just $z(t)$. One way to construct this is as follows. By (4.4), we have

$$\begin{aligned} \frac{d^i}{dt^i} P =& 0, \quad i = 0, 1, .., n-2, \\ \frac{d^j}{dt^j} Q =& 0, \quad j = 0, 1, .., n-1. \end{aligned} \tag{4.5}$$

We now view (4.5) as a homogeneous *linear* system of equations:

$$\sum_{l=0}^{2n-2} M_{kl}(z(t), .., z^{(2n-1)}(t)) \frac{d^l}{dt^l} g_1 = 0, \quad k = 0, .., 2n-2, \tag{4.6}$$

where each (M_{kl}) is the following $(2n-1) \times (2n-1)$ matrix:

$$\begin{pmatrix} c_0 & c_1 & .. & c_{n-2} & 0 & 1 & 0 & .. & 0 \\ c_0' & c_0 + c_1' & .. & c_{n-3} + c_{n-2}' & . & 0 & 1 & 0 & 0 \\ & \cdots & .. & \cdots & & & & .. & \\ c_0^{(n-2)} & (n-2)c_0^{(n-3)} + c_1^{(n-2)} & .. & \cdots & . & . & . & 0 & 1 \\ c_1 & 2c_2 & .. & (n-2)c_{n-2} & n & 0 & 0 & .. & 0 \\ c_1' & c_1 + 2c_2' & .. & (n-3)c_{n-3} + (n-2)c_{n-2}' & 0 & n & 0 & .. & 0 \\ & \cdots & .. & \cdots & & & .. & & \\ c_1^{(n-1)} & (n-1)c_1^{(n-2)} + 2c_2^{(n-1)} & .. & \cdots & . & & . & 0 & n \end{pmatrix} \tag{4.7}$$

More precisely if we define the 1st and n^{th} (n fixed) row vectors to be $(M_{1l}) = (c_0, c_1, .., c_{n-2}, 0, 1, 0, .., 0)$ and $(M_{nl}) = (c_1, 2c_2, .., (n-2)c_{n-2}, 0, n, 0, .., 0)$ respectively, then the matrix (M_{kl}) is given by the recursion relation:

$$M_{k+1,l} = M_{k,l-1} + M_{k,l}', \quad l = 1, .., 2n-1; \;\; k = 1, .., n-2, n, .., 2n-2. \tag{4.8}$$

Thus the (M_{kl}) depends rationally on $z(t), .., z^{(2n-1)}(t)$. Since g_1 is nonzero, it follows that

$$det\left(M_{kl}(z(t), .., z^{(2n-1)}(t))\right) = 0. \tag{4.9}$$

This is what we call the Schwarzian equation associated with (4.1). Note that by suitably clearing denominators, this becomes a $(2n-1)^{st}$ order polynomial ODE for $z(t)$ with constant coefficients. It is clear that this equation depends only on the data $q_i(z)$ we began with. In the case in which all the q_i are identically zero, we call the determinant in (4.9) the nth Schwarzian bracket $\{z(t), t\}_n$.

Despite having a general form of the Schwarzian equation, it is useful to see a few simple examples. As the first example, consider the case $n = 2$:

$$\frac{d^2}{dz^2} f + q_0(z) f = 0. \tag{4.10}$$

The eqns (4.4) become

$$\begin{aligned} {g_1}'' + c_0 g_1 &= 0 \\ 2{g_1}' &= 0 \end{aligned} \tag{4.11}$$

where

$$c_0(t) := q_0(z(t)) z'^2 - \frac{1}{2}\{z, t\}_2. \tag{4.12}$$

The corresponding linear system (4.6) has

$$(M_{kl}) = \begin{pmatrix} c_0 & 0 & 1 \\ 0 & 2 & 0 \\ 0 & 0 & 2 \end{pmatrix}. \tag{4.13}$$

Hence the associated Schwarzian equation (4.9) in this case is

$$det(M_{kl}) = 4c_0 = 2\left(2q_0 z'^2 - \{z,t\}_2\right) = 0 \tag{4.14}$$

which is the well-known classical Schwarzian equation.

For $n = 3$, we begin with the data

$$\left(\frac{d^3}{dz^3} + q_1(z)\frac{d}{dz} + q_0(z)\right) f = 0. \tag{4.15}$$

The transformed equation (4.3) in this case becomes

$$\left(\frac{d^3}{dt^3} + c_1(t)\frac{d}{dt} + c_0(t)\right) g = 0 \tag{4.16}$$

where

$$\begin{aligned} c_1(t) :=& q_1(z(t))\, z'(t)^2 + 2\{z(t),t\}_2 \\ c_0(t) :=& q_0(z(t))\, z'(t)^3 + q_1(z(t))\, z'(t)\, z''(t) + \frac{3\, z''(t)^3}{z'(t)^3} \\ &- \frac{4\, z''(t)\, z^{(3)}(t)}{z'(t)^2} + \frac{z^{(4)}(t)}{z'(t)}. \end{aligned} \tag{4.17}$$

The corresponding linear system (4.6) has:

$$(M_{kl}) = \begin{pmatrix} c_0 & c_1 & 0 & 1 & 0 \\ c_0' & c_0 + c_1' & c_1 & 0 & 1 \\ c_1 & 0 & 3 & 0 & 0 \\ c_1' & c_1 & 0 & 3 & 0 \\ c_1'' & 2c_1' & c_1 & 0 & 3 \end{pmatrix}. \tag{4.18}$$

Computing the associated Schwarzian equation, we get

$$det(M_{kl}) := 27c_0^2 + 4c_1^3 - 18c_1c_0' - 3{c_1'}^2 + 6c_1c_1'' = 0. \tag{4.19}$$

Substituting (4.17) into (4.19), we get the explicit form (3.14).

Now let's consider the case $n = 4$ which begins with

$$\left(\frac{d^4}{dz^4} + q_2(z)\frac{d^2}{dz^2} + q_1(z)\frac{d}{dz} + q_0(z)\right) f = 0. \tag{4.20}$$

We assume that this is the Picard-Fuchs equation for the periods of a 3 dimensional Calabi-Yau hypersurface. Then as pointed out earlier, there is a basis of solutions which takes the form (2.15). This implies that $q_1(z) = \frac{dq_2}{dz}$. The analogue of transformed equation (4.3) now becomes

$$\left(\frac{d^4}{dt^4} + c_2(t)\frac{d^2}{dt^2} + c_2'(t)\frac{d}{dt} + c_0(t)\right) g = 0 \tag{4.21}$$

where

$$\begin{aligned} c_2(t) :=& q_2(z)z'^2 + 5\{z(t),t\}_2 \\ c_0(t) :=& q_0(z)z'^4 + \frac{3}{2}\frac{dq_2(z)}{dz}z'^2 z'' - \frac{3}{4}q_2(z)z''^2 - \frac{135 z''^4}{16 z'^4} \\ &+\frac{3}{2}q_2(z)z'z^{(3)} + \frac{75 z''^2 z^{(3)}}{4z'^3} - \frac{15 {z^{(3)}}^2}{4z'^2} - \frac{15 z'' z^{(4)}}{2z'^2} + \frac{3z^{(5)}}{2z'} \end{aligned} \tag{4.22}$$

The associated linear system (4.6) in this case is 7×7. Computing the associated Schwarzian equation (4.9), we get

$$\begin{aligned} det(M_{kl}) :=& 16\, {c_2}^4\, c_0 - 128\, {c_2}^2\, {c_0}^2 + 256\, {c_0}^3 + 4\, {c_2}^3\, {c_2'}^2 \\ &+240\, c_2\, c_0\, {c_2'}^2 - 15\, {c_2'}^4 - 144\, {c_2}^2\, c_2'\, c_0' - 448\, c_0\, c_2'\, c_0' \\ &+256\, c_2\, {c_0'}^2 - 8\, {c_2}^4\, c_2'' + 128\, {c_0}^2\, c_2'' - 48\, c_2\, {c_2'}^2\, c_2'' \\ &+48\, c_2'\, c_0'\, c_2'' + 12\, {c_2}^2\, {c_2''}^2 - 48\, c_0\, {c_2''}^2 + 32\, {c_2}^3\, c_0'' \\ &-128\, c_2\, c_0\, c_0'' + 48\, {c_2'}^2\, c_0'' + 32\, {c_2}^2\, c_2'\, c_2^{(3)} + 64\, c_0\, c_2'\, c_2^{(3)} \\ &-96\, c_2\, c_0'\, c_2^{(3)} + 8\, c_2\, {c_2^{(3)}}^2 - 8\, {c_2}^3\, c_2^{(4)} + 32\, c_2\, c_0\, c_2^{(4)} - 12\, {c_2'}^2\, c_2^{(4)} = 0. \end{aligned} \tag{4.23}$$

Substituting (4.22) into (4.23), we get the 7th order ODE (3.17).

5. ODE for the Yukawa Coupling

We can give an analogous construction of an ODE for the Yukawa coupling. To be brief, we will instead explain an approach which will in the end results in a simple *characterization.*

Let us consider first a classical problem: given a pair of polynomials $r(y,z), s(y,z)$, let X be the intersection of their zero loci (in $\mathbf{C}^2$) and X_y, X_z the projections of X onto the y,z-directions. When can we construct (by quadrature) nontrivial polynomials $p(y), q(z)$ whose zero loci contain X_y, X_z respectively? The answer is simple: Hilbert's Nullstellensatz gives the following characterization. Namely $p(y), q(z)$ are constructible iff X is finite. Moreover there exists canonical choices for $p(y), q(z)$, namely the ones with the minimal degrees. Unfortunately, no similar general characterization is known in the case of *differential* polynomials (see however [21]). However in the cases we consider, we can formulate our problem of constructing an ODE for the Yukawa coupling in a similar spirit. We can also construct the analogues of the $p(y), q(z)$ above.

We begin with (4.21)and (4.22). By applying (4.21)to the four solutions of the form $u(t), u(t)t, u(t)F'(t), u(t)(2F(t) - tF'(t))$ (cf. eqn (2.15)), we get a system of four equations which is equivalent to:

$$\begin{aligned} u^{(4)} + c_2 u'' + c_2' u' + c_0 u &= 0 \\ 4u''' + 2c_2 u' + c_2' u &= 0 \\ F^{(3)}(6u'' + c_2 u) + 4F^{(4)} u' + F^{(5)} u &= 0 \\ 2F^{(3)} u' + F^{(4)} u &= 0. \end{aligned} \tag{5.1}$$

Solving the last equation gives $u = (F^{(3)})^{-1/2}$. The second equation is redundant because it is the derivative of the third one. Thus the above system reduces to just

$$\begin{aligned} A_2(z;t) - B_2(y;t) &= 0 \\ A_4(z;t) - B_4(y;t) &= 0 \end{aligned} \tag{5.2}$$

where $A_2(z;t), A_4(z;t), B_2(y;t), B_4(y;t)$ are differential rational functions defined by

$$\begin{aligned} A_2(z;t) :=& c_2(t) = q_2(z) z'^2 + 5\{z(t), t\}_2 \\ A_4(z;t) :=& c_0(t) = q_0(z) z'^4 + \frac{3}{2}\frac{dq_2(z)}{dz} z'^2 z'' - \frac{3}{4} q_2(z) z''^2 - \frac{135 z''^4}{16 z'^4} \\ &+ \frac{3}{2} q_2(z) z' z^{(3)} + \frac{75 z''^2 z^{(3)}}{4 z'^3} - \frac{15 {z^{(3)}}^2}{4 z'^2} - \frac{15 z'' z^{(4)}}{2 z'^2} + \frac{3 z^{(5)}}{2 z'} \\ B_2(y;t) :=& 2y'' - \frac{y'^2}{2} \\ B_4(y;t) :=& \frac{y^{(4)}}{2} + \frac{y''^2}{4} - \frac{y'' y'^2}{2} + \frac{y'^4}{16} \\ y(t) :=& Log\ F'''(t). \end{aligned} \tag{5.3}$$

The system (5.2) depends only on the two rational functions $q_2(z), q_0(z)$ via A_2, A_4, whereas B_2, B_4 are independent of such data. Observe that if we assign a weight 0 to z, y and weight 1 to $\frac{d}{dt}$, then the expressions A_2, B_2 (resp. A_4, B_4) are formally homogeneous of weights 2 (resp. 4). Thus we can speak of the weight of a quasi-homogeneous differential polynomial of these expressions. Finally, we note that the system of equations (5.2) should be regarded as a differential analogue of the system of polynomial equations $r(y,z) = 0 = s(y,z)$ considered above. With this analogy in mind, we now state the first of the two main results of this section:

Given a pair of rational function $(q_0(z), q_2(z))$ (which determines the Picard-Fuchs equation), there exists a differential polynomial P with the following properties:

(i) P quasi-homogeneous;

(ii) $P(A_2(z;t), A_4(z;t))$ is identically zero (see eqns. (5.3));

(iii) $P(B_2(y;t), B_4(y;t)) = 0$ is a nontrivial 7th order ODE in y which has a solution $y(t) = Log\ F'''(t)$ where $F(t)$ is the prepotential;

(iv) P is minimal, ie. every differential polynomial satisfying (i)–(iii) has weight no less than that of P.

(v) The polynomial ODE $P(B_2(y;t), B_4(y;t)) = 0$ is $SL(2, \mathbf{C})$ invariant.

We have observed in all the known examples that P is in fact characterized by the above properties, ie. P satisfying (i)-(v) is unique up to constant multiple. We conjecture that this is the case in general. Note that P depends on the data $(q_0(z), q_2(z))$ precisely via property (ii). The polynomial ODE $P(B_2(y;t), B_4(y;t)) = 0$ should be regarded as the differential analogue of $p(y) = 0$ considered above. In this analogy, the group $SL(2, \mathbf{C})$ plays the role for the *differential* polynomial $P(B_2(y;t), B_4(y;t))$ as the Galois group does for the *ordinary* polynomial $p(y)$. The simplest example of the above ODE for $F(t)$ is given by the considering the Picard-Fuchs equation for the complete intersection of 4 quadrics in $\mathbf{P}^7$, which is given by (2.25) after factorizing θ^4 as $\left[\theta^4 - 16^7(2\theta - 1)^4\right]\omega_\Gamma = 0$. To write the ODE for $F(t)$ down, we first define the notations:

$$\begin{aligned} \rho &= 100B_4 - 9B_2^2 - 30B_2'' \\ \chi &= -32\rho^2 B_2 - 45\rho'^2 + 40\rho\rho'' \\ \delta &= 5\chi\rho' - 2\chi'\rho. \end{aligned} \tag{5.4}$$

Then our ODE for the prepotential in this case is

$$
\begin{aligned}
&3783403212890625\,\chi^{18} + 52967644980468750\,\chi^{15}\,\delta^2 + 292835408677734375\,\chi^{12}\,\delta^4 \\
&+ 833559395864062500\,\chi^9\,\delta^6 + 1301823644717109375\,\chi^6\,\delta^8 + 1064406315612768750\,\chi^3\,\delta^{10} \\
&+ 357449882108765625\,\delta^{12} + 9097175898878906250\,\chi^{16}\,\rho^5 + 75543680906950781250\,\chi^{13}\,\delta^2\,\rho^5 \\
&- 551687817628209375 00\,\chi^{10}\,\delta^4\,\rho^5 - 1235933279927738437500\,\chi^7\,\delta^6\,\rho^5 \\
&- 262832882938824706875 0\,\chi^4\,\delta^8\,\rho^5 - 1669442421173622243750\,\chi\,\delta^{10}\,\rho^5 \\
&- 316395922222462973709375\,\chi^{14}\,\rho^{10} - 1041303693581386404075000\,\chi^{11}\,\delta^2\,\rho^{10} \\
&+ 1397061241390545045311250\,\chi^8\,\delta^4\,\rho^{10} - 978071752628929206000\,\chi^5\,\delta^6\,\rho^{10} \\
&+ 308896151588294552017312 5\,\chi^2\,\delta^8\,\rho^{10} + 7263752639215828135041225 00\,\chi^{12}\,\rho^{15} \\
&- 240196756730625798291889200 0\,\chi^9\,\delta^2\,\rho^{15} + 29319060393675698423999778 00\,\chi^6\,\delta^4\,\rho^{15} \\
&- 259200772973054831072975200 0\,\chi^3\,\delta^6\,\rho^{15} + 264772114318563252921325 00\,\delta^8\,\rho^{15} \\
&+ 7317735278685046995613249290 24\,\chi^{10}\,\rho^{20} - 13844538867915453829873316659 20\,\chi^7\,\delta^2\,\rho^{20} \\
&+ 1003786188392583028918031769600\,\chi^4\,\delta^4\,\rho^{20} \\
&- 92650299984331138894225408000\,\chi\,\delta^6\,\rho^{20} \\
&+ 2643799507163740354805575660339 20\chi^8\,\rho^{25} \\
&- 3236538849966783594155399027097 60\chi^5\,\delta^2\rho^{25} + \\
&\qquad + 10512210115205768202081722695680 0\,\chi^2\,\delta^4\,\rho^{25} \\
&\qquad + 4885370016741424964003892343865 3440\,\chi^6\,\rho^{30} \\
&\qquad - 3315342376066498968351383146725 3760\,\chi^3\,\delta^2\,\rho^{30} \\
&\qquad + 5281206792539883211563693244416 00\,\delta^4\,\rho^{30} \\
&\qquad + 4965538896010513223822010617996 247040\,\chi^4\,\rho^{35} \\
&\qquad - 1238934080748073699029086124292 177920\,\chi\,\delta^2\,\rho^{35} \\
&\qquad + 2650217711622663559007619458167 68184320\,\chi^2\,\rho^{40} \\
&\qquad + 5822406825670998196401392296588 763725824\,\rho^{45} = 0
\end{aligned} \tag{5.5}
$$

Note that since δ, χ, ρ are of weights 15,10,4 respectively, P is a quasi-homogeneous differential polynomial of weight 180. Each of the 37 terms in this polynomial corresponds to a partition of 180 by 15,10,4.

It turns out that there is a dual characterization for the Schwarzian equation (4.23)we have constructed:

There exists a differential polynomial Q with the following properties:

(i) Q is quasi-homogeneous;

(ii) $Q(B_2(y;t), B_4(y;t))$ is identically zero (see eqns. (5.3));

(iii) $Q(A_2(z;t), A_4(z;t)) = 0$ is a nontrivial ODE which has a solution given by the mirror map $z(t)$;

(iv) Q is minimal of weight 12, ie. every differential polynomial satisfying (i)–(iii) has weight at least 12;

(v) The polynomial ODE $Q(A_2(z;t), A_4(z;t)) = 0$ is $SL(2, \mathbf{C})$ invariant;

(vi) Q is universal, ie. it is independent of the data $(q_2(z), q_0(z))$ and it is characterized by (i)-(v) up to constant multiple;

(vii) $Q(A_2(z;t), A_4(z;t)) = 0$ coincides with $det(M_{kl}) = 0$ in (4.23).

We will defer the detailed proofs of the above results to our forthcoming papers.

Acknowledgement: Two of us (A. K. and S-T. Y.) would like to thank Shinobu Hosono and Stefan Theisen for collaboration on the matters discussed in section 2. B.H.L. thanks Gregg Zuckerman for helpful discussions.

References

[1] P. Candelas, X. della Ossa, P. Green, L. Parkes, *Nucl. Phys.* B359 (1991) 21

[2] P. Aspinwall and D. Morrison, *Commun. Math. Phys.* 151 (1993) 45

[3] E. Witten, *Commun. Math. Phys.* 118 (1988) 411; *Nucl. Phys.* B371 (1992) 191; and in *Essays on Mirror Manifolds*, Ed. S.-T. Yau, International Press, Hong Kong, 1992

[4] R. Rohm and E. Witten, *Ann. Phys.* 234 (1987) 454

[5] V. Batyrev and D. van Straten, *Generalized hypergeometric functions and rational curves on Calabi-Yau complete intersections in toric varieties*, Essen Preprint (1993)

[6] S. Hosono, A. Klemm, S. Theisen and S.-T. Yau, *Mirror symmetry, mirror map and applications to Calabi-Yau hypersurfaces*, HUTMP-93/0801, LMU-TPW-93-22 (hep-th/9308122), to be published in *Commun. Math. Phys.*

[7] S. Hosono, A. Klemm, S.Theisen and S. T. Yau, *Mirror symmetry, mirror map and applications to complete intersection Calabi-Yau spaces*, Preprint HUTMP-94-02, hep-th 9406055

[8] V. Batyrev, *Dual Polyhedra and the Mirror Symmetry for Calabi-Yau Hypersurfaces in Toric Varieties*, Univ. Essen Preprint (1992), to appear in *Journal of Alg. Geom.*

[9] R. Bryant and P. Griffiths, *Arithmetic and Geometry*, Progress in Mathematics **36** 77, Birkhäuser Boston, 1983

[10] P. Griffiths, Periods of integrals on algebraic manifolds, I and II., *Amer. Jour. Math.* vol. 90 (1968) 568; *Ann. Math.* 90 (1969) 460

[11] I. M. Gelfand, A. V. Zelevinky and M. M. Kapranov, Hypergeometric func-

tions and toral manifolds, *Functional Anal. Appl.* **23** 2 (1989) 12, English trans., 1994

[12] A. Strominger, *Commun. Math. Phys.* 133(1990)163; P. Candelas and X. della Ossa,*Nucl. Phys.* B355(1991), '455

[13] G. Tian, in *Mathematical Aspects of String Theory*, ed. S. T. Yau, World Scientific, Singapore, 1987

[14] M. Bershadsky, S. Ceccotti, H. Ooguri and C. Vafa *Kodaira-Spencer theory of gravity and exact results for quantum string amplitudes*, HUTP-93/A025, RIMS-946, SISSA-142/93/EP

[15] D. R. Morrison, in *Essays on mirror manifolds*, Ed. S.-T.Yau, International Press Singapore, 1992

[16] M. Yoshida, *Fuchsian Differential Equations*, Frier. Vieweg & Sohn, Bonn, 1987

[17] D. E. Sommervoll, *Rational Curves of low degree on a complete intersection Calabi-Yau threefold in* $P^3 \times P^3$, Oslo preprint ISBN 82-553-0838-5.

[18] E. Verlinde and N. Warner, *Phys. Lett.* 269 (1991) 96

[19] A. Klemm, S. Theisen and M. Schmidt , *Int. J. Mod. Phys.* A7 (1992) 6215

[20] A.O. Klemm, B.H. Lian, S.S. Roan and S.T. Yau, *Differential Equations from Mirror Symmetry I, II.*, in preparation.

[21] J.F. Ritt, *Differential Algebra*, Colloq. Publ. vol.33, AMS, Providence, 1950

A. Klemm
Theory Division
CERN
CH-1211 Geneva, 23 Switzerland
klemm@nxth21.cern.ch

B. H. Lian and S. T. Yau
Department of Mathematics
Harvard University
Cambridge, MA 02138, USA
lian@math.harvard.edu; yau@math.harvard.edu

S. S. Roan
Institute of Mathematics
Academia Sinica
Taipei, Taiwan
maroan@ccvax.sinica.edu.tw

Received October 1994

Progress in Mathematics

Edited by:

Progress in Mathematics is a series of books intended for professional mathematicians and scientists, encompassing all areas of pure mathematics. This distinguished series, which began in 1979, includes authored monographs and edited collections of papers on important research developments as well as expositions of particular subject areas.

We encourage preparation of manuscripts in some form of TeX for delivery in camera-ready copy which leads to rapid publication, or in electronic form for interfacing with laser printers or typesetters.

Proposals should be sent directly to the editors or to: Birkhäuser Boston, 675 Massachusetts Avenue, Cambridge, MA 02139, U. S. A.

1 GROSS. Quadratic Forms in Infinite-Dimensional Vector Spaces
2 PHAM. Singularités des Systèmes Différentiels de Gauss-Manin
3 OKONEK/SCHNEIDER/SPINDLER. Vector Bundles on Complex Projective Spaces
4 AUPETIT. Complex Approximation, Proceedings, Quebec, Canada, July 3-8, 1978
5 HELGASON. The Radon Transform
6 LION/VERGNE. The Weil Representation, Maslov Index and Theta Series
7 HIRSCHOWITZ. Vector Bundles and Differential Equations Proceedings. Nice, France, June 12-17, 1979
8 GUCKENHEIMER/MOSER/NEWHOUSE. Dynamical Systems, C.I.M.E. Lectures. Bressanone, Italy, June, 1978
9 SPRINGER. Linear Algebraic Groups
10 KATOK. Ergodic Theory and Dynamical Systems I
11 BALSEV. 18th Scandinavian Conferess of Mathematicians, Aarhus, Denmark, 1980
12 BERTIN. Séminaire de Théorie des Nombres, Paris 1979-80
13 HELGASON. Topics in Harmonic Analysis on Homogeneous Spaces
14 HANO/MARIMOTO/MURAKAMI/OKAMOTO/OZEKI. Manifolds and Lie Groups: Papers in Honor of Yozo Matsushima
15 VOGAN. Representations of Real Reductive Lie Groups
16 GRIFFITHS/MORGAN. Rational Homotopy Theory and Differential Forms
17 VOVSI. Triangular Products of Group Representations and Their Applications
18 FRESNEL/VAN DER PUT. Géométrie Analytique Rigide et Applications
19 ODA. Periods of Hilbert Modular Surfaces
20 STEVENS. Arithmetic on Modular Curves

21 KATOK. Ergodic Theory and Dynamical Systems II
22 BERTIN. Séminaire de Théorie des Nombres, Paris 1980-81
23 WEIL. Adeles and Algebraic Groups
24 LE BARZ/HERVIER. Enumerative Geometry and Classical Algebraic Geometry
25 GRIFFITHS. Exterior Differential Systems and the Calculus of Variations
26 KOBLITZ. Number Theory Related to Fermat's Last Theorem
27 BROCKETT/MILLMAN/SUSSMAN. Differential Geometric Control Theory
28 MUMFORD. Tata Lectures on Theta I
29 FRIEDMAN/MORRISON. Birational Geometry of Degenrations
30 YANO/KON. CR Submanifolds of Kaehlerian and Sasakian Manifolds
31 BERTRAND/WALDSCHMIDT. Approximations Diophantiennes et Nombres Transcendants
32 BOOKS/GRAY/REINHART. Differential Geometry
33 ZUILY. Uniqueness and Non-Uniqueness in the Cauchy Problem
34 KASHIWARA. Systems of Microdifferential Equations
35 ARTIN/TATE. Arithmetic and Geometry: Papers Dedicated to I. R. Shafarevich on the Occasion of His Sixtieth Birthday. Vol. 1
36 ARTIN/TATE. Arithmetic and Geometry: Papers Dedicated to I. R. Shafarevich on the Occasion of His Sixtieth Birthday. Vol. 1I
37 DE MONVEL. Mathématique et Physique
38 BERTIN. Séminaire de Théorie des Nombres, Paris 1981-82
39 UENO. Classification of Algebraic and Analytic Manifolds
40 TROMBI. Representation Theory of Reductive Groups
41 STANLEY. Combinatorics and Commutative Algebra; 1st ed. 1984 2nd ed., 1996
42 JOUANOLOU. Théorèmes de Bertini et Applications
43 MUMFORD. Tata Lectures on Theta II
44 KAC. Infitine Dimensional Lie Algebras
45 BISMUT. Large deviations and the Malliavin Calculus
46 SATAKE/MORITA. Automorphic Forms of Several Variables, Taniguchi Symposium, Katata, 1983
47 TATE. Les Conjectures de Stark sur les Fonctions L d'Artin en $s = 0$
48 FRÖLICH. Classgroups and Hermitian Modules
49 SCHLICHTKRULL. Hyperfunctions and Harmonic Analysis on Symmetric Spaces
50 BOREL ET AL. Intersection Cohomology
51 BERTIN/GOLDSTEIN. Séminaire de Théorie des Nombres, Paris 1982-83
52 GASQUI/GOLDSCHMIDT. Déformations Infinitesimales des Structures Conformes Plates
53 LAURENT. Théorie de la Deuxième Microlocalisation dans le Domaine Complexe
54 VERDIER/LE POTIER. Module des Fibres Stables sur les Courbes Algébriques: Notes de l'Ecole Normale Supérieure, Printemps, 1983
55 EICHLER/ZAGIER. The Theory of Jacobi Forms
56 SHIFFMAN /SOMMESE. Vanishing Theorems on Complex Manifolds
57 RIESEL. Prime Numbers and Computer Methods for Factorization
58 HELFFER/NOURRIGAT. Hypoellipticité Maximale pour des Opérateurs Polynomes de Champs de Vecteurs
59 GOLDSTEIN. Séminaire de Théorie des Nombres, Paris 1983–84
60 PROCESI. Geometry Today: Giornate Di Geometria, Roma. 1984

61 BALLMANN/GROMOV/SCHROEDER. Manifolds of Nonpositive Curvature
62 GUILLOU/MARIN. A la Recherche de la Topologie Perdue
63 GOLDSTEIN. Séminaire de Théorie des Nombres, Paris 1984–85
64 MYUNG. Malcev-Admissible Algebras
65 GRUBB. Functional Calculus of Pseudo-Differential Boundary Problems
66 CASSOU-NOGUES/TAYLOR. Elliptic Functions and Rings and Integers
67 HOWE. Discrete Groups in Geometry and Analysis: Papers in Honor of G.D. Mostow on His Sixtieth Birthday
68 ROBERT. Autour de L'Approximation Semi-Classique
69 FARAUT/HARZALLAH. Deux Cours d'Analyse Harmonique
70 ADOLPHSON/CONREY/GHOSH/YAGER. Analytic Number Theory and Diophantine Problems: Proceedings of a Conference at Oklahoma State University
71 GOLDSTEIN. Séminaire de Théorie des Nombres, Paris 1985–86
72 VAISMAN. Symplectic Geometry and Secondary Characteristic Classes
73 MOLINO. Riemannian Foliations
74 HENKIN/LEITERER. Andreotti-Grauert Theory by Integral Formulas
75 GOLDSTEIN. Séminaire de Théorie des Nombres, Paris 1986–87
76 COSSEC/DOLGACHEV. Enriques Surfaces I
77 REYSSAT. Quelques Aspects des Surfaces de Riemann
78 BORHO/BRYLINSKI/MACPHERSON. Nilpotent Orbits, Primitive Ideals, and Characteristic Classes
79 MCKENZIE/VALERIOTE. The Structure of Decidable Locally Finite Varieties
80 KRAFT/PETRIE/SCHWARZ. Topological Methods in Algebraic Transformation Groups
81 GOLDSTEIN. Séminaire de Théorie des Nombres, Paris 1987–88
82 DUFLO/PEDERSEN/VERGNE. The Orbit Method in Representation Theory: Proceedings of a Conference held in Copenhagen, August to September 1988
83 GHYS/DE LA HARPE. Sur les Groupes Hyperboliques d'après Mikhael Gromov
84 ARAKI/KADISON. Mappings of Operator Algebras: Proceedings of the Japan-U.S. Joint Seminar, University of Pennsylvania, Philadelphia, Pennsylvania, 1988
85 BERNDT/DIAMOND/HALBERSTAM/HILDEBRAND. Analytic Number Theory: Proceedings of a Conference in Honor of Paul T. Bateman
86 CARTIER/ILLUSIE/KATZ/LAUMON/MANIN/RIBET. The Grothendieck Festschrift: A Collection of Articles Written in Honor of the 60th Birthday of Alexander Grothendieck. Vol. I
87 CARTIER/ILLUSIE/KATZ/LAUMON/MANIN/RIBET. The Grothendieck Festschrift: A Collection of Articles Written in Honor of the 60th Birthday of Alexander Grothendieck. Volume II
88 CARTIER/ILLUSIE/KATZ/LAUMON/MANIN/RIBET. The Grothendieck Festschrift: A Collection of Articles Written in Honor of the 60th Birthday of Alexander Grothendieck. Volume III
89 VAN DER GEER/OORT / STEENBRINK. Arithmetic Algebraic Geometry
90 SRINIVAS. Algebraic K-Theory
91 GOLDSTEIN. Séminaire de Théorie des Nombres, Paris 1988–89
92 CONNES/DUFLO/JOSEPH/RENTSCHLER. Operator Algebras, Unitary Representations, Enveloping Algebras, and Invariant Theory. A Collection of Articles in Honor of the 65th Birthday of Jacques Dixmier

93 AUDIN. The Topology of Torus Actions on Symplectic Manifolds
94 MORA/TRAVERSO (eds.) Effective Methods in Algebraic Geometry
95 MICHLER/RINGEL (eds.) Representation Theory of Finite Groups and Finite Dimensional Algebras
96 MALGRANGE. Equations Différentielles à Coefficients Polynomiaux
97 MUMFORD/NORI/NORMAN. Tata Lectures on Theta III
98 GODBILLON. Feuilletages, Etudes géométriques
99 DONATO /DUVAL/ELHADAD/TUYNMAN. Symplectic Geometry and Mathematical Physics. A Collection of Articles in Honor of J.-M. Souriau
100 TAYLOR. Pseudodifferential Operators and Nonlinear PDE
101 BARKER/SALLY. Harmonic Analysis on Reductive Groups
102 DAVID. Séminaire de Théorie des Nombres, Paris 1989-90
103 ANGER /PORTENIER. Radon Integrals
104 ADAMS /BARBASCH/VOGAN. The Langlands Classification and Irreducible Characters for Real Reductive Groups
105 TIRAO/WALLACH. New Developments in Lie Theory and Their Applications
106 BUSER. Geometry and Spectra of Compact Riemann Surfaces
108 BRYLINSKI. Loop Spaces, Characteristic Classes and Geometric Quantization
108 DAVID. Séminaire de Théorie des Nombres, Paris 1990-91
109 EYSSETTE/GALLIGO. Computational Algebraic Geometry
110 LUSZTIG. Introduction to Quantum Groups
111 SCHWARZ. Morse Homology
112 DONG/LEPOWSKY. Generalized Vertex Algebras and Relative Vertex Operators
113 MOEGLIN/WALDSPURGER. Décomposition spectrale et séries d'Eisenstein
114 BERENSTEIN/GAY/VIDRAS/YGER. Residue Currents and Bezout Identities
115 BABELON/CARTIER/KOSMANN-SCHWARZBACH. Integrable Systems, The Verdier Memorial Conference: Actes du Colloque International de Luminy
116 DAVID. Séminaire de Théorie des Nombres, Paris 1991-92
117 AUDIN/LAFONTAINE (eds). Holomorphic Curves in Symplectic Geometry
118 VAISMAN. Lectures on the Geometry of Poisson Manifolds
119 JOSEPH/ MEURAT/MIGNON/PRUM/ RENTSCHLER (eds). First European Congress of Mathematics, July, 1992, Vol.I
120 JOSEPH/ MEURAT/MIGNON/PRUM/ RENTSCHLER (eds). First European Congress of Mathematics, July, 1992, Vol. II
121 JOSEPH/ MEURAT/MIGNON/PRUM/ RENTSCHLER (eds). First European Congress of Mathematics, July, 1992, Vol. III (Round Tables)
122 GUILLEMIN. Moment Maps and Combinatorial Invariants of T^n-spaces
123 BRYLINSKI/BRYLINSKI/GUILLEMIN/KAC. Lie Theory and Geometry: In Honor of Bertram Kostant
124 AEBISCHER/BORER/KALIN/LEUENBERGER/ REIMANN. Symplectic Geometry
125 LUBOTZKY. Discrete Groups, Expanding Graphs and Invariant Measures
126 RIESEL. Prime Numbers and Computer Methods for Factorization
127 HORMANDER. Notions of Convexity
128 SCHMIDT. Dynamical Systems of Algebraic Origin
129 DIJGRAAF/FABER/VAN DER GEER. The Moduli Space of Curves
130 DUISTERMAAT. Fourier Integral Operators
131 GINDIKIN/LEPOWSKY/WILSON. Functional Analysis on the Even of the 21st Century. In Honor of the Eightieth Birthday of I. M. Gelfand, Vol. 1
132 GINDIKIN/LEPOWSKY/WILSON. Functional Analysis on the Even of the 21st Century. In Honor of the Eightieth Birthday of I. M. Gelfand, Vol. 2
133 HOFER/TAUBES/WEINSTEIN/ZEHNDER. The Floer Memorial Volume